Claude Reignier Conder

The Survey of Western Palestine

Arabic and English name lists collected during the survey

Claude Reignier Conder

The Survey of Western Palestine
Arabic and English name lists collected during the survey

ISBN/EAN: 9783744755016

Printed in Europe, USA, Canada, Australia, Japan

Cover: Foto ©berggeist007 / pixelio.de

More available books at **www.hansebooks.com**

THE SURVEY

OF

WESTERN PALESTINE.

ARABIC AND ENGLISH NAME LISTS

COLLECTED DURING THE SURVEY BY

LIEUTENANTS CONDER AND KITCHENER, R.E.

TRANSLITERATED AND EXPLAINED BY

E. H. PALMER, M.A.,

Lord Almoner's Professor of Arabic in the University of Cambridge.

FOR
THE COMMITTEE OF THE PALESTINE EXPLORATION FUND,
1, ADAM STREET, ADELPHI, LONDON, W.C.
1881.

HARRISON AND SONS,
PRINTERS IN ORDINARY TO HER MAJESTY,
ST. MARTIN'S LANE, LONDON.

PREFACE.

THESE Lists were originally put together by Lieutenant Conder, but I have revised the Arabic spelling from careful comparison with the field books and the notes of the native scribe. For the English definitions I am myself responsible. The transliteration I have been obliged to leave untouched, as it was already finally engraved upon the Map before I took the matter in hand. The Committee had decided upon the adoption of Robinson's system, which has, however, only been partially followed, no difference being made between the equivalents for the cognate Arabic letters س and ص, ت and ط &c.; still it fairly represents the sound of the Arabic words in English; and as they are given in Arabic characters as well, this circumstance can occasion no difficulty to scholars. Many of the names were collected by the English members of the party, and submitted to the scribe afterwards, who seems to have written down some conjecturally, the delicate distinction between certain letters not being appreciable to an ordinary European ear. I have endeavoured by careful philological investigation of the words to settle their orthography, and believe that the names will, in most instances, be found correct, though it is quite possible that some may need revision on the spot by an investigator with a practised ear.

To determine the exact meaning of Arabic topographical names is by no means easy. Some are descriptive of physical features, but even these

are often either obsolete or distorted words. Others are derived from long
since forgotten incidents, or owners whose memory has passed away.
Others again are survivals of older Nabathean, Hebrew, Canaanite, and
other names, either quite meaningless in Arabic, or having an Arabic
form in which the original sound is perhaps more or less preserved, but
the sense entirely lost. Occasionally Hebrew, especially Biblical and
Talmudic names, remain scarcely altered : some of these I have indicated :
but I have only done so in obvious cases, avoiding mere speculation.
The lists will, nevertheless, be found to furnish better data for such
researches than any which have hitherto appeared, and future study of
them will no doubt reveal much that is of interest, both to philology
and Biblical topography.

The Arabic follows the Hebrew letter for letter, except that س‎ *s*
in the former generally represents either שׁ‎ *sh* or ס‎ *ṣ* in the latter :
while ش‎ *sh* stands for س‎ *s* : sometimes ע‎ *'ain* appears as ح‎ *ḥ* ;
כ‎ *k* and ח‎ *kh*, and occasionally ע‎ *'ain* and א‎ *alif* are also interchanged,
in Arabic, but these last are rare. In some instances the modern local
form represents the Hebrew with scarcely any change : in others the
Hebrew word is simply translated into Arabic, as Dan, "a judge."
which appears as *Tell Kâdhi*, "the judge's mound." In others, again,
the older form has grown into an intelligible Arabic word, similar in
sound, but quite different in meaning, and therefore inapplicable as a
descriptive epithet.

The topographical commonplaces, such as رأس‎, *râs*, "a hill-top or
headland," Heb. ראשׁ‎ *rosh* ; عين‎ *'ain*, "a spring," Heb. עין‎ *'en*, &c., are
for the most part alike in both languages ; while such Hebrew or Aramaic
words as בירה‎ *tireh*, "a fort," בירה‎ *bîreh*, "a castle," are often confounded
with the Arabic طير‎ *tair*, "a bird." and بئر‎ *bir*, "a well."

The reader will find in the following pages the meanings of the Arabic words, as far as I could ascertain them, with suggestions as to the more obvious Hebrew and Aramaic derivations. The letters p.n. (proper name) after a name, mean either that it is a common Arabic personal appellation, or that it is a word to which no meaning can be assigned; the former will be at once recognized by the Arabic scholar, the latter will form interesting problems for future investigation.

My previous work in the nomenclature of Sinai and the Tih, has given me the key to many difficulties in Arabic nomenclature, but many still remain unsolved. The names are not mere Arabic words to be translated, but, as I have said, are most varied forms containing multifarious elements, upon which philology alone does not afford us sufficient data to pronounce. All that I could hope to do was to state the certainties, offer suggestions as to the probabilities, and leave the rest to other hands.

E. H. PALMER.

July, 1881.

SURVEY OF WESTERN PALESTINE.

ARABIC AND ENGLISH NAME LISTS,

WITH

TRANSLITERATION AND EXPLANATION.

SHEET I.

عباسيه   Nb   *'Abbásíyeh.* From the proper name 'Abbás, the uncle of the Prophet, from whose family descended the Abbaside Khalifs of Bagdad.

عين ابى عمرو   Nb   *'Ain Abu 'Amr.* Abu Amr's spring. p.n.

عين العليليات   Nc   *'Ain el Alcilíyát.* 'The spring of successive draughts or torrents'; but it is probably a proper name.

عين عطاءالله   Mb   *'Ain 'Atáallah.* 'Atáallah's spring. p.n., meaning 'God's gift,' 'Theodore.'

عين الظلا   Mc   *'Ain edh Dhelat,* or لا زٍ *ezzalat,* which latter word means in the vulgar language 'pebbles,' and in the older language to 'move violently.'

عين فراوية   Mc   *'Ain Furáwiyat;* perhaps from فرِزٍ in the sense of 'waste-land.'

عين الغربية   Nc   *'Ain el Gharbíyeh.* The western spring.

عين الحبيشيه   Nb   *'Ain el Hubeishíyeh.* The spring of the Abyssinian. Sepp suggests that the name is connected with the Egypto-Ethiopian deity Abus, from which the Theban Temple of Medinet Abu and the colossal monuments of Aboo-Simbel are named. Justin, XLIV, 4, mentions the God-King of the Celtiberians, Habis, from whom the Phœnicians may have taken this deity. Abibaal is the name of a Tyrian king, and points to a similar connection.

B

عين ابعل   Nc   *'Ain Ib'âl*, vulgar pronunciation *Ib'âl*, pl. of *bâl*. 'Elevated land on which no water falls,' or 'unwatered vegetation.' The word may be connected with the name Baal.

عين جيلو   Nc   *'Ain Jilû.* The spring of Jilû, perhaps a corruption of *Jila*, or 'antimony ore.'

عين الجديدة   Mb   *'Ain el Judeideh.* The new spring; or, the spring of 'dykes,' *i.e.*, coloured streaks in the rocks.

عين كنيسة   Nc   *'Ain Kuneiseh.* The spring of the church.

عين كردية   Nc   *'Ain Kurdîyeh.* The spring of the Kurds.

عين القسيس   Nc   *'Ain el Kûsîs.* The (Christian) Priest's spring.

عين المعليا   Mc   *'Ain el Mâllîyeh.* The spring of 'successive draughts' or 'irrigations.'

عين المثنيات   Mc   *'Ain el Mûthniyât.* Spring of the lands twice turned over for sowing.

عين المدفنة   Mc   *'Ain el Medfeneh.* The spring of the burial-place.

عين شدغيث   Nb   *'Ain Shidghìth.* The spring of Shidghith. Cf. Heb. שָׂדֶה 'fields surrounding a city,' and עַיָּה Phœn. for עֵמֶק 'a vale.'

عين صور   Mb   *'Ain Sûr.* The well of the rock or, of Tyre.

عين تورد   Nb   *'Ain Torah.* The spring of running water.

عين ام العمل   Mb   *'Ain Umm el 'Aml.* The spring of the mother of work.

عين يارين   Nc   *'Ain Yârîn.* The spring of Yárin.

عين الزهيرية   Mc   *'Ain ez Zaheirîyeh* (properly *Zuheirîyeh*). The spring of little flowers; but this is probably a proper name.

عين ابي عبدالله   Nb   *'Ain Abu 'Abdallah.* Abu Abdallah's spring. p.n.

عقبة علي   Nb   *'Akabat 'Alî.* Ali's steep. Ali, the cousin and son-in-law of the Prophet Mohammed, is the great saint of the Metawileh and other Shiah sects.

عمود الاطرش   Mc   *'Amûd el Atrash.* The column (oil press) of the deaf man.

عمود البيروتي   Mc   *'Amud el Beirûty.* The column (oil press) of the Beirût man.

عرب النميرات المياضة(or الغياضة)   Mb   *'Arab en Numeirât el Kaiyâdeh.* The Arabs of the Numeirat el Kaiyâdeh tribe, descendants of Numeir, 'the little leopard'; *cf. Nimrim.*

ارض السواقى {*Mc* *nw*} *Ard es Sawáky.* The land of small water-courses.

ارض الزهيرية {*Mc* *nw*} *Ard ez Zaheiriyeh.* The land of little flowers; *see above,* *'Ain ez Zaheiriyeh.*

عريض الكبير {*Nb* *nw*} *'Arídh el Kebír.* The large broad space or mountain side

عواميد المنقلة Mb *'Awámíd el Menkeleh.* The columns of the Menkeleh, *i.e.,* a game played by Arab children with a long board containing 14 squares, 7 on each side, and each square containing 7 pebbles. The word also means 'a stage' (of a journey).

بحور Nb *Bahhûr.* Seas, or large tracts of land.

بازورية Nb *Bázûriyeh.* Producing pot-herbs; *see Khallet el Bázûriyeh.* p. 5.

بيت بيدون Mc *Beit Beidûn.* The house of Beidûn. p.n.

بيت هولى Nc *Beit Hûlei.* The house of Hûlei. p.n.

بيت شاكورد Mb *Beit Shákûrah.* The house of Shákûrah. p.n.

بستان العقاد *Mb* *Bestán el 'Akkád.* The garden of the 'Akkád, *i.e.,* of the maker of a *taákkud,* or 'the projecting portion of the interior stone casing of a well.' But perhaps from عقد the bud of a plant.

بدياس Nb *Bidiyás.* Bidiyás. p.n.

بئر العقاد Mb *Bir el 'Akkád.* The well of the 'Akkád; *see above.*

بئر بيت الصغير {*Nb* *nw*} *Bir Beit es Saghir.* The well of the little house.

بئر الجبلون Mb *Bir el Jebelûn.* The well of Jebelûn (a local form of word from *jebel,* a mountain).

بئر الواسع Mc *Bir el Wásih.* The well of ample contents.

بركة عين الزرقآ Ma *Birket 'Ain ez Zarká.* The pool of the blue spring.

بركة البقبق Mb *Birket el Bakbûk.* The pool of gurgling.

بركة البس Mb *Birket el Bass* (probably البسّ). The pool of crumbling, or the swamp.

بركة العسراوى *Mc* *Birket el Isráwy.* El Isráwiy's pool, p.n., meaning left-handed or ambidexter.

بركة الصفصافة *Mc* *Birket es Sûfsáfeh.* The pool of the osier.

بركة تل البحر Mb *Birket Tell el Bahr.* The pool of the mound of the sea

بياض بدياس   Nb   *Biyâdh Bidiâs.* The white place of Bidiâs.

برج الهوا   Mb   *Burj el Hâwa (hawâ).* The air tower.

برج القبلى   Mb   *Burj el Kibly.* The south tower.

برج رحال   Nb   *Burj Rahhâl.* The traveller's tower.

برج الشمالى   Mb   *Burj esh Shemâly.* The northern tower.

دير الربعيين   Nb   *Deir el Arba'în.* The 'convent of the Forty,' *i.c.*, the Forty Martyrs of Cappadocia, canonized by the Orthodox Greek Church.

دير قانون رأس العين   Mc   *Deir Kânûn Râs el 'Ain.* The convent of the Rule (canon) at Râs el 'Ain. So called to distinguish it from another village of the same name not far distant.

دير القبى   Nc   *Deir el Kubbeî.* The vaulted or domed convent.

ظهر حيدر   Nb   *Dhahr Heider.* Heider's ridge. *Heider*, 'lion,' is an epithet of Ali the Metawileh saint.

الدير   Nb   *Ed Deir.* The convent.

الحمادية   Mb   *El Hammâdiyeh.* The place of Hammâd. p.n.

العزية   Mb   *El 'Ezziyah.* Probably connected with *'izz*, 'flowing' water.

عزية التحتا   Mc   *'Ezziyat et Tahta (Uzziyet).* The lower 'Ezziyah.

حوطة الاعافرة   { Nb / mc }   *Hautat el Aâfreh.* The enclosure of the A'âfreh (family name).

ابن زكريا   Nc   *Ibn Zakariyeh.* The son of Zakariyah, *i.c.*, 'John the Baptist.' Perhaps one of his numerous shrines was to be found here.

حنوية   Nc   *Henawei.* The little bend (of a valley, &c.).

جبال البطم   Nc   *Jebâl el Butm.* The mountains of the Terebinth trees.

جبل العمود   Nc   *Jebel el 'Amûd.* The mountain of the column.

جبل النقب   Nb   *Jebel en Nûkb.* The mountain of the Pass.

جبل الصوان   Nc   *Jebel es Suwân.* The mountain of the Flint.

جسر العزية   Lc   *Jisr el 'Ezziyeh.* The 'Ezziyeh bridge ; *see 'Ezziyat et Tahtâ.*

جسر القاسمية   Nb   *Jisr el Kâsimiyeh.* The bridge of the Kasimiyeh ; *see Nahr el Kâsimiyeh.*

جب سويد   Mc   *Jubb Sûweid.* *Jubb* means 'a pit,' and *sûweid* 'blackish'; but the latter also means 'water.' and 'salt-tracts.'

جوار النخل   Mb   *Juwâr en Nûkhl.* The water-holes of the palms. The word *Nukhl* is however probably connected with נַחַל 'a watercourse'; *see Khûrbet Juwâr en Nûkhl.*

تبر حيرام   Nc   *Kabr Hirâm.* Hiram's tomb.

قآ صور   {Mc / Nb}   *Kada Sûr.* The district of Tyre; *see* "Memoirs," p. 44.

قانا   Nc   *Kânâ.* p.n. Cana.

الكنيسة   Nc   *El Kenîseh.* The church.

خلة ابى علول   {Mb / Nb / nw}   *Khallet Abu 'Allûl.* Abu Allûl's dell. p.n.; *cf. Ain el 'Aleilat.*

خلة ابى عمش   Nb   *Khallet Abu 'Amesh.* Abu 'Amesh's dell.

خلة عين ابى عمرو   Nb   *Khallet Ain Abu Amr.* The dell of the spring of Abu 'Amr.

خلة عطيطة   Nc   *Khallet 'Atîtah.* The dell of the rent or split.

خلة البازورية   {Mc / nw}   *Khallet el Bâzûriyeh.* The dell of the Bâzûriyeh. From بز 'seeds of leguminous plants,' or 'silkworms' eggs.'

خلة الدفعة   Nc   *Khallet Dafâh.* The dell of the rushing (torrent, &c.).

خلة حسن   Nb   *Khallet Hasan.* Hasan's dell. Hasan, the son of Ali, is another of the Metawileh saints.

خلة حاتم   Mc   *Khallet Hâtim.* Hâtim's dell. Hâtim et Tâiy was an Arab who lived about Mohammed's time, and was celebrated for his liberality.

خلة الكرار   Mc   *Khallet el Kerâr.* The dell of the store-house.

خلة الخروبة   Mb   *Khallet el Kharrûbeh.* The dell of the Carob (or locust) tree.

خلة المطحنة   Nb   *Khallet el Metahneh.* The dell of the water-mill.

خلة المزرعانى   Mc   *Khallet el Mezrâny.* The dell of the cultivator.

خلة الملاحة   Nb   *Khallet el Millâhah.* The dell of the salt-works; *see Khûrbet 'Akabet el Mellahah.*

خلة المعنى   {Nb / nw}   *Khallet el Muânna.* The dell of the captive. But the word may also mean 'land fertilized by showers and producing herbage.'

خلة المغارة   {Nc / nw}   *Khallet el Mughârat.* The dell of the cave.

خلة النبع   *Nc*   *Khallet en Nebâ.* The dell of the perennial spring.

خلة النقاب   {*Nb* / *nc*}   *Khallet en Nûkâb.* The dell of the passes.

خلة العليقة   *Nc*   *Khallet 'Olleikah.* The dell of the thorn bushes or brambles.

خلة الشمسية   *Nc*   *Khallet es Shemsîyeh.* *Shemsîyeh* means 'sunny.'

خلة الشوحة   *Nb*   *Khallet esh Shûhah.* The dell of the kite.

خلة السوق   *Nc*   *Khallet es Sûk.* The dell of the market.

خلة الزاوية   *Nc*   *Khallet ez Zâwiyeh.* The dell of the corner or hermitage.

خلة الزبل   *Nc*   *Khallet ez Zibl.* The dell of dung (manure).

خان القاسمية   *Ma*   *Khân el Kâsimîyeh.* The Kasimiyeh Khân or Caravanserai; see *Nahr el Kasimîyeh.*

خلال الصابون   *Mb*   *Khelâl es Sâbôn.* The dells of the soap-works.

الخربة   *Mc*   *El Khûrbeh.* The ruin.

خربة ابي الفرحات   *Nc*   *Khûrbet Abu Ferhât.* Abu Ferhât's ruin. Ferhât is a Syrian proper name.

خربة ابي عجنة   *Nc*   *Khûrbet Abu 'Ajneh.* The ruin of Abu 'Ajneh. The name may mean "the man with the fat cattle," but in Hebrew and Aramaic it has the sense of 'hedged in,' or 'debarred.'

خربة ابي صالح   *Nb*   *Khûrbet Abu Sâleh.* The ruin of Abu Saleh. p.n.

خربة علي صالح   *Mc*   *Khûrbet 'Alî Sâleh.* Ali Saleh's ruin. p.n.

خربة عقبة الملاحة   *Nb*   *Khûrbet 'Akabet el Mellâhah.* The ruin of the steep, or mountain road of the salt-works (but the last word may be a corruption from *Imlâheh,* water becoming sweet after being brackish).

خربة العمود   *Mc*   *Khûrbet el 'Amûd.* The ruin of the column.

خربة العواميد   *Mc*   *Khûrbet el 'Awâmîd.* The ruin of the columns.

خربة دبيش   *Nc*   *Khûrbet Dabbîsh.* The ruin of Dabbish; perhaps from دبش، which in vulgar Arabic means a 'clod.'

خربة الدير   *Mc*   *Khûrbet ed Deir.* The ruin of the convent.

خربة الدبس   *Nc*   *Khûrbet ed Dibs.* The ruin of grape-syrup; but this is probably connected with the Hebrew דבשה 'a hill.'

خربة دكمش	Mb	*Khŭrbet Dŭkmash.* The ruin of the purblind man. Vulgar for *dakash;* this also means a 'pole for raking out an oven,' or 'stirring a fire.'
خربة عزية الفوقآ	Mc	*Khŭrbet 'Ezziyât el Fôka.* The ruin of the upper 'Ezziyah.
خربة الفريوية	Nc	*Khŭrbet el Fureiwiyeh; see 'Ain el Fureiwiyeh.*
خربة حودة	Mc	*Khŭrbet el Hammûdeh.* The ruin of Hammûdeh. p.n. from حمد 'to praise.'
خربة الحدية	Mc	*Khŭrbet el Hanîyeh.* The ruin of the bend.
خربة جارودية	Mb	*Khŭrbet Jârûdiyeh.* The ruin of the Jârûdiyeh (a Shiah sect, followers of Abu'l Jârûd Ziyâd).
خربة جبل الكبير	Nc	*Khŭrbet Jebel el Kabîr.* The ruin of the great mountain.
خربة جنجيل	Nc	*Khŭrbet Junjeil.* Probably connected with or corrupted from *jundeil,* diminutive of *jandal,* a strong place.
خربة جوار النخل	Nb	*Khŭrbet Juwâr en Nŭkhl.* Jawâr means both 'abundant and deep water,' and an 'exterior courtyard.' *Nŭkhl* means a palm-tree; but is perhaps connected with the Hebrew נַחַל a water-course.
خربة قابو المعيصرة	Mc	*Khŭrbet Kabu el Ma'tsreh.* The ruin of the cellar or vault of the olive press.
خربة قبور الرصاص	Mc	*Khŭrbet Kabûr er Resâs.* The ruin of the leaden tombs.
خربة الخمسية	Nc	*Khŭrbet el Khamsîyeh.* The ruin of the Khamsiyeh (from خمسة five).
خربة خشنه	Nc	*Khŭrbet Khashnah.* The ruin of the coarse ground.
خربة خوتة	Mc	*Khŭrbet Khôta.* The ruin of Khota; perhaps from خوات the noise of a flowing torrent, &c.
خربة الكنيسة	Nc	*Khŭrbet el Kuneiseh.* The ruin of the church.
خربة القرين	Mc	*Khŭrbet el Kurein.* The ruin of the little horn, or peak.
خربة المعليا	Mc	*Khŭrbet el Mâllîyeh; see 'Ain el Mâllîyeh.*
خربة الملوحيه	Mc	*Khŭrbet el Malûhiyeh; see Khŭrbet el Mellâhah.*
خربة مشتا الاشحر	Nc	*Khŭrbet Mashta el As-her.* The ruin of the dusty-red winter quarters.
خربة المدفنة	Mc	*Khŭrbet el Medfeneh.* The ruin of the burial place.
خربة الملاحة	Nc	*Khŭrbet el Mellâhah; see 'Akabat Khŭrbet el Mellakha.*

خربة المرج   Mc   *Khŭrbet el Merj.* The ruin of the meadow or 'prairie.'

خربة مرج النسر   Nb   *Khŭrbet Merj en Nŭsr.* The ruin of the meadow of the eagle.

خربة ميماس   Mc   *Khŭrbet Mimâs.* The ruin of Mimâs. p n.

خربة المصبح   Nc   *Khŭrbet el Musebbah.* *Musebbah* means a man who travels till morning, or who gives a morning draught to another.

خربة المثنية   Mc   *Khŭrbet el Muthniyeh.* The ruin of the land twice turned over for sowing.

خربة رأس اللوزة   Mc   *Khŭrbet Ras el Lôzeh.* The ruin of the almond-tree head (fountain head or cape).

خربة شحور القنا   Mc   *Khŭrbet Shahûr el Kaná.* The conspicuous or notable parts of the aqueduct.

خربة السدين   Mb   *Khŭrbet Siddein.* The ruin of the two ramparts.

خربة الشرافيات   Nb   *Khŭrbet es Sherâfiyât.* The ruin of the cornices or parapets.

خربة شدغيث   Nb   *Khŭrbet Shidghîth.* See '*Ain Shidghîth,* p. 2.

خربة سويدية   Nc   *Khŭrbet Suweidîyeh.* The blackish ruin ; but *see Jubb Sûweid.*

خربة الطيبة   Mc   *Khŭrbet et Taiyebeh.* The ruin of sweet or wholesome water.

خربة طربة   Nb   *Khŭrbet Tarabiyeh.* The spring of the Tarabiyeh. The word means 'joy' or 'mirth.'

خربة تل القاضى   Nb   *Khŭrbet Tell el Kâdy.* The ruin of the Judge's mound.

خربة الورديانة   Nc   *Khŭrbet el Wardiâneh.* Either from *ward*, 'to go down to a well or spring for water,' or the modern Arabic word borrowed from the Italian, and meaning 'guardians.'

خربة يارين   Nc   *Khŭrbet Yárin.* The ruin of Yárin. p.n.

خربة الزهيرية   Mc   *Khŭrbet ez Zaheiriyeh.* The ruin of the little flowers. But probably a family name.

خربة الزعترية   Mc   *Khŭrbet ez Zàteriyeh* (properly *Zatheriyeh*). The ruin of thyme.

خريبة   Nc   *Khureibeh.* The little ruin.

الكنيسة   Nc   *El Kuneiseh.* The little church.

القرية   Nb   *El Kureih.* The little village.

كرم دابل Nb *Kurm Dâbil.* The vineyard of Dâbil ; perhaps from *dhabel,* 'withering.'

كرم مسرتن Nc *Kurm Muserten.* The vineyard of Muserten. p.n.

كرم السوقى Nb *Kurm es Sûky.* The vineyard of the market-man.

ليلة or ليلى Mc *Leileh.* Leileh, a female proper name. It means also 'night.'

مالكية Mc *Mâlkiyeh.* The royal (building). It may refer to the (Christian) Malekite sect, but is more probably a reminiscence of the worship of the Tyrian deity Melkarth, 'the city king.'

منصورة Mc *Mansûrah.* Mansûr's (building) ; *cf. Neby Mansûr.*

معشوق *Mâshûk.* Beloved ; *see Nebi Mâshûk.*

مرج قليلة Mc *Merj Kuleileh.* The meadow of the hillock.

مرج منصورة Mc *Merj el Mansûrah.* The meadow of Mansûrah. p.n.

مغارة جورة ادريس Nb *Mûghârat Jûrat Idrîs.* The cave of the water-hole of Enoch.

مغارة ابى ذكر {Mb / Nw} *Mughârat Abu Dhekr.* Abu Dhekr's cave. p.n.

مغارة اللواتين Mb *Mughârat el Lawatin* (properly *Lawwâtin*). The cave of the plasterers. The word has also an indecent meaning.

محمد نوف Nb *Muhammed Nûf.* Mohammed Nûf. p.n.

نهر القاسمية Nb *Nahr el Kâsimiyeh.* The river Kâsimiyeh.

The old and generally accepted interpretation of the word *Kâsimiyeh* as the 'boundary river,' from the Arabic root *Kasam,* 'to divide,' is inadmissible, the latter word meaning to 'apportion,' and not to 'divide,' in the sense alluded to. It is unquestionably named from *Sheikh* or *Neby Kâsim,* whose shrine is still the most venerated in the country, and who is to be identified with the Phœnician Cadmus, Cadmiel or Casmiel.

نبع العزية Mc *Nebâ el 'Ezziyeh.* The spring of 'El Ezziyeh ; q.v.

نبى عمران Mc *Neby 'Amrân.* The prophet Amrân. Neby means here a 'prophet's tomb' or 'shrine.' Amrân (or rather Imrân), who is *not* a Mohammedan prophet, is identical with the Amram of the Bible, the father of Moses. He is mentioned several times in the Kor'ân.

c

نبى الجليل Nc *Neby el Jelil.* The prophet El Jelil, 'the illustrious.' Probably a reminiscence of the name Galilee.

نبى قاسم Mb *Neby Kâsim.* The prophet Kâsim; *see Nahr el Kâsimîych.*

نبى منصور Mc *Neby Mansûr.* The prophet Mansûr (victor).

نبى معشوق Mb *Neby Mâshûk.* The prophet Mâshûk (beloved).

The word *Mâshûk* means 'beloved.' It is needless to say that no prophet of the name exists in either the Mohammedan or Christian hagiology. The word is probably a survival of the ancient title of the deity to whom the shrine was dedicated, namely the Tyrian Hercules, *Baal Moloch* or *Melkarth*, worshipped under the Egyptian name of Mi-amun or Memnon,' the Beloved of Ammon,' to whom, as Lucian distinctly tells us, an *Egyptian* temple existed at Tyre. The name may be compared with that of Abraham, *Khalil Allah*, 'the Friend of Allah,' and with that of the Cretan Zeus, called by the Greeks Meilichios, who is also identified with this deity.

رأس العين، Mc *Râs el 'Ain.* The head of the spring.

رميدية Nc *Rumeidîych.* The ashy or ash coloured.

رشيدية Mc *Rusheidîych.* The building of Rusheid Pasha.

رويسات عبد علي Mc *Ruweisat 'Abd 'Ali.* 'Abd 'Ali's 'heads,' or hill-tops.

رويسات البيارة Nc *Ruweisat el Beiyârah.* The hill-tops of the well-sinkers.

سهل معشوق Mb *Sahel Ma'shûk.* Ma'shûk's plain.

سماعية Mc *Semmââiych.* Semma'iych. p.n.

شعثيه Nc *Shâtîych.* 'Dishevelled,' 'straggling.'

شيخ داود Mb *Sheikh Dâûd.* Sheikh David. *Sheikh*, elder, is used for a saint, or saint's tomb. *Dâûd* is the Arabic form of the name David, but the shrine must not be confounded with that of the Hebrew Poet King. It is most probably another of the shrines of Melkarth, who, according to Syncellus, was called by the Phœnicians Dibdan, Διβδαν The name David itself is identical with *Mâshûk* in signification; *see Neby Mâshûk* above.

شقيف الاحمر Mc *Shakif el Ahmar.* The red cleft.

شقيف الدلافه Nc *Shakîf ed Dellâfeh.* The cleft of dripping water.

شرنى Mb *Shernei.* Shernei. p.n.

الصور Mb *Er Sûr.* Tyre.

تيردبّة Nb *Teir Dubbeh.* The fortress of the bear. The meaning of this word *teir* is very doubtful. The Arab scribe who wrote out the list of names has added a note, written in pencil, in Arabic, explaining that it 'means a beam or partition between two walled gardens, and is sometimes used as an equivalent for *tîh';* the latter word he interprets 'desert.' His explanation is, however, taken *verbatim* from the smaller Arabic dictionary by Bustani, published in Beirût (*Katr el Muhît*), and he has misunderstood the word *tîh*, which, as there used, means 'pride,' and is coupled with the synonym *kibr.* It is probably the Aramaic word בירה (طير) meaning 'fort,' or 'court.'

وادى الثغرة Mb *Tell eth Thoghrah.* The mound of the frontier road.

تورة Nb *Torah.* Flowing water.

وادى ابى حسن Nc *Wâdy Abu Hasan.* The valley of Abu Hasan. p.n.

وادى العتّاب Nc *Wâdy el 'Akkâb.* The valley of the 'Akkâb; probably عَتّاب the plural of 'Akabah, 'mountain roads,' or عَتّاب 'a mass of projecting rock,' or 'a projecting stone in a well.'

وادى العاور Nc *Wâdy el 'Awar.* The valley of the one-eyed man.

وادى فراويات {Mc Nc} *Wâdy Furâwîyât; see 'Ain el Furâwîyât.*

وادى الحبيشية Nb *Wâdy el Hubeishiyeh.* The valley of the Abyssinian; *see 'Ain el Hubeishîyeh.*

وادى الحمرانية Nb *Wâdy el Hamrânîyeh.* The valley of the Hamrâni Arabs.

وادى سليم Nc *Wâdy Islim.* Selim's valley. p.n.

وادى الجزائر Nc *Wâdy el Jezâir.* The valley of the islands.

وادى الجسر Mc *Wâdy el Jisr.* The valley of the bridge.

وادى الجديدة *Wâdy el Judeiyideh.* The valley of the dykes. (Geological term, meaning streaks in the rock.)

وادى الكنيسة Nc *Wâdy el Kenîseh.* The valley of the church.

وادى كردية Nc *Wâdy Kurdîyeh.* The valley of the Kurds.

وادى المعليا Mc *Wâdy el Mâllîyeh.* The valley of el Mâllîyeh; *see 'Ain el Mâllîyeh.*

وادى المنصور	Mc	*Wâdy el Mansûr.* Mansûr's valley; *see Neby Mansûr.*
وادى المزيرع	Mc	*Wâdy el Mezeirâh.* The valley of sown land.
وادى الموارخة	Nc	*Wâdy el Muweirikheh.* The valley of flowing waters.
وادى الرصاص	Nc	*Wâdy er Rasâs.* The valley of lead.
وادى السماعية	Mc	*Wâdy el Semmâäiyeh.* The valley of Semmâäiyeh.
وادى الشمالية	Mc	*Wâdy es Shemâliyeh.* The northern valley.
وادى تورة	Nb	*Wâdy Tôrah.* The valley of flowing water; *see 'Ain Torah.*
وادى الوعر	Mc	*Wâdy el Wâr.* The valley of rugged ground.
وادى يانوح	Nb	*Wâdy Yânûh.* The valley of Yânûh. The Janoah of the Bible (2 Kings xv, 29).
وادى اليهودية	Ma	*Wâdy el Yehûdiyeh.* The valley of the Jews.
وادى الزاروب	Nc	*Wâdy ez Zârûb.* The valley of the water-pipe.
زاروب نمار	Nb	*Zârûb nimmâr.* The pipe of abundant water.

SHEET II.

العبابيد Qc *El Ababid.* The diverging roads (this has no connection with the word *'Abbad*, 'devotee'; see *Sheikh Abbad*).

آبل Qb *Abl.* Probably the Abel Beth Maachah of 2 Sam. xx, 14–22; 1 Kings xv, 20.

ابريخا Pb *Abrikha.* Abrikha. p.n.

العبسيه Rc *'Absiyeh.* Absiyeh. p.n.

اخصاص الحلبية Qc *Ahsas el Halabiyeh.* A mistake for *Akhsas el Halabiyeh.* q.v.

عيديب Pb *'Atdib.* A local form connected with عدوب 'much sand.'

عين ابى غيبة Ob *'Ain Abu Gheibeh.* The spring of the low ground.

عين ابى كسية Nc *'Ain Abu Kuseiyeh.* Abu Kusaiyeh's spring. p.n.

عين ابى نحله Ob *'Ain Abu Nahleh.* The spring of the bees, but *nahleh* in its Hebrew form means 'stream.'

عين ابى سدون Pb *'Ain Abu Sudun.* Abu Sudun's spring. The word *sudun* means the coverings of a camel-litter. It may, however, be a local form or plural of *sadu*, the office of attendant at the Kaabeh, which was hereditary in the family of *Abu dar*, a mosque to whose honour is found in this sheet. Abu Sudun would then be identical with the name Abu Dar itself.

عين عيديب Pb *'Ain 'Aidib.* The well of 'Aidib.

عين عقبة القندوله Pb *'Ain Akabet el Kandoleh;* see *Akabet el Kanduliyeh.*

عين العلق Rb *'Ain el 'Alak.* The spring of the leeches.

عين العريض Oc *'Ain el Arid.* The spring of the broad mountain-side.

عين بانياس *'Ain Banias.* The spring of Banias. q.v.

عين الباردة Nb *'Ain el Bardeh.* The spring of cool water.

عين باريش Ob *'Ain Barish.* The spring of the place abounding in herbage.

عين بيبعان Pc *'Ain Beib'an.* The spring of Beib'an. p.n.

عين البنيه Pb *'Ain el Beneiyeh.* The spring of the little building.

عين برعشيت	Pc	'Ain Berâshît. The spring of Berâshit. p.n.
عين بروخيه	Ob	'Ain Berûkhei. The spring of Berûkhei. p.n.
عين بدياس	Nb	'Ain Bidiâs. The spring of Bidiâs. p.n.
عين بركة صوان	Pc	'Ain Birket Suwân. The spring of the Flint Pool.
عين بطيطا	Oc	'Ain Buteitah. The spring of the little duck (or 'oil bottle').
عين الداليه	Ob	'Ain ed dâlieh. The spring of the hanging vine.
عين الضيعة	Nb	'Ain ed Deiâh. The spring of the farm.
عين دير النوقا	Ob	'Ain Deir el Fôka. The upper convent spring.
عين دير قانون	Nb	'Ain Deir Kanûn. The spring of the convent of the Rule (canon).
عين دير كيفا	Ob	'Ain Deir Kifa. The spring of the convent of Kifa. p.n.
عين دير ميماس	Qb	'Ain Deir Mîmâs. The spring of the convent of Mimâs. p.n.
عين دلافه	Qc	'Ain Dellâfeh. The trickling spring.
عين الدردارة	Qb	'Ain ed Derdârah. The spring of the Derdârah (elm) tree.
عين الذهب	Qc	'Ain edh Dhaheb. The golden spring.
عين الدروز	Oc	'Ain ed Drûz. The Druze's spring.
عين الفقيعه	Nb	'Ain el Fakâieh. The spring of the truffle.
عين النوقا	Qc	'Ain el Fôka. The upper spring.
عين الفوار	Nb	'Ain el Fuwâr. The spring of the fountain.
عين الغنم	Qb	'Ain el Ghanem. The spring of the flocks.
عين الغزار	Nb	'Ain el Ghazzâr. The spring of abundant water.
عين الغسيل	Ob	'Ain el Ghûsîl. - The spring of washing (perhaps corrupted from ghasûl, 'marsh-mallows').
عين الحجل	Qb	'Ain el Hajl. The spring of the partridge.
عين الحجر	Qb	'Ain el Hajr. The spring of the rock.
عين حرب	Qc	'Ain Harb. The spring of war.
عين الحوش	Qb	'Ain el Hosh. The spring of the fenced-in enclosure.
عين الحمه	Pc	'Ain el Hǔmma. The hot spring.

عين الحمرآ	Ob	'Ain el Hûmra. The red spring.
عين الحمرآ الفوقآ	Ob	'Ain el Hûmra el fôka. The upper red spring.
عين الحمرآ التحتآ	Ob	'Ain el Hûmra et tahta. The lower red spring.
عين حور	Nb	'Ain Hûr. The spring of the 'water-hole,' or plane-tree.
عين جديدد	Ob	'Ain Jedîdeh. The new spring.
عين جهير	Ob	'Ain Jehir. The spring of Jehir. The meaning is 'conspicuous'; but it is probably misspelt, and connected with juhr, جحر 'a den.'
عين الجبنان	Ob	'Ain el Jenân. The spring of the gardens.
عين جنّاتا	Nb	'Ain Jennâta. Connected with جنّة jennah, 'a garden.'
عين الجوزد	Nc	'Ain el Jozeh. The spring of the almond tree.
عين جبيب	Ob	'Ain el Jubeib. The spring of the little pit.
عين القنطرد	Pb	'Ain el Kantarah. The spring of the arch.
عين كفره	Oc	'Ain Kefrah. The spring of the village.
عين كفر كلا	Ob	'Ain Kefr Kila. The spring of the village on the valley side.
عين قرزون	Ob	'Ain Kerzôn. The spring of Kerzon. p.n.
عين الخان	Pb	'Ain el Khân. The spring of the Caravanserai.
عين القنية	Pc	'Ain el Kinîyeh. The spring of the aqueduct.
عين القبى	Nb	'Ain el Kubbei. The spring of the domed (building).
عين اللدان	Re	'Ain el Leddân. The spring of the (River) Leddân.
عين لوبية	Qb	'Ain Lûbiyeh. The spring of Lubiyeh; a kind of bean.
عين مرنبة	Ob	'Ain Marnabah. The spring of Marnabah. p.n.
عين معروب	Ob	'Ain Marûb. The spring of abundant water.
عين المطمورد	{Nb}{Oc}	'Ain Matmûrah. The spring of the 'Matamore,' i.e., a subterranean granary.
عين الميسة	Oc	'Ain el Meiseh. The spring of the meiseh tree.
عين الملح	Qb	'Ain el Mellâh. The spring of the salt-worker, or 'mariner.'
عين المتاولة	Qb	'Ain el Metâwileh. The spring of the Metawileh, a sect of the Shiah Mohammedans; probably deriving their origin from the Hashshâshin (Assassins).

عين المزراب {Oc} {Pc} 'Ain el Mezrâb. The spring of the water-pipe, or spout.

عين المزرعة Nb 'Ain el Mezrâh. The spring of the corn-land.

عين الموسى Qc 'Ain el Mûsa. The spring of the razor.

عين المسمار Oc 'Ain el Musmar. The spring of the nail.

عين نائب Qb 'Ain Nâïb. The spring of the 'Deputy:' ناب nâb is also used of wild beasts drinking at a stream.

عين الندا (ندى) Qb 'Ain en Nedâ. The spring of moisture.

عين نويا Ob 'Ain Nuweiya. The spring of Nuweiya; perhaps diminutive of نو nau', 'a storm,' 'rain.'

عين الراموح Ob 'Ain er Râmûh. The spring of wild barley.

عين الريحانة Nb 'Ain er Rîhâneh. The spring of sweet-smelling herbs (sweet basil).

عين رويحينة Qc 'Ain Rûeihîneh. Diminutive form of rîhâneh.

عين الرميلة Nb 'Ain er Rumeileh. The sandy spring.

عين الصبيح Nb 'Ain es Sabbîh. The spring of 'watering camels at dawn.'

عين صفد Pc 'Ain Safed. The spring of Safed. q.v.

عين السهيد Ob 'Ain es Sahîa. Perhaps for صيبة, a booth set up by desert Arabs near to water as a shelter.

عين صريفا Ob 'Ain Sarîfa. The spring of Sarifa. q.v.

عين السفلى {Nc} {Qc} 'Ain es Sifla. The lower spring.

عين صقور Qc 'Ain Sukûr. The spring of hawks or kites.

عين الطافورة Pb 'Ain et Tâfûrah. From tafara, 'to leap over a brook or rivulet'; also 'to spring up.'

عين الطيبة Qb 'Ain et Taiyebeh. The goodly spring.

عين طلحة Qc 'Ain Talhah. The spring of the Talha tree.

عين تير قلسية Nb 'Ain Teir Filsich; see Teir Filsich.

عين تير سمحات Ob 'Ain Teir Samhât; see Teir Samhât.

عين تير زنبة Oc 'Ain Teir Zinbeh; see Teir Zinbeh.

عين التين {Nc} {Oc} 'Ain et Tîn. The spring of the fig.

عين التينة Rb 'Ain et Tineh. The spring of the fig-tree.

عين ام عكوش   Qb   *'Ain Umm 'Okûsh.* The spring of the thicket.

عين الوردة   Pc   *'Ain ed Werdeh.* The spring of the going down to water.

عين اليانوح   Nb   *'Ain el Yânûh.* The spring of Yanûh (the Biblical Janoah).

عين الزرقآ   Oc   *'Ain ez Zerkâ.* The spring of the blue glittering water.

عيتيت   Nc   *'Aitît.* Aitit. p.n.

عيتا الزط   Oc   *'Aita ez Zutt.* The high mountain of the Zutt. *Zutt* is the name by which a certain tribe of Indian gipsies, the *Jats*, are known in Syria.

عقبة علي   Nb   *'Akabat 'Aly.* Aly's steep, or mountain road.

عقبة القندولية   Pc   *Akabat al Kandûlieh.* The steep or mountain road of the Kandûlieh (a thorny tree).

عقبة القنطرة   Oc   *'Akabat al Kantarah.* The steep or mountain road of the arch.

عقبة الخرار   Ob   *'Akabat al Kharrar.* The steep or mountain road of rushing water.

عقبة المقاتيل   Pc   *'Akabat al Makâtîl.* The steep or mountain road of the slain.

عقبة الزقية   Pb   *Akabat ez Zûkkiyeh.* The steep or mountain road of the Zukkiyeh. Connected with زق a waterskin, زق 'a diver' (waterfowl), or زقاق a narrow road or lane.

اخصاص الحلبية   Qc   *Akhsâs el Halabiyeh.* The silk-worm frames of the Aleppines.

علمان   Pb   *Almân.* Almân. Perhaps from علم 'a sign-post' or 'a mountain.' Hebrew עלמון

ارض الابيض   Pb   *Ard el Abiad.* The white land.

ارض ابى مكنا   Qb   *Ard Abu Mukanná.* The land of Abu Mukanna. p.n.

ارض ابى رماد   Nb   *Ard Abu Ramád.* The ashy or ash-coloured land.

ارض عاصى المعلاوى   Qc   *Ard 'Asy el Mâllâwy.* The land of the 'Asy el Mâllâwy. p.n.

ارض البوزيه   Qc   *Ard el Bûzîyeh.* The land of hawks.

ارض دفنه   Rc   *Ard Dufneh.* The land of Daphne (Oleander). In Arabic the word would also mean 'burial.'

ارض حرشاح   Oc   *Ard Hirshâh*. The land of Hirshâh. p.n.

ارض القنطرد   Pb   *Ard el Kantarah.* The land of the arch.

ارض راج   Pb   *Ard Râj.* The land of Râj. p.n.

ارض الرميديه   Rc   *Ard er Rumeidich.* The land of ashy ground.

ارض شوتا   Rc   *Ard Shoka.* The land of Shoka. شوتآ is the feminine of اشوق 'tall.'

ارض الصدور   Pc   *Ard es Sudûr.* The land of prominent places.

ارض التتابيع   Rc   *Ard et Tetâbiâ.* The land of successive followings.

عريض الابيض   Pb   *'Arîd el Abyad.* The wide white mountain side.

عريض الكعقور   Rc   *'Arîd el Kâkûr.* The broad mountain side of the cairn.

عريض الكندوليه   Pc   *'Arîd el Kandôliyeh.* The broad mountain side of the Kanduliyeh.

عريض المدوّرد   Nb   *'Arîd el Mudauwarah.* The circular mountain side.

عريض المزيبة   Sc   *'Arîd el Muzeibileh.* The mountain side of the little dung heap.

عريض شومر   Qb   *'Arîd Shomar.* The mountain side of Shomar. Shomar is the name of a kind of herb (wild fennel), and also of several Arab heroes.

عتشيش   Pb   *'Atshîsh.* 'Atshish. p.n.

عيون العجّال   Qc   *'Ayûn el 'Ajjâl.* The springs of the herd of cattle.

عيون الغزلان   Qb   *'Ayûn el Ghûzlân.* The springs of the gazelles.

عيون الجديديه   Qc   *'Ayûn el Judeidiyeh.* The springs of the dykes (streaks in rock).

عيون الخان   Pc   *Ayûn el Khân.* The springs of the Caravanserai.

عيون المغر   Qc   *Ayûn el Mughr.* The springs of the caves.

بانلى   Ob   *Bâflet.* Baflai. p.n.

بانياس   Sc   *Bâniâs.* Banias, the ancient Panium or Panias.

باريش   Ob   *Bârîsh.* Abounding in herbage.

باب الثنية   Rb   *Bab eth Theniyeh.* The gate of the mountain road

بائكة فرنسيس   Rc   *Baikat Fransîs.* Francis' cattle-shed.

بين الرؤس   Nc   *Bein er Rûs* (*ru-ûs*). The space between the hill-tops.

البياض   Nc   *El Beiyâdh.* The white spot.

بياض صريفا   Ob   *Beiyâd Sarîfa.*

بلاد بشارة   {Oc} {Pc}   *Belâd Beshâra.* The Beshârah country. The Beshârah family are a noble and once very powerful house in Syria.

بنو حيان   Pc   *Beni Haiyân.* The Beni Haiyân, an Arab tribe.

بَرَشِيت   Pc   *Berâshit.* Berâshit. p.n.

بستان القيطية   Rc   *Bestân el Keitîyeh.* The Keitîyeh's garden. p.n.

بيار دير سريان   Qb   *Biâr Deir Suriân.* The wells of the Syrian convent.

بئر بيت الشيب   Nc   *Bîr beit esh Shaib.* The well of the house of the old man.

بئر الحاج يونس   Oc   *Bîr el Hâjj Yûnis.* The well of the pilgrim Jonas.

بئر الحواش   Rb   *Bîr el Hôwâsh.* The well of the fenced-in enclosure.

بئر كنيسة   Qb   *Bîr Kuneiseh.* The well of the little church.

بئر ميتون   Oc   *Bîr Meitûn.* The well of Meitûn. p.n. from ميت 'dead.'

بئر المصلبات   Qc   *Bîr el Musellabât.* The well of the crossed (roads).

بئر الساتفة   Nc   *Bir es Sâkfeh.* The roofed-over well.

بئر شبلى   Oc   *Bîr Shibly.* Shibly's well. Probably from Abu Bekr Shibly, a descendant of Ali, and founder of a sect of dervishes.

بركة بنى حيان   Pc   *Birket Beni Haiyân.* The pool of the Beni Haiyân (an Arab tribe).

بركة دير السريان   Pb   *Birket deir es Suriân.* The pool of the convent of the Syrians.

بركة الغُرز   Qb   *Birket el Gharz.* The pool of the rushes.

بركة حدائا   Oc   *Birket Haddâtha.* The pool of Haddâtha. p.n.

بركة الحجر   Qc   *Birket el Hajr.* The pool of the rock.

بركة الحمرآ   Sb   *Birket el Hamrâ.* The red pool.

بركة اكساف   *Birket Iksâf.* The pool of Iksâf.

بركة الجلمة   Nc   *Birket el Jelameh.* The pool of the hill. جلمة *jelameh*, is properly written جلبمة *jelhameh*.

بركة القرقاف   Pc   *Birket el Kárkáf.* The pool of the Karkáf (a small bird).

بركة نقيّة   Pc   *Birket Nakíyeh.* The pool of the sand-worms.

بركة راج   Pb   *Birket Ráj.* The pool of Ráj. p.n.

بركت الطيبه   *Birket et Taiyebeh.* The pool of sweet water.

         Ob   *Bistáth.* p.n. The Arabic form of this is not given in the lists; it might be either بسطاث بستاث or بسطاث.

برج العلوى   Ob   *Burj el Alawei.* The tower of the Alawei. p.n.

دحروج   Qc   *Dahrûj.* 'Rolling.'

الدير   {Oc Pc}   *Ed Deir.* The convent.

دير ابى دى   Ob   *Deir Abu Deî.* The convent of Abu Dei. p.n.

دير عال   Oc   *Deir 'Ál.* The lofty convent.

دير عامص   Ne   *Deir 'Ámis.* The convent of 'Ámis. p.n.

دير عقرب   Oc   *Deir 'Akrab.* The convent of the scorpion.

دير دنيّا   Ob   *Deir Dughîya.* The convent of Dughíya. p.n.

دير قنطار   Oc   *Deir Kantár.* The convent of arches.

دير قانون   Nb   *Deir Kánûn.* The convent of the Rule (canon).

دير كيفا   Ob   *Deir Kîfa.* The convent of Kifa. p.n.

دير ميماس   Qb   *Deir Mîmás.* The convent of Mimás. p.n.

دير السريان   Pb   *Deir es Suriân.* The convent of the Syrian.

ظهر عين الجنان   Ob   *Dhahr 'Ain el Jenân.* The ridge of the spring of gardens.

ظهر الاصفر   Ob   *Dhahr el Asfar.* The yellow ridge.

ظهر الحاصبانى   Rb   *Dhahr el Hásbány.* The Hasbány ridge.

ظهر قبر النصرانى   Nb   *Dhahr Kabr en Nusrání.* The ridge of the Christian's tomb.

ظهر الكرم   Nb   *Dhahr el Kurm.* The ridge of the vineyard.

ظهر السيد   Ob   *Dhahr es Seiyid.* The ridge of the lord.

ظهر السوق   Nb   *Dhahr es Sûk.* The ridge of the market.

ظهر يعقوب   Ob   *Dhahr Ya'kûb.* Jacob's ridge.

ظهرة البياض   *Qb*   *Dhahret el Beiyâd.* The ridges of the white place.

دبعال   Oc   *Dib'âl.* Dib'âl. p.n.

فرون   Pb   *Furûn.* The ovens or reservoirs.

الغجر   Rb   *El Ghajir.* Perhaps the Jazer (Gazoros) of 1 Maccab. v, 8.

الغزارد   {Nb / Qc}   *El Ghazzârch.* Abundant.

حجر المقدوح   Sc   *Hajr el Makdûh.* The scooped-out stone.

حلوسية   Nb   *Hallûsîyeh.* Hallûsîyeh. Perhaps from حلس abundant herbage.

حمام صفد   *Hammam Safed.* The hot baths of Safed.

حنوى   *Hanawei.* Hanawei. p.n. حنو the bend of a valley.

حريس   Oc   *Harîs.* Guarded.

حا هورة   *Qb*   *Hima Hûra.* The preserve of Hûra. p.n. A *Hima* was a parcel of ground set apart for the sole pasturage of the cattle belonging to the king or chief.

الحولة   Qc   *El Hola.* Properly الحوّل, dark green herbage.

حصن تبنين   Oc   *Hosn Tibnîn.* The fort of Tibnîn.

حميرة   Ob   *Humeireh.* The red patch.

هونين   Qc   *Hûnîn.* Hunîn. p.n.

جامع ابى در   Qc   *Jamià Abu Dar.* The mosque of Abu Dar. (One of the companions of Mohammed, and a prominent character in the early history of Islam.)

جواوير   Rb   *Jawâwîr,* pl. of جورة A hollow in the ground for water.

جبل ابى غيب   Ob   *Jebel Abu Gheibeh.* Abu Gheibeh's mountain.

جبل ابى قطائا   Ob   *Jebel Abu Katât.* The craggy mountain.

جبل العين   Ob   *Jebel el 'Ain.* The mountain of the spring.

جبل عين الشقفان   Ob   *Jebel 'Ain esh Shûkfân.* The mountain of the spring of clefts.

جبل بعيج   Oc   *Jebel Buâij.* The mountain at the wide part of the valley.

جبل البرج   Pb   *Jebel el Burj.* The mountain of the tower.

جبل عسيه   Nb   *Jebel 'Esîa.* Mount 'Esîa. p.n.

جبل حونين   Qc   *Jebel Hûnîn.* Mount Hûnin. p.n.

جبل حسين   Qc   *Jebel Husein.* Mount Husein. p.n.

جبل جمة   Oc   *Jebel Jumleh.* The mountain of Gamala (mentioned in the Talmud as a Galilean site).

جبل القلعة   Qc   *Jebel el Kŭlàh.* The mountain of the castle.

جبل ميدان   Qb   *Jebel Meidân.* The mountain of the plain.

جبل المروج   Qc   *Jebel el Merûj.* The mountain of the prairies.

جبل المشاتى   Ob   *Jebel el Meshâty.* The mountain of the winter quarters.

جبل المصلبات   Qc   *Jebel el Musellabât.* The mountain of the cross (roads).

جبل الرويس   Qc   *Jebel er Rûeis.* The mountain of the hill-tops.

جبل الشيخ   Nb   *Jebel esh Sheikh.* The mountain of the Sheikh. A sheikh is an 'elder,' the chief of an Arab tribe, encampment or village; the word is also sometimes used for a 'saint.'

جبل الصيرة   Sb   *Jebel es Sîrah.* The mountain of the fold.

جبل الشحيل   Sb   *Jebel esh Shuheil.* The mountain of Shuheil. p.n. Perhaps شُهَيل 'green-eyed.'

جبل السلطان   Ob   *Jebel es Sultân.* The mountain of the Sultan.

جبل تير سمحات   Ob   *Jebel Teir Samhât; see Teir Samhât.*

جبل التينة   Ob   *Jebel et Tineh.* The mountain of the fig-tree.

جبل الثور   Ob   *Jebel et Tor. Tor* also signifies mountain.

جبل ام على   Rb   *Jebel Umm Ali.* Mount Umm 'Ali, Aly's mother.

جبل وردد   Qc   *Jebel Werdeh.* The mountain of going down for water.

جبل اليعقوص   Qb   *Jebel el Yâkûsah.* The mountain of the pathless sand.

جناتا   Nb   *Jennâta.* Jennâta, from *Jennat,* 'a garden.'

الجزيرة   Nb   *El Jezîreh.* The island.

جيلو   Nc   *Jîlû.* See '*Ain Jîlû*, p. 1.

جسر قعقعية   Pb   *Jisr Kâkâiyeh.* The bridge of Kâkâiyeh. The word means 'rumbling,' 'clanking'; also a small piebald bird.

جسر القبلى   Sc   *Jisr el Kibly.* The southern bridge.

جسر المشرع   Rc   *Jisr el Mishrà.* The bridge of the straight way.

البميجمه Pc *El Jumeijmeh.* 'The little sculls'; but *Jamâjim* pronounced Syrian fashion, *Jamaijim*, also means the piece of wood at the end of which the ploughshare is fastened.

جورة عبدالله Se *Jûrat 'Abdallah.* Abdallah's water-hole. p.n.

جورة مسيعد Pb *Jûrat Museiâd.* Museiâd's water-hole. p.n.

جوار دير قانون Nb *Juwâr Deir Kânûn.* The neighbourhood or water-holes of the convent of the Rule (canon).

جويا Nc *Juweiya.* Juweiyâ. p.n.

قبر عجائب Rc *Kabr 'Ajâib.* The tomb of wonders.

قدا قنيطره   *Kada Kuneitrah.* The Kuneitrah (little arch) district.

قلوى Ob *Kalawei.* The ancient Kelabo.

قناية صندينه Rb *Kanâyet Sandîneh.* The aqueduct of Sandina. p.n.

القنطره {Pb Oc} *El Kantarah.* The arch.

كفره Oc *Kefrah.* The village.

كفر دونين Oc *Kefr Dûnîn.* The village of Dûnin. p.n.

كفر كلا Qb *Kefr Kila.* The village of the pasturage.

القيطية Rc *El Keitîyeh.* El Keitiyeh. p.n.

كنيسة مار جرجس Ob *Keniset Mâr Jirjis.* The church of St. George.

كنيسة مار توما Ob *Keniset Mâr Tûmâ.* The church of St. Thomas.

كروم العريض Qc *Kerûm el 'Arîd.* The vineyards of the mountain side.

قرزون Ob *Kerzon.* Kerzon. p.n.

الخالصة Qc *El Khâlisah.* Pure, sincere; the name appears in the south of Palestine, and is there identical with the Greek Elusa.

خلة العبد Qb *Khallet el 'Abd.* The dell of the slave.

خلة العين Pc *Khallet el 'Ain.* The dell of the spring.

خلة ابى عيسى Rb *Khallet 'Abu 'Aîsa.* Abu Aisa's dell.

خلة عين الغسيل Ob *Khallet 'Ain el Ghûsîl.* The dell of washing (or of marshmallows).

خلة عين جناتا Nb *Khallet 'Ain Jennâta.* The dell of the spring of gardens.

خلة عين الجوزة   *Nc*   *Khallet 'Ain el Jozeh.* The walnut spring.

خلة عين السنية   *Pb*   *Khallet 'Ain Suneiyeh.* The dell of the brilliant spring.

خلة العلماني   *Pc*   *Khallet el Alamâny.* The dell of the Alamâny.   p.n.

خلة علول   *Ob*   *Khallet 'Allûl.* 'Allûl's dell.   p.n.

خلة الوينه   *Oc*   *Khallet Alwineh.* Alwineh's dell.   p.n.

خلة العمود   *Ob*   *Khallet el 'Amûd.* The dell of the column.

خلة عصينة   *Nb*   *Khallet 'Asîfeh.* The dell of the wheat-ears.

خلة العوجآ   *Ob*   *Khallet el 'Aujah.* The crooked dell.

خلة العويدد   *Qb*   *Khallet el 'Aweideh.* 'Aweideh's dell.   p.n.

خلة عويزد   *Qb*   *Khallet 'Awîzeh.* 'Awizeh's dell.   p.n.

خلة البيت   *Oc*   *Khallet el Beit.* The dell of the house.

خلة برقيل   *Ob*   *Khallet el Berkîl.* The dell of bullets; but perhaps it should be بركيل, which in the vulgar Arabic means a short thick black snake.

خلة البئر   *Qc*   *Khallet el Bîr.* The dell of the well.

خلة الدحروج   *Qc*   *Khallet ed Dahrûj.* The dell of the rolling (stone).

خلة الدالية   *Pc*   *Khallet ed Dâlieh.* The dell of the hanging vine.

خلة الدروز   *Oc*   *Khallet ed Drûz.* The dell of the Druzes.

خلة الضبعة   *Rc*   *Khallet ed Dŭbâh.* The dell of the hyæna.

خلة الدوارنة   *Nb*   *Khallet ed Duwârneh.* The dell of the duwarneh (lotus) tree.

خلة الفراشة   *Qc*   *Khallet el Ferrâsheh.* The dell of the moths.

خلة الغربية   *Oc*   *Khallet el Gharbîyeh.* The western dell.

خلة حدائا   *Oc*   *Khallet Haddâtha.* The dell of Haddâtha.   p.n.

خلة الحجة   *Pb*   *Khallet el Hajjeh.* The dell of pilgrimage. The original meaning of the word is 'to repair to a place.'

خلة الحمارة   *Nb*   *Khallet el Hammareh.* The dell of asses.

خلة الحما   *Oc*   *Khallet el Hima.* The dell of the preserve.

خلة حسين   *Pc*   *Khallet Husein.* Husein's dell.

خلة الجبصينة   *Qc*   *Khallet el Jibsîneh.* The dell of gypsum.

خلة القباره	Nb	*Khallet Kabbârah.* The dell of tall trees.
خلة كروم العتاق	Qc	*Khallet Kerûm el 'Atâk.* The dell of old vineyards.
خلة الخنازير	Nc	*Khallet el Khanâzîr.* The dell of swine.
خلة الكرسى	Qb	*Khallet el Kursy.* The dell of the throne.
خلة اللوز	Pc	*Khallet el Loz.* The dell of the almond.
خلة الملول	Pb	*Khallet el Mellûl.* The dell of the oak trees.
خلة المذبح	Qc	*Khallet el Medbah.* The dell of the sacrificial altar.
خلة المزروره	Pb	*Khallet el Mezrûrah.* The narrow dell (lit. buttoned close).
خلة مينا	Qb	*Khallet Mîna.* The dell of Minâ. p.n.
خلة المغاره	{Qb / Nb}	*Khallet el Mughârah.* The dell of the cane.
خلة المرشد	Oc	*Khallet el Murshid.* The dell of the director or guide.
خلة النهير	Qc	*Khallet en Naheir.* The dell of the little river.
خلة النبع	{Qc / Qb}	*Khallet en Nebâ.* The dell of the perennial spring.
خلة النقع	Ob	*Khallet en Nǔkâ.* The dell of the marsh.
خلة عريق	Oc	*Khallet 'Oreik.* 'Oreik signifies 'a small mountain path,' also 'exudation.'
خلة الريحانه	Nb	*Khallet er Rîhâneh.* The dell of sweet-smelling herbs.
خلة سفرجا	Oc	*Khallet Saferja,* probably *Safarjel.* The dell of quinces.
خلة الساقية	Pc	*Khallet es Sâkiyeh.* The dell of the water-wheel.
خربة سلوم	Qc	*Khallet Sellûm.* The dell of Sellûm. p.n.
خلة السماقه	Pb	*Khallet es Semmâkah.* The dell of the Semmakah (Sumach) tree.
خلة الشبيه	Oc	*Khallet esh Shubeih.* The dell of the little *shabah* (a thorny plant used in medicine).
خلة السيد	Qc	*Khallet es Sîd.* The dell of the lord.
خلة الست	Qb	*Khallet es Sitt.* The dell of the lady.
خلة السوق	Pc	*Khallet es Sûk.* The dell of the market.
خلة التينة	Pb	*Khallet et Tîneh.* The dell of the fig-tree.
خلة التون	{Qc / Nb}	*Khallet et Tûn.* The dell of the lime-kiln.

E

خلة الزراقطا   Qc   *Khallet ez Zerâkit.* The dell of wasps.

خلة زريق   Pc   *Khallet Zereik.* The bluish dell.

الخان   {Nc} {Pc}   *El Khân.* The Caravanserai.

خان الدوير   Rc   *Khân ed Duweir.* The Caravanserai of the little convent or circle.

الخيام   Rb   *El Khiâm.* The tents.

خرب السبان   Rb   *Khurab es Sebban.* (Perhaps الصبان.) The ruins of the soapmaker.

الخربة   Qb   *El Khurbeh.* The ruin.

خربة ابي سركن   Oc   *Khŭrbet Abu Sirkîn.* Abu Sirkin's ruin. p.n.

خربة عيا   Nc   *Khŭrbet 'Aiya.* The ruin of 'Aiya. p.n.

خربة ارياق   Qb   *Khŭrbet Ariâk.* The ruins of Ariâk. p.n.

خربة العرصآء   Qb   *Khŭrbet el 'Arsa.* The ruin of the open court.

خربة العزية   Qb   *Khŭrbet el 'Azzîyeh.* The ruin of el Azziyeh. p.n.

خربة بقيرة   Qb   *Khŭrbet Bakeirah.* The ruin of the little cow.

خربة بلعبين   Nc   *Khŭrbet Balâbîn.* The ruin of Balâbin. p.n.

خربة بريش   Nb   *Khŭrbet Berbish.* The ruin of Berbish. p.n.

خربة برياش   Nc   *Khŭrbet Berîâsh.* The ruin of Berîâsh. p.n.

خربة بروخى   Ob   *Khŭrbet Berûkhei.* The ruin of Berukhei. p.n.

خربة البركة   Pb   *Khŭrbet el Birkeh.* The ruin of the pool.

خربة دير عبدو   Nc   *Khŭrbet Deir 'Abdu.* The ruin of Abdu's convent. Abdu, 'His servant,' a Syrian proper name.

خربة دفنة   Rc   *Khŭrbet Dafnah.* The ruin of Daphne (Oleander).

خربة دير شمعون   Nc   *Khŭrbet Deir Shem'ûn.* The ruin of the convent of Simeon.

خربة الظهر الصغير   Rb   *Khŭrbet edh Dhahr es Saghîr.* The ruin of the little ridge.

خربة دورة   Oc   *Khŭrbet Dûrah.* The ruin of Dùrah. Perhaps *daura,* 'a circular enclosure.'

خربة الامير يحيى   Pb   *Khŭrbet el Emîr Yahia.* The ruin of Emir Yahya (John).

خربة حارة   Pc   *Khŭrbet Hârah.* The ruin of the quarter (in the sense of 'quarter' or 'district' of a town).

| خربة الحيط | Rc | *Khŭrbet el Heit.* The ruin of the wall. |

خربة الحيط Rc *Khŭrbet el Heit.* The ruin of the wall.

خربة الحميره Ob *Khŭrbet el Humeirah.* The reddish ruin.

خربة الحمرآ *Khŭrbet el Hŭmra.* The red ruin.

خربة حور Nb *Khŭrbet Hŭr.* Probably حَوْر the ruin of the plane tree.

خربة هورآ Qb *Khŭrbet Hŭra.* The ruin of the crumbling bank.

خربة اكساف Pb *Khŭrbet Iksâf.* The ruin of Iksâf (Achsaph).

خربة الجلمة Oc *Khŭrbet el Jelameh.* The ruin of the large hill جُلَمَة.

خربة جمول Rb *Khŭrbet Jammŭl.* The ruin of Jammŭl. p.n.

خربة جولة Qb *Khŭrbet Jola.* The ruin of Jola. p.n.

خربة جملة Oc *Khŭrbet Jumleh.* The ruin of Gamala.

خربة جنيجل Pb *Khŭrbet Juneijil.* The ruin of Juneijil, diminutive of جنجيل; *see Khŭrbet Junjail, p. 7.*

خربة كفر ناى Ob *Khŭrbet Kefr Nay.* The village of Nai. p.n.

خربة الكنيسة Qb *Khŭrbet el Kuneiseh.* The ruin of the little church.

خربة كرم العواميد Nb *Khŭrbet Kurm el 'Awâmid.* The ruin of the vineyard of the columns.

خربة كرم الحلو Nb *Khŭrbet Kurm el Helu.* The ruin of the sweet vineyard.

خربة القصير Oc *Khŭrbet el Kuseir.* The ruin of the little castle or palace.

خربة اللقيقة Oc *Khŭrbet el Lakeikah.* The ruin of the narrow-mouthed hole.

خربة اللوبية Qb *Khŭrbet el Lûbieh.* The ruin of the Lubieh (a sort of bean).

خربة اللويزية Rb *Khŭrbet el Luweiziyeh.* The ruin of the Luweiziyeh. The word is derived from *loz*, an almond; but is probably the name of a dervish sect.

خربة المنصورد Oc *Khŭrbet el Mansŭrah.* The ruin of el Mansurah. q.v.

خربة المنارة Qc *Khŭrbet el Menârah.* The ruin of the lighthouse.

خربة المزارع Sb *Khŭrbet el Mezâra.* The ruin of the sown land.

خربة المغائر Nc *Khŭrbet el Mughair.* The ruin of the caves.

خربة المسجد Nc *Khŭrbet el Museijid.* The ruin of the little mosque.

L 2

خربة نيحا   Qb   *Khŭrbet Nîha*. The ruin of Niha. p.n.

خربة النخيلة   Rb   *Khŭrbet en Nukheileh*. The ruin of the little palm-grove.

خربة العزيزيات   Rc   *Khŭrbet el Ozeiziyât*. The ruin of the Ozeiziyât. p.n.

خربة راج   Pb   *Khŭrbet Râj*. The ruin of Râj. p.n.

خربة روبحينه   Qc   *Khŭrbet Rûeihîneh*. The ruin of the sweet-smelling herbs.

خربة السنبرية   Rc   *Khŭrbet es Sambarîyeh*. The ruin of the Sambarîyeh; probably the name of a dervish order.

خربة السبوغية   Ob   *Khŭrbet es Sebbûghîyeh* (probably الصبوغية). The ruin of the dyers.

خربة سلم   Pc   *Khŭrbet Selem*. Selem's ruin. p.n.; also the name of a plant.

خربه سردأ   Rb   *Khŭrbet Serada*. The ruin of Serada. p.n.

خربة شاغورى   Nb   *Khŭrbet Shâghûry*. The ruins of Shâghûry (Shâgûr, for Shanghûr, the Senigora סניגורא of the Talmud).

خربة صقور   Qc   *Khŭrbet Sukûr*. The ruin of hawks.

خربة طلحة   Qc   *Khŭrbet Talhah*. The ruin of the talh tree.

خربة تل الناعم   Rc   *Khŭrbet Tell en Nâăm*. The ruin of the mound of en Nâăm. p.n. signifying 'pleasant,' 'comfortable,' 'soft.'

خربة التربة   Pb   *Khŭrbet et Turbeh*. The tomb.

خربة تريثة   Rc   *Khŭrbet Tŭrrîtha*. The ruin of Turritha. p.n.

خربة ام العمود   Nc   *Khŭrbet Umm el Amûd*. The ruin with the columns.

خربة اليادون   Oc   *Khŭrbet el Yâdhûn*. The ruin of Yâdhûn. p.n.

خربة الزبيب   Qc   *Khŭrbet ez Zabîb*. The ruin of raisins.

خربة زوق الحج   Qc   *Khŭrbet Zûk el Hâj*. The pilgrim's town or village. *Zûk* is a local Syriac word.

الخضر   Rc   *El Khŭdr*. El Khŭdhr, 'the green old man,' is a personage who occurs in Mohammedan legends, and is identified sometimes with Elias and sometimes with St. George; *see* my translation of the Kor'ân, Vol. II, p. 23, note 3.

القبع   Pc   *El Kubâ*. The scull-cap.

قلعة الدبة   Pc   *Kŭlăt ed Dubbeh*. The bear's castle.

قلعة القط   Pc   *Kŭlăt ed Kott*. The cat's castle.

قلعة مارون   Ob   *Kŭlât Mârûn.* Maron's castle. Named after St. Maron, the founder of the Maronite sect.

قلعة الشقيف   Qb   *Kŭlât esh Shŭkif.* The castle of the cleft or crevasse.

قلعة صبيبة   Sb   *Kŭlât Subeibeh.*

قلعة نمرود   „   *Kŭlât Nimrûd.* } The castle of Subeibeh, or Nimrod's castle.

قلعة تبنين   Oc   *Kŭlât Tibnîn.* The castle of Tibnin. The Toron of the Crusaders.

القليعة   Qb   *El Kuleiâh.* The little castle.

الكنيسة   Qc   *El Kuneiseh.* The little church.

كرسى ابى مكنا   Qb   *Kursy Abu Mukenna.* Abu Mukenna's throne or chair.

القصير   Pb   *El Kuseir.* The little palace.

القصر   Nc   *El Kŭsr.* The palace.

قصر الجورة   Nc   *Kusr el Jŭrah.* The palace of the hollow in the ground.

النزحية   Nb   *El Kŭzhîych.* p.n. 'sprouting' or 'squirting.'

اللزازة   Rc   *El Lazâzeh.* p.n. 'fastening.'

محرونة   Oc   *Mahrûneh.* 'carded' (as cotton).

المنصورة   Rc   *El Mansŭrah.* Mansûr's place. (Mansûr = Victor.)

منزلة العسكر   {Rb}{Sb}   *Manzalet el 'Asker.* The halting place of the army.

مراح ابى على   Sc   *Marâh Abu 'Aly.* Abu 'Aly's resting place.

معركة   Nb   *Mârakeh.* The battle field.

مرنبه   Pb   *Marnabeh.* Marnabeh. p.n.

معروب   Ob   *Mârûb.* Abounding in water.

مارون   Ob   *Mârûn.* St. Maron.

مزار الاربعين   Rb   *Mazâr el Arbâin.* The shrine of the Forty (Martyrs of Cappadocia).

الميادين   Qb   *El Meiâdin.* The *meidâns, i.e.,* plains or exercise grounds.

ميس   Qc   *Meis.* Name of a tree.

مجدل اسلم   Pc   *Mejdel Islim.* Islim's watch-tower. p.n.

ملعب الغزلان   Sc   *Melâb el Ghuzlân.* The play-ground of the gazelles.

عيون مرج (Qb) (Qc) (Pc) *Merj 'Ayûn.* The meadow, or prairie of springs.

فاس مرج Nb *Merj Fâs.* The meadow of the hatchet.

احمص مرج Qc *Merj el Hummus.* The meadow of chick peas.

الجبل مرج Qc *Merj el Jebel.* The meadow of the mountain.

قيطيه مرج Rc *Merj Keitiych.* Keitiych's meadow.

اللزازة مرج Rc *Merj el Lazâzeh.* The meadow of Lazâzeh. p.n.

النعامة مرج {Rc} {Oc} *Merj en Nââmeh.* The meadow of soft ground.

نيحا مرج Qb *Merj Nîhâ.* The meadow of Nihâ. p.n.

عديثا مرج Qc *Merj 'Odeitha.* The meadow of Odeitha. p.n.

الشمالى مرج Qc *Merj esh Shemâly.* The northern meadow.

الحاج زوق مرج Rc *Merj Zûk el Hâj.* The meadow of the pilgrims' village.

التحتا زوق مرج Rc *Merj Zûk et Tahta.* The meadow of the lower village. Zûk is a local Syriac word.

مرسبة *Merseheh.* The place with pillars. It also means a place where water subsides or sinks into the ground.

مركبة Qc *Merkebeh.* p.n. from ركب to 'ride' or 'to lie,' as one thing on the top of another.

مزرعة Nc *Mezrâh.* The sown land.

نقار مغارة Pb *Mughâret en Nukkâr.* The cave of the pecker.

الشحل مغارد Pb *Mughâret esh Shahl.* The cave of esh Shahl. p.n.

الشبعانة مغارة Rb *Mughr esh Sheb'âneh.* The cave of the satiated one.

البزاز ام مغارد Nc *Mughâret Umm el Bezâz.* The cave with the cloths.

العلالى مغائر Oc *Mûghâir el 'Alâly.* The caves of the upper chambers.

المغيرة Nb *El Mugheirah.* The little cave.

نيحا مغراقة Ob *Mughrâket Niha.* The inundated land of Niha. p.n.

شلمى مغر Qc *Mughr Shalamei.* The caves of Shalamei. p.n.

مجيدل Oc *Mujeidil.* The little watch-tower. Perhaps Migdal, Josh. xix, 38.

المصيطبة Qc *El Museitibeh.* The little stone bench.

المطلة Qb *El Mutalleh.* 'Belle Vue.' From طل ' to overlook.'

النعامة Rc *En Na'âmeh.* The soft soil.

نهر بانياس Rc *Nahr Bániâs.* The river of Banias.

نهر بريغيت Qc *Nahr Bareighît.* The river of fleas.

نهر الحاصبانى {Rb} {Rc} *Nahr el Hâsbânî.* The River Hasbany.

نهر القاسمية {Pb} {Ob} {Nb} {Qb} *Nahr el Kâsimiyeh.* The River Kâsimiyeh.

نهر اللدان Rc *Nahr el Leddân.* The River Leddân.

نهر الليطانى Qb *Nahr el Lîtâni.* The River Litany.

نهر النيله Rc *Nahr en Nîleh.* The Blue River. (*Nîleh*, 'indigo'; *cf.* The Nile.)

النبى محمد بن يعقوب Nb *En Neby Mohammed ibn Yâkûb.* The prophet Mohammed, the son of Jacob.

النبى منذر Qc *El Neby Mundhir.* The prophet Mundhir, 'the warner.'

نبعة البياض Pc *Nebât el Beiâdh.* The perennial spring of the white place.

نبعة الجندى Pc *Nebât el Jindi.* The perennial spring of the soldier.

النبى عويدد Qb *En Neby 'Âweidah.* The prophet 'Aweidah.

النبى الصديق Oc *Neby es Saddîk* (properly *Enneby es Siddîk*). The Truthful Prophet.

النبى سارى Qb *En Neby Sâry.* The prophet Sâry (from *sara*, to journey by night). Perhaps an allusion to the celebrated 'Night Journey,' or Mohammed's 'vision of his ascent into Heaven'; *see* Kor'ân, ch. XVII, v. I.

النبى سياح Nc *En Neby Seiyâh.* The prophet Seiyâh, *i.e.*, the traveller or wanderer.

النفاخية Ob *En Neffâkhiyeh.* The blowers. Probably the name of a dervish order.

نيحا Ob *Niha.* Niha. p.n.

نورية نيحا Qb *Nûriyet Nîha.* The gypsies of Nîha. p.n.

عديثة الفوقا Qb *'Odeithat el Foka.* The upper 'Odeitha.

عديثة التحتا Qb *'Odeithat et Tahta.* The lower 'Odeitha.

الرفيد Ob *Er Rafîd.* The milk-bowl.

الرملية   Oc   *Er Ramliyeh.* The sandy place.

رأس ابى نذ   Oc   *Rás Abu Nedd.* The hill-top of Abu Nedd. p.n.

راس الاصفر   Ob   *Rás el Asfar.* The yellow hill-top.

رأس الظهر   Oc   *Rás edh Dhahr.* The top of the ridge.

رأس ام قبر   Nc   *Rás Umm Kabar.* The hill-top with the tomb.

رش كنانين   Nc   *Resh Kenânîn.* p.n.

رب ثلاثين   Qc   *Rubb Thcláthin.* Rubb means 'syrup'; *Thcláthin,* 'thirty.' Perhaps *rubb* may be a mistake for *rabb,* 'lord' (Rabbi).

الرويس   Ob   *Er Ruweis.* The little hill-top.

رويسات نيحا   Ob   *Ruweisât Niha.* The little hill-tops of Niha. p.n.

صفد البطيخ   Pb   *Safed el Battikh.* Safed of the Melons. The word *battikh* (melons) is used by the vulgar in the sense of 'nothing.' Thus if asked whether anything grows in a place which he knows to be barren, an Arab will reply ironically, 'melons'! This mistake has given rise to a statement in Robinson that a certain bare valley in the Tih is productive of 'melons'!

سهل الخان   Pc   *Sahel el Khân.* The level plain of the Caravanserai.

صريفا   Ob   *Sarîfa.* Sarifa. p.n.

شفا ابى مكنا   Qb   *Shcfa Abu Mukenna.* The edge or marge of Abu Mukenna. p.n.

شيخ عباد   Qc   *Sheikh 'Abbâd.* The sheikh Abbâd (devotee).

شيخ مرزوق   Rc   *Sheikh Marzûk.* The sheikh Marzûk (provided for by God).

شيخ محمد رشيد   Nb   *Sheikh Mohammed Rashîd.* The sheikh Mohammed Rashid.

شجور   Ob   *Shuhûr.* Mud walls, or conspicuous parts.

شقرا   *Shakra.* Grey.

شقيف القط   Qc   *Shakîf el Kott.* The cleft of the Cat.

شقيف الشعبان   Pc   *Shâkîf esh Shâbân.* The cleft of the satiated.

السد   Pc   *Es Sidd.* The barrier or dam.

صديقين   Nc   *Siddîkîn.* The truthful ones.

سيد هودا بن يعقوب   Rc   *Seiyid Hûda Ibn Y'kûb.* Lord Juda, the son of Jacob.

| سهم الخان | Rc | *Sihm el Khân.* The beam of the Caravanserai. |

سلعة   Ob   *Silâh.* The crevasse.

صيرة   Sb   *Sîreh.* The fold.

سلطان ابراهيم   Sc   *Sultân Ibrâhîm.* Sultân Ibrâhim.

سوق العصر   Qb   *Sûk el 'Asir.* The afternoon market.

السوائد   {Pb}{Pc}   *Es Suwâneh.* The flinty place.

طاحونة العبسية   Rc   *Tahûnet el 'Absîyeh.* The mill of the Abs tribe.

طاحونة احمد الموسى   Rc   *Tahûnet Ahmed el Musa.* The mill of Ahmed el Musa. The name is so written by the Arab scribe, but the article *el* appears incorrect.

طاحونة الكوكية   Qb   *Tahûnet el Kôlakîyeh.* The mill of the wooden ring.

طاحونة المنصورة   Rc   *Tahûnet el Mansûrah.* The mill of the Mansûrah. q.v.

طاحونة الست   Rc   *Tahûnet es Sitt.* The lady's mill.

طاحونة التل   Rc   *Tahûnet et Tell.* The mill of the mound.

الطيبة   Qb   *Et Taiyibeh.* The good, sweet, or wholesome (water).

طلوسة   Pc   *Tallûseh.* Tallûsch. p.n. Perhaps $\theta\alpha\rho\sigma\eta$ = Tirza, 1 Kings xv, 33. Jos. Ant. viii, 6.

تير فلسيه   Ob   *Teir Filsieh.* The fortress of the scales or small coins.

تير سمحات   Ob   *Teir Samhât.* The fortress of Samhât. The word is probably a Syrian family name, from *Samh*, 'to be liberal'; *cf. Farhât. Samhât* might also mean wide, open spaces.

تير زنبه   Oc   *Teir Zinbeh.* The fortress of Zinbeh. *Zinbeh* means 'hummocks in a valley,' and also a place from which to shoot partridges.

تل عجول   Qb   *Tell 'Ajûl.* The mound of 'Ajûl. *'Ajûl* means a calf.

تل علا   Sc   *Tell 'Alla.* Perhaps for علا 'high.' The high mound.

تل العروس   Rb   *Tell el 'Arûs.* The bride's mound.

تل البلان   {Sb}{Sc}   *Tell el Bellân.* The mound of the thorn-bush.

تل الحية   Oc   *Tell el Haiyeh.* The mound of the snake.

تل جريدد   Nb   *Tell Jureideh.* The bare mound, or perhaps from جرادة 'locust.'

تل القاضى   Rc   *Tell el Kâdy.* The mound of the Judge. The Arabic word *Kâdy* is equivalent to the Hebrew *Dan.*

تل القصعه   Qb   *Tell el Kasâh.* The mound of the wooden bowl.

تل شيخ يوسف   Rc   *Tell Sheikh Yûsuf.* The mound of the Sheikh Joseph.

تبنين   Oc   *Tibnîn.* Tibnin. p.n.

توليـن   Pc   *Tûlîn.* Tûlin. p.n.

تويره   Ob   *Tuweirah.* The little hill or fort.

ام ابراهيم   Rb   *Umm Ibrahim.* Abraham's mother. p.n.

وادى ابى نحله   Nb   *Wady Abu Nahleh.* The valley of the bee (Heb. stream).

وادى ابى كوير   Qb   *Wady Abu Kuweir.* The valley with the little beehive.

وادى العين   Ob   *Wady el 'Ain.* The valley of the spring.

وادى عين جحر   Ob   *Wady 'Ain Jehir.* The valley of the spring of the den.

وادى عين السعيد   Oc   *Wady 'Ain Saîd.* The valley of the spring of Saîd. p.n. (Felix.)

وادى عين الثافورة   Pb   *Wady 'Ain et Tâfûrah.* The valley of 'Ain et Tâfûrah. q.v.

وادى عين التين   Oc   *Wady 'Ain et Tîn.* The valley of the spring of the fig.

وادى عين التورة   Nb   *Wady 'Ain et Torah.* The valley of 'Ain et Torah. q.v.

وادى عيتيت   Nb   *Wady 'Aitît.* The valley of 'Aitit. p.n.

وادى العيزقانه   Qb   *Wady el 'Aizakânch.* The valley of 'Aizakânch (from عزق 'to dig up the ground with a mattock').

وادى عقبة القندوله   Pb   *Wady 'Akabet el Kandoleh.* The valley of the steep mountain path of the kandoleh (thorny plant).

وادى عصفور   Nb   *Wady 'Asfûr.* The valley of the sparrow.

وادى العسل   Rc   *Wady el 'Asl.* The valley of honey.

وادى عاشور   {Nc}{Oc}   *Wady 'Âshûr.* The valley of Âshûr (name of the 10th day of the month Moharram in the old Arab calendar).

وادى عطوه   Oc   *Wady 'Atweh.* 'Atwah's valley. p.n.

وادى العزية   Qb   *Wady el 'Azzîyeh.* The valley of el 'Azziyeh. p.n.

وادى الباغة   Oc   *Wady el Báá.* The valley of the courtyard.

وادى بافلى   Ob   *Wa.ly Báflei.* The valley of Báflei. p.n.

وادى البذ   Nb   *Wa.ly el Bedd.* The valley of el Bedd. p.n.   Perhaps بذ 'idol temple.'

وادى بيعان   Pc   *Wady Beib'án.* The valley of Beib'án. p.n.

وادى البياض   Pc   *Wady el Beiyád.* The valley of the white place.

وادى برشيت   Pc   *Wady Beráshit.* The valley of Beráshit.

وادى بريك   Pb   *Wady Burcik.* The valley of the little pool.

وادى البرج   Ob   *Wady el Burj.* The valley of the tower.

وادى داجع   Pb   *Wady Dámij.* The intricate valley.

وادى دير كينا   Ob   *Wady Deir Kifa.* The valley of the convent of Kifa. p.n.

وادى دير قنية   Pc   *Wady Deir Kinich.* The valley of the convent of Kinich. p.n.

وادى دير التينة   Nc   *Wady Deir et Tinch.* The valley of the convent of the fig-tree.

وادى الدبة   Pc   *Wady ed Dubbch.* The valley of the bear.

وادى الفقعانة   Qb   *Wady el Fak'ánch.* The valley of the truffle.

وادى الفيق   {Nb / Oc}   *Wady el Feik.* The valley of Feik. p.n.

وادى الفرون   Pb   *Wady el Furún.* The valley of the ovens, or reservoirs.

وادى الفوار   Ab   *Wady el Fúwár.* The valley of the fountain.

وادى الغميق   Ob   *Wady el Ghamik.* The deep valley.

وادى الغار   {Sb / Qc}   *Wady el Ghár.* The valley of the hollow.

وادى غدران   Nb   *Wady el Ghadrán.* The valley of the swamp, or pools.

وادى حذيه   Pc   *Wady Haddeih* (perhaps حدآ). The valley of the kite.

وادى حذره   Nc   *Wady Hadra.* The valley of descent.

وادى الحجير   {Pb / Pc}   *Wady el Hajeir.* The valley of the little rock.

وادى الحميد   Ob   *Wady el Humeid.* The valley of Humeid. p.n.

وادى هونين   Qc   *Wady Húnín.* The valley of Húnín.

وادى ارزون   Ob   *Wady Irzón.* The valley of Irzón. p.n.

وادى اصطبل	Pc	*Wady Istabl.* The valley of the stable.
وادى الجمل	Pc	*Wady el Jemel.* The valley of the camel.
وادى جيلو	Nc	*Wady Jîlu.* The valley of Jilu. q.v.
وادى الجوز	Pc	*Wady el Joz.* The valley of the nut.
وادى جملة	Oc	*Wady Jumleh.* The valley of Gamala.
وادى جويك	Pc	*Wady Juweik.* The devious valley.
وادى قاسيون	Nc	*Wady Kâsiûn.* The valley of Kâsiûn. p.n.
وادى كفر ناى	Ob	*Wady Kefr Nai.* The valley of the village of Nai.
وادى القيسية	Pc	*Wady el Keisîyeh.* The valley of the Keis (Arab tribe).
وادى الكريم	Oc	*Wady el Kerîm.* The valley of the generous or noble one.
وادى قرزون	Ob	*Wady Kerzôn.* The valley of the Kerzôn. p.n.
وادى الخندق	Ob	*Wady el Khandûk.* The valley of the fosse or trench.
وادى الخانوق	Pc	*Wady el Khânûk.* The valley of the gorge.
وادى الخنزير	Pb	*Wady el Khanzîr.* The valley of the swine.
وادى الخرار	Qb	*Wady el Kharrar.* The valley of murmuring water.
وادى الخشابة	Rc	*Wady el Khashâbeh.* The valley of the timber-cutters.
وادى خربة سردد	Rb	*Wady el Khûrbet Serada.* The valley of the ruin of Serada. p.n.
وادى قلعة دورة	Oc	*Wady Kulât Dûra.* The valley of Castle Dûra.
وادى قلعة مارون	Ob	*Wady Kûlât Mârûn.* The valley of Maron's Castle. (St. Maron, the founder of the Maronite sect.)
وادى الكواخ	Pc	*Wady el Kuwâkh.* The valley of the huts made of branches of trees.
وادى الماء	Oc	*Wady el Mâ.* The valley of the waters.
وادى الملوة	Pc	*Wady el Mallûleh.* The valley of the oak trees.
وادى المعروب	Nb	*Wady el Mârûb.* The valley of abundant water.
وادى الميس	Pb	*Wady el Meis.* The valley of the meis tree.
وادى المشاتى	Ob	*Wady el Meshâty.* The valley of the winter quarters (for cattle).
وادى المغاير	Nc	*Wady el Mughair.* The valley of the caves.

وادى المغارة	Pc	*Wady el Mágharah*. The valley of the cave.
وادى المغيسيل	Qb	*Wady el Mugheisil*. The valley of washing.
وادى النحل	Pc	*Wady en Nahl*. The valley of the bees (Heb. water-course).
وادى النهر	Pb	*Wady en Nahr*. The valley of the river.
وادى ناصر	Oc	*Wady Násir*. The valley of Nâsir. p.n. signifying 'helper.'
وادى النمور	Rb	*Wady en Nimûr*. The valley of the Leopards. But this is the name of an Arab tribe; and in Hebrew it also signifies 'running water.'
وادى النقيب	Sc	*Wady en Nukíb*. The valley of the warden.
وادى نصف رتل	Qb	*Wady Nusf Rotl*. The valley of 'half a rotl' (a weight).
وادى العرقانى	Qc	*Wady el 'Orkâny*. The valley of the cliffs.
وادى الراج	Pb	*Wady er Ráj*. The valley of er Raj. p.n.
وادى الراموح	Nb	*Wady er Rámâh*. The valley of Rámûh. p.n.
وادى الرصيف	Pc	*Wady er Rasíf*. The valley of the paved water channel.
وادى الصديقين	Nc	*Wady es Siddikín*. The valley of the truthful ones.
وادى صفوة	Ob	*Wady Safweh*. The valley of pure (water.)
وادى سغليب	Ob	*Wady Saghlib*. The valley of Saghlib. p.n.
وادى الساقفة	Nc	*Wady es Sákfeh*. The valley of the roofed-over (well).
وادى الساقطة	Oc	*Wady es Sáktah*. The valley of the falling stone.
وادى السعارى	Sc.	*Wady es S'áry*. The valley of es Saâry. p.n.
وادى السواقى	Nc	*Wady es Sawáky*. The valley of the small streams.
وادى سلم	{O} {Pc}	*Wady Selem*. The valley of the Selem (mimosa flava), a thorny tree.
وادى سلوكية	Pc	*Wady Selûkieh*. The valley of the hunting dogs (sleuth-hounds, *Seleucidæ*).
وادى شدغيث	Nb	*Wady Shidghíth*; see *'Ain Shidghíth*, p. 2.
وادى شق العبوز	Rb	*Wady Shukk el 'Ajûz*. The valley of the old woman's crevasse or cleft. The 'old woman' is probably 'Queen Dellooka' successor of Pharaoh according to the Arab legends. Nearly all the specimens of ancient architectural or engineering skill are attributed by the Arabs either to Pharaoh or the 'old woman.'

وادى السد	Oc	*Wady es Sidd.* The valley of the barrier.
وادى سلعة.	Oc	*Wady Silåh.* The valley of the 'fissure.'
وادى السوق	Oc	*Wady es Sûk.* The valley of the market.
وادى السلطانية.	Oc	*Wady es Sultâniyeh.* The valley of the king's highway.
وادى السنية.	Pb	*Wady es Suneiyeh.* The barren valley.
وادى الصرارة	Qc	*Wady es Sûrarah.* The valley of pebbles.
وادى الصوان	Pc	*Wady es Sûwân.* The valley of the flint.
وادى التقليد	Oc	*Wady et Taklîd.* The valley of Taklid (the word means imitation).
وادى التركمان	Qc	*Wady et Turkmân.* The valley of the Turkomans.
وادى الوحش	Nb	*Wady el Wahsh.* The wild valley, or valley of the wild beast.
وادى الزاروب	Ob	*Wady ez Zârûb.* The valley of the water-course.
وادى الزواريب	Nb	*Wady ez Zawârîb.* The valley of the water-courses; plural of *Zârûb.*
وادى الزاوية	Oc	*Wady ez Zâwieh.* The valley of the corner, or hermitage.
وادى الزبيه	Nc	*Wady Zubeih.* The valley of the pitfall.
يانوح	Nb	*Yânûh.* The Yanoah of the Bible (2 Kings, xv, 39).
اليهودية	Oc	*El Yehûdîyeh.* The Jews, or Jewish woman.
زور التحتانى	Rc	*Zôr et Tahtâni.* The lower Zor; perhaps meaning 'meandering'; *cf. ez Zaurâ* (a name of the Tigris).
زوق التحتا	Rc	*Zûk et Tahta.* The lower Zûk. *Zûk* is a Syriac word meaning 'town' or 'village.'

SHEET III.

ابو عتبة‌ Lf *Abu '-Itabeh.* Having a lintel.

ابو سنان Mf *Abu Senân.* Producing pasturage, especially such plants as خبض 'sorrel.'

عين علما Md *'Ain '-Alma.* The spring of 'Almâ. p.n.

عين العنكليس Me *'Ain el 'Ankalîs.* The spring of the eels.

عين العسل Le *'Ain el 'Asl.* The spring of honey.

عين عوينات Nd *'Ain 'Aweinât.* The spring of 'Aweinât. Either from حوان, 'land watered by rain'; or حوانة, 'a sand worm.'

عين البتة Le *'Ain el Basseh.* The spring of the marsh.

عين البيضآء Ld *'Ain el Beida.* The white spring.

عين الدبشة Nf *'Ain ed Dubsheh.* The spring of clods or chips of stone.

عين دير الاسد Nf *'Ain Deir el Asad.* The spring of the convent of the lion.

عين الدم Me *'Ain ed Dumm.* The spring of collyrium or pigment. The word also signifies the earth that stops up a jerboas hole. The original root implies 'gushing' or 'flowing.'

عين الفوارة Le *'Ain el Fûwârah.* The spring of the fountain.

عين الغفر Ld *'Ain el Ghufr.* The spring of the escort. So called from a fort which was used as a station at which toll or escort-money was paid.

عين الحمرآ Md *'Ain el Hamra.* The red spring.

عين حامول Ld *'Ain Hâmûl.* The Hammon of Joshua xix. v. 28.

عين خور Me *'Ain Haur.* The spring of the white poplar, or plane-tree.

عين اسكندرونة Ld *'Ain Iskänderûneh.* The spring of Iskanderûn. p.n. (Alexandroskene.)

عين الجوزة Me *'Ain el Jôzeh.* The spring of the walnut-tree.

عين كركرة Me *'Ain Kerkera.* The bubbling spring.

عين القلعة Me *'Ain el Kŭläh.* The spring of the castle.

عين لبونا Ld *'Ain Lebbûna.* The spring of Lebbûna ; from the Hebrew *Lebona,* 'white.'

عين المجنونة	Me	*'Ain el Majnûneh.* The spring of the mad woman ; or it may mean 'haunted by jinns' or demons.
عين المزراب	Me	*'Ain el Mezrâb.* The spring of the spout.
عين ميماس	Lf	*'Ain Mîmâs.* The spring of Mimâs.
عين المشيرفة	Le	*'Ain el Musheirefeh.* The spring of the look out, or high place. The Misrephoth-maim of Joshua xi, v. 8.
عين المحدورة		*'Ain el Mahdûra.* The spring of the descent.
عين العليقة	Me	*'Ain el 'Olleikah.* The spring of the bramble bush.
عين شمع	Md	*'Ain Shemâ.* The spring of the candle.
عين سيريا	Nf	*'Ain Siria.* The spring of Siria. p.n.
عين الست	Kf	*'Ain es Sitt.* The spring of the lady.
عين الصفصافة	Ld	*'Ain es Sûfsâfeh.* The spring of the osier.
عين التنور	Nd	*'Ain et Tannûr.* The spring of the reservoir.
عين وازع	Ne	*'Ain Wâziâ.* The spring of the beehive.
عين يارين	Md	*'Ain Yârîn.* The spring of Yârin.
عين الزرورة	Me	*'Ain ez Zârûrah.* The spring of the hawthorn tree.
عين الزيتونة	Me	*'Ain ez Zeitûneh.* The spring of the olive tree.
عين الزيب	Le	*'Ain ez Zîb.* The spring of ez Zîb. (Achzib.)
عقبة القاضى	Ne	*'Akabet el Kâdy.* The steep ascent or mountain-road of the judge.
عكا	Kf	*'Akka.* Acre. The Biblical Akko. (Ptolemais.)
اقرث	Ne	*Akrith.* Akrith. p.n.
علما الشعب	Md	*'Alma esh Shâub.* *'Almâ* means 'a coat of mail'; but the word is most probably a form of علم ' a mountain,' 'sign-post,' or 'way mark.' *Shâub,* 'mountain spurs.'
عمقا	Lf	*'Amka.* From عمق 'deep.' (Bethemek.)
عرب العرامشة	{Ld} {Md}	*'Arab el 'Arâmsheh.* The Arabs of the 'Arâmsheh tribe.
عرب الغوارنة	Lf	*'Arab el Ghawârneh.* The Arabs of the Ghor, or low ground, singular of *Ghôr.*
عرب الكليتات (؟)	Nd	*'Arab el Kleitât.* The Kleitat Arabs. (The name might be written four ways in Arabic. The scribe has not written it down.)

عرب السمنية	Mc	'*Arab es Semnîyeh.*	The Semnîyeh Arabs.
عرب السَوَيْتَات	Mc	'*Arab es Suweitât.*	The Suweitât Arabs.
عرب التُكِيين	{Ld Md}	'*Arab et Tukîyyin.*	The Tukkiyîn Arabs.
ارض الملك	Mc	*Ard el Mulk.*	The freehold land.
ارض سحماتا	Nf	*Ard Sahmâta.*	The land of Sahmâta. p.n.; probably from سحم ' black.'
ارض الزُوينيتا	Mc	*Ard ez Zuweinîta.*	The land of ez Zuweinîtâ. p.n.
عريض الدولة	Md	'*Arûd ed Daulch.*	The broad space or mountain side of the Government.
اشلول عبده	Mc	*Ashlûl 'Abdeh.*	Cascades of 'Abdeh. p.n. (Eboda.)
عيون اقرث	Nc	'*Ayûn 'Akrith.*	The springs of Akrith.
عز الدين	Kf	'*Azz ed Dîn.*	p.n.; meaning the might of the faith.
البهجة	Lf	*El Bahjeh.*	' The beautiful,' applied to land (Lane).
البعنه	Nf	*El Bâneh.*	El Bâneh. p.n.
البتة	Lc	*El Basseh.*	The marsh.
بيوت السيد	Md	*Beiût es Seiyid.*	The houses of the lord.
البياضة	Ld	*El Beiyâdah.*	The white spot.
بيار عرنوس	Nc	*Biâr 'Arnûs.*	The wells of 'Arnús. p.n.
بئر ابى طاروح	Mf	*Bir Abu Târûh.*	The well of Abu Târûh. p.n.
بئر الكليل	Mf	*Bîr Iklîl.*	The well of the garland.
بئر كفر نبيذ	Lc	*Bir Kefr Nebîd.*	The well of the village of beer.
بئر المدقق	Md	*Bîr el Mudakkak.*	The well of pounded or crushed (corn).
بئر النصف	Mc	*Bir en Nusf.*	The well of the half.
بئر الزويس	Nd	*Bîr er Ruweis.*	The well of the little headland.
بئر الشرق	Nf	*Bir esh Sherk.*	The eastern well.
بئر الزيتون	Mf	*Bîr ez Zeitûn.*	The well of olive.
بركة حجر	Nd	*Birket Hajr.*	The pool of the stone or rock.
بركة مفشوخ	Lc	*Birket Mefshûkh.*	The pool of the split (rock, &c.).

G

بركة الرحراح　Nd　*Birket er Rahráh.* The pool of the rahráh, a small white bird so called. The word also means 'wide and shallow.'

بركة راميا　Nd　*Birket Rámia.* The pool of Rámia (perhaps Rama of Naphthali).

بركة ريشة　Nd　*Birket Rîsheh.* The pool of the feather.

بركة الرجمين　Md　*Birket er Rujmein.* The pool of the two heaps of stones.

بركة سروح　Nc　*Birket Surâh.* The pool of the plain or level ground.

بركة تيريبخا　Nc　*Birket Terbîkha,* p.n.

بركة طاراعه　Nc　*Birket Tetárámah.* The pool of Tetárámah. p.n.

برج البياضة　Ld　*Burj el Beiyâdah.* The tower of the blank (*i.e.* uncultivated) spot.

برج الغفر　Ld　*Burj el Ghúfr.* The tower of the escort.

برج مصر　Ne　*Burj Músr.* The tower of Egypt.

دار عبدالله باشا　Lf　*Dâr 'Abdallah Básha.* The house of Abdallah Pasha.

دار الجباخانجى　Le　*Dâr el Jebâkhânjy.* The house of Jebákhânji; a proper name.

دار سرسق　Lf　*Dâr Sursuk.* The house of Sursuk. p.n.

دير الاسد　Nf　*Deir el Asad.* The convent of the lion.

ظهر الضبعة　Md　*Dhahr ed Dúbâh.* The ridge of the hyena.

ظهر حمار　Le　*Dhahr Humâr.* The ridge of the ass.

ظهر الجمل　Nd　*Dhahr el Jemel.* The ridge of the camel.

ظهر المدفون　Ld　*Dhahr el Medfân.* The ridge of the buried (person or thing).

ظهرة ام مقدوح　Ne　*Dhahret Umm Makdâh.* The ridges with the hollowed-out rocks.

الغابسية　Lf　*El Ghâbsîyeh.* From *ghabus,* 'dusky ashen grey.'

حقول زهير　Mc　*Hakûl Zaheir.* The field of ez Zuheir. p.n. *Hukúl* is the plural of *Hakl,* which means a piece of land without trees, but fit for sowing.

جامع العابسية　Lf　*Jámiâ el Ghâbsîyeh.* The mosque of el Ghâbsiyeh; q.v.

جامع الجزار　Kf　*Jámiâ el Jezzâr.* The mosque of Jezzâr Pasha.

جبل ابى قيراط Nf *Jebel Abu Kirât.* The mountain of the Kirât (a land measure).

جبل بلاط Nd *Jebel Belât.* The mountain of the flat stone slab.

جبل حانوتا {Ld / Md} *Jebel Hânûta.* The mountain of the tavern. (*Hanuta* in Talmud.)

جبل الكسار Ld *Jebel el Kisâr.* The mountain of broken bits.

جبل لبونا Ld *Jebel Lebbûna.* Mount Lebbuna; from the Hebrew *Lebonah,* 'milk white.'

جبل مبلية Mf *Jebel Mibilich.* Mount Mibilich. The term in ancient Arabic was applied to a camel tied and left to starve at its master's grave.

جبل المُشَقَّح Ld *Jebel el Mushakkah.* The word signifies the change of colour in ripening fruit.

جبل الرؤس Mf *Jebel er Rûs.* The mount of the hill-tops.

جبل الطويل Mf *Jebel et Tawîl.* The tall mountain.

جلمة البعنة Nf *Jelamet el Bânch.* (جلمة properly جلیمة) 'a large hill'; the hill of el Bânch.

جلمة النحف Nf *Jelamet en Nuhf.* The hill of en Nehf. p.n. (Nef.)

جت Mf *Jett.* Perhaps جَت 'elevated ground'; or Heb. Gath.

جسر المدفون Ld *Jisr el Medfûn.* The bridge of the buried one.

الجبين Md *El Jubbein.* The two pits.

الجديدة Lf *El Judeiyideh.* The word *judeiyida* is the diminutive of *jedîdeh,* a 'dyke,' or coloured streak in a mountain side.

جولس Mf *Jûlis.* Jûlis. p.n. (Julius.)

قاعدة المعصر Ne *Kââdet el M'âser.* Foundation of wine press.

الكابرى Le *El Kâbry.* The bridge (in Turkish); perhaps the *Kabartha* of the Talmud.

قبو جمرية Nd *Kabu Jamrîyeh.* Vault of pebble or rubble.

القاعدة Lf *El Kâdeh.* The foundation. The word also means 'the capital of a column.'

قفزة المهر Md *Kafzet el Muhr.* The foal's leap (trad.).

القهوة Le *El Kahweh.* The coffee-shop.

قناية ابى لزيق Le *Kanâit Abu Lizzik.* The water-courses of Abu Lizzik. p.n.

كفر سميع   Nf   *Kefr Sumeid.* The village of Sumeia.

كفر ياسف   Lf   *Kefr Yásif.* The village of Yásif.

قصر القوداءه   *Kŭsr el Kŭdámeh.* The castle or mansion of the Kŭdámeh ; a family name.

خلة ابي العراقي   Nd   *Khallet Abu el 'Aráky.* The dell with the river bank.

خلة ابي غيث   Md   *Khallet Abu Gheith.* The dell with the fertile soil, or where the showers fall.

خلة ابي محمد   Md   *Khallet Abu Muhammed.* Abu Mohammed's dell. p.n.

خلة ابي سعد   Le   *Khallet Abu Sâd.* Abu Saad's dell. p.n.

خلة ابي تركة   Md   *Khallet Abu Turkeh.* Abu Turkeh's dell. p.n.

خلة العدس   {Mf / Md}   *Khallet el 'Adas.* The dell of lentils.

خلة العين   Ne   *Khallet el Aïn.* The dell of the spring.

خلة علم   Mf   *Khallet 'Alam.* The dell of the conspicuous mountain, or way mark.

خلة آسة   Me   *Khallet Ásch.* The myrtle dell.

خلة بعيش   Nf   *Khallet Bâîsh.* The dell of Bäïsh. p.n.

خلة بالوة   Nf   *Khallet Bâlŭa.* The dell of the water-hole, or sunk well.

خلة بورا   Ne   *Khallet Bûra.* The dell of unsown land.

خلة الضعون   Nd   *Khallet ed Dâûn.* The dell of Dâûn (perhaps ظعون 'a camel of burden').

خلة فارس   Ne   *Khallet Fáris.* The dell of the knight or rider.

خلة الغور   Me   *Khallet el Ghôr.* The dell of the lowland.

خلة الحافور   *Khallet el Hafûr.* The dell of the digging or excavation.

خلة حمدان   Md   *Khallet Hamdhân.* Hamdhân's dell ; perhaps a mistake for حماض, from حمض wild sorrel, a favourite pasture for camels.

خلة حاروئة   Ne   *Khallet Hárûneh.* The dell of Hárûneh ('refractory' ; also 'cotton carding').

خلة هاشم   Md   *Khallet Háshem.* Háshem's dell. p.n. Háshem was the founder of Mohammed's family.

خلة الحصان   Ld   *Khallet el Hisân.* The dell of the horse.

خلة حَارَة Ld *Khallet Humârah.* The dell of the stones set round a trough.

خلة جِيش Me *Khallet Jeish.* The dell of the army.

خلة الجمادية Nd *Khallet el Jemâdiyeh.* The dell of dry or unfruitful ground.

خلة جوخة Md *Khallet Jôkhah.* The dell of the excavation or of the crumbling banks.

خلة الجوزانة Md *Khallet el Jôzâneh.* The dell of the walnut trees.

خلة جورة الضبعة Md *Khallet Jûrat ed Dûbâa.* The dell of the hyena hollow.

خلة الخَص Md *Khallet el Khuss.* The dell of the lattices, huts, or silk-worm frames; *see Aksâs*, of which it is the singular.

خلة كرم التين Md *Khallet Kurm et Tîn.* The dell of the fig orchard.

خلة المدور Md *Khallet el Madauwar.* The round dell.

خلة مقتل العبد Me *Khallet Maktal el 'Abd.* The dell of the place where the slave was slain.

خلة مقتلة النمر Nf *Khallet Maktalet en Nimr.* The dell of the place where the leopard was slain.

خلة مارون Md *Khallet Mârûn.* (St.) Maron's dell.

خلة الطحنة Ne *Khallet el Mat-haneh.* The dell of the mill.

خلة مطيلة Mf *Khallet Mattîlah.* Perhaps from مطيلة water remaining at the bottom of a tank.

خلة المعازب Md *Khallet el M'âzib.* The dell of the herdsmen who pasture their cattle at a distance.

خلة محمد حيدر Nd *Khallet Muhammed Heider.* Mohammed Heider's dell. p.n.

خلة المربعة Md *Khallet el Murabbâh.* The square dell.

خلة النجاس Ne *Khallet en Nejâs.* The dell of pears.

خلة العليق Ne *Khallet el 'Olleik.* The dell of the bramble.

خلة الرجية Nd *Khallet er Raheiyeh.* The dell of the 'hillock,' or of 'the soft ground'; dimin. of رجوة.

خلة السيد Md *Khallet es Seiyid.* The dell of the lord.

خلة سلوم Ne *Khallet Sellûm.* Probably a local form from *selem*, a thorny tree (the *mimosa flava*).

خلة الشريف	Me	Khallet esh Sherîf. The dell of the noble.
خلة السيح	Md	Khallet es Sîh. The dell of the stream.
خلة السوق	Me	Khallet es Sûk. The dell of the market.
خلة السماقة	Md	Khallet es Summâkah. The dell of the Sumach tree (Rhus Coriaria).
خلة الطاقة	Ne	Khallet et Tâkah. The dell of the spray, or of the arch.
خلة تير حرفا	Md	Khallet Teir Harfa. The dell of the fort of Harfa. p.n.
خلة التين	{Ne} {Ld}	Khallet et Tîn. The dell of figs.
خلة التون		Khallet et Tûn. The dell of the lime-kiln.
خلة ام ظروف	Md	Khallet Umm Dhurûf. The dell with the vessels. p.n.
خلة ام الرّب	Md	Khallet Umm er Rubb. The syrup-producing dell.
خلة وعول	Nd	Khallet Wûûl. The dell of the wild goats (ibex).
خلة الواوى	Me	Khallet el Wâwy. The dell of the jackal.
خلة يوسف	Md	Khallet Yûsef. Joseph's dell.
خلة الزيتون	Mf	Khallet ez Zeitûn. The dell of the olive.
خلة زرنيخ	Ne	Khallet Zernîkh. The dell of orpiment (yellow arsenic), or of sandarac (red arsenic).
خلة زبقين	Nd	Khallet Zubkîn. Probably from زبق 'to bind' or 'confine.'
خان ابى هدا	Le	Khân Abu Heda. Abu Heda's Caravanserai. p.n.
خلال مصلح	Me	Khelâl Masleh. Masleh's dells. p.n.
خلال عقدة	Md	Khelâl 'Okdeh. The dells of pasturage or fruitful ground.
الخضر	Le	El Khŭdr. See p. 28.
الخضر ابو عباس	Nf	El Khŭdr Abu 'Abbâs. El Khŭdr, father of 'Abbâs. p.n.
خراب جب سويد	Md	Khŭrâb Jubb Suweid. The ruins of the black pit.
خربة عباسية	Le	Khŭrbet 'Abbâsîyeh. The ruins of Abbâsiyeh. p.n.; see p. 1.
خربة عبدة	Le	Khŭrbet 'Abdeh. The ruin of Abdeh, from عبد 'a slave.' The ancient Abdon. The name occurs in Arabia Petraea, and is written in the older itineraries 'Εβόδα (Eboda).
خربة ابى شاش	Md	Khŭrbet Abu Shâsh. The ruin of Abu Shâsh. p.n.
خربة عين حور	Me	Khŭrbet 'Ain Haur. The ruin of the plane or poplar tree.

خربة أيتيم Le   *Khŭrbet Aiteiyim.* The ruin of Aiteiyim. p.n.

خربة العجليات Nd   *Khŭrbet el 'Ajliyât.* The ruin of el Ajliyât (from عجل 'haste,' or عجل 'a calf').

خربة عكروش Mf   *Khŭrbet 'Akrŭsh.* The ruin of 'Akrŭsh; from a plant called 'Akrish.

خربة علّيا Ne   *Khŭrbet 'Alia.* The ruin of the high place.

خربة العمود Le   *Khŭrbet el 'Amŭd.* The ruin of the column.

خربة عربين Me   *Khŭrbet 'Arubbîn.* The ruin of Arubbîn. p.n.

خربة عوينات Nd   *Khŭrbet 'Aweinât. See 'Ain 'Aweinât.*

خربة البالوع Nf   *Khŭrbet el Bâlûâ.* The ruin of the water-hole or sunk well.

خربة بيت ايريا Me   *Khŭrbet Beit Iria.* The ruin of Beit Iria. p.n.

خربة البياض Md   *Khŭrbet el Beiyâd.* The ruin of the white (blank or uncultivated) place.

خربةبياض الجواني Md   *Khŭrbet Beiyâd el Juwâny.* The ruin of the inner white place.

خربة بلاط Nd   *Khŭrbet Belât.* The ruin of the slab of stone; *see* "Memoirs," p. 173.

خربة بلاطون Me   *Khŭrbet Belâtûn.* A local form of the preceding.

خربة بنا Me   *Khŭrbet Benna.* The ruin of the builder.

خربة البوبرية Le   *Khŭrbet el Bôbrîyeh.* Also called *Khŭrbet el Menawât.* المنوات.

خربة بودا Lf   *Khŭrbet Bûda.* The ruin of Bûda. p.n.

خربة البطيشية Md   *Khŭrbet el Buteishîyeh.* The ruin of those who assault.

خربة الدبشة Ne   *Khŭrbet ed Dabsheh.* The ruin of ed Dabsheh. q.v.

خربة دانيان Ld   *Khŭrbet Dâniân.* The ruin of Dâniân, the Danjaan of 2 Sam. xxvi. 6.

خربة ظهور الحارة Md   *Khŭrbet Dhahûr el Hârah.* The ruin of the ridges of the quarter.

خربة الغريب Le   *Khŭrbet el Ghureib.* The ruin of the little raven.

خربة الحبان Mf   *Khŭrbet el Habâi.* The ruin of Habâi. p.n.

خربة الحمرآ Md   *Khŭrbet el Hamra.* The red ruin.

خربة حمسين Le   *Khŭrbet Hamsîn.* The ruin of Hamsîn. p.n.

خربة حنبلية   Mf   *Khŭrbet Hanbaliyeh.* The ruin of the Hambaliyeh, a Mohammedan sect.

خربة حانوتا   Ld   *Khŭrbet Hânûta.* The ruin of the tavern.

خربة حارونة   Ne   *Khŭrbet Hârûneh.* The ruin of Hârûneh; *see Khallet Hârûneh,* p. 44.

خربة الحما   Nd   *Khŭrbet el Hima.* The ruin of the preserve.

خربة ادمث   Me   *Khŭrbet Idmith.* The ruin of Idmith. p.n.

خربة اكليل   Mf   *Khŭrbet Iklîl.* The ruin of the garland.

خربة إنعيلة   Me   *Khŭrbet Inàîleh.* The ruin of the horseshoes.

خربة اسكندرونة   Ld   *Khŭrbet Iskanderûneh.* The ruin of Alexandroskene.

خربة جعثون   Me   *Khŭrbet Jâthûn.* The ruin of Jâthûn. *Jàthûn* or *jàthûm* means in Arabic a large *penis;* it may be a form of *jî'thin* or *jî'thim,* meaning the roots of the leguminous plant called سليان *silliyân* (the *Gathûn* of the Talmud).

خربة جوهرة   Ld   *Khŭrbet Jauharah.* The ruin of the jewel.

خربة جليل   Me   *Khŭrbet Jelil.* The ruin of Jelil; *see Neby Jelil,* p. 10.

خربة جحجين   Md   *Khŭrbet Jijjin.* The ruin of Jijjin. p.n.

خربة الجوق   Ne   *Khŭrbet el Jôk.* The ruin of el Jôk. (From جوق to be crooked.)

خربة الجبين   Md   *Khŭrbet el Jubbein.* The ruin of the two pits.

خربة جب رحيج   Nf   *Khŭrbet Jubb Ruheij.* The ruin of the pit for rain water.

خربة جديدة   Le   *Khŭrbet Judeideh.* The ruin of the streaks, or dykes.

خربة جرديه   Md   *Khŭrbet Jurdeih.* The ruin of the barren land.

خربة القبارصة   Lf   *Khŭrbet el Kabârsah.* The ruin of the Cypriotes.

خربة كابرى   Nf   *Khŭrbet Kábra.* The ruin of Kabara (Gabara). Jos. Vita. 37.

خربة كفكفا   Md   *Khŭrbet Kafkafa.* The word *kafkafa* means 'a camel's beard'; the original root signifies 'drying up' (of vegetables), or 'shivering.'

خربة قرحاثا   Ne   *Khŭrbet Karhâtha.* The ruin of Karhâtha. p.n.

خربة كرمث   Md   *Khŭrbet Kermith.* The ruin of Kermith. p.n.

خربة كروم الحميض   Ne   *Khŭrbet Kerûm el Hummeid.* The ruin of the sorrel orchard.

خربة خلة الواوى   Me   *Khŭrbet Khallet el Wâwy.* The ruin of the jackal's dell.

خربة الخشم   Md   *Khŭrbet el Khashm.* The ruin of the mouth or outlet; but perhaps from خشام 'a mountain having a thick prominence.

الخريبة   Md   *El Khŭreibeh.* The little ruin.

خربة قصيقس   Ne   *Khŭrbet Kuscikis.* قصاقس means thick or stout.

خربة القصير   Md   *Khŭrbet el Kŭseir.* The ruin of the little castle.

خربة القطعة   Nd   *Khŭrbet el Kŭtâh.* The ruin of the cutting.

خربة قزيزية   Me   *Khŭrbet Kuzîzîyeh.* The ruin of the glass-works.

خربة المدور   Md   *Khŭrbet el Madauwar.* The round ruin.

خربة محّوز   Nf   *Khŭrbet Mahhûz.* The ruin of Mahhûz (name of a plant).

خربة مروحين   Nd   *Khŭrbet Marwahîn.* Either from مروحة 'a place where the wind blows, effacing the traces of dwellings,' or from مروحة 'a fan.'

خربة معصوب   Le   *Khŭrbet Mâsûb.* The ruin of Mâsûb. p. n.; signifying 'bound' or 'compact.'

خربة المطمورة   Md   *Khŭrbet el Matmûrah.* The ruin of the underground granary.

خربة مازى   Ld   *Khŭrbet Mâzî.* The ruin of Mâzî. p.n.

خربة مناويثة   Mf   *Khŭrbet Menâwîtha.* The ruin of Menâwîtha. p.n.

خربة منحتة   Me   *Khŭrbet Menhatah.* The ruin of the quarry.

خربة مرج البينة   Me   *Khŭrbet Merj el Beineh.* The ruin of the meadow of the conspicuous object.

خربة ميليا   Mf   *Khŭrbet Mibilich; see Jebel Mibilich, p. 43.*

خربة ميريامين   Na   *Khŭrbet Miriâmîn.* The ruin of Miriâmîn (the name of a plant).

خربة المونة   Mf   *Khŭrbet el Mûneh.* The ruin of provisions.

خربة المشيرفة   Ld   *Khŭrbet el Musheirefeh.* The ruin of the high places.

خربة المشمش   Me   *Khŭrbet el Mushmush.* The ruin of the apricot.

خربة مصلح   Lf   *Khŭrbet Muslih.* Muslih's ruin. p.n.

خربة ناصر   Md   *Khŭrbet Nâsr.* Nâsir's ruin. p.n.

خربة نيفد   Nd   *Khŭrbet Neifed.* Perhaps for نافذ 'penetrating.'

خربة العمرى   Le   *Khŭrbet el 'Omry.* The ruin of el 'Omary. p.n.

خربة رمح　Me　*Khŭrbet Rŭmeh*. Perhaps from *rumh*, 'a lance.'

خربة الرويس　Ld　*Khŭrbet Ruweis*. The ruin of the hill-top.

خربة رويسات　Ne　*Khŭrbet er Ruweisât*. The ruin of the hill-tops.

خربة صمخ　Me　*Khŭrbet Semakh*. The ruin of the Semakh (in the vulgar language = صمغ 'gum').

خربة السيح　Ne　*Khŭrbet es Sîh*. The ruin of the stream.

خربة الشقارة　Nf　*Khŭrbet esh Shakkârah*. The ruin of the Shakkâra; properly شقارى 'a kind of clover.'

خربة الشفية　Me　*Khŭrbet esh Shefeiyeh*. The ruin of the little brink or margin.

خربة شومرية　Me　*Khŭrbet Shômerîyeh*. Either from Shomer (p.n.), or from the herb wild fennel.

خربة الشبيكة　Me　*Khŭrbet esh Shubeikeh*. The ruin of the net.

خربة سروح الفوقا　Ne　*Khŭrbet Surûh el Fôka*. The ruin of the upper Surûh. q.v.

خربة صوانة　Ne　*Khŭrbet Sûwânch*. The ruin of flint.

خربة الصوانة　Me　*Khŭrbet es Sûwânch*. The ruin of the flint.

خربة سويجيرة　Le　*Khŭrbet Suweijîreh*. The ruin of the dog-collar; but in the local dialect a tree like a willow is so called.

خربة الطبلة　Me　*Khŭrbet et Tableh*. The ruin of the tray, or tambourine.

خربة تل فرخة　Me　*Khŭrbet Tell Ferkhah*. The ruin of the chicken.

خربة الترجمان　Md　*Khŭrbet et Terjemân*. The ruin of the dragoman or interpreter.

خربة طظارامة　Ne　*Khŭrbet Tetárâmah*. The ruin of Tetárâmeh. p.n.

خربة طبريا　Me　*Khŭrbet Tibrîa*. The ruin of Tibria. p.n.; *see Tabariyeh*.

خربة طبق الحنة　Le　*Khŭrbet Tâbk el Henna*. The ruin of the terrace of Henna (a plant used by Arab women to dye their finger tips. Written in name-book *Tabak*).

خربة ام العمود　Ld　*Khŭrbet Umm el 'Amûd*. The ruin with the columns.

خربة ام الفرج　Le　*Khŭrbet Umm el Ferj*. The ruin with the gap or chink.

خربة ام عفية　Ld　*Khŭrbet Umm 'Ofeiyeh*. Probably from عفو 'land with no traces of dwellings.'

خربة ام الرب　Md　*Khŭrbet Umm er Rubb*. The syrup-producing ruin.

خربة ام التوتة   Md   *Khŭrbet Umm et Tûtch.* The ruin with the mulberry trees.

خربة ام الزينات   Nd   *Khŭrbet Umm ez Zeinât.* The ruin with the ornaments.

خربة الوزية   Mf   *Khŭrbet el Waziych.* The ruin of mud walls.

خربة يارين   Md   *Khŭrbet Yârîn.* The ruin of Yârîn. p.n.

خربة زلوطية   Md   *Khŭrbet Zallûtiych.* The ruin of pebbles.

خربة الزاوية   Ne   *Khŭrbet ez Zawiyeh.* The ruin of the corner or hermitage.

خربة زبد   Md   *Khŭrbet Zebed.* The ruin of Zebed. *Zebed* means 'froth,' or 'civet,' but is probably connected with the Hebrew proper name Zebedee, &c. The Talmud (Beresch Rab. s. 56) mentions a Zebud in Gallilee in the neighbourhood of Tyre.

خربة زبدية   Le   *Khŭrbet Zubdiyeh.* The ruin of Zubdiyeh ; *see* last paragraph.

خربة زوينيتا   Me   *Khŭrbet Zuweinita.* The ruin of Zuweinita. p.n.

كسرا   Nf   *Kisra.* The fragment.

قلعة جدين   Mf   *Kŭlàt Jiddîn.* The castle of Jiddîn. p.n.

قلعة القرين   Me   *Kŭlàt el Kurein.* The castle of the little horn or peak.

قلعة شمع   Md   *Kŭlàt Shemà.* The castle of Shemà. *Shemà* means wax, but is probably connected with the name *Shimeon.*

قلعة الطوفانية   Nf   *Kŭlàt et Tûfàniych.* Castle of the flood or inundation.

قرن حنوى   Mf   *Kŭrn Hennâwy.* The horn (peak) of Henawei. q.v.

كرم حنوس   Md   *Kurm Hanûs.* The vineyard of Hanûs. p.n.

قصر الاحمر   Md   *Kŭsr el Ahmar.* The red palace.

قصر الحمار   Lf   *Kŭsr el Hammâr.* The palace of the stones set round a trough.

قصر القوادمة   Md   *Kŭsr el Kŭdâmeh.* The palace of the Kŭdâmeh family.

قصر محمد بك   Lf   *Kŭsr Muhammed Bek.* The palace of Mohammed Bey. p.n.

قصر الطويل   Md   *Kŭsr et Tawîl.* The long palace.

القصيرات   Md   *El Kŭseirât.* The little palaces.

الكويكات   Lf   *El Kuweikât.* Perhaps كواخات 'huts'; *see* Wady el Kuweikh, p. 30.

لبونا   Ld   *Lebbûna.* 'White,' from the Hebrew.

المدوّر   *Nd*   *El Madauwer.* The circular.

عُلّيا   *Ne*   *Màlia.* The high place.

مدينة النحاس   *Nd*   *Medinet en Nehás.* The city of brass.

المجدل   *Md*   *El Mejdel.* The watch-house. Heb. Migdal.

مجد الكروم   *Nf*   *Mejd el Kerûm.* Probably for الكروم مجدل 'the watch-house of the vineyard.'

المكّ   *Lf*   *El Mekr.* El Mekr. p.n.

المنارة   *Md*   *El Menárah.* The lighthouse.

المَنْشية   *Lf*   *El Menshîyeh.* El Menshiyeh. p.n.

مرج العبد   *Mf*   *Merj el 'Abd.* The plain of the slaves, or the bare desert.

مرج عكبرا   *Ne*   *Merj 'Akbara.* The meadow or prairie of the (male) jerboa.

مرج البينة   *Me*   *Merj el Beineh.* The meadow or prairie of the conspicuous object.

مرج البُقَيْعة   *Nf*   *Merj el Bŭkeiah.* The meadow or prairie of the little vale.

مرج حارونة   *Ne*   *Merj Hárûneh.* The meadow or prairie of *Hárûneh.* q.v.

مرج كفر سميع   *Nf*   *Merj Kefr Sumeià.* The meadow or prairie of the village of Sumeià. p.n.

مرج راميا   *Nd*   *Merj Rámia.* The meadow or prairie of Ramia. p.n. (Ramah).

مرج شمع   *Md*   *Merj Shemà.* The meadow or prairie of Shemà (Simeon).

المروج   *Nd*   *El Merûj.* The meadows.

مزرعة اسكندرونة   *Md*   *Mezràt Iskanderûneh.* The sown land of Alexandroschene.

المزرعة   *Lf*   *El Mezràh.* The sown land.

مغارة ابى الجراجمة   *Me*   *Mŭghàret Abu el Jeràjimeh.* The cave of the skulls. p.n.

مغارة عترة   *Md*   *Mŭghàret 'Atrah.* The cave of the 'Atrah; name of a plant, perhaps the marjoram.

مغارة البياضة   *Mŭghàret el Beiyâdah.* The cave of the white or blank space.

مغارة خاطر   *Md*   *Mŭghàret Khàter.* The cave of danger, or 'of him who risks.'

مغارة المرصوفة	Ld	*Mûghâret el Mersûfeh.* The cave of the dammed-up stream.
مغر الطاقات	Ld	*Mûhgr et Tâkât.* The caves of the arches.
نهر مفشوخ	Lf	*Nahr Mefshûkh.* The River Mefshûkh; *see Birket Mefshûkh,* p. 41.
الناقورة	Ld	*En Nâkûrah.* The horn or trumpet. This name rises apparently from a misconception on the part of the Arab-speaking inhabitants, as the name صر Tyre means in Arabic a 'horn' or 'trumpet'; therefore *Ras Sûr* (the 'headland' or 'ladder of Tyre') is rendered by *Nakûra,* the synonym for *Sûr.* The word is also connected with نقر 'to peck' or 'perforate.'
النبى روبين	Ne	*Neby Rûbîn.* The prophet Reuben.
النبى الصديق	Mf	*Neby es Saddîk.* The truthful prophet.
النبى الصالح	Mf	*Neby Sâleh.* The righteous prophet.
النبى شمع	Md	*Neby Shemâ.* The prophet Shemâ; perhaps Simeon.
النبى زكريآ	Mf	*Neby Zakarîya.* The prophet Zackarias.
رأس العزاقة	Me	*Râs el 'Azzâkeh,* or *Ruweisat el Kenîseh.* The head or hill-top of el Azzâkeh. p.n.; or the heads or hill-tops of the church.
رأس الظهر	Nf	*Râs ed Dhahr.* The top of the ridge.
رأس كلبان	Mf	*Râs Kelbân.* The head of Kelbân. p.n.
رأس المصرى	Md	*Râs el Mûsry.* The hill-top or headland of the Egyptian.
رأس الناقورة	Ld	*Râs en Nâkûrah.* 'The headland of the trumpet'; *see en Nâkûrah.*
رأس النبع	Ne	*Râs en Nebâ.* The hill-top or headland of the spring.
رأس النمرة	Md	*Râs en Nimreh.* The hill-top or headland of the abundant water.
رأس تمرون	Me	*Râs Tamrûn.* The head or hill-top of Tamrûn. p.n. (from تمر fruit).
رويسات العين	Nf	*Ruweisât el 'Ain.* The hill-tops of the spring.
رويسات كنعان	Md	*Ruweisât Kan'ân.* The hill-tops of Canaan.
سهلة لجونة	Me	*Sahlet Lejûnah.* The level plain of Lejûnah; Latin *legiones.*
صلحانة	Nd	*Salhâneh.* Salhâneh. p.n. *cf.* صالح *Sâleh* (righteous).

سنجق عكا {Lf / Lc} *Sanjak 'Akka.* The province of Akka ; *see* "Memoirs," p. 44.

سميرية Lf *Semeiriyeh.* The word means in Arabic 'tawny,' ' brown'; but the name is probably a survival of the Hebrew שמירון מראון *Shimron meron* (Joshua xii, 20). It is the Casale Somelaria Templi of the mediaeval writers.

شيخة مباركة Nc *Sheikhah Mubârakeh.* The shrine of the female sheikh Mubârakeh (blessed).

شيخ عبد الرحيم Nd *Sheikh 'Abd er Rahîm.* The shrine of Sheikh 'Abd er Rahîm ('slave of the merciful'). p.n.

شيخ عبد الرسول Nd *Sheikh 'Abd er Rasûl.* The shrine of Sheikh 'Abd er Rasûl ('slave of the apostle'). p.n.

شيخ علي ابي سعد Nc *Sheikh 'Aly Abu Sâd.* The shrine of Sheikh Abu Sâd ('father of auspiciousness'). p.n.

شيخ علي فارس Mf *Sheikh 'Aly Fâris.* The shrine of Sheikh 'Ali Fâris. p.n.

الشيخ دنون Lf *Esh Sheikh Dannûn.* The shrine of Sheikh Dannûn. p.n.

شيخ داود Lf *Sheikh Dâûd.* The shrine of Sheikh David ; *see Sheikh Dâûd,* p. 10.

شيخ حسين Mc *Sheikh Husein.* The shrine of Sheikh Husein. p.n. Husein was the son of 'Ali, murdered at Kerbela, and one of the chief saints of the Metawileh.

الشيخ ابراهيم السعدى Lc *Esh Sheikh Ibrahîm es Sâdy.* The shrine of Sheikh Ibrahim es Sâdi. p.n.

شيخ قويقس Nc *Sheikh Kûeikis.* The shrine of Sheikh Kuweikis. p.n.

شيخ مجاهد Nc *Sheikh Mejâhed.* The shrine of Sheikh Mejâhed (one who fights for the faith). p.n.

شيخ موسى Nc *Sheikh Mûsa.* The shrine of Sheikh Musa (Moses).

الشيخ شحادد Mc *Esh Sheikh Shehâdeh.* The shrine of Sheikh Shehâdeh (begging). p.n.

شيحين Nd *Shîhîn.* Shihin. p.n.

سحماتا Nc *Suhmâta.* Suhmata ; perhaps from سحم ' black.'

سروح Nd *Surûh.* Surûh. p.n. The root of the word in Arabic means 'to flow freely' (water) or 'to pasture at large' (cattle).

طاحونة الدرابسة Mc *Tâhûnet ed Darâbsah.* The mill of the Darâbseh. p.n.

طاحونة الفرج *Lc* *Tâhûnet el Ferj.* The mill of the hole or chink.

طاحونة الجميزة *Lc* *Tâhûnet el Jimmeizeh.* The mill of the sycamore tree.

طاحونة مفشوخ *Lc* *Tâhûnet Mefshâkh.* The mill of Mefshûkh; *see Birket Mefshûkh,* p. 41.

طاحونة المناوات *Lc* *Tâhûnet el Menâwât.* The mill of Menâwât (perhaps from مُنًا safety).

طاحونة ام الفار *Lc* *Tâhûnet Umm el Fâr.* The mill with the mouse.

طاحونةام الحجرين *Lc* *Tâhûnet Umm el Hajrein.* The mill of the two stones.

طاحونة حامود *Ld* *Tâhûnet Hâmûd.* The mill of Hâmûd. p.n.

طاحونة حميمة *Lc* *Tâhûnet el Humeimeh.* The mill of the little bath or pigeon.

طلعة البلان *Lf* *Talât el Bellâneh.* The mountain path or ascent of bellân (*Poterium spinosum*).

تير حرفا *Md* *Teir Harfa.* The fort of Harfa. p.n.

تير شيحة *Nc* *Teir Shîha.* Teir Shiha. *Teir,* 'a fortress.' *Shîh* is a fragrant herb.

التل *Lc* *Et Tell.* The Mound.

تل عمران *Md* *Tell 'Amrân.* 'Amrân's mound.

تل العريس *Mc* *Tell el 'Areis.* The mound of the bride.

تل العودة *Nc* *Tell el 'Aûdeh.* 'Aûdeh's mound. p.n.

تل باسيل *Nd* *Tell Bâsil.* Basil's mound (perhaps the Latin Basilius). p.n.

تل بلاطا *Nc* *Tell Belât.* The mound of the stone slab.

تل ظهور الحارة *Tell Dhuhûr el Hârah.* The mound of the ridges of the enclosure.

تل الفقيه *Md* *Tell el Fakieh.* The mound of the jurisprudist.

تل الفخار *Lf* *Tell el Fokhkhâr.* The mound of pottery.

تل فطيس *Nc* *Tell Futeis.* The mound of Futeis; an animal that has not been legally slaughtered, and is therefore unfit for food (braxy). (افطس also means 'flat-nosed.')

تل الغياضة *Mc* *Tell el Gheiyâdah.* The mound of the thicket.

تل الحامى قرصه *Md* *Tell el Hâmy Kurseh.* The mound of 'him who protects his loaf or disk.'

تل الهوا *Lc* *Tell el Hawa.* The mound of the air.

تل الحسبة Md *Tell el Hisbch.* The mound of calculation.

تل الخوارة Me *Tell el Hŭwárah.* The mound of the white marl.

تل الكشك Md *Tell el Kishk.* The mound of Kishk; a dish made of (برغل ولبن) parched corn and soured milk.

تل اللقية Me *Tell el Lekîych. Lekiych* means 'anything thrown.'

تل المشنقة Me *Tell el Mashnakah.* The mound of the gallows.

تل ميماس Lf *Tell Mîmâs.* The mound of Mimâs.

تل الصفا Ne *Tell es Safa.* The mound of clear water.

تل السميرية Lf *Tell es Semeirîych. See es Semeirîych,* p. 54.

تل الشعتون Md *Tell esh Shâtûn.* The mound of Shâtûn; perhaps from شعث to be dusty and dishevelled.

تل الشوبة Me *Tell esh Shôbch.* The mound of deceit. The root of the word in Arabic implies mixing; in Hebrew, to be white or hoary.

تل شبيب Le *Tell Shubeib.* The mound of the youth.

تل الزعتر Le *Tell ez Zátcr.* The mound of thyme.

تل الزبدية Le *Tell ez Zubdîych.* The mound of Zubdiych; *see Khŭrbet Zebed,* p. 51.

تريبخا Ne *Terbikha.* Terbikha. p.n.; perhaps *Teir Bikha,* the fortress of Bikha.

طول مردا Me *Tûl Merda.* The length of Merda. p.n.

وادى ابى الذهب Lf *Wâdy Abu 'dh Dhaheb.* Abu 'dh Dhaheb's valley. Abu Daheb was the name of a Mamlûk who plundered the convent on Carmel in 1775.

وادى ابى لباد Nd *Wâdy Abu Lebbâd.* The valley of Abu Lebbâd. p.n. signifying 'felt maker.'

وادى ابى محمد Nd *Wâdy Abu Muhammcd.* The valley of Abu Muhammed. p.n.

وادى ابى ناصر Ld *Wâdy Abu Nâsr.* The valley of Abu Nâsr. p.n. signifying 'helper.'

وادى ابى سنان Lf *Wâdy Abu Scnân.* The valley with the serrated ridge.

وادى ابى طبيخ Lf *Wâdy Abu Tabîkh.* The valley of Abu Tabikh; lit. 'father of cooked meat.'

وادى ابى ترك Md *Wâdy Abu Tŭrkch.* The valley of Abu Tŭrkch. *Tark,* 'to leave' or 'abandon.'

وادى عيدة Ld *Wâdy 'Aîdeh.* The valley of 'Aideh. p.n.

وادى العين {Md Ld} *Wâdy el 'Ain.* The valley of the spring.

وادى عين الدم Me *Wâdy 'Ain ed Dumm.* See *'Ain ed Dumm,* p. 39.

وادى علوية Md *Wâdy 'Alawîyeh.* The valley of 'Alawiyeh. p.n. 'Ala-wiyeh is either from the root علا *álâ*, 'to be lofty,' or from the proper name علي *'Alí,* the chief saint of the Shiah and Metawileh sects.

وادى العيون Mf *Wâdy el 'Ayûn.* The valley of springs.

وادى البق Na *Wâdy el Bakk.* The valley of the bug.

وادى البئر Le *Wâdy el Bîr.* The valley of the well.

وادى الدفين Me *Wâdy ed Dafîn.* The valley of the buried.

وادى داروما Mf *Wâdy Dârûma.* From the Hebrew *Daroma*, 'dry land.'

وادى الدلم Me *Wâdy ed Delam.* The valley of ed Delam; the root *Delam* means to be intensely black (a rock), and is also a name for the male of the attagen, francolin, or rail (bird).

وادى الدولة Md *Wâdy ed Daulch.* The valley of the government.

وادى الديكية Ld *Wâdy ed Dîkîyeh.* From *dik*, 'a cock.'

وادى الضبعة Md *Wâdy ed Dûbâh.* The valley of the hyena.

وادى العزية Md *Wâdy el 'Ezzîyeh.* The valley of el 'Ezziyeh; *cf. el 'Ezzîyah,* p. 4.

وادى غابسية Lf *Wâdy Ghâbsiyeh.* The valley of Ghâbsiyeh. q.v.

وادى الغضب Md *Wâdy el Ghadab.* The valley of anger. *Ghadab* also means 'a hard rock,' and 'anything very red.'

وادى الغراب Ld *Wâdy el Ghurâb.* The valley of the raven.

وادى الحافور Me *Wâdy el Hâfûr.* The valley of the excavation.

وادى الحاج حسين {M Nd} *Wâdy el Hâjj Husein.* The valley of the pilgrim Husein.

وادى حامول Ld *Wâdy Hâmûl.* The valley of Hâmûl; perhaps the Hammon of Joshua xix, 28.

وادى حسن Md *Wâdy Hasan.* The valley of Hasan. Hasan and Husein, sons of 'Ali, as has been before remarked, are the great Metawileh and Shiah saints.

وادى الحميرا Lf *Wâdy el Humeira.* The reddish-coloured valley.

I

وادى جعثون   Mc   *Wâdy Jâthûn.* The valley of Jâthûn ; *see Khûrbet Jâthûn,* p. 48.

وادى الجمل   Nc   *Wâdy el Jemel.* The valley of the camel.

وادى الجزازة   Lf   *Wâdy el Jezzâzeh.* The valley of shearing.

وادى جدّين   {Mf / Lf}   *Wâdy Jiddîn.* The valley of Jiddin. p.n.

وادى جولس   Mf   *Wâdy Jûlis.* The valley of Jûlis. p.n. (Julius).

وادى التراقع   Lf   *Wâdy el Kerâkâ.* The valley of croaking, or of frogs.

وادى كواشين   Mf   *Wâdy Kuwâshîn.* The valley of the summer houses ; sing. *Kâshân.* It is a Persian word.

وادى كركرة   Mc   *Wâdy Kerkera.* The valley of the bubbling water.

وادى قرمت   Md   *Wâdy Kermith.* The valley of Kermith. p.n.

وادى قبالى   Mf   *Wâdy Kibâly.* The southern valley.

وادى تيراط   Mc   *Wâdy Kîrât.* The valley of the Kirât (a measure).

وادى الخشنة   Mf   *Wâdy el Khashneh.* The valley of rough ground.

وادى الخرب   Nc   *Wâdy el Kherûb.* The valley of ruins.

وادى القرن   Lc   *Wâdy el Kûrn.* The valley of the horn or peak.

وادى قطعية   Ld   *Wâdy Kutâiyeh.* The valley of the fragment.

وادى قزيزية   Mc   *Wâdy Kuzîzîyeh.* The valley of Kuziziyeh. p.n.; perhaps ‘glass-makers.’

وادى محوز   Nf   *Wâdy Mahhûz.* The valley of Mahhûz (a sweet-smelling herb).

وادى معيسلة   Mf   *Wâdy Mâtsleh.* The valley of the place abounding with honey.

وادى المجنونة   {Lf / Mf}   *Wâdy el Majnûneh.* The valley of the mad woman, or haunted by genii.

وادى المربط   Nd   *Wâdy el Marbat.* The valley of the place for tying up horses.

وادى المعشوق   Mc   *Wâdy el Mâshûk.* The valley of Mashûk ; *see Neby Mashûk,* p. 10.

وادى المدفون   Ld   *Wâdy el Medfûn.* The valley of the buried one.

وادى المفخوتة   Md   *Wâdy el Mefkhûteh.* Perhaps from فخت ‘round perforations in a roof.’

وادى مناويثا   Mf   *Wâdy Menâwîtha.* The valley of Menâwitha. p.n.

وادى مرج العبد   Lf   *Wâdy Merj el 'Abd.* The valley of the prairie of the slave.

وادى المشاحر   Md   *Wâdy el Meshâhir.* The valley of the charcoal furnaces.

وادى المتقع   Ne   *Wâdy el Meskâ.* The valley of the cold or frosty place.

وادى المناظر   Ld   *Wâdy el Munâtir.* The valley of watch-towers.

وادى المشاوش   Md   *Wâdy el Musheiwish.* The valley of water-pits.

وادى النكاتة   Ld   *Wâdy en Nekkâteh.* The valley of en Nekkâta. p.n. meaning 'spotted.'

وادى النظارة   Nd   *Wâdy Nettârah.* The valley of the watcher or 'look-out.'

وادى الرخم   Mf   *Wâdy er Rakham.* The valley of the Egyptian vulture.

وادى رجمين   Md   *Wâdy Rujmein.* The valley of the two stone heaps.

وادى الركبة   Ld   *Wâdy er Rukbeh.* The valley of the riders.

وادى الرويسات   Ld   *Wâdy er Ruweisât.* The valley of the hill-tops.

وادى سلحانة   Nd   *Wâdy Salhâneh.* The valley of Salhâneh. p.n.; see p. 53.

وادى الصعاليك   Le   *Wâdy es Sâlîk.* The valley of the dervish or poor wanderer.

وادى السميرية   {Kf / Lf}   *Wâdy es Semeiriyeh.* The valley of Semeiriyeh; *see* Semeiriyeh, p. 54.

وادى السماك   Mf   *Wâdy es Semmâk.* The valley of the fisher.

وادى السروات   Md   *Wâdy es Serâwât.* The valley of the cypresses.

وادى الشاغور   Ne   *Wâdy esh Shâghûr.* The valley of the Shâghûr, which is the name of the district. q.v.

وادى شمع   Md   *Wâdy Shemâ.* The valley of Shemâ; *see Neby Shemâ.* p. 53.

وادى الشمالى   Mf   *Wâdy esh Shemâly.* The northern valley.

وادى الشناتى   Ld   *Wâdy esh Shenâty.* Probably from *Shanath = shathin.* 'to weave.'

وادى شبحين   Md   *Wâdy Shîhîn.* The valley of Shihin. p.n.

وادى شويتا   Lf   *Wâdy Shûeit.* Shûeit is the diminutive of *shaut,* and means 'a place between two tracts of high ground in which water collects.'

وادى صبح   Na   *Wâdy Subh.* The valley of the dawn.

وادى سحماتا   Ne   *Wâdy Suhmâta.* See Suhmâta, p. 54.

وادى السُّقاق — Me — *Wâdy es Sukâk.* Probably from *sakfkeh*, a kind of herb.

وادى سروح — Ne — *Wâdy Surûh.* See *Surûh*, p. 54.

وادى صوانه — Ne — *Wâdy Shwâneh.* The valley of flints.

وادى السويطية — Lf — *Wâdy es Suweitîyeh.* The valley of the Suweitîyeh, so called from the Arabs of that name.

وادى الوقية — Md — *Wâdy el Wakîyeh.* The valley of the *Wakîyeh*, the vulgar pronunciation of *ukîyeh*, an ounce (weight). The root *waka* also signifies protecting.

وادى الزعارير — Md — *Wâdy ez Zârîr.* The valley of the hawthorn.

وادى زمزم — Nd — *Wâdy Zemzem.* The valley of Zemzem. Zemzem is the name of the well at Mecca—supposed to be that which was shown by the angel to Hagar. The word means 'murmuring water,' and is applied to any sweet or abundant spring.

وادى زريق — {M Nd} — *Wâdy Zereik.* The azure valley.

وادى الزرقآ — Md — *Wâdy ez Zerka.* The valley of blue water.

يانوح — Mf — *Yânûh.* See SHEET II.

يركا — Mf — *Yerka.* Yerka. p.n.

الزيب — Le — *Ez Zîb.* The ancient Achzib.

زبقين — Nd — *Zubkîn.* Zubkin. p.n.; see *Khallet Zubkîn*, p. 46.

SHEET IV.

انو بانك Pe *Abu Bálek.* Abu Bálek. p.n.

ابى شوارب Qb *Abu Shúárib.* The father (possessor) of moustaches, or 'drinkers.'

عين عباد Rf *'Ain 'Abbád.* The spring of 'Abbád (a devotee). p.n.

عين عبضية Of *'Ain 'Abbadiych.* The spring of the ابانية 'Ibadiych, a puritanical Mohammedan sect.

عين العافية Pf *'Ain el 'Áfich.* The spring of health or soundness. The root signifies the effacing of vestiges of habitation, culture, &c., from land.

عين عكبرة Ne *'Ain 'Akbara.* The exact form does not occur in Arabic, but *ákbar* means a male jerboa, and *'ikbir* a viscous substance collected by bees. It is the Hebrew עכברה mentioned several times in the Talmud, and called in Josephus (*B. J.* II, 25, 6) the rock of the Achaberi.

عين العلوية Pe *'Ain el 'Alawiyeh.* The spring of the 'Alawiyeh; *see Wâdy 'Alawiyeh,* p. 57.

عين العلمانية Re *'Ain el 'Almániych.* The spring of 'Almániych. p.n. *cf.* Heb. עלמון as a proper name of a place, Josh. xxi, 18.

عين البخرد Ne *'Ain el Bakhrah.* The spring of vapour or steam.

عين البلاط Pe *'Ain el Balát.* The spring of the stone slabs.

عين البلاطة Qd *'Ain el Balátah.* The spring of the stone slab.

عين البانية Pf *'Ain el Bánich.* The built-up spring.

عين البربير Qd *'Ain el Barbir.* Perhaps from *barbár,* a bucket that makes a noise in the water.

عين الباردة Pe *'Ain el Bárdeh.* The cold spring.

عين البيضآ· {Qe Pf} *'Ain el Beida.* The white spring.

عين البلانة Pf *'Ain el Bellánch.* The spring of bellánch (a thorny plant, *Poterium spinosum*).

عين بيريا Pf *'Ain Biria.* The spring of Biria. From *Bir,* 'a well,' the Hebrew בירי. Talmud, Baba Meziah, 84*b*; Pessachim, 51.

عين البرانية {Pf Nf} *'Ain el Búrrániych.* The outer spring.

عين الدرة | Oc | 'Ain el Durrah. The spring of maize.

عين فرتم | Qf | 'Ain Ferâm. The spring of Ferâm. p.n.

عين الفوقآ | {Od} {Pe} | 'Ain el Fôka. The upper spring.

عين غباطى | Oc | 'Ain Ghabbâtî. Perhaps from ghabît, 'low-lying land.'

عين الغنم | Of | 'Ain el Ghanem. The spring of flocks.

عين الغاورداى | Ne | 'Ain el Ghâwardâi. The spring of the place where truffles abound.

عين الغدران | Od | 'Ain el Ghudrân. The spring of the swamps.

عين الحمرة الفوقآ | Pf | 'Ain el Hamrat el Fôka. Properly Humrat. The upper spring of the reddish muddy water.

عين الحمرة التحتآ | Pf | 'Ain el Hamrat el Tahta. The lower spring of the reddish muddy water.

عين حنين | Od | 'Ain Hânîn. The spring of Hânin. p.n.

عين الحارة | Pd | 'Ain el Hârah. The spring of the enclosed space.

عين حرمون | Oc | 'Ain Haramûn. The spring of Hermon (doubtful).

عين الحاصل | Pf | 'Ain el Hâsel. The spring of produce.

عين حوضين | Of | 'Ain Haudein. The spring of the two cisterns.

عين الحوش | Pf | 'Ain el Hôsh. The spring of the enclosure.

عين حميمة | Oe | 'Ain Humeimeh. The spring of the little bath.

عين الحمام | Re | 'Ain el Hummâm. The spring of the (warm) bath.

عين الحرية | Od | 'Ain el Hurrîyeh. The spring of freedom.

عين حوارة | Pe | 'Ain Huwârah. The spring of white meal or soft chalk.

عين ابل | Od | 'Ain Ibl. The spring of camels.

عين جاعونة | Qf | 'Ain Jâûneh. Either connected with ju'ûneh, a short, fat man; or with jî'ân, hungry.

عين جحولة | Qd | 'Ain Jahûlah. The spring of the large rock.

عين الجامع | Pf | 'Ain el Jâmiâ. The spring of the mosque.

عين الجديده | {Oe} {Of} {Pf} | 'Ain el Jedîdeh. The new spring.

عين الجمل | Of | 'Ain el Jemel. The spring of the camel.

عين الجندية   Nd   '*Ain el Jenadiyeh*. The spring of the military. Syria was formerly divided into five *ajnâd* (sing. *jund*) or military divisions. There is also an Arab tribe called el Jenâdy.

عين جنان   Nf   '*Ain Jenân*. The spring of gardens.

عين الجرب   Pf   '*Ain el Jerab*. The spring of the plantations.

عين الجن   Pf   '*Ain el Jinn*. The spring of the jinn (genie or demon).

عين الجش   Pe   '*Ain el Jish*. The spring of el Jish (Giscala). *See el Jish*, p. 76.

عين الجوزة   Pd   '*Ain el Jôzeh*. The spring of the walnut tree.

عين الجديدة   Pf   '*Ain el Judeiyideh*. The spring of el Judeiyideh, diminutive of *jedideh*, 'a dyke'; a coloured streak on a mountain side.

عين الجوران   Pf   '*Ain el Jurân*. The spring of the hollow stone.

عين الجرون   Of   '*Ain el Jurûn*. The spring of the place for storing wheat or drying raisins.

عين قديثا   Pe   '*Ain Kaddîtha*. The spring of Kadditha.

عين القاضى   Qf   '*Ain el Kâdy*. The spring of the judge.

عين القنطرة   Pe   '*Ain el Kantarah*. The spring of the arch.

عين تطمون   Oe   '*Ain Katamûn*. The spring of Katamûn. The root قطم means 'to quarry'; *cf. Jebel el Mukattem*, in Egypt.

عين الكبيرة   Oe   '*Ain el Kebîreh*. The big spring.

عين كفرة   Od   '*Ain Kefrah*. The spring of the village.

عين الكروم   Qf   '*Ain el Kerûm*. The spring of the vineyards.

عين الخروبة   Of   '*Ain el Kharrûbeh*. The spring of the carub or locust tree.

عين خطّارة   Pf   '*Ain Khattârah*. Probably from خطرة a shower, or patch of herbage growing after a shower.

عين الخربة   Pe   '*Ain el Khûrbeh*. The spring of the ruin.

عين القصيبة   {Qe / Pf}   '*Ain el Kûseibeh*. The spring of the little reed.

عين اللبوة   Rf   '*Ain el Lebweh*. The spring of the lioness.

عين اللوز   Rf   '*Ain el Lôz*. The spring of the almond.

عين اللوزية   Qe   '*Ain el Lôziyeh*. The spring of almonds.

عين ماروس   Qe   *'Ain Mârûs.* The spring of Mârûs.

عين ميرون   Pf   *'Ain Meirôn.* The spring of Meirôn (Merom).

عين الملاحة   Qd   *'Ain el Mellâhah.* The spring of sweetened water; *see* SHEET I, p. 3.

عين المرج   Od   *'Ain el Merj.* The spring of the meadow.

عين مرج شرش الخبب   Qf   *'Ain Merj Shersh el Khubb.* The spring of the meadow of Shersh el Khubb. *Shersh* (properly *shers*) means small thorny trees on which camels pasture; *khubb* means low ground.

عين المزاريب   Oe   *'Ain el Mezârîb.* The spring of the channels.

عين المغار   Qf   *'Ain el Mâghâr.* The spring of the caves.

عين المقيصبة   Pe   *'Ain el Mukeisibeh.* The spring of the reedy ground.

عين المخلد   Qf   *'Ain el Mukhallid.* The perpetual spring.

عين النحيلة   Qf   *'Ain en Naheileh.* The spring of the little bee. (But *cf.* the Hebrew נַחַל 'a water-course' or 'wady.')

عين النبرتين   Qe   *'Ain en Nebratein.* The spring of the two elevated spots.

عين نبي حانيا   Qe   *'Ain Neby Hâniya.* The prophet Hâniya's spring. p.n.

عين النمرة   Oe   *'Ain en Nimreh.* The spring of the leopard. (It also means abundant water.)

عين النوم   Of   *'Ain en Nôm.* The spring of sleep.

عين النحف   Nf   *'Ain Nühf.* The spring of Nühf. p.n. The *Nef* of the Crusaders.

عين النسورة   Pf   *'Ain en Nûsûrah.* The spring of the eagles.

عين الراهب   Pe   *'Ain er Râhib.* The spring of the monk.

عين الريحانة   Pf   *'Ain er Rîhâneh.* The spring of basil, a sweet herb.

عين الرميلة   Pf   *'Ain er Rumeileh.* The spring of the sandy ground.

عين الرشيدة   Pf   *'Ain er Rusheideh.* The spring of the Rusheideh.; *cf. er Rusheidîyeh,* p. 10.

عين رشيف   Od   *'Ain Rusheif.* The spring of the swamp.

عين الصغيرة   {Oe Of}   *'Ain es Saghîreh.* The small spring.

عين السهلة   Of   *'Ain es Sahleh.* The spring of the level ground

عين الصخرة   Pf   *'Ain es Sakhrah.* The spring of the rock.

عين الصالح Pf 'Ain es Sâleh. The spring of es Sâleh. p.n. (righteous).

عين سمورة Pf 'Ain Samûrah. The brown spring.

عين سيجور Of 'Ain Seijûr. The spring of the Seijûr. 'A dog collar,' also 'red turbid water.'

عين السموئية Pf 'Ain es Semûäieh. The spring of es Semûäieh. p.n.

عين الشعير Oc 'Ain esh Shâîr. The spring of barley.

عين الشيخ Of 'Ain esh Sheikh. Spring of the Sheikh (elder or saint).

عين سدرة اللهبية Rf 'Ain Sidret el Lehebiyeh. The spring of the lote-tree of the Lehebiyeh (Arabs).

عين صفرة Oe 'Ain Sûfra. The yellow spring.

عين صوف Pe 'Ain Sûf. The spring of wool.

عين الصفصافة Rf 'Ain es Sûfsâfeh. The spring of the osier willow.

عين السلطان Pe 'Ain es Sultân. The Sultan's spring.

عين الصرار Of 'Ain es Sûrâr. Sûrâr means high places over which water does not flow.

عين سربين Od 'Ain Surubbîn. The fountain of Surubbîn. p.n.

عين الطبل {Oc / Pf} 'Ain et Tabil. The spring of the drum.

عين التحتا Pe 'Ain et Tahta. The lower spring.

عين الطارة Of 'Ain et Târah. The spring of the rim or fringed border.

عين طيريا Nf 'Ain Tiria. The spring of Tiria. p.n.

عين طوبى Rf 'Ain Tôba. The goodly spring. طوبى is the name of a tree in Paradise, according to the Kor'ân.

عين التفاح Oe 'Ain et Tuffâh. The spring of the apple.

عين ام قادوس Od 'Ain Umm Kâdûs. The spring of the mother of the holy one.

عين ام القرع Rf 'Ain Umm el Kûrâ. The gourd-producing spring.

عين ام طحين Pf 'Ain Umm Tahin. The flour-producing spring.

عين الورقة Pf 'Ain el Warakah. The spring of the leaf.

عين يعتر Nd 'Ain Yâter. The spring of Yâter. p.n.

عين الزيتون Pf 'Ain ez Zeitûn. The spring of the olive.

عين الزرقا Pf 'Ain ez Zerka. The spring of azure, glittering water.

K

عينيثا I'd *'Ainitha.* 'Ainitha. p.n.

عيطآ· الشعب Nd *'Aita esh Shâub.* The tall mountain of the spur.

عيثرون Pd *'Aitherûn.* 'Aitherûn. p.n.

العقربة I'e *El 'Akrabeh.* The (female) scorpion.

عقبة البئر Qd *'Akabet el Bîr.* The steep or mountain road of the well.

عقبة اللبن Qf *'Akabet el Leben.* The steep or mountain road of *leben.* *Leben* means 'sour, curded milk,' but the name is probably connected with the Hebrew לבנה 'white.'

عقبة المرج Pd *'Akabet el Merj.* The steep or mountain road of the meadow.

عكبرة Pf *'Akbara.* See *'Ain 'Akbara,* p. 61.

علما I'e *'Alma.* 'Alma. p.n.; *see Almân,* p. 17 ; and *Almanîyeh,* p. 61.

العمارة Qd *El 'Amârah.* The edifice.

عموقة Qe *'Ammûkah.* From عمق 'deep.'

عرب الاكراد {Qe / Re} *'Arab el 'Akrâd.* The Kurdish Arabs.

عرب الغوارنة Qd *'Arab el Ghawârneh.* The lowland Arabs.

عرب الحمدون {Qd / Qe} *'Arab el Hamdûn.* The Hamdûn Arabs.

عرب الحسنية {Qe / Re} *'Arab el Hasanîyeh.* The Hasanîyeh Arabs.

عرب الجنادى Qd *'Arab el Jenâdy.* The Jenády Arabs.

عرب الخرانبة Pf *'Arab el Kharanbeh.* The Kharanbeh Arabs.

عرب الخوابى Pf *'Arab el Khawâby.* The Khawâby Arabs.

عرب القديرات Pf *'Arab el Kudeirât.* The Kudeirât Arabs.

عرب لهيب العيني {Pe / Qe} *'Arab Luheib el 'Aithy.* The Luheib el 'Aithy Arabs. p.n.

عرب لهيب المريدات {Ne / Of} *'Arab Luheib el Mureidât.* The Luheib el Mureidât Arabs. p.n.

عرب لهيب Oe *'Arab Luheib er Rusâtmeh.* The Luheib er Rusâtmeh Arabs. p.n.

عرب المعاسى {Ne / Oe} *'Arab el Mûâsy.* The Mûâsy Arabs.

| عرب السويطات | {Nc Od} | *'Arab es Suweitât.* The Suweitât Arabs. |

عرب السويطات {Nc Od} *'Arab es Suweitât.* The Suweitât Arabs.

عرب الزنغرية Qf *'Arab ez Zenghariyeh.* The Zenghariyeh Arabs. p.n.

عرب الزبيد Qe *'Arab ez Zubeid.* The Zubeid Arabs; cf. *Khŭrbet Zebed*, p. 51.

ارض دواميس Qe *Ard ed Dawâmîs.* The land of the dome trees.

ارض الدرجة Oe *Ard ed Derajeh.* The land of the steps.

ارض الحمرة Od *Ard el Hamra.* The reddish land.

ارض حريثة Qe *Ard Harîtheh.* The ploughed land.

ارض المعقور Of *Ard el Kâkûr.* The land of the cairn (stone heap).

ارض كريسفة Oe *Ard Kersifa.* The land of Kersifa. Perhaps كرسوف 'cotton.'

ارض الخيط Re *Ard el Kheit.* The land of the thread or string.

ارض الخضرآء Pf *Ard el Khûdra.* The land of verdure.

ارض المغراقة Oe *Ard el Mughrâkah.* The inundated land.

ارض الناقة Pd *Ard en Nâkah.* The land of the she-camel.

ارض النمورة Oe *Ard en Nimûrah.* The watered land.

ارض الرحراح Pe *Ard er Rahrâh.* The land of the Rahrâh, a small white bird.

ارض السيارة Re *Ard es Seiyârah.* The land of the traveller.

ارض الشقفان Pf *Ard esh Shŭkfân.* The land of the clefts.

ارض السحيمية Qd *Ard es Suheimiyeh.* The land of the Suheimiyeh. p.n.

ارض السواد Qe *Ard es Suwâd.* The land of black basalt.

ارض اليابس Od *Ard el Yâbis.* The dry land.

ارض الزنار Pd *Ard ez Zennâr.* The land of the girdle.

عريض ابى شنان Oe *'Arîd Abu Shennân.* The broad space with the dispersed water. (Perhaps اشنان, meaning plants burnt to make alkali.)

عريض عين الصفرء Oe *'Arîd 'Ain es Sûfra.* The broad place of the yellow spring.

عريض عقبية Oe *'Arîd 'Akabiyeh.* The broad place of the steep or mountain road.

عريض الصبية Qd *'Arîd es Sûbiyeh.* The broad place of the girl.

تريض السيجور	Nf	*'Arîd es Seijûr.* The broad place of the *seijûr; see Ain Seijûr,* p. 65.
تريض الشمس	Oe	*'Arîd esh Shems.* The broad place of the sun.
تريض التون	Qd	*'Arîd et Tûn.* The broad place of the lime-kiln.
تريض البطم	Qf	*'Ayûn el Butm.* The springs of the terebinth tree.
عيون المالحة	Pf	*'Ayûn el Mâlhah.* The salt springs.
عيون العقيبة	Pf	*'Ayûn el 'Okeibeh.* The springs of the little steep or mountain road.
عيون الوقاص	Qe	*'Ayûn el Wakkâs.* The spring of the broken-necked one.
عيون الورد	Pf	*'Ayûn el Werd.* The spring of the rose, or at which one goes down to water.
باب خلة حية	Oe	*Bâb Khallet Haiyeh.* The gate of the serpent's dell.
باب الخانوق	Qf	*Bâb el Khânûk.* The gate of the gorge.
باب الصديق	Pe	*Bâb es Saddîk.* The gate of the truth-teller.
بحيرة الحولة	Re	*Baheiret el Hûleh.* The lake of Hûleh.
بين الجبلين	Pf	*Bein el Jebelein.* Between the two mountains.
بيت جن	Of	*Beit Jenn.* The house of the genie: but probably *Beth gann,* the garden house.
بيت ليف	Nd	*Beit Lîf.* The house of *lîf* (palm-fibre).
بيت ياحون	Od	*Beit Yâhûn.* The house of Yâhûn. p.n.
البلاط	Qd	*El Balât.* The pavement or slab of rock.
بلاد بشارة	{Nd} {Od}	*Belâd Beshârah.* The country of the Beshârah, a noble and influential Syrian family.
بليدة	Qd	*Belîdeh.* The little village.
بنات يعقوب	Of	*Benât Yâkûb.* The daughters of Jacob.
بيار الجرمق	Of	*Biâr el Jermuk.* The wells of the Jermuk.
بيار السكر	Od	*Biâr es Sukker.* The wells of sugar.
بيت ام جبيل	Pd	*Bint umm Jubeil.* The daughter of the mother of the little mountain.
بئر العبد	Qf	*Bîr el 'Abd.* The well of the slave.
بئر العيون	Od	*Bîr el 'Ayûn.* The well of the springs.
بئر البياض	Nd	*Bîr el Beiyâd.* The well of the blank space.

بئر الحيات	Qf	*Bîr el Haiyât.* The well of the snakes.
بئر الخشب	Nf	*Bîr el Khashab.* The well of the woodman.
بئر المقاطع	Od	*Bîr el Makatîa.* The well of the cuttings or quarries.
بئر المزرعة	Qd	*Bîr el Mezrâh.* The well of the sown land.
بئر شيخ الخرابى	Pf	*Bîr Sh. Huzâby.* The well of Sheikh Huzâby. p.n.
بئر الشيح	Pf	*Bîr esh Shîh.* The well of Shih, a certain aromatic herb.
بئر السكر	Of	*Bîr es Sukker.* The well of sugar ; see *Biâr es Sukker.*
بئر التل	Pe	*Bîr et Tell.* The pool of the mound.
بئر الثنية	Nd	*Bîr eth Theniyeh.* The well of the mountain road.
بئر يوشع	Qd	*Bîr Yûshâ.* Joshua's well.
بئر زبود	Of	*Bîr Zebûd.* The well of Zebûd ; see *Khûrbet Zebed,* p. 51.
بئر زحلتى	Qf	*Bîr Zûhlûk.* The well of rolling down.
بيريا	Pf	*Bîria. See 'Ain Bîria,* p. 61.
بركة عيتا	Nd	*Birket 'Aîta.* The pool of the high mountain.
بركة عوبا	Be	*Birket 'Aûba.* The pool of Aûba. p.n.
بركة الدير	Ae	*Birket ed Deir.* The pool of the monastery.
بركة الدجاج	Qd	*Birket ed Dejâj.* The pool of the fowls.
بركة دلاتا	Pe	*Birket Delâta.* The pool of Delâta. p.n.
بركة دبل	Od	*Birket Dibl.* The pool of manure.
بركة الحافور	Pd	*Birket el Hâfûr.* The pool of the excavation.
بركة حامد	Of	*Birket Hâmed.* The pool of Hâmed. p.n.
بركة حامل	Oe	*Birket Hâml.* The pool of the bearer.
بركة الجش	Pe	*Birket el Jish.* The pool of Giscala.
بركة جب يوسف	Qf	*Birket Jubb Yûsef.* The pool of Joseph's pit.
بركة قديثا	Pe	*Birket Kadditha.* The pool of Kadditha. p.n.
بركة كعود	Rf	*Birket Kaûd.* The pool of Kaûd. p.n. ; also called بركة سلك *Birket Silak.* The pool of beetroot.
بركة القيومة	Pf	*Birket el Keiyûmeh.* The peerless or subsisting pool.

بركة كونين   Pd   *Birket Kûnîn.* The pool of Kûnîn. p.n.; perhaps from كنين *Kanîn,* 'concealed,' or from the Phen. כנן 'a post.'

بركة القصب   Qf   *Birket el Kŭsab.* The pool of reeds.

بركة المالكية   Pd   *Birket el Mâlkîyeh.* The royal pool, or the pool of the Malechites; *see Mâlkîyeh,* p. 9.

بركة النبع   Qf   *Birket en Nebâ.* The pool of the fountain.

بركة النقيز   Od   *Birket en Nukkeiz.* The pool of the sparrow.

بركة النصيبة   Of   *Birket en Nuseibeh.* The pool of small sacrificial stones, or 'high places.'

بركة رنجيغية   Pe   *Birket Ranjighia.* The pool of Ranjighia. p.n.

بركة سعسع   Oe   *Birket Sâsâ.* The pool of Sâsâ; *sâsâ* is the sound used in driving goats.

بركة شلعبون   Od   *Birket Shelâbûn.* The pool of Shelâbûn. p.n.

بركة شورا   Qf   *Birket Shôra.* The pool of Shôra. شورى a tree growing near the sea.

بركة شفنين   Of   *Birket Shufnîn.* The pool of the dove.

بركة طيطبا   Pe   *Birket Teitaba.* The pool of Teitabâ. p.n.; *cf.* ديدبا *Deidabâ,* 'a watch-tower.'

بركة ترجم   Qd   *Birket Terjem.* The pool which is pelted with stones.

البقيعة   Nf   *El Bukeiâh.* The little valley (between mountains).

برك عيثرون   Pd   *Burak 'Aitherûn.* The pools of Aitherûn. p.n.

برك علما   Pe   *Burak 'Alma.* The pools of 'Almâ; *see 'Ain 'Almânîyeh,* p. 61.

برك الشحم   Pf   *Burak esh Shahm.* The pools of fat, or perhaps of low-lying ground. شحم

البرج   {Pf/Oe}   *El Burj.* (2) The tower.

دبشة النبي   Qd   *Dabshet en Neby.* The prophet's hill. Heb. דבשה. In Arabic the word means 'a clod.'

الدبادب   Pe   *Ed Debâdib.* Debâdib. p.n. Perhaps from دبدبة 'clattering of hoofs.'

دبة تل حمار   Pe   *Debbet Tell Hamâr.* The plain of the asses' hill.

الدير   Pe   *Ed Deir.* The convent.

دير الغابية   Pd   *Deir el Ghábieh.* The convent of el Ghábieh (the thicket).

دير حبيب   Qd   *Deir Habib.* The convent of Habib. p.n.; signifying 'the friend.'

دير قلنسود   Nd   *Deir Kŭlŭnsawy.* The convent of the conical cap.

دير مار بطرس   Pe   *Deir Már Butrus.* St. Peter's Monastery.

دير وادى القاسى   Ne   *Ed Deir Wady el Kásy.* The convent of Wady el Kásy. q.v.

ديشون   Qe   *Deishûn.* Deishûn. p.n.

دلاتا   Pe   *Delâta.* Delâta. p.n.

الدرجة   Pf   *Ed Derajeh.* The steps.

الظاهرية الفوقآ   Pf   *Edh Dháheríyeh el Fôka.* The upper village on the ridge.

الظاهرية التحتآ   Pf   *Edh Dháheríyeh et Tahta.* The lower village on the ridge.

ظهر عليا   Pf   *Dhahr 'Alia.* The ridge of the height.

ظهر ارض الغربية   Pe   *Dhahr Ard el Gharbíyeh.* The ridge of the western land.

ظهر بيعبان   Pd   *Dhahr Beib'ân.* The ridge of Beib'ân. p.n.

ظهر البياض   Pe   *Dhahr el Beiyád.* The ridge of the white place.

ظهر برعشيت   Pd   *Dhahr Beráshít.* The ridge of Berâshit.

ظهر البّاره   Od   *Dhahr el Bârah.* The ridge of the well-sinkers.

ظهر الفطنين   Pf   *Dhahr el Fatnîn.* The ridge of Fatnîn. p.n.; from فطن to be intelligent.

ظهر الحزاريم   Pe   *Dhahr el Hazarîm.* The ridge of the round holes. (Phen. חור to be round).

ظهر الجمل   Nd   *Dhahr el Jemel.* The ridge of the camel.

ظهر خلة الشيح   Pe   *Dhahr Khallet esh Shîh.* The ridge of the dell of Shîh 'an aromatic plant, something like thyme.'

ظهر شقيف الغراب   Pe   *Dhahr Shúkíf el Ghuráb.* The ridge of the raven's cleft.

ظهر التوته   Qf   *Dhahr et Tûteh.* The ridge of the mulberry tree.

ظهر الوسطانى   Pe   *Dhahr el Wustâny.* The central ridge.

دبل   Od   *Dibl.* Dibl. p.n. ('manure'). Probably the Diblath of Ezekiel vi, 14.

الدوارة   Pd   *Ed Duwârah.* The circle.

الدوير Oe   *Ed Duweir.* The little monastery.

فارع Pe   *Fâráh.* The highest parts of a mountain.

فسُوطة Ne   *Fassûtah.* Fassûtah. p.n.

فرِم Qf   *Ferâm.* Ferâm. p.n.

فراضية Pf   *Ferrâdich.* Ferrâdich. p.n.

الفوارة Pf   *El Fûwârah.* The fountain.

الغابية Pf   *El Ghâbich.* El Ghâbich. ' the thicket.'

حبل العظيم Od   *Habl el 'Adhîm.* The grand terrace.

الحدب Pe   *El Hadab.* The hump.

حذاثا Od   *Haddâtha.* Haddâtha. p.n.

حجر الدم Qf   *Hajr ed Dûmm.* See *'Ain ed Dumm,* p. 39; but it may be *Hajar ed Dam.* The rock of blood.

حجر الحبلى Pf   *Hajr el Hûblch.* The pregnant stone.

حجر المكسور Pe   *Hajr el Maksûr.* The broken rock.

حجر منيقع Pf   *Hajr Muneika.* The rock of stagnant water.

حقل جنان Pf   *Hakel Jenân.* The field of gardens.

حانين Od   *Hânîn.* Hânin. p.n.

حريق الجنادية Nd   *Harîk el Jenadîych.* Burning of the Jenâdy Arabs.

حزور Od   *Hazzûr.* Hazzûr. p.n. Hazor.

الحجاج Pe   *El Hijâj.* The pilgrims.

الحولة {Od Rd}   *El Hûlch.* El Hûleh. p.n.

حمام بنات يعقوب Re   *Hümmâm Benât Yâkûb.* The hot-bath of the daughters of Jacob.

حرفيش Oe   *Hurfeish.* Hurfeish. p.n. Perhaps from حريش ' a snake.'

ابن يعقوب Qe   *Ibn Yâkûb.* The son of Jacob.

جاعونة Qf   *Jâûneh.* Jâ'ûneh. p.n.; meaning ' short and fat.'

جامع الاحمر Pf   *Jamiâ el Ahmar.* The red mosque.

جامع السويقة Pf   *Jâmiâ es Suweikah.* The mosque of the little market.

جبل ابى ثر Oe   *Jebel Abu Ghumr.* The mountain of the flood.

جبل ابی حجر *Og* *Jebel Abu Hajr.* The mountain with the rock.

جبل عدائر *Oe* *Jebel 'Adather.* Mount 'Adather. p.n.

جبل عین الشعیر *Of* *Jebel 'Ain esh Shâir.* The mountain of the spring of barley.

جبل العاصی *Pd* *Jebel el 'Âsy.* The mountain of el 'Âsy. p.n.; 'rebellious.'

جبل بیت ابی ذیاب *Pf* *Jebel Beit Abu Dhiâb.* The mountain of Abu Dhiâb's house. p.n.; signifying the fame of the Wolf: an Arab tribal name.

جبل برکوم *Oe* *Jebel Berkûm.* The mountain of Berkûm. p.n.

جبل بلعویل *Pd* *Jebel Belâwil.* Mount Belâwil. p.n.

جبل بیریا *Pf* *Jebel Biria.* See *'Ain Biria.*

جبل البقیعة *Nf* *Jebel el Bukeiâh.* The mountain of the little vale.

جبل الدیدبة *Of* *Jebel ed Deidebeh.* The mountain of the look-out.

جبل الظاهریة *Jebel edh Dhâheriyeh.* The mountain of the ridge.

جبل الضو *Oe* *Jebel ed Dô.* The mountain of light.

جبل غباطی *Oe* *Jebel Ghabbâtî.* See *Ain Ghabbâtî.*

جبل الغابیة *Pd* *Jebel el Ghâbieh.* The mountain of the thicket.

جبل غدماثا *Pd* *Jebel Ghudmâtha.* Mount Ghudmâtha.

جبل حضیرة *Pe* *Jebel Hadîreh.* The mountain of the fold.

جبل الحمرآ *Pf* *Jebel el Hamra.* The brown mountain.

جبل حمران *Pe* *Jebel Hamrân.* The red mountain.

جبل الهنبلی *Pf* *Jebel el Hanbely.* The mountain of el Hanbely. p.n.; 'lame.'

جبل هرمون *Oe* *Jebel Haramûn.* Mount Hermon.

جبل حیدر *Of* *Jebel Heider.* Mount Heider (the Lion; a title of Ali, the Shiah saint).

جبل حمید *Od* *Jebel Humeid.* The mountain of Humeid. p.n.; from حمد 'praise.'

جبل حریة *Pf* *Jebel Hureiyeh.* The mountain of freedom.

جبل جاءونة *Qf* *Jebel Jâûneh.* Mount Jâ'ûneh; *see Jâ'ûneh.*

L.

جبل جبيل   Pf *Jebel Jebeil.* The mountain of Jebail (small mountain).

جبل الجراد   Pf *Jebel el Jerâd.* The mountain of locusts.

جبل البريش   Of *Jebel el Jermûk.* Mount Jermûk (Jermocha).

جبل كنعان   Of *Jebel Kan'ân.* The scribe has written this *Jebel Kan'ân,* 'Mount Canaan,' but has interpreted it in a pencil note as if written جَبّ الكنعان 'the pits of Canaan.' The word is doubtful.

جبل قيسون   Qe *Jebel Keisûn.* Keisûn, a sort of herb.

جبل القيومة   Pf *Jebel el Keiyûmeh.* The mountain of 'the peerless,' 'up-right' or 'subsisting' one.

جبل خلة البئر   Qd *Jebel Khallet el Bir.* The mountain of valley of well.

جبل خلة الحد   Pe *Jebel Khallet el Hadd.* The mountain of the dell of the frontier.

جبل كحيل   Pd *Jebel Koheil.* The mountain of manganese, or antimony ore.

جبل كوكار   Pd *Jebel Kôkar.* The mountain of Kôkar. p.n.

جبل الكنيسة   Od *Jebel el Kuneiseh.* The mountain of the small church.

جبل اللوباني   Pf *Jebel el Lûbâny.* The mountain of el Lûbâny; perhaps from Heb. לבנה.

جبل مارون   Pd *Jebel Mârûn.* Mount Maron.

جبل المعصرة   Qd *Jebel el Mâserah.* The mountain of the oil press.

جبل المعتق   Od *Jebel el Mâtek.* The mountain of Mâtek. p.n.; meaning 'emancipated.'

جبل الملكي   Qd *Jebel Melky.* The royal mountain.

جبل المنابع   Of *Jebel el Menâbâ.* The mountain of the perennial springs.

جبل مغارة شهاب   Pf *Jebel Mûghâret Shehâb.* The mountain of the Shehâb's cave. p.n. Shehâb ed Din, the name of a Syrian chief.

جبل مخبّا   Qe *Jebel Mukhabbah.* The mountain of hiding place.

جبل المربع   Pf *Jebel el Murabbâ.* The square mountain.

جبل النحيلة   Qf *Jebel en Naheileh.* The mountain of the little bee; if from Hebrew, 'watercourse.'

جبل النطاح   Qe *Jebel en Nattâh.* The mountain of en Nattâh. p.n.; 'butting.'

جبل عبيد   Oe   *Jebel 'Obeid.* The mountain of the little stone.

جبل سعد   Ne   *Jebel Sâd.* Mount Sâd. p.n.; Felix.

جبل صفد   { Of / De / Pe / Pf }   *Jebel Safed.* Mount Safed. p.n.; *see* p. 32.

جبل سبلان   Oe   *Jebel Sebelân.* Mount Sebelân. Name of a Sheikh venerated by the Druzes.

جبل سليم   Od   *Jebel Selim.* Mount Selim. p.n.

جبل سراج   Ne   *Jebel Serâj.* Mountain of the lamp.

جبل السرج   Pe   *Jebel es Serj.* The mountain of the saddle.

جبل الشحار   Od   *Jebel esh Shahhâr.* The mountain of blackish ground.

جبل الشكارة   Qd   *Jebel esh Shakârah.* Mountain of charity land.

جبل الشيخ ربيعه   Qf   *Jebel esh Sh. Rabiâh.* The mountain of Sheikh Rabiâh.

جبل الصوانة   Of   *Jebel es Sûwâneh.* The mountain of the flint.

جبل تنورية   Oe   *Jebel Tannûriyeh.* The mountain of the reservoirs.

جبل الطويل   { Pf / Oe }   *Jebel et Tawîl.* The long mountain.

جبل الثعلب   Gd   *Jebel eth Thâleb.* The mountain of the fox.

جبل طبقات الزبعين   Pf   *Jebel Tûbakat el Arbâin.* The mountain of the stone slabs of the Forty, *i.e.*, of the shrine of the Forty Martyrs of Cappadocia.

جبل ام هرون   Pf   *Jebel Umm Harûn.* The mountain of Harûn's mother. p.n.

جبل ام التراى   Qd   *Jebel Umm el Kerâmy.* The mountain with the stumps.

جبل وعر موسى   Pf   *Jebel Wâr Mûsa.* The mountain of the rugged ground of Moses.

جبل الزبائك   Of   *Jebel ez Zebâik.* The mountain of Zebâik. p.n.; *see* Zubkin.

جبلة العروس   Of   *Jebelet el 'Arûs.* The single mountain of the bride.

جبلة الصفائف   Oe   *Jebelet es Sefâif.* The mountain of rows.

الجيرمك   Of   *El Jermûk.* Jermocha.

جمال الحصن   Od   *Jemâl el Hosn.* The beautiful fortress; literally, 'the beauty of the fortress,' an expression often used in Arabic for 'the beautiful fortress.'

الجش   Pe   *El Jish.*   Giscala (Gush Chaleb); *see* Memoirs, vol. I, p. 225.

جسر بنات يعقوب   Re   *Jisr Benât Yâkûb.*   The bridge of Jacob's daughter.

جباب القصب   Of   *Jubâb el Kŭsab.*   The pits of the reeds.

جور العنكليك   Of   *Jûr el 'Ankalik.*   The hollows of el 'Ankalik.   p.n.

جوران الذهب   Jurân edh Dhaheb.   The troughs of gold.

جورة الست مكة   Pd   *Jûrat es Sitt Mikkeh.*   Lady Mikkeh's Hollow.   p.n.

قبعة منصور   Pd   *Kabât Mansûr.*   Mansûr's skull cap.

قباعة   Qf   *Kabbâah.*   Kabbâah.   The word means 'large-headed.'

الكبّاش   Pe   *El Kabbâsh.*   The owner of rams.

قبور الشراسكة   Pf   *Kabûr es Serâsikah. (Sherâsikah.)*   The graves of the Circassians.

تدّيثا   Pe   *Kaddîtha.*   Kaddîtha.   p.n.

قديس   Qd   *Kades.*   Kadesh.

Pe   *Kasâret Butrus.*   This name is not written in Arabic in any of the lists. I am therefore not certain of the orthography; it is probably قصارة بطرس Peter's Hut.

كفر عنان   Of   *Kefr 'Anân.*   The village of 'Anân.   Perhaps the Kefar Chanan כפר חנן of the Talmud.

كفر برعم   Oc   *Kefr Birîm.*   The village of Birîm.   p.n.

كروم الميادين   Pd   *Kerûm el Meiâdîn.*   Vineyards of open places.

خلة العبد   Od   *Khallet el 'Abd.*   The dell of the slave.

خلة ابواب الهوا   Nc   *Khallet Abûâb el Hawa.*   The dell of the wind gates.

خلة ابى قتلة   Nf   *Khallet Abu Kŭtleh.*   The dell of slaughter.

خلة ابى منصور   Pf   *Khallet Abu Mansûr.*   Abu Mansûr's dell.   p.n.; meaning 'Victor's Father.'

خلة ابى الثلوج   Oc   *Khallet Abu 'th Thellûj.*   The snowy dell.

خلة ابى زيد   Oc   *Khallet Abu Zeid.*   Abu Zeid's dell.   p.n.

خلة العدس   Qd   *Khallet el 'Adas.*   The dell of lentils.

خلة العين   (Nd / Ne / Od)   *Khallet el 'Ain.*   The dell of the spring.

خلّة عين داما   Pe   *Khallet 'Ain Dáma.* The dell of '.Ain Dáma, 'The spring of Dáma.' p.n.

خلّة عين البند   Pd   *Khallet 'Ain el Hind.* The dell of the spring of *el Hind. Hind* without the article *el* was a common female name amongst the ancient Arabs.

خلّة عين الصفرآ   Oe   *Khallet 'Ain es Sŭfra.* The dell of the yellow spring.

خلّة عين الصغيرة   Oe   *Khallet 'Ain es Saghíreh.* The dell of the small spring.

خلّة عيسى   Pe   *Khallet 'Aîsa.* The dell of 'Aîsa, properly 'Isã, ' Jesus.'

خلّة عياد   Of   *Khallet 'Aiyád.* The dell of 'Aiyád. p.n.

خلّة العجرمى   Pd   *Khallet el 'Ajramy.* The dell of the 'Ajram (a plant).

خلّة عمشا   Pf   *Khallet 'Amsha.* Amsha's dell. p.n.

خلّة عطية   Od   *Khallet 'Atiyeh.* 'Atiyeh's dell. p.n.

خلّة البدارنة   Pd   *Khallet el Badárneh.* p.n. The dell of the Badarneh (family name).

خلّة البياض   Od   *Khallet el Beiyád.* The dell of the white place.

خلّة بيوض   Od   *Khallet Beiyûd.* The dell of the white place.

خلّة بئر الثنية   Nd   *Khallet Bîr eth Theníyeh.* The dell of the mountain road.

خلّة البرج   Pe   *Khallet el Burj.* The dell of the tower.

خلّة برّة   Oe   *Khallet Burrah.* The dell of wheat.

خلّة دقيقة   Oe   *Khallet Dakíkah.* The dell of flour.

خلّة الدالية   Pf   *Khallet ed Dálich.* The dell of the trailing vine.

خلّة دراج   Od   *Khallet Darráj.* The dell of the francolm or rail (a bird).

خلّة الضيعة   Od   *Khallet ed Deiáh.* The dell of the farm.

خلّة ديوس   Qf   *Khallet Deiyûs.* p.n. (Deiyûs is a term of abuse, meaning a willing cuckold.)

خلّة دردح   Nd   *Khallet Derdeh.* The dell of the elm tree.

خلّة الديك   Oe   *Khallet ed Dîk.* The dell of the cock.

خلّة الدير   Ne   *Khallet ed Deir.* The dell of the convent.

خلّة الدوم   Oe   *Khallet ed Dôm.* The dell of the Dom palm or lotus tree.

خلّة الدود   Pe   *Khallet ed Dûd.* The dell of the worm.

خلة الدوير   *Od*   *Khallet ed Duweir.*   The dell of the little monastery.

خلة فائق   *Od*   *Khallet Fâik.*   The dell of Fâik. p.n.

خلة فانس   *Ne*   *Khallet Fânis.*   The dell of Fânis. p.n.; signifying 'a pauper.'

خلة الفقيه   *Oe*   *Khallet el Fikich.*   The cloven valley.

خلة فضة   *Pd*   *Khallet Fuddah.*   The dell of silver.

خلة الغميس   *Pd*   *Khallet el Ghamîk.*   The deep dell.

خلة غزال   *Pe*   *Khallet Ghŭzâl.*   The dell of the gazelle.

خلة غزالة   *Qd*   *Khallet Ghŭzâleh.*   The dell of gazelles.

خلة غزالية   *Pf*   *Khallet el Ghŭzâlîyeh.*   The dell frequented by gazelles.

خلة حدايا   *Pf*   *Khallet Hadâia.*   The dell of the kite (bird).

خلة الحد   *Pe*   *Khallet el Hadd* (2).   The dell of the boundary.

خلة الحية   *Oe*   *Khallet el Haiyeh.*   The dell of the snake.

خلة الحجة   *Pd*   *Khallet el Hajjeh.*   The dell of the pilgrimage (or 'footpath ').

خلة الحجر   *Pe*   *Khallet el Hajr.*   The dell of the rock.

خلة حمود سيد   *Oe*   *Khallet Hammûd Seid.*   The dell of Hammûd Seid. p.n.

خلة حرمند   *Pd*   *Khallet Haramand.*   The dell of Haramand. p.n.

خلة الهوا   *Pd*   *Khallet el Hawa.*   The dell of the wind.

خلة حلاوة   *Pf*   *Khallet Helâweh.*   The dell of sweetness. p.n.

خلة حسيكة   *Pf*   *Khallet Huseikeh.*   The dell of thistles.

خلة حسين   *Nf*   *Khallet Husein.*   Husein's dell. p.n.; the Shiah saint and martyr.

خلة جميل   *Oe*   *Khallet Jemîl.*   The dell of Jemil. p.n.; meaning 'handsome.'

خلة الجوبا   *Nd*   *Khallet el Jûba.*   The dell of the watering trough; perhaps for جبى pl. of جبوة 'a watering trough.'

خلة الجبيب   *Pd*   *Khallet el Jubeib.*   The dell of the little pit.

خلة القاضى   *Pf*   *Khallet el Kâdy.*   The judge's dell.

خلة كاكيش   *Pf*   *Khallet Kâkish.*   The dell of Kâkish. p.n.

خلة كونة *Oe* *Khallet Kammūnch.* The dell of Cummin (plant).

خلة قامعة *Pe* *Khallet Kammāmeh.* The dell of the church or meeting-house.

خلة الكراسفة *Oe* *Khallet el Kerāsifeh.* The dell of the Kerāsifeh (a family name) ; singular, Kersifeh ; see *Ard Kersīfah*, p. 67.

خلة كرك *Od* *Khallet el Kerekeh.* The dell of the fortress.

خلة الخنزيرة *Qe* *Khallet el Khanzīreh.* The dell of the swine.

خلة خزنة *Qf* *Khallet Khaznch.* The dell of the treasure.

خلة خير الدين *Pf* *Khallet Kheir ed Dīn.* Kheir ed Din's dell. p.n.

خلة الخربة *Qd* *Khallet el Khŭrbeh.* The dell of the ruin.

خلة الخبازة *Od* *Khallet el Khobbāzeh.* The dell of the mallow plant.

خلة القسيس *Pd* *Khallet el Kūssis.* The dell of the (Christian) priest.

خلة كحيل *Pd* *Khallet Koheil.* The dell of manganese.

خلة القط *Od* *Khallet el Kott.* The dell of the cat.

خلة التوزح *Nd* *Khallet el Kōzāh.* The dell of the hill-top or height.

خلة المغابيط *Oe* *Khallet el Maghābīt.* Valley of naturally irrigated land.

خلة منصور *Oe* *Khallet Mansūr.* Mansur's dell. p.n. (victor).

خلة المراح *Pd* *Khallet el Marāh.* Valley of the fold.

خلة المحافر *Qd* *Khallet el Mehāfir.* The dell of the excavation.

خلة المل *Od* *Khallet el Mell.* The dell of ash-coloured earth.

خلة مشحرة *Oe* *Khallet Mesh-harah.* Dell of charcoal burning.

خلة المطحنة *Nd* *Khallet el Met-hanch.* The dell of the place of grinding corn.

خلة المغارة *Qd* *Khallet el Mŭghārah.* The dell of the cane.

خلة المغر *Qd* *Khallet el Mughr.* The dell of caves.

خلة محمد حيدر *Nd* *Khallet Muhammed Heider.* Muhammed Heider's dell. p.n.

خلة محيشم *Od* *Khallet Muheishim.* The dell of the torrents that carry trees, &c., with them.

خلة المتقص *Od* *Khallet el Mukaskas.* The cut-up valley.

خلة المكيسبة *Od* *Khallet el Mukeisibeh.* The valley of gains.

خلة الملك	Oe	*Khallet el Melek.* The king's dell.
خلة المنشار	Oe	*Khallet el Mŭnshâr.* The dell of the saw.
خلة موسى	Nf	*Khallet Mûsa.* Mûsa's (Moses) dell. p.n.
خلة المتفح	Pd	*Khallet el Mussaffâh.* The spacious dell.
خلة المشعبة	Oe	*Khallet el Mushâbeh.* The dell of the mountain paths.
خلة النعت	Of	*Khallet en Nâseh.* The dell of en Nâseh. p.n. ; signifying 'helper.'
خلة النخاشية	Qd	*Khallet en Nakhkhâsîyeh.* The dell of the drovers or cattle sellers.
خلة الناقورية	Pe	*Khallet en Nâkûrîyeh.* The dell of the perforators.
خلة ناصر	Bd	*Khallet Nâsr.* The dell of Nâsir. p.n.
خلة النجار	Oe	*Khallet en Nejjâr.* The dell of the carpenter.
خلة النجاصة	{Oe / Pd}	*Khallet en Nejâseh* (2). The dell of the pear trees.
خلة النيارة	Nd	*Khallet en Niârah.* The dell of the yoke.
خلة العبيد	{O / Pd}	*Khallet 'Obeid.* The dell of Obeid. p.n. ; meaning ' little slave.'
خلة العليق	Od	*Khallet el 'Olleik.* The dell of the brambles.
خلة العنق	Of	*Khallet er 'Onk.* The dell of the neck (of land).
خلة الراهب	Ne	*Khallet el Râhib.* The dell of the monk.
خلة راجح	Oe	*Khallet Râjih.* The dell of Râjih. p.n. ; meaning ' grave,' ' with a well-balanced mind.'
خلة رجب	Pe	*Khallet Rejeb.* The dell of Rejeb. p.n. ; but perhaps from رجيب bricks, &c., used to prop up a tree when too heavily laden with fruit.
خلة الرمانة	Pe	*Khallet er Rummâneh.* The dell of the pomegranate. (Heb. Rimmon.)
خلة رشيف	Od	*Khallet Rasheif.* The dell of the swamp.
خلة الصعد	Oe	*Khallet es Safed.* The dell of Safed. p.n. see p. 32.
خلة صافي	Oe	*Khallet Sâfy.* The dell of pure (water).
خلة سعيد	Oe	*Khallet Sâid.* The dell of Sâid. p.n. ; Felix.
خلة سعسع	Oe	*Khallet Sâsâ.* The dell of Sâsâ. *Sâsâ* is the noise used in calling goats.

خلة الصقعانة *Pd* *Khallet es Sek'âneh.* The frosty dell.

خلة الشيخ *Qe* *Khallet esh Sheikh.* The dell of the elder.

خلة الشلال *Nd* *Khallet esh Shellâl.* The dell of the cascades.

خلة الشرقية *Nd* *Khallet esh Sherkîyeh.* The eastern dell.

خلة الشومر *Pf* *Khallet esh Shômer.* The dell of wild fennel.

خلة شحيتا *Oe* *Khallet Shuheitah.* The dell of the mountain pine.

خلة شقير *Pe* *Khallet Shukeir.* The grey dell.

خلة السيقيع *Od* *Khallet es Sikya.* The dell of the hoar frost.

خلة الصبيات *Qf* *Khallet es Sûbiyât.* The dell of the girls.

خلة سريج *Pd* *Khallet Surîj.* The dell of the small lamp.

خلة التون $\left\{\begin{matrix} Pf \\ Od \end{matrix}\right\}$ *Khallet et Tûn* (3). The dell of the lime-kiln.

خلة تونة *Od* *Khallet et Tûneh.* The dell of the lime-kiln.

خلة التوت *Qe* *Khallet et Tût.* The dell of the mulberry.

خلة أمثاث *Nf* *Khallet Umthâs.* The dell of sprinkling.

خلة واوى *Pe* *Khallet Wâwy.* The dell of the jackal.

خلة الوسطى *Pe* *Khallet el Wusta.* The middle dell.

خلة ياسكى *Qe* *Khallet Yâsky.* The dell of Yâsky. p.n.

خلة الزاغ *Od* *Khallet ez Zâgh.* The dell of the crow.

خلة الزقزوق *Oe* *Khallet ez Zakzûk.* The dell of the linnet.

خلة الزاروب *Oe* *Khallet ez Zârûb.* The dell of the water-pipe.

خلة زوان *Oe* *Khallet Zuwân.* The dell of tares.

خلة الزيتون *Pd* *Khallet ez Zeitûn.* The dell of the olive.

خلة الزويلة *Od* *Khallet ez Zuweileh.* The dell of falcons.

خلة الزويتينة *Od* *Khallet ez Zûeitîneh.* The dell of the small olive tree.

خلة الخالصة *Of* *El Khâlisah.* The dell of Khálisah ('pure,' 'clear'), the Greek Elusa.

خلة خان $\left\{\begin{matrix} Pf \\ Re \end{matrix}\right\}$ *Khallet el Khân.* The dell of Caravanserai.

خان جب يوسف *Qf* *Khân Jubb Yûsef.* The Caravanserai of Joseph's Pit.

الخانوق	Oe	*El Khânûk.* The gorge.
الخرشومية	Od	*El Kharshûmiych.* The mountain spurs.
الخشم	Qf	*El Khashm.* The mouth, outlet.
الخوانيق	Qd	*El Khawânîk.* The gorges.
خلال الدرة	Nc	*Khelâl edh Dhurah (durrah).* The dells of maize.
خلال حضيرة	Qe	*Khelâl Hadîreh.* The dells of the fold.
خلال قطاعين	Qd	*Khelâl el Katâ'in (Kattâ'in).* The dells of the highway robbers.
خلال شحرور	Pe	*Khelâl Shahrûr.* The dells of the Grackle (bird).
خلال شيريا	Oe	*Khelâl Shîria.* The dells of Shiria. p.n.
خلال سبيع	Oe	*Khelâl Subeiâ.* The dells of the wild beast.
خلايل اللوز	Qf	*Khelayil el Lôz.* The small dells of the almond.
الخربة	Pe	*El Khûrbeh* (2). The ruin.
خربة عباد	Nf	*Khûrbet 'Abbâd.* The ruin of the devotee.
خربة ابى لوزة	Rf	*Khûrbet Abu Lôzeh.* The ruin with the almond tree.
خربة ابى شبا	Of	*Khûrbet Abu esh Sheba.* The ruin with the chickweed.
خربة ابى زلفة	Qf	*Khûrbet Abu Zelefeh.* The ruin with the cistern.
خربة عين البطم	Qf	*Khûrbet 'Ain el Butm.* The ruin of the fountain of the terebinth tree.
خربة عكبرا	Ne	*Khûrbet 'Akbara.* The ruin of Akbara ; *see 'Ain 'Akbara,* p. 61.
خربة العلوية	Pe	*Khûrbet el 'Alawîyeh.* The ruin of the 'Alawîyeh. p.n. ; meaning ' of the family of 'Ali.'
خربة العليا	Qf	*Khûrbet el 'Alîeh.* The upper ruin.
خربة علمانية	Qe	*Khûrbet 'Almânîyeh.* See Ain Almânîyeh, p. 61.
خربة العسلية	Qf	*Khûrbet el 'Asalîyeh.* The ruin of the 'Asalîyeh. (Jews are so called from their being formerly obliged to wear a turban of the colour of honey ['Asal].)
خربة عصيلة	Od	*Khûrbet 'Assîleh.* The ruin of the oleander.
خربة عوبا	Pd	*Khûrbet 'Aûba.* The ruin of 'Aûba. p.n.

خربة بدية	Oe	*Khŭrbet el Bedîyeh.* The ruin of Bediyeh. Either from بَدْو 'a desert,' بَدِيَّة 'the side of a valley,' or 'بدى 'truffles.
خربة البلانة	Pf	*Khŭrbet el Bellâneh.* The ruin of Bellâneh. From بَعْلَة a thorny plant.
خربة بنات يعقوب	Re	*Khŭrbet Benât Yâkûb.* The ruin of the Daughters of Jacob.
خربة بنيت	Qf	*Khŭrbet Benît.* The ruin of Benit. p.n.
خربة برزة	Ne	*Khŭrbet Berza.* The ruin of the mountain road.
خربة البيار	Od	*Khŭrbet el Biâr.* The ruin of the wells.
خربة البيارة	Ne	*Khŭrbet el Biâreh.* The ruin of the well-sinker.
خربة الدواجية	Pe	*Khŭrbet ed Dâwajîyeh.* The ruin of ed Dâwajiyeh, family name; meaning 'Bedesman.'
خربة الدوير	Od	*Khŭrbet ed Duweir.* The ruin of the little monastery.
خربة فانس	Ne	*Khŭrbet Fânis.* The ruin of Fânis. p.n.; meaning 'pauper.'
خربة فصل دانيال	Ne	*Khŭrbet Fasil Dâniâl.* The ruin of Daniel's judgment.
خربة غباطى	Oe	*Khŭrbet Ghabbâti.* The ruin of Ghabbâti; *see 'Ain Ghabbâti.*
خربة غزالة	Qf	*Khŭrbet Ghŭzâleh.* The ruin of the gazelles.
خربة الحجار	Od	*Khŭrbet el Hajâr.* The ruin of the rocks.
خربة الحمرآ	Pf	*Khŭrbet el Hamra.* The red ruin.
خربة حرة	Qd	*Khŭrbet Harrah.* The ruin of Harreh. p.n.
خربة الحسنية	Qe	*Khŭrbet el Hasanîyeh.* The ruin of el Hasaniyeh. p.n.; meaning family or followers of el Hasan, the son of 'Ali, and grandson of the Prophet.
خربة الحصيرة	Od	*Khŭrbet Hazîreh.* The ruin of the fold.
خربة الحقاب	Pf	*Khŭrbet el Hekâb.* The ruin of the Hekâb. حِقَاب 'an ornamented thing which a woman binds on her waist.'— LANE. But it may be a corruption of حِقَاف 'curved tracts of sand'; or perhaps of عِقَاب 'an overhanging rock.'
خربة حينة	Od	*Khŭrbet Hîneh.* The ruin of Hineh. p.n.; the word means 'milking time.'
خربة الحمام	Rf	*Khŭrbet el Hŭmmâm.* The ruin of the hot bath.

M 2

خربة امسية   Od   *Khŭrbet Imsieh.* The ruin of Imsieh. p.n.

خربة جافا   Pe   *Khŭrbet Jâfa.* The ruin of Jâfa. Either from جائفة 'a deep hollow' (watercourse), or from جفا 'rubbish, &c., cast up by a torrent.'

خربة جفتلك   Qe   *Khŭrbet Jeftelek.* The ruin of Jeftelek. p.n.

خربة جندية   Nd   *Khŭrbet el Jenadîyeh.* The ruin of the Jenâdiyeh Arabs. p.n.

خربة جب يوسف   Qf   *Khŭrbet Jubb Yûsef.* The ruin of Joseph's Pit.

خربة جول   Of   *Khŭrbet Jûl.* The ruin of Jûl. p.n.

خربة تطمون   Oe   *Khŭrbet Katamŭn.* The ruin of Katamûn; perhaps from قطم, 'to quarry'; cf. جبل مقطم *Jebel Mokattem,* near Cairo.

             Qf   *Khŭrbet Katanah.* Either قطنة cotton, or كتنة flax. The Arabic form is not given in the lists.

خربة كفر ابنين   Od   *Khŭrbet Kefr Ibnîn.* The ruin of the village of Ibnin (sons).

خربة تيسون   Qe   *Khŭrbet Keisûn.* The ruin of Keisûn; name of a herb.

خربة القيومة   Pf   *Khŭrbet el Keiyûmeh.* The erect, peerless, or existent ruin.

خربة كرسيفة   Od   *Khŭrbet Kersîfa.* The ruin of Kersifa; a family name; perhaps from كرسوف = 'cotton.'

خربة الخمارة   Nd   *Khŭrbet el Khammârah.* The ruin of the wine-tavern.

خربة الخضرآ   Ne   *Khŭrbet el Khŭdra.* The green ruin.

خربة القلنسوى   Nd   *Khŭrbet el Kŭlŭnsawy.* The ruin of the dervish's cap.

خربة الكورة   Od   *Khŭrbet el Kŭrah.* The ruin of the hole.

خربة القرية   Od   *Khŭrbet el Kureiyeh.* The ruin of little village.

خربة لوزية   Qe   *Khŭrbet Lôzîyeh.* The ruin of the almond tree.

خربة المنصورة   {Od / Ne}   *Khŭrbet el Mansûrah (2).* The ruin of Mansûra. p.n.; = Victoria.

خربة ماروس   Qe   *Khŭrbet Mârûs.* The ruin of Marûs. p.n.

خربة معصرة   Qd   *Khŭrbet el Mâserah.* The ruin of the oil press.

خربة الحفائر   Qd   *Khŭrbet el Mehâfir.* The ruin of the excavated water-pits.

خربة المجدل   Qd   *Khŭrbet el Mejdel.* The ruin of the watch-tower.

خربة المنارة   Oe   *Khŭrbet el Menárah.* The ruin of the light-house.

خربة المرج   Od   *Khŭrbet el Merj.* The ruin of the meadow or prairie.

خربة معظمية   Pe   *Khŭrbet Muáddemiyeh.* The ruin of (el) Muáddemiyeh. Probably from el Melik el Muáddhem, one of the Aiyúbite sultans of Egypt.

خربة المنطار   Rf   *Khŭrbet el Mŭntár.* The ruin of the watch-tower.

خربة المشيرفة   Rf   *Khŭrbet el Musheirefeh.* The ruin of high places or 'eminences.'

خربة المزيبلات   Nd   *Khŭrbet el Muzeibelát.* The ruin of the dung-hills.

خربة النبرة   Qe   *Khŭrbet en Nebrah.* The ruin of the high-place.

خربة النبرتين   Qe   *Khŭrbet en Nebratein.* The ruin of the two high-places.

خربة النظارة   Oe   *Khŭrbet en Netárah.* The ruin of the watch.

خربة النصيبة   Pe   *Khŭrbet en Nuseibeh.* The ruin of the erected stones (نصب a stone set up as an object of worship in the Pagan times.)

خربة العتيبة   Pf   *Khŭrbet el 'Okeibeh.* The ruin of the little steep or mountain road.

خربة عكيمة   Qf   *Khŭrbet 'Okeimeh.* The ruin of the عكم *i.e.,* the pulley of a well. But *see Wády el Okeimeh,* p. 102.

خربة ربيص   Pe   *Khŭrbet Rabbis.* The ruin of the ground irrigated in summer.

خربة الرندة   Oe   *Khŭrbet er Randeh.* The ruin of the Randeh (aloes) tree.

خربة الرجم   Od   *Khŭrbet er Rŭjm.* The ruin of the stone-heap.

خربة الرشيدة   Pf   *Khŭrbet Rusheideh.* The ruin of Rusheideh. p.n.; diminutive of Rashid (orthodox).

خربة الرويس   Ne   *Khŭrbet er Ruweis.* Ruin of the little head or hill-top.

خربة الصعابنة   Of   *Khŭrbet es S'ábneh.* The ruin of the S'ábneh. p.n.; from صعب 'difficult.'

خربة السهلة   Of   *Khŭrbet es Sahleh.* The ruin of the plain.

خربة سمورة   Pf   *Khŭrbet Samûrah.* The ruin of Samûra. p.n.; from اسمر brown.

خربة السيارة	Rf	*Khŭrbet es Seiyârah.* The ruin of the wanderer.
خربة سموخية	Oc	*Khŭrbet Semmûkhich.* The ruin of the Semmûkich. p.n. (خماج = خماع and means the first sprouting of herbage.)
خربة الشعرة	Pf	*Khŭrbet esh Shârah.* The ruin of the thick foliage (literally 'hair').
خربة السنية	Qf	*Khŭrbet es Sentuch.* The ruin of 'grit,' or of 'elevated tracts of sand.'
خربة شلعدون	Od	*Khŭrbet Shelâbûn.* The ruin of Shelâbûn. p.n.
خربة شمع	Pf	*Khŭrbet Shemâ.* The ruin of the candle.
خربة شرتة	Od	*Khŭrbet Sherta.* The ruin of Sherta. p.n. ('scarifying')
خربة شورا	Qf	*Khŭrbet Shôra.* The ruin of Shôra. Perhaps شورى a tree that grows near the sea.
خربة شنفين	Of	*Khŭrbet Shufnîn.* The ruin of the dove.
خربة شويطا	Nd	*Khŭrbet Shuweit.* The ruin of the morass or bog.
خربة سيريا	Nf	*Khŭrbet Stria.* The ruin of Siria. Perhaps from سرى 'a rivulet,' or from the Aramaic שור 'a wall.'
خربة سرطبة	Of	*Khŭrbet Sŭrtŭba.* The ruin of Sŭrtŭba. p.n.; perhaps from سرداب a subterranean vault. The 'Belle Vue' of the Crusaders.
خربة الطاحونة	Of	*Khŭrbet et Tâhûnch.* The ruin of the mill.
خربة تير حرمة	Nd	*Khŭrbet Teir Hirmeh.* The ruin of the fort of Hirmeh. *cf.* حرام 'a sanctuary.'
خربة طيرطيرة	Pd	*Khŭrbet Teirtireh.* The ruin of the Fort of Tireh. p.n.; also meaning 'fortress.'
خربة التليل	Nf	*Khŭrbet et Teleil.* The ruin of the small mound.
خربة الطبقة	Od	*Khŭrbet et Tûbakah.* The ruin of the terrace.
خربة ام علي	Qd	*Khŭrbet Umm 'Aly.* The ruin of 'Ali's mother. p.n.
خربة ام الهموم	Qe	*Khŭrbet Umm el Humûm.* The ruin of the 'mother of cares.'
خربة اميّة	Od	*Khŭrbet Ummich.* The ruin of 'Ommaiych. Family name of the caliphs who succeeded Muhammed.
خربة وقاص	Qe	*Khŭrbet Wakkâs.* The ruin of the 'man with a broken neck.'

خربة زبود   Of   *Khŭrbet Zebŭd.* The ruin of Zebûd. p.n. ; *see Khŭrbet Zebed,* p. 51.

الخريبة   Qe   *El Khŭreibeh.* The little ruin.

القبلية   Oe   *El Kiblïyeh.* The southern (place) or 'frontage.'

كتف العاصى   Pd   *Kitf el 'Asy.* The shoulder of el 'Âsy. p.n. ; 'the rebel.'

الكوزح   Xd   *El Kôzah.* The height or top.

القلعة   {Pf / Oe}   *El Kŭlah* (2). The castle.

قلعة حديه   Of   *Kŭlat Hiddeiyeh.* The boundary castle.

قلعة المرج   Pd   *Kŭlat el Merj.* The castle of the meadow.

قلعة الراهب   Ne   *Kŭlat er Râhib.* The monk's castle.

قلعة شيريا   *Kŭlat Shiria.* The castle of Shiria. p.n.

كونين   Pd   *Kŭnïn.* Kunîn. p.n.; *see Birket Kŭnîn,* p. 70.

كرم ابى غبار   Oe   *Kurm Abu Ghabâr.* The vineyard of Abu Ghabâr. p.n. : meaning 'dusty.'

كرم العتيق   Oe   *Kurm el 'Atïk.* The vineyard of 'Atik. p.n. ; meaning 'noble,' 'emancipated.'

كرم برغوث   Oe   *Kurm Barghût.* The vineyard of the flea.

كرم البركة   Pd   *Kurm el Birkeh.* The vineyard of the pool.

كرم نو   Pf   *Kurm Dô* (properly *Dhau*). The vineyard of light.

كرم الدب   Pf   *Kurm ed Dubb.* The vineyard of the bear.

كرم الغورانة   Of   *Kurm el Ghôrany.* Vineyard of the Arab of the Ghawârineh tribe, *i.e.,* the dwellers in the Ghor, or 'lowlands.'

كرم حامد   Pf   *Kurm Hâmed.* The vineyard of Hâmed. p.n.

كرم الحجيلاوى   Oe   *Kurm el Hujeilâwï.* The vineyard of el Hujeilâwi. p.n.

القرمية   Of   *El Kurmïyeh.* The stumps.

قصر عترة   Re   *Kŭsr 'Atra.* The mansion. p.n.

قصر مروش   Od   *Kŭsr Marrûsh.* The mansion of Marrûsh. p.n.

القطعة   Pf   *El Kŭtâh.* The cutting.

قطوع رمسية‎ Qe *Kutûâ Rumsîyeh.* The quarries of Rumsiyeh. Perhaps from رمس‎, the dust or earth over a grave that has been levelled.

الغيثة‎ Pf *El Maghîteh.* The pasture irrigated with rain.

حفرة البياض‎ Od *Mahferet el Beiyâd.* The white diggings.

مهوا الجمال‎ Pe *Mahwa el Jemâl.* The precipice of the camels.

مجنونة‎ Qd *Majnûneh.* The mad (woman); or the place haunted by *jinns* (demons or 'genii').

المثاني‎ Pf *El Makâthy.* The cucumber plots.

مخاضة السيارة‎ Rf *Makhâdet es Seiyârah.* The wanderer's ford.

مخاضة الطشطاش‎ Re *Makhâdet et Tushtâsh.* The plashing ford.

مقسم القهوة‎ Pf *Maksam el Kahweh.* The allotment of the coffee-shop.

المالكية‎ Qd *El Mâlkîyeh.* El Malkiyeh. From Malek, 'to possess,' or 'reign'; see *Mâlkiyeh,* p. 9.

ملولة عين الدرب‎ Qf *Mallûlet 'Ain ed Derb.* Oaks of spring of road.

المراح‎ {Oe} {Od} *El Marâh* (2). The place for resting at night (a fold or cattle-shed).

مارون الرأس‎ Pd *Mârûn er Râs.* Mârûn of the head (a headland). p.n.

المصاطب‎ Pe *El Masâtib.* The benches.

معاصر العقيبة‎ Pf *M'âser el 'Okeibeh.* The oil or wine-press of the little steep, or mountain-road.

مزار الوقاص‎ Qe *Mazâr el Wakkâs.* The shrine of the broken-necked one.

ميرون‎ Pf *Meirôn.* Merom.

مرج العبد‎ Rf *Merj el 'Abd.* The meadow of the slave.

مرج العين‎ Ne *Merj el 'Ain.* The meadow of the spring.

مرج علما‎ Pe *Merj 'Alma.* The meadow of 'Alma. p.n.; see *'Alma esh Shâub,* p. 40.

مرج العرائس‎ Pd *Merj el 'Arâis.* The meadow of the brides.

مرج بيسمون‎ *Merj Beisamûn.* The meadow of Beisamûn (a sort of tree).

مرج بليدة‎ Qd *Merj Belîdeh.* The meadow of the little town. The word also means 'a foolish woman.'

مرج البصل   Oe   *Merj el Básl.*   The meadow of onions.

مرج فوعاثا   Oe   *Merj Fûk'átha.*   The meadow of Fûk'átha.   p.n.; perhaps from فقع a sort of truffle.

مرج ديشون   Qe   *Merj Deishûn.*   The meadow of Deishûn.   Perhaps from دشن a new and as yet uninhabited house.

مرج دبل   Od   *Merj Dibl.*   The meadow of manure.

مرج الدم   Ne   *Merj ed Dumm.*   The meadow of blood; but *see 'Ain ed Dumm*, p. 39.

مرج الحظيرة   Pe   *Merj el Hadireh.*   The meadow of the fold.

مرج الحافور   Pd   *Merj el Háfûr.*   The meadow of the excavation or 'the hoof.'

مرج هلال   Ne   *Merj Halál.*   The meadow of the crescent.

مرج الحمرآ   Qd   *Merj el Hamra.*   The red meadow.

مرج الحمرة   Pf   *Merj el Hamrah.*   The red meadow.

مرج حمام الصبايا   Od   *Merj Hûmmâm es Sûbâya.*   The meadow of the girls' bath.

مرج الحمص   Qf   *Merj el Hummûs.*   The meadow of the chick-peas.

مرج الجش   Pe   *Merj el Jish.*   The meadow of el Jish.   q.v.

مرج قديس   Qd   *Merj Kades.*   The meadow of Cadesh.

مرج القطع   Pe   *Merj el Katâ.*   The meadow of the cutting or quarry.

مرج القبلة   Qd   *Merj el Kibleh.*   The south meadow.

مرج قوزح   Od   *Merj Kôzah.*   The meadow of the height or top.

مرج المالكية   Pd   *Merj el Málkiyeh.*   The meadow of Málkiyeh.   q.v., p. 39.

مرج لحافر   Pd   *Merj el Mehâfir.*   The meadow of the excavated water-pits.

مرج النقيز   {Nd} {Od}   *Merj en Nakkeiz.*   Meadow of the *nakkeiz* (name of a small bird).

مرج النطاح   Pe   *Merj en Nattâh.*   The meadow of the one who butts.

مرج العليق   Oe   *Merj el 'Olleik.*   The meadow of the bramble.

مرج راميجيا   Pe   *Merj Ramjighia.*   The meadow of Ramjighia.   p.n.

مرج رميش   Oe   *Merj Rumeish.*   The meadow of scanty herbage.

N

مرج الرمانة   *Od*   *Merj er Rummánch.* The meadow of the pomegranate-tree.

مرج سعسع   *Oe*   *Merj Sásá.* The meadow of Sásá. q.v.

مرج سلمى   Pd   *Merj Selmy.* The meadow of Selmy. p.n. ; perhaps from سلم name of a plant.

مرج شرش النخب   *Rf*   *Merj Shersh el Khubb.* (Properly *shers.*) The meadow of the small thorny trees of the marsh.

مرج سدرة اللهبية   *Rf*   *Merj Sidret el Lehebíych.* The meadow of the lote tree of the Lehebiych Arabs.

مرج سدر الشيح   Pe   *Merj Sidr esh Shíh.* The lote trees of the *Shíh.* *Shih* in Arabic means a certain aromatic plant; but it is perhaps from the Aramaic שיח 'a cleft' or pit.

مرج السنديانة   Qe   *Merj es Sindiánch.* The meadow of oaks.

مرج الصفصاف   Pe   *Merj es Súfsáf.* The meadow of the osier willow.

مرج تل السنجق   *Qf*   *Merj Tell es Sanjak.* The meadow of the mound of the flag.

مرج طوفة   Qd   *Merj Túfch.* The meadow of Túfch. p.n.; perhaps from طاف 'to go round.'

مرج يارون   Pd   *Merj Yárún.* The meadow of Yárún. p.n.

مرج الزيتون   Pd   *Merj ez Zeitún.* The meadow of the olives.

مسيل الدباغة   Qe   *Mesíl ed Dabbághah.* The stream of the tanners.

الموبرة   Of   *El Móbarah.* The quarry.

موبرة الحما   Qf   *Móbarat el Hima.* The quarry of the royal preserves.

المغار   Qf   *El Mughâr.* The caves.

مغارة الديوس   *Mughârat ed Deiyús.* The cave of the cuckold.

مغارة الدب   Nd   *Mughârat ed Dubb.* The cave of the bear.

مغارة الغدارة   Pf   *Mughârat el Ghúddárah.* The cave of the traitors.

مغارة الحمدة   Qe   *Mughârat el Hamdeh.* The cave of the Hamdeh. p.n.

مغارة الحلس   Qe   *Mughârat el Hilis.* The cave of verdure.

مغارة قمامة   Pe   *Mughârat Kammâmah.* The cave of the temple or church.

مغارة خير الدين   Pf   *Mughârat Kheir ed Din.* Kheir ed Din's cave. p.n.

مغارة المجلس   Pe   *Mughârat el Mejlis.* The cave of the assembly or council.

مغارة المشبّك   Pf   *Mŭghârat el Mushebbak.* The cave of the 'labyrinth' or of the 'lattice work.'

مغارة نحلية   Pd   *Mŭghârat Nahleiya.* The cave of the bees.

مغارة النورية   Of   *Mŭghârat en Nurîyeh.* The gipsies' cave.

مغارة عمير   Oe   *Mŭghârat 'Omeir.* The cave of 'Omair. p.n.

مغارة سبلان   Of   *Mŭghârat Sebelân.* The cave of Sebelân; *see Neby Sebelân,* a Druse sheikh (perhaps Zabulon).

مغارة التنور   Pe   *Mŭghârat et Tannûr.* The cave of the reservoir.

مغارة الطبّة   Od   *Mŭghârat et Tŭbakah.* The cave of the terrace.

مغارة الزاغ   Od   *Mŭghârat ez Zâgh.* The cave of the crow.

مغراقة   Qd   *Mughrâkah.* Flooded land.

الحيشه،   Pe   *El Mŭheisheh.* The enclosure.

المقطعة   Rf   *El Mŭktâh.* The cutting (quarry).

المناطر   Oe   *El Mŭnâtir.* The watch towers.

المنيشير   Od   *El Mŭneishîr.* The saws.

المنطرة   Pf   *El Mŭntarah.* The watch tower.

منطرة القطن   Pe   *Mŭntaret el Kotn.* The watch tower of the cotton.

نخاشية   Qd   *Nakhkhâsiyeh;* see p. 80.

الناقورية   Pe   *En Nâkûrîyeh;* see p. 80.

نبع عوبا   Pe   *Nebâ 'Aûba.* The perennial spring of 'Aûba.

نبعة الذياب   Pf   *Nebât edh Dhîâb.* The perennial spring of the wolves. (Also an Arab tribal name.)

نبعة دبل   Od   *Nebât Dibl.* The perennial springs of Dibl. q.v., p. 71.

نبعة الجوزات   Pf   *Nebât el Jôzât.* The perennial springs of the walnut trees.

نبعات وادى فارة   Pe   *Nebât Wâdy Fârâh.* The perennial springs of Wâdy Fârâh. q.v., p. 99.

نبى ابو هليون   Ne   *Neby Abu Haliûn.* The perennial springs where the asparagus grows.

نبى حنيا   Qe   *Neby Hânîya.* The prophet Hânîya. p.n.

نبى حيدر   Of   *Neby Heider.* The prophet Heider, 'the Lion'; a name of 'Ali.

N 2

نبى محيبيب   Qd   *Neby Muheibib.* The prophet Muheibib, 'beloved.'

نبى سبلان   Oe   *Neby Sebelân.* The prophet Sebelân ; perhaps Zebulon.

نبى يوشع   Qd   *Neby Yûshâ.* The prophet Joshua.

النمورة   Oe   *En Nimmûrah.* The abundant water.

نحف   Nf   *Nûhf.* Nûhf. p.n. ; see p. 64.

النقار   Pe   *Nûkkâr.* Perforations.

عش الشوحة   Qd   *'Osh esh Shûhah.* The kite's nest.

رخصون   Oe   *Rakhasûn.* Rakhasûn. p.n. ; perhaps from رخص, soft.

الراعة   Of   *Er Râmeh.* Er Râmeh.   p.n. ; a common Arabian topographical name, the Hebrew רמה, meaning 'elevated,' 'lofty.' (Josh. xix, 36.)

راعيا   Nd   *Râmia.* Râmia.   p.n. ; connected with the above in meaning.

رارا   Pd   *Râra.* Rârâ. p.n.

رأس الاحمر   Pe   *Râs el Ahmar.* The red head, or hill-top.

رأس البدندى   Nd   *Râs el Bedendy.* The 'head' or hill-top of el Bedendy. p.n.

رأس الغربية   Nd   *Râs el Gharbîyeh.* The western 'head.'

رأس غصون   Pd   *Râs Ghusûn.* The 'head' of branches.

رأس ابنيف   Qf   *Râs Ibnîf.* The 'head' of Ibnîf. p.n.

رأس القبلة   Pf   *Râs el Kibleh.* The south 'head.'

رأس الكبس   Pf   *Râs el Kibs.* The head of el Kibs ('earth rammed round a well' ; from كبس 'to press').

رأس الكتف   Pf   *Râs el Kitf.* The 'head' of the shoulder.

رأس القبة   Râs el Kubbeh.* The head or hill-top of the dome.

رأس الكرسى   Oe   *Râs el Kursî.* The 'head' or hill-top of the chair or throne.

رأس النخاشية   Qd   *Râs en Nakhkhâsîyeh.* See p. 80.

رأس النبع   Pf   *Râs en Nebâ.* The head of the perennial spring.

رأس الشرقية   {N Od}   *Râs esh Sherkîyeh.* The eastern head.

رأس الطيرة   Od   *Rás et Tirch.* The head of el Tirch, 'The Fort.'

رأس الوعر   Pf   *Rás el Wár.* The head of the rugged ground.

رويش   Oe   *Rumeish.* Scanty herbage.

رشيف   Od   *Rusheif.* Morass or bog.

رويس الجاموس   Od   *Ruweis el Jámús.* The little head or hill-top of the buffalo.

صفد   Pf   *Safed.* Safed; *see* p. 32.

الصديق   Pe   *Es Saddík.* (Properly *Siddík.*) The truthful one.

صلحا   Pe   *Salhah.* Salhah. p.n.; meaning 'righteous.'

الصالحية   Rd   *Es Salihiyeh.* p.n. This name is elsewhere attached to buildings or establishments founded by Saláh ed din (Saladin).

سعسع   Oe   *Sásá.* Sásá. p.n.; *see* p. 70.

سجور   Of   *Seijúr.* Seijúr; *see* pp. 50 and 65.

السموعية   Pf   *Es Semúáieh.* Es Semúaieh. p.n.

السرير   Pf   *Es Serír.* The sarcophagus, or 'bedstead.'

الشاغور   {Nf} {Of}   *Esh Shághúr.* *See* p. 28.

شيخ عبدالله   Qf   *Sheikh 'Abdallah.* Sheikh Abdallah. p.n.

شيخ ابى بيت   Of   *Sheikh Abu Beit.* Sheikh Abu Beit. p.n. 'The elder, the father, or owner of the house.'

شيخ احمد القاسم   Pe   *Sheikh Ahmed el Kásim.* Sheikh Ahmed el Kásim. p.n.

شيخ بنيت   Qf   *Sheikh Benít.* Sheikh Benít. p.n.

شيخ حسين الحكابى   *Sheikh Husein el Hekáby.* Sheikh Husein el Hekáby. p.n.

شيخ الكويس   Pf   *Sheikh el Kuweiyis.* The pretty sheikh.

شيخ منصور   Pe   *Sheikh Mansúr.* Sheikh Mansúr. p.n. = victor.

شيخ مرزوق   Nd   *Sheikh Marzúk.* Sheikh Marzúk. p.n. (= 'provided for').

شيخ محمد العجمى   Pf   *Sheikh Muhammed el 'Ajamy.* Sheikh Mohammed the Persian.

شيخ محمد الحديد   Qf   *Sheikh Muhammed el Hadíd.* Sheikh Mohammed el Hadíd (iron).

| شيخ النطاح | Qe | *Sheikh en Nattâh.* Sheikh en Nattâh (the butting one). |

| شيخ محمدالقطري | Nd | *Sheikh Muhammed el Kutry.* Sheikh Mohammed el Kutry. p.n. |

| شيخ ربع | Nf | *Sheikh Rabiá.* Sheikh Rabiá. p.n. |

| شيخ ربعة | Qf | *Sheikh Rabiâh.* Sheikh Rabiâh. p.n. |

| شيخ وهيب | Oe | *Sheikh Waheib.* Sheikh Waheib. p.n. |

| شجرة بنات يعوب | Re | *Shejerât Benât Yâkûb.* The tree of Jacob's Daughters. |

| شمعة العتيقة | Pf | *Shemât el 'Atikah.* The old candlestick (a pointed stone). |

| شكارة فارس | Pf | *Shkâret Fâris.* Fâris' plot. Shakhârat means land ploughed by the rest of the villagers for one who for any reason cannot plough himself, as a priest, a craftsman, &c. |

| شكارة جمول | Pf | *Shkâret Jemmûl.* Jemmûl's (p.n.) plot. |

| شتيل | Pd | *Shukeil.* Shukeil. p.n. (perhaps شاقول a levelling or measuring pole). |

| شقفان البركة | Pf | *Shŭkfân el Birkeh.* The clefts of the pool. |

| شقيف العين | Pe | *Shŭkîf el 'Ain.* The cleft of the spring. |

| شقيف العلوية | Pe | *Shŭkîf el 'Alawîyeh.* The cleft of el 'Alawîyeh; see Wâdy 'Alawîyeh, p. 57. |

| شقيف حسن بركات | Pf | *Shŭkîf Hasan Barakât.* Cleft of Hasan; Barakât. These are both proper names; the first being the common Moslem name signifying 'beautiful,' the second meaning 'blessings.' |

| شقيف احتاب | Pf | *Shŭkîf el Hekâb.* The cleft of the Hekâb; see Khŭrbet el Hekâb, p. 63. |

| شقيف النمرة | Od | *Shŭkîf en Nimreh.* The cliff of the leopard (or of abundant water). |

| شقيف صلحا | Pe | *Shŭkîf Salha.* The cleft of Salha. |

| شقيف وادى عوبا | Pe | *Shŭkîf Wady 'Aûba.* The cleft of Wady 'Aûba. p.n. |

| شونة الناقة | Pd | *Shûnet en Nâkah.* The barn of the she-camel. |

| سيد احمد المنتار | Qf | *Sîd Ahmed el Muntâr.* Lord Ahmed of the watch-tower. |

| سيد المشارف | Pe | *Sîd el Meshârif.* The Lord of the eminent places. |

| سدرة اللهبية | Rf | *Sidret el Lehebîyeh.* Lotus tree of the Lehebiyeh Arabs. |

صفصاف   Pe   *Sûfsâf.* The osier willow.

السحيمية   Qd   *Es Suheimiyeh.* Es Suheimiyeh. Perhaps from سحم 'black,' or from سحم a sort of tree so called.

صربين   Od   *Surubbîn.* Surubbin. p.n.

طاحونة البطيطة   Pf   *Tâhûnet el Buteitah.* The mill of the 'little duck,' or 'caterpillar.'

طاحونة دبوسية   Pf   *Tâhûnet Dabbûsîyeh.* The mill of the Dabbûsiyeh. p.n.; from *dabbûs*, a 'knobstick' or 'cudgel.'

طاحونة الغدارة   Pf   *Tâhûnet el Ghaddârah.* The mill of the traitor.

طاحونة الهندى   Re   *Tâhûnet el Hindy.* The mill of the Indian.

طاحونة العسراوى   Pf   *Tâhûnet el 'Isrâwy.* The mill of el 'Isrâwy. p.n.; see p. 3.

طاحونة الجبيلية   Pf   *Tâhûnet el Jebeilîyeh.* The mill of the Jebeiliyeh. p.n.; from Jebeil, a town in Syria.

طاحونة الجسر   Re   *Tâhûnet el Jisr.* The mill of the bridge.

طاحونة الملاحة   Qe   *Tâhûnet el Mellâhah.* The mill of the salt-works; see *Khŭrbet Akabet el Mellâhah,* p. 6.

طاحونة المعلق   Pf   *Tâhûnet el Muallak.* The suspended mill.

طاحونة السيارة   Re   *Tâhûnet es Seiyârah.* The mill of the wanderer.

طاحونة السموئية   Pf   *Tâhûnet es Semmûâîyeh.* The mill of Semarûaiyeh. p.n.

طاحونة سوق العصر   Re   *Tâhûnet Sûk el 'Aser.* The mill of the afternoon market.

طاحونة ام الحديد   Pf   *Tâhûnet Umm el Hadîd.* The mill with the iron.

طاحونة ام جوزة   Pf   *Tâhûnet Umm Jôzeh.* The mill with the walnut tree.

طاحونة ام المشمش   Pf   *Tâhûnet Umm el Mushmusheh.* The mill with the apricot tree.

طاحونة ام التوت   Pf   *Tâhûnet Umm et Tût.* The mill with mulberry.

طواحين عوبا   Pe   *Tawâhîn 'Aûba.* The mills of 'Aûba. p.n.

طواحين فرائية   Pf   *Tawâhîn Ferrâdieh.* The mills of Ferrâdiyeh p.n.

طواحين سيرين   Pf   *Tawâhîn Sîrîn.* The mills of Sirin. p.n.

طيطبا   Pe   *Teitaba.* Teitaba; perhaps the same as طديده, 'a watch-tower.' Sepp suggests that this is the place from which Elijah derived his patronymic of the Tishbite, the ת and ש being often interchanged in Aramaic dialects.

التليل   Re   *Et Teleil.* The small mound.

تل ابليس   Re   *Tell Abális.* The mound of the devils.

تل ابى بابين   Ne   *Tell Abu Bábein.* The mound with the two doors.

تل ابى جراد   Nf   *Tell Abu Jeráđ.* The mound where locusts abound.

تل ابى معابى   Pd   *Tell Abu M'áby.* The mound of Abu M'áby. p.n.

تل عيصلان   Qd   *Tell 'Aísalân.* The mound of the oleanders.

تل الاحا   Pe   *Tell el Ahma.* The mound of el Ahma. From حى either superlative of حامى 'defender,' or plural of حامية 'stone casing for a well.'

تل العنقور   Pe   *Tell el 'Ankûr.* (Properly عُنقُر) 'The mound of sprouting rushes.'

تل عارا   Pe   *Tell 'Âra.* The mound of fissures.

تل البياض   Nd   *Tell el Beiyâd.* The white mound.

تل بعراجة   Oe   *Tell Bi'rájeh.* The mound of Bi'rájeh. p.n.

تل الدور   Pf   *Tell ed Dûr.* The mound of the circle.

تل الغوردية   Od   *Tell el Ghawardiyeh.* The mound of 'cane booths' or of 'truffles.'

تل الغزار   Nd   *Tell el Ghazzâr.* The mound of rushes.

تل احلس   Qd   *Tell el Hilis.* The mound of dark green herbage.

تل القواس   Qd   *Tell el Kôwâs.* The mound of the archer. The word is now used for a guard or policeman.

تل القصب   Qf   *Tell el Kûsab.* The mound of reeds.

تل معطنة   Pe   *Tell Matanah.* The mound of the Matanah; i.e., a place round which camels, &c., lie down while waiting to drink.

تل الميدان   Pe   *Tell el Meidân.* The mound of the plain or exercise ground.

تل الملاحة   Qd   *Tell el Mellâhah.* The mound of salt works.

تل المريمغة   Pf   *Tell el Mureimighah.* The mound of the tanners or skin dressers.

تل العريمة   Re   *Tell el 'Oreimeh.* The mound of the heap or dam.

تل الراهب   Ne   *Tell er Râhib.* The monk's mound.

تل الرمان   Rf   *Tell er Rummân.* The mound of the pomegranate.

تل الرمّانة Qd *Tell er Rummânch* (a wady). The mound of the pomegranate tree.

تل الصّفا Re *Tell es Safa.* The clear mound.

تل السّنجق Qf *Tell es Sanjak.* The mound of the flag.

تل الزّاغ Od *Tell ez Zágh.* The mound of the crow.

تل زكّار Oe *Tell Zakkâr.* The mound of the leathern wine bottle-maker.

تلّة النّسورة Qf *Tellet en Nûsûrah.* The eagles' mound.

تلّة النّبعة Qf *Tellet en Neb'ât.* The mound of the perennial spring.

تلول القائمات Pf *Tellûl el Káimât.* The mound of the standing column, &c.

تلول القطّاعين Qd *Tellûl el Katâ'in.* The mound of the highwaymen.

ثغيرة الهوا Of *Thogheiret el Hawa.* The little frontier of the wind.

الطّيرة Od *El Tîreh.* The fortress.

طرفا Pe *Turfa.* The tamarisks.

ام سويط Oe *Umm Suweit.* The place with stagnant water.

وادى ابى علي Of *Wâdy Abu 'Aly.* The valley of Abu 'Aly. p.n.

وادى ابى غار Pe *Wâdy Abu 'Amâr.* The valley with the habitations.

وادى ابى هرّة Pf *Wâdy Abu Hirreh.* The valley where wild cats abound.

وادى ابى جاجة Oe *Wâdy Abu Jâjeh.* The valley of Abu Jâjeh. p.n.; Jâjeh means 'a head,' and in the vulgar dialect as spoken at Damascus is equivalent to دجاجة 'a hen.'

وادى ابى لوزة Rf *Wâdy Abu Lôzeh.* The valley with the almond tree.

وادى ابى ملاعق Pd *Wâdy Abu Melâ'ik.* The valley with the spoons.

وادى ابى شقيف Od *Wâdy Abu Shukeifeh.* The valley with the little cleft.

وادى ابى طبل Pe *Wâdy Abu Tabil.* The valley of Abu Tabil. p.n.; meaning the man with the 'drum' or 'round tray.'

وادى العين Of *Wâdy el 'Ain.* The valley of the spring.

وادى عين البرّانية {Pf / Nf} *Wâdy 'Ain el Bárrânîyeh.* The valley of the outer spring.

وادى عين جنان Nf *Wâdy 'Ain Jenân.* The valley of the spring of the gardens.

o

وادى عين الجرانى	Pf	*Wâdy 'Ain el Jurâny.* The valley of the spring of the troughs.
وادى عين القنطرة	Qe	*Wâdy 'Ain el Kantarah.* The valley of the spring of the arch.
وادى عين الخطارة	Pf	*Wâdy 'Ain el Khattârah.* The valley of the spring of 'Ain el Khattârah ; see p. 63.
وادى عين القصيبة	Qe	*Wâdy 'Ain el Kûseibeh.* The valley of the spring of rushes.
وادى عين النسورة	Pf	*Wâdy 'Ain en Nûsûrah.* The valley of the spring of eagles.
وادى عين صوف	Pe	*Wâdy 'Ain Sûf.* The valley of the spring of wool.
وادى عيثرون	Pd	*Wâdy 'Aitherûn.* The valley of 'Aitherûn. p.n.
وادى دلاتينا	Pe	*Wâdy 'Alâkîna.* The valley of 'Alâkîna. p.n.
وادى العموقة	Qe	*Wâdy el 'Ammûkah.* The valley of 'Ammûka. p.n. : from عمق 'to be deep.'
وادى عمورية	Of	*Wâdy 'Ammûrich.* The valley of 'Ammûrich. p.n.
وادى عروس	Qd	*Wâdy 'Arûs.* The valley of the bride.
وادى اشليل	Pe	*Wâdy Ashlîl.* The valley of the cascade.
وادى العسل	Of	*Wâdy el 'Asl.* The valley of honey.
وادى عصيلة	Od	*Wâdy 'Assîleh.* The valley of the oleander.
وادى العاصى	Pd	*Wâdy el 'Âsy.* The valley of the 'Âsy. p.n.; meaning 'rebellious.'
وادى العتبة	Qd	*Wâdy el 'Atabeh.* The valley of the threshold.
وادى عوبا	Pe	*Wâdy 'Aûba.* The valley of 'Aûba. p.n.
وادى العيون	Nd	*Wâdy el 'Ayûn.* The valley of the springs.
وادى عظائم	Ne	*Wâdy 'Azâim.* The valley of 'Azâim. The word عظائم is the plural of عظيمة 'a great crime or misfortune'; it may be connected with عظم 'a bone.'
وادى بقيس	Od	*Wâdy Bakîs.* The valley of the box tree.
وادى البقر	Qf	*Wâdy el Bakr.* The valley of the cows.
وادى البدية	Oe	*Wâdy el Bediyeh.* See خربة بدية , p. 83.
وادى بيت ياحون	Pd	*Wâdy Beit Yâhûn.* The valley of the house of Yâhûn. p.n.

وادى البستان	Pd	*Wâdy el Bestân.* The valley of the garden.
وادى بئر الخربة	Qf	*Wâdy Bîr el Khûrbeh.* The valley of the well of the ruin.
وادى بئر الشيخ	{Pe, Pf}	*Wâdy Bîr esh Sheikh.* The valley of the Sheikh's well.
وادى الظل	Pe	*Wâdy edh Dhûl.* The valley of shade.
وادى الدب	Nd	*Wâdy ed Dubb.* The valley of the bear.
وادى فارة	Pe	*Wâdy Fârah.* The valley of the mouse. But *see Fârah,* p. 72.
وادى فرم	Qf	*Wâdy Ferâm.* The valley of Ferâm. p.n.
وادى غبائى	Oe	*Wâdy Ghabbâtî.* The valley of Ghabbâti; *see 'Ain Ghabbâtî,* p. 62.
وادى غدير الذبان	{Oe, Of}	*Wâdy Ghadîr edh Dhubbân.* The valley of the pond of the flies.
وادى الغميق	{Pf, Pf, Of}	*Wâdy el Ghamîk (3).* The deep valley. غميق *ghamîk* being a vulgar Syrian form for عميق *âmak.*
وادى الغربية	{Pf, Qf}	*Wâdy el Gharbîyeh.* The western valley.
وادى الغزار	Nd	*Wâdy el Ghazzâr.* The valley of rushes.
وادى غصون	Qd	*Wâdy Ghusûn.* The valley of branches.
وادى الحبيس	Ne	*Wâdy el Habîs (2).* The valley of the religious bequest. حبس to 'confine, or restrict,' is used in the technical sense of bequeathing a thing, as an endowment to a religious establishment.
وادى الحجية	Od	*Wâdy el Hajjeh.* The valley of the pathway.
وادى الهلال	Od	*Wâdy el Hallâl.* The valley of the crescent.
وادى حام	Ne	*Wâdy Hamâm.* The valley of the dove.
وادى حامد	Of	*Wâdy Hâmed.* The valley of Hâmed. p.n.
وادى حمرا	Pf	*Wâdy el Hamra.* The red valley.
وادى حرة	Qd	*Wâdy Harrah.* The valley of Harrah. p.n.
وادى الحجاج	Pe	*Wâdy el Hijâj.* The valley of the Pilgrims.
وادى حنداج	Oe	*Wâdy Hindâj.* The valley of Hindâj, *i.e.,* sandy soil with grass growing upon it.
وادى حيمة	Oe	*Wâdy Humeimeh.* The valley of 'the small bath,' or 'the little dove.'

وادى جاعونة	Qf	*Wâdy Jââûneh* (2). See '*Ain Jââûneh*, p. 62.
وادى الجحيف	Of	*Wâdy el Jahif.* Stripped or spoilt valley.
وادى جهنم	Qd	*Wâdy Jehennam.* The valley of Hell.
وادى جسيرد	Od	*Wâdy Jessireh.* The bridged valley.
وادى الجن	Pf	*Wâdy el Jinn.* The valley of the Jinn (demons or 'genii').
وادى الجش	Pe	*Wâdy el Jish.* The valley of el Jish (Giscala).
وادى الجوق	Pf	*Wâdy el Jôk* (2). The devious valley.
وادى الجوز	Pd	*Wâdy el Jôz.* The valley of the walnut.
وادى جديدة	Pf	*Wâdy Judeiyideh.* The valley of dykes; see *Wâdy el Judeiyideh*, p. 11.
وادى الجوانية	Of	*Wâdy el Jûwâniyeh.* The innermost valley.
وادى قطمون	Oe	*Wâdy Katamûn.* The valley of Katamûn; see '*Ain Katamûn*, p. 63.
وادى كفر عنان	Pf	*Wâdy Kefr 'Anân.* The valley of Kefr Anan. q.v. p. 76.
وادى قيسون	Qe	*Wâdy Keisûn.* The valley of Keisûn (a sort of plant).
وادى الكروم	Nf	*Wâdy el Kerûm.* The valley of the vineyards.
وادى خلة السوق	Pe	*Wâdy Khallet es Sûk.* The valley of the dell of the market.
وادى الخانوق	Od	*Wâdy el Khânûk.* The valley of the gorge.
وادى الخرار	Pf	*Wâdy el Kharrâr.* The valley of murmuring water.
وادى الخشب	Nf	*Wâdy el Khashab.* The valley of timber.
وادى الجلال	Oe	*Wâdy el Khelâl.* The valley of dells.
وادى الخوخ	Of	*Wâdy el Khôkh.* The valley of plums.
وادى خزانة	Of	*Wâdy Khûzâneh.* The valley of treasure.
وادى كحالة	Pf	*Wâdy Kohâleh.* The valley of manganese.
وادى كونين	Pd	*Wâdy Kûnîn* ; see *Kûnîn.*
وادى الكورة	Od	*Wâdy el Kûrah.* The valley of the hole.
وادى كرم الجوانى	Nf	*Wâdy Kurm el Juwâny.* The valley of the inner vineyard.

وادى الكوبّس Pf *Wády el Kuweiyis.* The pretty valley.

وادى اللّوز Qf *Wády el Lóz.* The valley of the walnut.

وادى لوزية Qe *Wády Lóziyeh.* The valley of (the) walnut tree.

وادى الوزية Qf *Wády el Lóziyeh.* The valley of the walnut-tree.

وادى المعتى Od *Wády el Mátek.* The valley of the emancipated one.

وادى مقطع المعاصرة Qd *Wády Maktà el Máserah.* The valley of the quarry of the wine press.

وادى المقتول Pf *Wády el Maktúl.* The valley of the slain.

وادى ميرون Pf *Wády Meirôn.* The valley of Merom.

وادى المنابع Oe *Wády el Menábà.* The valley of perennial springs.

وادى المرج Qe *Wády el Merj.* The valley of the meadow.

وادى معظمية Pe *Wády Muáddemiyeh.* The valley of Muáddemiyeh ; *see Khúrbet Muáddemîyeh,* p. 85.

وادى المظلم Nd *Wády el Mudhlim.* The valley of the oppressor ; but perhaps from ظلمة 'darkness.'

وادى المغار Qf *Wády el Mughár.* The valley of the caves.

وادى المخيبة Pe *Wády el Mukeisibeh.* The valley of the reed beds.

وادى المركة Of *Wády el Mûrikeh.* The valley of the pommel of the saddle.

وادى مشيرفة Re *Wády Musheirefeh* (2). The valley of the high places (eminences).

وادى نحلة Od *Wády Nahleh.* The valley of the bee ; but *see Khúrbet Juwâr en Nukhl,* p. 7.

وادى نحلية Pd *Wády Nahleiya.* The valley of the bees ; *see* last paragraph.

وادى النكّاتى Pf *Wády en Nakkáty.* The valley of en Nakkáty. p.n.

وادى الناشف {Re / Pf} *Wády en Náshef* (2). The dry valley.

وادى نبع البلاط Pe *Wády Nebà el Balát.* The valley of the spring of stone slabs or pavement.

وادى النجاعى Rf *Wády en Nejâs.* The valley of the pears.

وادى نطارة Nd *Wády Nettárah.* The valley of the scarecrow.

وادى النَمِر　Of　*Wády en Nimr.* Valley of abundant water.

وادى النوم　Of　*Wády en Nôm.* The valley of sleep.

وادى النَقعة　{Pd / Pf}　*Wády en Nŭkáh.* Valley of stagnant water.

وادى النَقّار　Pe　*Wády en Nŭkkar.* The valley of the perforator; from نَقَر to perforate or peck.

وادى عَكيمة　Qf　*Wády 'Okeimeh.* The valley of 'Okeimah. p.n.; signifying the 'pulley of a well'; but it should probably be written عَقيمة from عَقَم، pl. عَقوم, which in the Bedawi dialect means an embankment.

وادى العش　Qf　*Wády el 'Ash.* The valley of the nest.

وادى عش الغراب　Qf　*Wády 'Osh el Ghuráb.* The valley of the raven's nest.

وادى رأس الأحمر　Pe　*Wády Rás el Ahmar.* The valley of the red headland or hill-top.

وادى الرُميلة　Pf　*Wády er Rumeileh.* The valley of the sandy tracts.

وادى سعيد　Of　*Wády Saîd.* The valley of Saîd; p.n. (= Felix).

وادى صلحا　Pe　*Wády Salhah.* Valley of Salhah. p.n. (Righteousness).

وادى السقا　Qd　*Wády es Sakka.* The valley of the swan or pelican.

وادى السكّارة　Pe　*Wády es Sakkárah.* The valley of the sugar makers.

وادى سرطبا　Of　*Wády Sŭrtŭba.* The valley of Sŭrtŭba. p.n.

وادى السيسبان　Rf　*Wády es Seisebán.* Valley of Seisabân (a flowering shrub).

وادى السيارة　Rf　*Wády es Seiyárah.* The valley of the wanderers.

وادى شعبان　Pe　*Wády Shában.* The valley of Shábân. Shábân is the name of the 8th Arabian month; also of a kind of locust. The word is perhaps connected with شعب a mountain ravine or spur.

وادى الشاغور　{Ne / Pe}　*Wády esh Shághúr* (2). The valley of esh Shághúr. The word means 'land left without a guardian'; see *Khŭrbet Shághúry*, p. 28.

وادى الشبابيك　{Pe / Qe}　*Wády esh Shebâbík* (2). The valley of the windows, lattices, or mazes. The word is also applied to any intricate system of irrigation.

وادى الشَخيتى　Pf　*Wády esh Shekhety.* The valley of esh Shekhety. p.n.; meaning 'lank,' 'stingy.'

وادى شفنين　Of　*Wády Shufnín.* The valley of the dove.

وادى سيرين　Pf　*Wády Sírín.* The valley of Sirin.　p.n.

وادى السرار　Pe　*Wády es Súrár ; see 'Ain es Súrár,* p. 65.

وادى صربين　Od　*Wády Surubbín.* The valley of Surubbin.　p.n.

وادى التنور　Nd　*Wády et Tannúr.* The valley of the reservoir.

وادى الطراش　Pd　*Wády et Tarrásh.* The valley of the white-washer.

وادى الطواحين　Pf　*Wády et Tawáhín.* The valley of the mills.

وادى طيطبا　Pe　*Wády Teitaba.* The valley of Teitaba ; *see Teitaba,* p. 95.

or وادى تيربخا
تربخا　Nd　*Wády Terbíkha.* The valley of Terbikha ; *see Terbíkha,* p. 56.

وادى الطولى　Qf　*Wády et Túla.* The long valley.

وادى ام الهموم　Qe　*Wády Umm el Hamúm.* The valley with the well-filled wells ; or it may be a female proper name, signifying ‘mother of cares.’

وادى الوقاص　Qe　*Wády el Wakkás.* The valley of the broken-necked man.

وادى الوعول　Ne　*Wády el Wáál.* The valley of the ibexes or wild mountain goats.

وادى اليابس　Pf　*Wády el Yábis.* The dry valley.

وادى يارون　Od　*Wády Yárún.* The valley of Yárún.　p.n.

وادى يعتر　Wd　*Wády Yáter.* The valley of Yater.　p.n.

وادى زبود　Of　*Wády Zebúd.* The valley of Zebúd ; *see Khúrbet Zebed,* p. 51.

وادى الزقاق　Pd　*Wády ez Zekák.* The valley of the street or lane.

وادى الزرب　Of　*Wády ez Zerb.* The valley of the water channel.

وادى زحلق　Rf　*Wády Záhlúk.* The slippery valley.

وعرة باب القصيب　Pe　*Wáret Báb el Kadíb.* The rugged place of the door of the reeds.

وعرة باب الخانوق　Qf　*Wáret Báb el Khánúk.* The rugged place of the entrance to the gorge.

وعر المدورة　Pd　*Wár el Madáwerah.* The rugged place of the circular (rock).

ولى المنتار   Kf   *Welî el Muntâr.* The tomb or shrine of el Muntâr, the 'watchman.' *Weli* properly signifies a saint, but is used in Palestine for a saint's tomb.

الوسطى   Pf   *El Wustah.* The middle.

الوستانى   Qd   *El Wustâny.* The middle part.

يارون   Oe   *Yârûn.* Yârûn. p.n.; perhaps the Iron of Josh. xix, 38.

يعتر   Nd   *Yâter.* Yâter. p.n.

يوسف الغريب   Of   *Yûsef el Gharîb.* Joseph, the poor, or stranger.

الزعيترى   Pe   *Ez Zâitery.* For زعترى ; *i.q.*, زعتر , abounding in marjoram (the herb).

الزاروب   Oe   *Ez Zârûb.* The channels.

زيتون الحربص   Of   *Zeitûn el Hurbus.* The olive of the irrigated ground.

زقاق الغار   Pf   *Zekâk el Ghâr.* The street or lane of the hollow.

SHEET V.

عبلين Mh *'Abellin.* p.n.

العبهرية Li *El 'Abhariyeh.* The place of the mock orange (*Styrax officinalis*).

ابطون Lh *Abtûn.* Abtûn. p.n.

ابو الهيجا Mg *Abu el Heija.* p.n.; Father of the fight, a mukâm or shrine.

ابو قراد Ng *Abu Kerâd.* Abounding with ticks.

ابو سويد Jh *Abu Suweid.* Father of blackness; but *see Jubb Suweid,* p. 5.

عيلوط Ni *'Ailût.* Perhaps from عليط, a species of tree.

عين عافية Mh *'Ain 'Âfieh.* The wholesome spring; or the spring of the water drawers.

عن العليق Mi *'Ain el 'Aleik.* The spring of the bramble.

عين الطاروق Mi *'Ain Atârûk.* Perhaps connected with the word طريق, 'a road,' as two roads here diverge.

عين العولطة Mi *'Ain el 'Awaltah.* The spring of 'Awaltah. p.n.; *cf. 'Ain 'Ailût* above, with which the word is probably connected.

عين البيضا Mi *'Ain el Beida.* The white spring.

عين الغفر Li *'Ain el Ghûfr.* The spring of the escort or toll.

عين حميدة Li *'Ain Hamîdeh.* Hamideh's spring. p.n., from حمد, 'to praise.'

عين حوض Vi *'Ain Haud.* Spring of the cistern.

عين الحيق Ki *'Ain el Heik.* The spring of Heik, an aromatic herb (resembling *Shîh*), used for the table.

عين الحلو Mi *'Ain el Helu.* The sweet spring.

عين إسحق Li *'Ain Is-hak.* Isaac's spring; so called from Neby Is-hak, at the place called *el Is-hakiyeh.*

عين الجديدة Mi *'Ain el Judeideh.* The spring of dykes; *see Wâdy el Judeiyideh,* p. 11.

عين القبوة Li *'Ain el Kabweh.* The spring of the little vault or cellar.

P

عين الخشبة.   Li   *'Ain el Khashabeh.* The timber spring.

عين المدوّرة   Mi   *'Ain el Madadwarah.* The round spring.

عين المالحة   Mi   *'Ain el Mâlhah.* The salt spring.

عين الرسمك   Mi   *'Ain er Resmek.* The spring of Resmek. p.n.

عين السعادة   Kh   *'Ain es S'âdeh.* The spring of happiness.

عين شنا عمر   Lh   *'Ain Shefa 'Amr.* The spring of Shefa 'Amr. q.v.

عين الشيخة   Mi   *'Ain esh Sheikhah.* The spring of the female elder or saint.

عين شتوية   Mh   *'Ain Shittawîyeh.* The winter spring.

عين السيح   Jh   *'Ain es Sîh.* The spring of the stream.

عين الصفصافة   Ni   *'Ain es Sufsâfeh.* The spring of the osier willow.

عين صفنا عادي   Lh   *'Ain Sufta 'Ády.* The spring of Sufta 'Ády. q.v.

عين الطيبة   Mh   *'Ain et Taiyibeh.* The good spring.

عين التينة   Lh   *'Ain et Tîneh.* The spring of the fig-tree.

عين ام الفاروج   Jh   *'Ain Umm el Fârûj.* The spring with the crevices.

عرب الهنادي   Mh   *'Arab el Henâdy.* The Henâdy Arabs; see Section D.

عرب المواسي   *'Arab el Mûâsy.* The Arabs of the Mûâsy tribe.

عرب المحميدات   Kh   *'Arab el Muhammeidât.* The Arabs of the Mohammeid (dimin. of Mohammed) tribe.

العراق   Jh   *El 'Arâk.* The shore.

اربع استنبولي   Ji   *Arbà Stambûly.* The spring pastures of the Stamboulee.

أرض البراغيث   Ji   *Ard el Baraghîth.* Land of fleas; a plot of ground.

اشلل الحية   jh   *Ashlûl el Haiyeh.* Waterfalls of the serpent. The name applies to the cascades, from *'Ain es Sîh.*

اشلول الواوي   Kh   *Ashlûl el Wâwy.* Cascades of the jackal.

عثليت او عتليت   Ji   *'Athlît or 'Atlît.* Athlit. p.n.

عيون العافي   Li   *'Ayûn el 'Afy.* The springs of the water-drawer.

عيون البص   Lg   *'Ayûn el Bass.* Springs of the marsh.

عيون ام حميد   Mh   *'Ayûn Umm Humeid.* The springs of Mother of Humeid; p.n.

عيون كوكب Mh *'Ayûn Kawkab.* The springs of Kawkab ; q.v.

عيون الورد Kh *'Ayûn el Werd.* The spring of the rose. ورد also means 'to go down for water,' as opposed to صدر 'to come up from a watering place.'

باب الهوا Mh *Bâb el Hawa.* The gate of the wind.

البتّة Ji *El Bassah.* The marsh.

بيت لحم Mi *Beit Lahm.* The house of meat ; in Hebrew ב. לחם signifies the 'house of bread.'

بيت الملح Ji *Beit el Milh.* House of salt (an old salt-pan).

بلد الشيخ Kh *Belled esh Sheikh.* The town of the Sheikh.

البلّو Ji *El Bellu.* El Bellu. p.n. ; either from بلو 'worn,' or بلّ 'moist.'

بنات يعقوب Lh *Benât Yâkûb.* The daughters of Jacob.

السورة Mq *El Berweh.* The well of the bangle.

بستان Ji *Bestân.* The garden.

بستان ابو لبده Kq *Bestân Abu Libdeh.* The garden of Abu Libdeh. p.n.

بئر ابو زيد Ni *Bîr Abu Zeid.* The well of Abu Zeid ; *i.e.,* 'father of Zeid,' a common Arab name.

بئر العيّاديّه Lg *Bîr el 'Aiyâdîyeh.* The well of the 'Aiyadiyeh (or Keiyudiyeh) Arabs.

بئر دعوك Lq *Bîr Dâûk.* The well of Dâûk. p.n.

بئر الدستره Li *Bîr ed Dustrah.* The well of Dustrey ; *see* "Memoirs," p. 288.

بئر ابدويه Ji *Bîr Ebdawîyeh.* The well of the Bedawi woman.

بئر الامير Ni *Bîr el Emîr.* The well of the prince.

بئر الغربي Lg *Bîr el Gharby.* The western well.

بئر الحناني Mg *Bîr el Hanâny.* The well of Hanâny. p.n., meaning 'yearning.' But it may be either from حنان = Hennâ, Egyptian paint used for staining the finger-tips ; or from حنانه, which in vulgar Arabic means a waterwheel which turns with a murmuring noise, and also a trough.

بئر حوشه Lh *Bîr Hûsheh.* The well of Hûsheh. p.n. The אושה of the Talmud.

بئر الجاحوش Lg *Bîr el Jâhûsh.* Well of the ass.

بئر كيسان Lq *Bîr Keisân.* The well of treason.

بئر الكنيسة Jh *Bîr el Kenîseh.* The well of the church.

بئر المكسور Mh *Bîr el Maksûr.* The broken well.

بئر المالح Mi *Bîr el Mâleh.* The salt well.

بئر المنصورة Li *Bîr el Mansûrah.* The well of el Mansûrah. p.n.; from Mansûr (victor).

بئر المغاير Mg *Bîr el Mûghair.* The well of the caves.

بئر الصفا Mg *Bîr es Sûfa.* The clear well.

بئر الطيبة Mh *Bîr et Taiyibeh.* The well of et Taiyibeh, 'the goodly.'

بئر الطيرة Lq *Bîr et Tîreh.* The well of the pit.

بئر وادي عبلين Mh *Bîr Wâdy 'Abellîn.* The well of the valley of 'Abellîn. p.n.

بئر يعنين Mg *Bîr Yânîn.* The well of Yânîn. p.n.

بئر اليزك Ji *Bîr el Yezek.* The well of the sentinel (Turkish).

بركة الصفرا Lh *Birket es Sûfra.* The yellow pool.

بركة الشويكاني Ji *Birket esh Shuweikâny.* The thorny pool.

بركة سخنين Ng *Birket Sukhnîn.* The pool of Sukhnin; name of a village (hot); *see Sukhnîn,* p. 116.

بيارة الغوارنة Kg *Biyâret el Ghawâruch.* The wells of the Arabs of the Ghor or 'Lowlands'; singular Ghûri.

البرج Jh *El Burj.* The tower.

برج السهل Lh *Burj es Sahel.* The tower of the plain.

البوابة Ji *El Bûwâbeh.* The fortress or 'portals.'

الدبات Ki *Ed Dabbât.* The sandhills or sandy tracts (pl. of دبة), applies to two knolls near the ruin ed Duweimin.

دالية الكرمل Ki *Dâliet el Kurmûl.* The hanging vine of Carmel, a village; possibly the Hebrew ידאלה Idalah.

الدعون Mg *Ed Damûn.* Ed Damûn. p.n.

الدير Jh *Ed Deir.* The monastery.

ظهر بئر ابو خرازة Ji *Dhahr (Bîr) Abu Kherâzeh.* Ridge of the well of Abu Kherâzeh. p.n.; خراز means 'sewing skins or hides.'

دوبل	Ki	*Dûbil.* Dûbil. p.n. ; *see Dibl,* p. 71.
دويمين	Ki	*Duweimîn.* The little dôm trees, a kind of *Zyziphus.*
عسفيا	Ki	*'Esfia.* The devious (road).
فرش اسكندر	Jh	*Fersh Iskander.* Alexander's bed (plot of ground).
الفوارة	Kh	*El Fiwârah.* The fountains or jets of water.
حيفا	Jh	*Haifa.* p.n. Hebrew חיפה, from חוף 'shore.'
حيفا العتيقة	Jh	*Haifa el 'Atîkah.* Old Haifa.
البجلي	Ji	*El Hajly.* The low-lying land.
الحنانة	Ji	*El Hanâneh.* See *Bîr el Hanâny.*
الهربج	Lh	*El Harbaj.* p.n. ; it means work (such as masonry) that is badly done.
الحارثية	Li	*El Hârithîyeh or el Hâritheh.* The ploughed land.
الحوارة	Mi	*El Hurârah.* The white marl.
الإسحقية	Li	*El Is-hakîyeh or Neby Is-hak.* The place of Isaac, or 'the prophet Isaac.'
جباتا	Mi	*Jebâta.* Jotapata ; *see* "Memoirs," p. 289.
جبل ابو عاق	Ng	*Jebel Abu 'Ak.* The mountain of Abu 'Âk ; perhaps from عوق the bend of a valley.
جبل عقارا	Ki	*Jebel 'Akkâra.* The mountain of 'Akkârah. Either from عقار medicinal herbs, or from عقر 'to be barren.'
جبل الديدبة	Nh	*Jebel ed Deidebeh.* Mountain of the watch-tower ; *cf.* Aramaic דדבאות.
جبل كفسى	Ni	*Jebel Kafsy.* The wry mountain.
جبل قانا	Ng	*Jebel Kâna.* See *Kh. Kâna.*
جبل خنزيرة	Ng	*Jebel Khanzîreh.* The mountain of the swine.
جبل كرمل	{Jh Kh Ji Ki}	*Jebel Kûrmŭl.* Mount Carmel. Hebrew הכרמל.
جديدا	Li	*Jeida.* Jeidd. p.n. In Arabic it means 'a long-necked.'
جلمة	Li	*Jelameh.* The hill. *See* بركة الجلمة p. 19.
جلمة المنصورة	Li	*Jelamet el Mansûrah.* The hill of Mansûrah.

المجنّادية Li   *El Jennadîyeh.* El Jennadiyeh. p.n. ; *see 'Ain Jenadîyeh,* p. 63.

جزيرة العجّال Ji   *Jezîrat el 'Ajjâl.* The island of the hastener.

جزيرة النفّاخة Ji   *Jezîrat en Nefâkhah.* The island of the blower ; *see* p. 31.

جزيرة الوادي Ji   *Jezîiat el Wâdy.* The island of the valley.

جدرو Lh   *Jidrû.* Jidrû. p.n.

الجمّيزة Ji   *El Jimmeizeh.* The sycamore, a tree south of 'Athlit, near *Birket el Shuweikâny.*

كابول Mg   *Kâbûl.* p.n. Heb. כָּבוּל Cabul.

قضاء حيفا {Jh / Ki} *Kada Haifa.* The district of Haifa.

قضا ناصره Mi   *Kada Nâsirah.* The district of Jaffa.

القنيطرة Mg   *El Kaneitrah.* The little arch.

كوكب Mh   *Kaukab.* Kaukab. p.n. ; meaning 'star.'

كفر اتّا Lh   *Kefr Etta.* The village of Etta. p.n.

كفر مندا Nh   *Kefr Menda.* The village of Menda. p.n.

الخلّدية Mh   *El Khalladîyeh.* Family name ; meaning 'perpetual.'

خلّة عبا Mh   *Khallet 'Aba.* The dell of Aba. p.n. ; meaning 'dull,' 'heavy.'

خلّة علّوش Mg   *Khallet 'Allûsh.* The ruin of the jackal or wolf.

خلّة النوري Kh   *Khallet en Nûry.* The flowery or light dell, or the dell of the gipsy.

خلّة الرمّانة Ji   *Khallet er Rummâneh.* The dell of the pomegranate tree.

خلّة الطاف Ng   *Khallet et Tâf.* The dell of et Tâf. p.n. ; perhaps connected with طاف, 'to go round in circuit,' though no noun of the form exists.

خلّة الزيبونة Ji   *Khallet ez Zeitûneh.* The dell of the olive tree.

خان بديوية Nh   *Khân Bedeiwîyeh.* The Caravanserai of the Bedawi woman, or family.

الخندق Ji   *El Khanduk.* 'The fosse' at 'Athlit.

الخشّاش Ni   *El Khashâsh.* 'The entrance' of the pass leading to Nazareth.

الخوانيق	Ji	*El Khawânîk.* The gorges (sing. *Khânûk*).
الخزيرقة	Ki	*El Khazîrkah.* The confined spot.
الخضر	Jh	*El Khûdr.* The Green Old Man ; a mythical personage in Moslem writings, sometimes identified with Elias, and sometimes with St. George.
خربة عبلين	Mh	*Khûrbet 'Abellîn.* The ruin of 'Abellin. p.n.
خربة ابى مسلسل	Lh	*Khûrbet Abu Musilsil.* The ruin of Abu Musilsil ; meaning ' father of concatenation.'
خربة العيطاوية	Lg	*Khûrbet el 'Aitâwîyeh.* The ruin of 'Aitawiyeh. p.n.
خربة العياضية	Lg	*Khûrbet el 'Aiyâdiyeh.* The ruin of the 'Aiyâdiyeh Arabs ; elsewhere written قيانية.
خربة العسافنة	Li	*Khûrbet el 'Asâfneh.* The ruin of the people of 'Esfia. q.v.
خربة اليسي	Jh	*Khûrbet Atteisy.* The ruin of Attisi. p.n.
خربة البيفا	Li	*Khûrbet el Beida.* The white ruin.
خربة البزوتية	Mg	*Khûrbet el Bezewâiyeh.* The ruin of el Bezewáiyeh. p.n. ; diminutive of بزيع, 'a fair youth.'
خربة البئر	Li	*Khûrbet el Bîr.* The ruin of the well.
خربة دعوك	Lg	*Khûrbet Dâûk.* The ruin of Dâûk. p.n.
خربة دسطري	Ji	*Khûrbet Dustrey.* The ruin of Dŭstrey ; a corruption of the old French ' *Destroit,*' ' pass' ; *see* "Memoirs," p. 288.
خربة الدويبة	Ki	*Khûrbet ed Duweibeh.* Ruin of the beast or reptile.
خربة الحميرة	Lh	*Khûrbet el Humeireh.* The reddish ruin.
خربة حوشة	Lh	*Khûrbet Hûsheh.* The ruin of Hûsheh ; the אושה of the Talmud.
خربة الجاحوش	Lg	*Khûrbet el Jáhûsh.* Ruin of the ass.
خربة جلّون	Mg	*Khûrbet Jallûn.* The ruin of Jallûn. Heb. גַּלִּים, ' heaps.'
خربة جفات	Ng	*Khûrbet Jefât.* The Jotapata of Josephus (Vit. 46 and 51) ; *see* "Memoirs," p. 289. Heb. יוֹדְפַתָה.
خربة جلمة	Lh	*Khûrbet Jelameh.* The ruin of the hill.
خربة القبو	Ng	*Khûrbet el Kabu.* The ruin of the vault.

خربة قانا Nh *Khûrbet Kâna.* The ruin of Cana ; from قنا, a channel or water-pipe. Heb. קנה ; *see Kefr Kenna,* "Memoirs," p. 391.

خربة كفر السمير Jh *Khûrbet Kefr es Samîr.* Ruin of the village of es Semir ; the word means 'one who converses by night,' but may be connected with the Samaritans.

خربة الكنيسة Jh *Khûrbet el Kenîseh.* The ruin of the church.

خربة الكرك Ki *Khûrbet el Kerek.* The ruin of the fort.

خربة القزاز Ng *Khûrbet el Kezâz.* The ruin of glass.

خربة الخضيرة Mi *Khûrbet el Khûdeirah.* The ruin of the little green patch.

خربة كردانة Lg *Khûrbet Kurdâneh.* Ruin of Kurdâneh. p.n.

خربة المكسور Mh *Khûrbet el Maksûr.* The ruin of the broken one.

خربة ماحلة Ji *Khûrbet Mâlhah.* The salt ruin.

خربة مثلية Ji *Khûrbet Mithilia.* The ruin of Mithilieh ; from مثل, 'an example' ; Heb. משל ; in Phœnician, 'an image' ; in old Arabic it means 'traces of a dwelling which are becoming effaced.'

خربة المشيرفة Mi *Khûrbet el Musheirefeh.* Ruin of the little high-place ; also called صور الطرابلسية *Sûr et Trâblesîyeh,* rock of the people of Tripoli.

خربة مسرارة Li *Khûrbet Mûsrârah.* The ruin of the Mûsrârah. The word may either mean 'place of gladness,' 'place where a child's naval-string is cut' (*i.e.,* birthplace ; *cf.* أسد, مسقط). or 'where sweet herbs grow' ; but it is perhaps مصرار 'pebbly.'

خربة الرختية Ki *Khûrbet er Rakhtîyeh.* *See er Rakhtîyeh.*

خربة راس الظهر Ng *Khûrbet Râs edh Dhahr.* Ruin of the top of the ridge.

خربة الرجم Lh *Khûrbet er Rujm.* Ruin of the stone-heap.

خربة رومة Nh *Khûrbet Rûmeh.* The ruin of Rûmeh. Heb. רומא 'height.'

خربة ساسا Lh *Khûrbet Sâsâ.* The ruin of Sâsâ ; *see Khallet Sâsâ,* p. 80.

خربة سماقا Ki *Khûrbet Semmâkâ.* Ruin of the sumach-tree (*Rhus Coriaria,* Linn.).

خربة الشلقية Ki *Khûrbet esh Shelkîyeh.* The ruin of esh Shelkîyeh. p.n. ; شلق means 'to beat' or 'flog.'

خربة شلّالة Ki *Khŭrbet Shellâleh.* Ruin of the waterfall.

خربة شرتا Lh *Khŭrbet Sherta.* The ruin of Sherta. p.n.; *see Wâdy el Ashert*, p. 118.

خربة شيحة Ji *Khŭrbet Shîhah.* Ruin of the Shîh, an aromatic plant (*Artimisia Judaica*). But *see Merj Sidr esh Shîh*, p. 90.

خربة السيح Ng *Khŭrbet es Siyeh.* *cf. 'Ain es Sîh.* The spring of the stream.

خربة صفتا عادي Lh *Khŭrbet Sufta 'Ady.* *See 'Ain Sufta Ády*, p. 106.

خربة سليمان Lh *Khŭrbet Suleimân.* The ruin of Solomon.

خربة الطيبة Mh *Khŭrbet et Taiyibeh.* The goodly ruin.

خربة الطيرة Lg *Khŭrbet et Tîreh.* The ruin of the fortress.

خربة يعنين Mq *Khŭrbet Yânîn.* The ruin of Yânîn. p.n.; probably the Hebrew הדביאל.

خربة طيرة الغزاز Ng *Khŭrbet Tiret el Kezâz.* Ruin of the fortress of glass.

خربة يونس Ji *Khŭrbet Yûnis.* The ruin of Jonas.

الخريبة Kh *El Khŭreibeh.* The little ruin.

قسقس Li *Kuskus.* Kuskus. p.n. = 'mincemeat.'

قصر الزير Mi *Kusr ez Zîr.* The palace of ez Zîr. p.n.

قسطال صفورية Ni *Kŭstâl Seffûrieh.* The castle of Seffûriyeh (a garden). It may be the crusading term *Casale*, which is applied to country villages (*cf.* Will. of Tyre).

ليّة زحلوق Li *Leiyet Zahlûk.* The slippery dry ground; *leiyeh* means land distant from water. The place is also called تل عمر *Tell 'Amr*, the mound of 'Amr (a common Arabic proper name). *cf. Nâhiet Shefa 'Amr*, p. 114.

المحرقة Ki *El Mahrakah.* The place of burning.

المقبية Ni *El Makbiyeh.* The vaulted place.

مقطعة عثليت Ji *Maktiyet 'Athlît.* The quarry of 'Athlit.

معلول Mi *Malûl*, p.n.; meaning 'caused or given to drink a second time.'

المطبع Li *El Matbâ.* The press (place where anything is stamped).

Q

المحفرة	Lh	*El Mehāferah.* The excavation ; also called *Birket es Sūfra.* q.v.
ميدان الزير	Mi	*Meidān ez Zīr.* The open place of ez Zīr.
المجدل	Lh	*El Mejdel.* The watch-tower. Heb. Migdol.
الملاحة	Kj	*El Mellāhah.* The salt-pan ; a place where salt is prepared.
مرج الذهب	Lh	*Merj edh Dhaheb.* The meadow of gold.
مرجات الدالية	Ki	*Merjāt ed Dālieh.* The meadows of the trailing vine ; *see* p. 77.
معار	Mg	*Māīr.* Mi'âr. p.n. ; perhaps from معر ' bare.'
المفتلة	Ki	*El Miftelah.* The twist ; from فتل to twist round, like yarn, &c., in spinning. Applied to a ridge which curves round.
المقلة	Ji	*El Miklch.* El Miklch. p.n. The word means a frying-pan, or a boy's toy.
المقتلة	Ki	*El Miktelch.* The place of slaughter.
المشريع	Ji	*El Misheriā.* The drinking-place (a spring and pool).
مغارة الجهنم	Li	*Mŭghâret el Jehennum.* The cave of Hell.
مغارة السمك او جزيرة العالى	Ji	*Mŭghâret es Semak.* The cave of the fish. Also called *Jezîrat el 'Aly,* the island or promontory of the lofty (thing).
مغارة الشيح	Li	*Mŭghâret es Sîh.* The cave of the shreik ; but it should perhaps be مغارة السيح the cave of the stream.
مغارة الوادي	Ji	*Mŭghâret el Wâdy.* The cave of the valley.
المغرشة	Ki	*El Mŭghrushah.* El Mŭghrashah. p.n. ; perhaps for عترشة ' barren land.'
المجيدل	Mi	*El Mujeidil.* The little watch-tower.
ناحية شفا عمر	LMh	*Nâhiet Shefa 'Amr.* The Shefa 'Amr commune ; *see* " Memoirs," p. 44.
نهر المنتنة	Kh	*Nahr el Mautuch.* The stinking river.
نهر المقطع	{Kh} {Li }	*Nahr el Mukŭttâ.* The river of the ' cut-up one.' The name is not uncommon in Syria, and may mean one to whom a portion of land has been allotted.
نهر نعمين	Lg	*Nahr Nâmein.* The river of Nâmein. p.n. ; Naaman.

الناصرة	Ni	*En Nâsirah.* Nazareth.
الناطف	Ji	*En Nâtef.* The (rock) dripping with water.
نبي هوشان	Lh	*Neby Hûshân.* The prophet Hûshân; cf. *Khŭrbet Hûsheh*, p. 111.
نبي سعين	Ni	*Neby Sâîn.* The prophet Sâîn; apparently 'Isaiah.'
العريسة	Kg	*El 'Orbîseh.* The level land.
الرختية	Ki	*Er Rakhtîyeh.* The place of winnowing corn. (Heb. רחת a winnowing fan.)
رأس العين	Mh	*Râs el 'Ain.* The fountain-head.
رأس الاقرع	Ji	*Râs el Akra.* The bald hill-top.
رأس براد	Mg	*Râs Barâd.* The cold hill-top.
رأس الظهر	Ng	*Râs edh Dhahr.* The top of the ridge.
رأس الكروم	Jg	*Râs el Kerûm.* The promontory of the vineyards.
رأس الخوانيق	Ji	*Râs el Khawânîk.* The hill-top of the gorges.
رأس الملة	Ng	*Râs el Mallch.* The hill-top of the oaks (*Quercus Ægilops*).
رأس المشاهر	Ki	*Râs el Meshahir.* The conspicuous top.
رأس مثليا	Ji	*Râs Mithilia.* See *Khŭrbet Mithilich*, p. 112.
رأس المهلل	Ki	*Râs el Muhellel.* The crescent-shaped top.
رأس طمرة	Mg	*Râs Tŭmrah.* The hill-top of Tŭmrah. q.v., p. 117.
رأس ام الشقف	Ki	*Râs Umm esh Shŭkf.* The hill-top with the cleft.
رأس الزالقة	Kh	*Râs ez Zelâkah.* The slippery hill-top.
الرويس	Mg	*Er Rúeis.* The little hill-top or headland.
رشميا	Jh	*Rushmia.* Rushmia. p.n.
سهل البطوف	Nh	*Sahel el Buttauf.* The plain of el Buttauf. p.n.
سهل صفورية	Nh	*Sahel Seffûrich.* Plain of Sepphoris.
سهل الطيبة	Mh	*Sahel et Taiyibeh.* The plain of et Taiyibeh (the goodly).
صفورية	Nh	*Seffûrich.* Sepphoris; Heb. צפורי.
سمونية	Mi	*Semûnieh.* Simoniah. p.n.; Heb. סימוניה.
شبانة	Mi	*Shabâneh.* Shabbâneh. p.n; 'a bridesmaid.'

Q 2

شعب   Mg   *Shâib.* The spur.

شنا عر   Lh   *Shefa 'Amr.* The margin or edge of 'Amr. Locally and erroneously supposed to mean 'the healing of 'Amer' (ed Dhâher), but really a corruption of the Hebrew Shafram שברעם.

شجرة ابو سُقر   Ki   *Shejeret Abu Sŭkr.* The tree with the falcons.

الشيخ عبدالله العروري   *Sheikh 'Abdallah el 'Arûry.* Sheikh Abdallah el 'Arûry (possessed by demons).

الشيخ ابراك   Ji   *Sheikh Abrâk.* Sheikh Abrâk. p.n.; perhaps from برك 'to bless.'

شيخ ابريك   Li   *Sheikh Abreik.* Sheikh Abreik. p.n. Really the same word as the last, but pronounced with the *imâleh.*

الشيخ احيا   Ji   *Sheikh Ahia.* Sheikh Ahia ; St. John (of Tyre). St. John of Tyre is mentioned in this direction in 1187 (*see* Du Vogüé, "Eglises," p. 445 ; 'La Citez de Jherusalem'). Ahia (properly Yahya يحيى) is the native name for St. John the Baptist.

سخنين   Ng   *Sukhnin.* Sukhnin. p.n.; perhaps from Heb. סיכנין.

السورية   Kh   *Es Sûriyeh.* Es Sûriyeh. p.n.; Syria.

طاحونة كردانة   Lg   *Tâhûnet Kurdâneh.* The mill of Kurdâneh. p.n.

طاحونة المرفوقة   Lh   *Tâhûnet el Merfûkah.* The mill of el Merfûkah. p.n.; 'escorted.'

طاحونة الراهب   Lh   *Tâhûnet er Râhib.* The monk's mill (belonging to the Carmel Convent).

طواحين النعمين   Lq   *Tawâhîn en Nâmein.* The mills of the River Nâmein ; *see* Nahr en Nâmein, p. 114.

طبيحة الفرس   Lh   *Teihet el Faras.* The grazing-ground of the horses.

تل ابو حوام   Kh   *Tell Abu Hûwâm.* The mound of the flocks of wildfowl.

تل ابو مدور   Kh   *Tell Abu Madaûwar.* The circular mound.

تل العلي   Li   *Tell el 'Aly.* The high mound.

تل بدوية   Nh   *Tell Bedeiwîyeh.* The mound of the Bedouins.

تل الفار   Lh   *Tell el Fâr.* The mound of the mouse.

تل غلطة   Li   *Tell Ghaltah.* The mound of the mistake.

تل الغربي Lg *Tell el Gharby.* The western mound.

تل كيسان Lg *Tell Keisân.* The mound of treachery.

تل الخشنا Mh *Tell el Khashna.* The mound of rough ground; but perhaps خشنآ a certain leguminous plant.

تل الخيار Kh *Tell el Khiâr.* The mound of the cucumbers.

تل كردانة Lg *Tell Kurdâneh.* The mound of Kurdâneh. p.n.

تل القسيس Li *Tell el Kûssîs.* The mound of the (Christian) priest.

تل المرسان Ng *Tell el Merasân.* The mound of myrtle.

تل المواجه Li *Tell el Muwâjeh.* The facing mound (opposite to *Tell esh Shemmân*).

تل النهل Kh *Tell en Nahl.* Mound of drinking, near springs.

تل سرج الونة Mh *Tell Saraj Alaunneh.* The mound of the saddle of Alaunneh. p.n.; perhaps *al-launî*, 'coloured.'

تل السمك Jh *Tell es Semak.* The mound of the fish.

تل السمن Lh *Tell es Semn.* The mound of butter.

تل الشمام Li *Tell esh Shemmâm.* The mound of the melon or colocynth.

تل الصوبات Kh *Tell es Sûbât.* The mound of heaps.

تل الطنطور Lg *Tell et Tantûr.* The mound of the peak. (The silver horn worn by women on their heads in some parts of Syria is so called.)

تل الوعر Li *Tell el Wâr.* The mound of rugged rocks; also called تل الزعتر *Tell ez Zâter.* The mound of thyme.

تل الواويات Nh *Tell el Wâwîyât.* The mound of the jackals.

تذاني Jh *Tinâny.* Tinâny. p.n.

الطيرة Jh *El Tîreh.* The fortress. Aramaic בירתה.

طمرة Mg *Tûmrah.* Tûmrah. p.n.; from طمر to make a pit for storing corn, &c.

طبعون Li *Tubâûn.* From Tabâ, 'to press' or stamp. Aramaic טבעין

أم العمد Li *Umm el 'Amed.* The place with the columns.

أم الشقف   Ki   *Umm esh Shŭkf.* The mother of the cleft.

وادي عبلين   Lg   *Wâdy 'Abellîn; see 'Abellîn (Mh),* p. 105.

وادي ابو عفن   Lh   *Wâdy Abu 'Afen.* The valley with the stinking soil.

وادي ابو حيّة   Ki   *Wâdy Abu Haiyeh.* The valley of the serpent.

وادي ابو مدور   Kh   *Wâdy Abu Madauwar; see Tell Abu Madanwar,* p. 116.

وادي العدسيّة   Lh   *Wâdy el 'Adasŷch.* The valley of lentils.

وادي العين   Ji   *Wâdy el 'Ain.* The valley of the spring.

وادي الاخرس   Ji   *Wâdy el Akhras.* The valley of the dumb man. Perhaps from Heb. חרס 'rugged.'

وادي عمر   Jh   *Wâdy 'Amir.* The cultivated valley; or p.n., 'Amr.

وادي الاشرت   Mh   *Wâdy el Ashert.* The valley of Ashert. p.n.; perhaps from Heb. שרת 'to hollow out.'

وادي البلاط   Mg   *Wâdy el Balât.* The valley of slabs (of rock).

وادي بدوية   Mh   *Wâdy Bedeiwîyeh; see Tell Bedawîyeh.*

وادي فلّاح   Ji   *Wâdy Fellâh.* The valley of the peasant or agriculturist.

وادي الفوار   Lh   *Wâdy el Fŭwâr.* The valley of fountains.

وادي الغابية   Ng   *Wâdy el Ghâbich.* The valley of the thicket.

وادي الغميق   Jh   *Wâdy el Ghamîk.* The deep valley.

وادي الحجير   Mg   *Wâdy el Hajîr.* The rocky valley.

وادي الحجلي   Ki   *Wâdy Hajly.* The valley of el Hajly. q.v., p. 109.

وادي الحلزون   Lg   *Wâdy el Halzûn.* The valley of the snail. Valley runs to Nahr N'âmlin, near which the murex חלזון used to be caught. (*cf.* Reland, "Pal.," p. 720.)

وادي الحراميّة   Lh   *Wâdy el Harâmîyeh.* The valley of the robbers.

وادي حواصة   Kh   *Wâdy Hawâsah.* The narrow valley.

وادي { اليهود / اليبود }   Lh   *Wâdy el Hûd,* or *Wâdy el Yehûd.* The valley of the Jews.

وادي حوشة   Lh   *Wâdy Hûsheh.* The valley of Hûsheh; see p. 111.

وادي الجاموس   Ki   *Wâdy el Jâmûs.* The valley of the buffalo.

وادي كابول   Mg   *Wâdy Kâbûl.* The valley of Kâbûl; see Kâbûl, p. 110.

وادي كوكب Nh *Wâdy Kaukab.* The valley of Kaukab; *see Kaukab,* p. 110.

وادي الكرك Mh *Wâdy el Kerek.* The valley of el Kerek (the fortress).

وادي قرف او غرف Mh *Wâdy Kerf* or *Gharf.* The valley of Kerf or Gharf. The first means 'excoriation' or 'aversion'; the second a tree used in tanning.

وادي الخليل Lh *Wâdy el Khalcil.* The valley of the friend, or 'thoroughfare.'

وادي الخلدية Mh *Wâdy el Khalladiyeh.* The valley of el Khalladiyeh. p.n. (signifying perpetual).

وادي القصب Jh *Wâdy el Kûsab.* The valley of reeds.

وادي الملك Lh *Wâdy el Melek.* The king's valley; *see Mâlkîyeh,* p. 9.

وادي المزيرة Mh *Wâdy el Mezeireh.* Valley of the little shrine.

وادي المفتلة Ki *Wâdy el Miftelah.* The valley of the twist; *see* p. 114.

وادي المغارة Ji *Wâdy el Mûghârah.* The valley of the cave.

وادي النحل Ki *Wâdy en Nahl.* The valley of the bee; but perhaps Heb. נַחַל 'watercourse.'

وادي النبك Ji *Wâdy en Nebek.* The valley of the hillock.

وادي عبيب Ng *Wâdy 'Obeib.* The valley of the little torrent.

وادي الريحان Kh *Wâdy er Rîhân.* The valley of sweet herbs (basil).

وادي الرجم Lh *Wâdy er Rujim.* The valley of the stone heap.

وادي رشميا Kh *Wâdy Rushmia.* The valley of Rushmia. p.n.

وادي الصقيعة Lh *Wâdy es Sakîâh.* The valley of the frosty ground.

وادي صفورية Nh *Wâdy Seffûrieh.* The valley of Seffûrieh (Sepphoris).

وادي سلمان Kh *Wâdy Selmân.* The valley of Selmân. p.n.

وادي سماقا Ki *Wâdy Semmâka.* The valley of the Semmâk tree; *see Khûrbet Semmâkâ,* p. 112.

وادي شعب Mg *Wâdy Shâib.* The valley of the mountain spur.

وادي الشاغور Mg *Wâdy esh Shâghûr.* The valley of esh Shâghûr; *see* p. 28.

وادي شلالة Ki *Wâdy Shellâleh.* The valley of cascades.

وادي شرقا Lh *Wâdy Sherka.* The eastern valley.

وادي الشمارية	Ki	*Wâdy esh Shomârîyeh.* The valley of the Shomarieh; family name, from شمر, a common name among the ancient Arabs. The word شمر also means 'wild fennel.'
وادي شويكة	Lh	*Wâdy Shuweikeh.* The valley of thorns.
وادي الصباوي	Mg	*Wâdy es Sûbâwy.* The valley of es Sŭbâwy. p.n.
وادي الصفا	Ng	*Wâdy es Sŭfa.* The valley of clear water or smooth stones.
وادي التابل	Kh	*Wâdy et Tâbil.* The valley of seeds (such as cummin, &c.) used as condiments for food.
وادي طبرية	Mh	*Wâdy Tŭbarîya.* Valley of Tiberias.
وادي طمرة	Mg	*Wâdy Tŭmrah.* *See* Tŭmrah, p. 117.
وادي الوزية	Lg	*Wâdy el Wazîyeh.* The valley of the geese.
وادي الزريب	Lh	*Wâdy ez Zureib.* Valley of the water-pipe.
وعرة صفورية	Lg	*Wâret Seffûrieh.* The rugged rocks of Sepphoris.
وعرة شنا	Mg	*Wâret Shenna.* Rocks of Shennah. p.n.
الواسط	Ji	*El Wâsit.* The middle.
الوسطاني	Kg	*El Wustâny.* The central.
يافا	Ni	*Yâfa.* Yâfa. Heb. יָפִיעַ Japhia.
ياجور	Kh	*Yâjûr.* Yâjûr. p.n.
زبدة العلالي	Mi	*Zebdat el 'Alâly.* Zebdat el 'Alâly. p.n.
الزور	Ji	*Ez Zôr.* The low ground.

SHEET VI.

العبيدية Qi *El 'Abeidiyeh.* El 'Abeidiyeh. p.n.; from *'Obeid,* 'a little slave' (but perhaps connected with the Biblical name Obadiah).

عيلبون Og *'Ailbûn.* 'Ailbun. p.n.; from علب 'hard, rocky ground.'

عين ابى حامد Qh *'Ain Abu Hâmed.* Abu Hâmed's spring. p.n.

عين ابى زينة Rg *'Ain Abu Zeineh.* The spring of Abu Zeineh. p.n.; meaning 'adorned.'

عين عيلبون Og *'Ain 'Ailbûn.* The spring of 'Ailbûn; see above.

عين علم Pi *'Ain 'Aulem.* The spring of 'Aulem. p.n.; from علم 'a mountain,' 'signpost,' &c.; cf. *Alma, Almaniyeh,* &c.

عين البيضآء Pi *'Ain el Beida.* The white spring.

عين بصوم Pi *'Ain Bessûm.* The spring of Bessûm. p.n.; perhaps from بقة for بقة 'a marsh'; see *'Ain el Basseh,* p. 39.

عين الداية Oi *'Ain ed Dâyeh.* The spring of the nurse.

عين الدالفة Ng *'Ain ed Dellâfeh.* The spring of dropping water.

عين الضيعة Og *'Ain ed Diâh.* The spring of the farm.

عين ايوب Qg *'Ain Eyûb.* The spring of Job.

عين الفولية البردة Qh *'Ain el Fûliyeh.* The spring of the bean-field; also called *'Ain el Bardeh,* the cold spring.

عين الحمام Ph *'Ain el Hamâm.* The spring of the pigeons.

عين حصونة Qi *'Ain Hassûneh.* The spring of the Hassûn family. p.n. Hassûn is a name borne by several Armenian families in Syria.

عين البختة Ni *'Ain el Jikleh.* The spring of the jackal.

عين الجنان Ni *'Ain el Jinnân.* The spring of the gardens.

عين الجيزان Oi *'Ain el Jîzân.* The spring of the walnuts.

عين الجوزة Ni *'Ain el Jôzeh.* The spring of the walnut-tree.

R

عين الجواني   Pi   '*Ain el Júâby.* The spring of the hollows or depressions.

عين قانا   Ni   '*Ain Kâna.* The spring of Kâna; *see Khûrbet Kânâ,* p. 112.

عين كتب   Ph   '*Ain Katab.* The spring of Katab. كتب means 'he wrote'; but it should probably be كثب *Ketheb,* 'a hillock or sand-heap.'

عين كفركنا   Ni   '*Ain Kefr Kenna.* The spring of Kefr Kenna. q.v.

عين الكلب   Qi   '*Ain el Kelb.* The spring of the dog.

عين الكلبة   Qh   '*Ain el Kelbeh.* The spring of the bitch.

عين كرازة   Qg   '*Ain Kerâzeh.* The spring of Kerâzeh. p.n.; perhaps Chorazin.

عين كحالة   Pg   '*Ain Kohâleh.* The spring of manganese.

عين لافى   Pi   '*Ain Lâfy.* The spring of Lâfy. p.n.; from لفا 'to defraud.'

عين المدى   Oi   '*Ain el Mady.* The spring of the Mada, a tree like the *ghadhâ;* the word also means 'limit,' 'range.'

عين ماحل   Oi   '*Ain Mâhil.* The spring of the barren land.

عين المنصورة   Og   '*Ain el Mansûrah.* The spring of el Mansûrah. q.v., p. 9.

عين المدورة   Qg   '*Ain el Mudauwerah.* The round spring.

عين المغرى   Pi   '*Ain el Mughrah.* The spring of cold water.

عين المتطشطش   Ph   '*Ain el Mutatŭshtŭsh.* The spring of sprinkling.

عين نصر الدين   Qh   '*Ain Nasr ed Din.* The spring of Nasr ed Din. p.n.; meaning 'the defender of the faith.'

عين النبى شعيب   Ph   '*Ain en Neby Shâib.* The spring of the prophet Shuáib (Jethro).

عين النجيمية   Og   '*Ain en Nejeimiyeh.* The spring of the Nejeimiyeh. Family name, from نجم 'a star.'

عين الربضية   Pg   '*Ain er Rŭbŭdiyeh.* The spring of the lair.

عين السمالية   Ni   '*Ain esh Shemâliyeh.* The northern spring.

عين السفلى   Pi   '*Ain es Sifla.* The lower spring.

عين السخنة   Ph   '*Ain es Sokhneh.* The warm spring.

عين الصفصافة   Pi   '*Ain es Sûfsâfeh.* The spring of the osier-willow.

عين الصرار   Ph   *'Ain es Sŭrár*. The spring of pebbles.

عين الطبل   Og   *'Ain et Tabil*. The spring of the drum.

عين التينة   {Pg Qg}   *'Ain et Tineh*. The spring of the fig-tree.

عين تورعان   Oh   *'Ain Tŏrán*. The spring of Torán. Perhaps from ترعة 'an outlet for water.' The form is Syriac.

عنين   Qi   *'Anin*. 'Anin. p.n.

عرب بني فهيد   {Oi Pi}   *'Arab Beni Feheid*. The Arabs of the sons of Feheid (the little panther).

عرب الدلايكة   {Ph Pi Qh Qi}   *'Arab ed Delaikeh*. The Arabs of ed Delàikeh. p.n.

عرب لهيب الشمالية   {Qg Rq}   *'Arab Luheib esh Shemálneh*. The northern branch of the Arabs of the Luheib tribe.

عرب السمكية   Qg   *'Arab es Semakiyeh*. The Semakiyeh (fisher) Arabs.

عرب السعيد   Og   *'Arab es Suáâid*. The Suââiyed Arabs. p.n.

عرب الصبيح   Oi   *'Arab es Subeih* (2). The Subeih Arabs. p.n.

عرب السويد   Qg   *'Arab es Suiyad*. The Suiyad Arabs. p.n.

عرب السميرية   Qg   *'Arab es Sumeiriyeh*. The Sumeiriyeh Arabs. p.n.

عرب التلاوية   {Pg Qg}   *'Arab et Tŭlláwiyeh*. The Tŭllâwiyeh Arabs. p.n.

عرب الوهيب   Pg   *'Arab el Waheib*. The Wuheib Arabs.

عرب الزنغرية   Qg   *'Arab ez Zenghariyeh*. The Zingari (gipsy) Arabs.

ارض العريض   Pi   *Ard el 'Arid*. The land of the broads.

ارض البوبرية   Pi   *Ard el Bŭberiyeh*. The land of the Bŭberiyeh. Family name; perhaps from بر 'a lion.'

ارض البرنس   Ph   *Ard el Bŭrnŭs*. The land of the prince. The name by which Count Renaud, of Kerek, was known to the Saracens.

ارض الدارون   Oi   *Ard ed Dárŭn*. The land of withered herbage; perhaps from Heb. דרום 'the south.'

ارض الكرك   Qi   *Ard el Kerak*. The land of the fortress.

ارض المنصورة   Qi   *Ard el Mansúrah*. The land of el Mansúrah. q.v., p. 9.

ارض المرادم	Pi	*Ard el Marâdem.* The land of ramparts, barriers, or demolitions.
ارض المغائر	Pi	*Ard el Mugheiyir.* The land of the caves.
عرابة البطوف	Oq	*'Arrâbet el Buttauf.* The steppe or plateau of the Buttauf. p'n.
العروى	Pi	*El 'Arûry.* El 'Arûry. p.n.; signifying 'mad,' or 'possessed by devils.'
عيون البيجان	Pi	*'Ayûn el Bijân.* The spring of Bijân. p.n.; meaning 'one who cannot keep a secret.'
عيون البساس او البصاص	Pi	*'Ayûn el Busâs.* The spring of the swamps.
عيون البخريع او تل الناعم	Qi	*'Ayûn el Kharâa.* The springs of the castor-oil plant; also called *'Ayûn Tell en Nââm.* The springs of the mound of the soft soil.
عيون موسى	Ni	*'Ayûn Mûsa.* The springs of Moses.
عيون العشة	Rg	*'Ayûn el 'Oshsheh.* The springs of the nest.
عيون الشعين	Oi	*'Ayûn esh Shâin.* The springs of Shâin. p.n.; perhaps Shihon.
عيون يما	Qi	*'Ayûn Yemma.* The spring of Yemma. p.n.
العزير	Nh	*El 'Azeir.* El 'Azeir. p.n.; Ezra is so called in Arabic.
باب البوابة	Qh	*Bâb el Buwâbeh.* The gate of the gate-keepers.
باب المرصد	Og	*Bâb el Mersed.* The gate of the observatory.
باب النقب	Pg	*Bâb en Nukb.* The gate of the pass.
باب التم	Qi	*Bâb et Tumm.* The gate of the mouth (*tumm,* vulgar Arabic for *fumm*).
البعينة	Oh	*El Bâineh.* El Bâineh. p.n.; the באינה of the Talmud.
بحر طبرية	Qd Rg Rh Ri	*Bahr Tûbarîya.* The lake of Tiberias.
بين التلول	Qg	*Bein et Telûl.* Between the mounds.
بيت جن	Pi	*Beit Jenn.* The house of the Genii (but it is probably Hebrew, meaning gardens).
بئار السبيل	Ph	*Biâr es Sebîl.* The wells by the way-side. *Sebîl* also means a way-side drinking fountain.

بئر العبيرة	Qg	*Bîr el 'Ajra.* The well of the projection.
بئر العرجا	Oh	*Bîr el 'Arja.* The well of the lame woman.
بئر بيين	Oi	*Bîr Beiyîn.* The well of Beiyin. p.n. (conspicuous).
بئر اربد	Ph	*Bîr Irbid.* The well of Irbid (Arbela).
بئر كبشانى	Oi	*Bîr Kibshâny.* The well of Kibshâni. p.n.; from كبش 'a ram.'
بئر مسلاخيت	Oh	*Bîr Mûslâkhit.* The well of Mûslâkhit. p.n.; perhaps from سلخ 'to strip,' 'to flay.'
بئر المزقة	Ph	*Bîr el Muzekka.* The well of Muzekka. p.n. (torn or rent in pieces).
بئر النويرية	Pg	*Bîr en Nûeirîyeh.* The well of the gipsies.
بئر سبانا	Pg	*Bîr Sebâna.* The well of Sebâna. p.n.
بركة على الظاهر	Qg	*Birket 'Aly edh Dhâher.* The pool of 'Aly ed Dhâher. p.n.; the son of the celebrated Dhâher el 'Amr.
بركة عرابة	Ng	*Birket Arrâbeh.* The pool of the carriage.
بركة البيضآء	Og	*Birket el Beida.* The white pool.
بركة الكرم	Ph	*Birket el Kûrm.* The pool of the vineyard.
بركة المرج	Og	*Birket el Merj.* The pool of the meadow.
برج نيات	Ph	*Burj Neiât.* The tower of Neiât. p.n.; meaning 'intentions.'
دامية	Pi	*Dâmieh.* Damieh. p.n. (Adami).
دبورية	Oi	*Debûrieh.* Deburiyeh. p.n. (Dabaritha, Daberath).
دبة المغر	Oh	*Debbet el Mughr.* The tract or plain of the caves, or of red ochre.
دبة الزعتر	Og	*Debbet ez Zâter.* The tract or plain of thyme.
الدير	Pi	*Ed Deir.* The convent.
دير حنا	Og	*Deir Hanna.* The convent of St. John.
ظهرة كبشانى	Oi	*Dhahret Kibshâny.* The ridge of Kibshâni. p.n.; see above, *Bir Kibshâny.*
ظهر السور	Ph	*Dhahr es Sûr.* The ridge of the wall.
الغوير	Qg	*El Ghuweir.* The little lowland, depression, or hollow.

الحدثة   Pi   *El Hadetheh.* El Hadetheh. p.n.; meaning 'new.'

حجر النملة   Qh   *Hajr en Nemleh.* The ant's stone.

حجارة النصارى   Ph   *Hajâret en Nusâra.* The Christian's stone.

حقول المغارة   Ph   *Hakûl el Mŭghârah.* The fields of the caves.

حنانة القسيس   Qh   *Hannânet el Kŭssîs.* The cistern of the priest; *see Bîr el Hanâny*, p. 107.

حطين   Ph   *Hattîn.* Hattin. p.n.

حام ابرهيم باشا   Qh   *Hŭmmâm Ibrahîm Bâsha.* The bath of Ibrahim Pasha.

حام سيدنا سليمان   Qh   *Hŭmmâm Sîdna Suleimân.* Bath of our lord Solomon.

اكسال   Ni   *Iksâl.* Iksal. p.n. (Chesulloth).

جبل عبهرية   Og   *Jebel 'Abharîyeh.* The mountain of the mock orange (*Styrax officinalis*).

جبل ابى عدوّر   {Nq} {Ph}   2. *Jebel Abu Mudauwer.* The mountain with the round place.

جبل ابى جمال   Ng   *Jebel Abu Jemâl.* The hut of Abu Jemâl. p.n.; meaning father of beauty.

جبل البلانة   Pg   *Jebel el Bellâneh.* The mountain of Bellâneh; a thorny plant (*Poterium spinosum*).

جبل حبقوق   Pg   *Jebel Habkûk.* The mountain of Habbakuk.

جبل حزور   Og   *Jebel Hazzûr.* The mountain of Hazzûr. p.n.; *cf.* Heb. חצר 'a court.'

جبل الكمانة   Og   *Jebel el Kummâneh.* The mountain of el Kemmâneh; perhaps from كمين 'an ambuscade.'

جبل الموبرة   Og   *Jebel el Môberah.* The mountain of the quarry.

جبل نمرين   Oh   *Jebel Nimrîn.* The mountain of Nimrin. (Heb. *Nimrim*, abundant water.)

جبل الرجوة   Nh   *Jebel er Rahweh.* The mountain with the knoll.

جبل سعد   Pg   *Jebel Sâd.* The mountain of Sâd. p.n.

جبل الشجرة   Oh   *Jebel esh Shejerah.* The mountain of the tree.

جبل السيح   Ni   *Jebel es Sîh.* The mountain of the stream.

جبل طويل   Pg   *Jebel Tawîl.* The high mountain.

جبل الطيارات   Og   *Jebel et Teiyárát.* The mountain of the tops (boy's play-thing).

جبل الطور   Oi   *Jebel et Tôr.* The mountain of ' the mount ' (Tabor).

جبل تورعان   Oh   *Jebel Tôr'án.* See *'Ain Tôr'án,* p. 123.

جسر السد   Qi   *Jisr es Sidd.* The bridge of the dam.

جورة العين   Qg   *Jûrat el 'Ain.* The hollow of the spring.

الجحيفة   Qg   *El Juheifeh.* The torrent, or 'water remaining at the bottom of a cistern.'

قبر يعقوب الصديق   Og   *Kabr Yákûb es Saddik.* The tomb of St. James the Just.

قآء ناصرة $\begin{Bmatrix} Ni \\ Oi \\ Oh \end{Bmatrix}$ *Kada Násirah.* The district of Nazareth ; *see* " Memoirs," p. 4.

قآء طبرية   Ph   *Kada Tûbariya.* The district of Tiberias.

تعتمية   Qh   *Kákáiyeh.* Place where magpies abound. The word also means 'rumbling,' 'clanking.'

القنقوزة   Ph   *El Kankûzah.* El Kankûzeh. p.n.

كفر كما   Pi   *Kefr Kama.* The village of truffles.

كفر كنا   Oi   *Kefr Kenna.* The village of Kenna. p.n.

كفر سبت   Pi   *Kefr Sabt.* The village of Sabbath. (Caphar Sobthi. Tal.)

الكرخانة   Qh   *El Kerkhâneh.* The warehouse or workshop.

الخلدية   Oi   *El Khalladiyeh.* El Khalladiyeh. p.n. ; from خلد 'to be perpetual.'

خليل جعفر   Oh   *Khaleil Jáfr.* Jáfr's little dell. *Jaafer* is a common Arabic name ; it means 'a rivulet.'

خلة الجريبة   Qg   *Khallet el Jureibeh.* The dell of sown land, or of the little leathern bag.

خلة النجار   Pg   *Khallet en Nejjâr.* The carpenter's dell.

خلة النصارى $\begin{Bmatrix} Pg \\ Qg \end{Bmatrix}$ *Khallet en Nására.* The Christian's dell.

خلة سعدية   Ph   *Khallet Sádiyeh.* The dell of the Sádiyeh. p.n ; of an order of dervishes.

خلة السماك   Qg   *Khallet es Semmák.* The dell of the fishmonger.

الخان   Ph   *El Khân.* The Caravanserai.

خان منيا   Qg   *Khân Minia.* The Caravanserai of Minia.

خان التجار   Oi   *Khân et Tujjâr.* The Caravanserai of the merchants; also called *Sûk el Khân*, market of the Caravanserai.

الحانوق   Ng   *El Khânûk.* The gorge.

الخزامة   Qg   *El Khazâmeh.* The (camel's) nose ring; but perhaps it is الخزامى, the name of a red wild flower. (*cf.* سوق الخزامين *Sûk el Khazâmain*, at Mecca.)

الخربة   Oh   *El Khûrbeh.* The ruin.

خربة العيثة   Ph   *Khûrbet el 'Aitch.* The ruin of the lofty (mountain).

خربة عريثا   Oi   *Khûrbet 'Arbîtha.* The ruin of 'Arbitha. p.n.

خربة ابى شوشة   Qg   *Khûrbet Abu Shûsheh.* The ruin of Abu Shûsheh. p.n.; meaning 'father of,' wearing 'a top knot.'

خربة ابى زينة   Rg   *Khûrbet Abu Zeineh.* The ruin of Abu Zeineh. p.n.; meaning 'father of adornment'; also called *Shûnet esh Shemâlneh*, the granary of the Shemâlneh Arabs.

خربة بيين   Oh   *Khûrbet Beiyîn.* The ruin of Beiyin. p.n.

خربة بصوم   Pi   *Khûrbet Bessûm.* The ruin of Bessûm; from بصة a marsh. Heb. בצה.

خربة فليح   Oh   *Khûrbet Felih* (2). The ruin of Felih. p.n.; root فلح 'to cultivate.'

خربة حزور   Og   *Khûrbet Hazzûr.* The ruin of Hazzûr (Hazor); *see Hazzûr*, p. 72.

خربة الحسينة   Ng   *Khûrbet el Hoseinîyeh.* The ruin of the Hoseiniyeh, family name for Husein, son of 'Ali.

خربة اربد   Ph   *Khûrbet Irbid.* The ruin of Irbid. p.n.; the ancient Arbela.

خربة قديش   Qi   *Khûrbet Kadîsh.* The ruin of Cadesh.

خربة قيسرية   Pg   *Khûrbet Kaisârîyeh.* The ruin of Cesaraea.

خربة قيسرون   Oh   *Khûrbet Kaisarûn.* The ruin of Cesareum.

خربة القنيطرة   Qh   *Khûrbet el Kaneitreh.* The ruin of the little arch.

خربة قسطة   Oi   *Khûrbet Kastah.* The ruin of el Kastah. Perhaps from قسطا *Kastâ*, the traditional name of the son of St. Luke.

خربة كنا   Nh   *Khûrbet Kenna.* The ruin of Kenna. p.n.

خربة كرازة   Qg   *Khŭrbet Keràzeh.* The ruin of Keràzeh. p.n. Chorazin.

خربة الكرك   Qi   *Khŭrbet el Kerak.* The ruin of the fortress.

خربة كبشاني   Oi   *Khŭrbet Kibshàny.* The ruin of Kibshàny. p.n.

خربة الكور   Qg   *Khŭrbet el Kûr.* The ruin of the forge; but perhaps for *Kuwar,* village.

خربة مدين   Ph   *Khŭrbet Madîn.* The ruin of Madîn; perhaps Madon.

خربة مامليا   Pg   *Khŭrbet Màmelia.* The ruin of Màmelia. p.n.

خربة المنصورة   Qi   *Khŭrbet el Mansûrah.* The ruin of Mansûrah; *see* p. 9.

خربة مسكنة   Oh   *Khŭrbet Meskeneh.* The ruin of the habitation.

خربة منيا   Qg   *Khŭrbet Minia.* The ruin of Minia. p.n.

خربة مغير   Oi   *Khŭrbet Mugheiyir.* The ruin of the little cave.

خربة مشتاة   Ng   *Khŭrbet Mushtâh.* The ruin of the winter houses (for cattle).

خربة مسلاخيط   Oh   *Khŭrbet Mŭslàkhît.* The ruin of Mŭslàkhit. p.n.; *see* p. 125.

خربة المزقة   Ph   *Khŭrbet el Muzekka.* The ruin of el Muzekka; torn, rent in pieces.

خربة نتف   Og   *Khŭrbet Nàtef.* The ruin of plucking. نتف pl. of نتفة, herbs plucked with the fingers. But *see en Nàtef,* p. 115.

خربة نجيمية   Og   *Khŭrbet Nejeimîyeh.* The ruin of the little star.

خربة النويرية   Pg   *Khŭrbet en Nûeirîyeh.* The ruin of the gypsies.

خربة العريمة   Qg   *Khŭrbet el 'Oreimeh.* The ruin of the heap or dam.

خربة عشة   Rg   *Khŭrbet 'Oshsheh.* The ruin of the nest.

خربة ربضية   Pg   *Khŭrbet Rŭbŭdîyeh.* The ruin of the lair.

خربة رويس الحمام   Ng   *Khŭrbet Ruweis el Hamàm.* The head-land or hill-top of the pigeon.

خربة سعد   Ph   *Khŭrbet Sàd.* The ruin of Saad; *see Jebel Sàd,* p. 126.

خربة سارة   Pi   *Khŭrbet Sàrah.* The ruin of Sàrah. p.n.

خربة سبانة   Pg   *Khŭrbet Sebàna.* The ruin of Sebàna. p.n.

خربة سيادة   Qi   *Khŭrbet Seiyâdeh.* The ruin of the Saiyids, *i.e.,* descendants of Fatima, daughter of Mohammed and wife of 'Ali.

S

خربة سلامة	Og	*Khŭrbet Sellâmeh.* The ruin of Sellâmeh. p.n.
خربة سرجونية	Pi	*Khŭrbet Serjûnieh.* The ruin of Serjûnieh. p.n.
خربة شعرة	Pi	*Khŭrbet Sh'ârah.* The ruin of Sh'ârah. Thick foliage, perhaps the Beth Shearim of the Talmud.
خربة شمسين	Qi	*Khŭrbet Shemsîn.* The ruin of Shemsin.
خربة سيرين	Pg	*Khŭrbet Sîrîn.* The ruin of folds.
خربة ام العلق	Pi	*Khŭrbet Umm el 'Alak.* The ruin with the leeches.
خربة ام العمد	Oh	*Khŭrbet Umm el 'Amed.* The ruin with the columns.
خربة ام الغنم	Oi	*Khŭrbet Umm el Ghanem.* The ruin with the sheep.
خربة ام جبيل	Oi	*Khŭrbet Umm Jebeil.* The ruin with the little mount.
خربة الوريدات	Ph	*Khŭrbet el Wereidât.* The ruin of the little roses, or of going down to water.
خربة الخريبة	Qg	*El Khŭreibeh.* The little ruin.
القلعة	{Qi / Qh}	*El Kŭlâh.* The castle.
قلعة الغول	Qi	*Kŭlât el Ghûl.* The castle of the ghoul (demon).
قلعة ابن معن	Ph	*Kŭlât Ibn Mân.* The ruin of Ibn Mân; the celebrated Syrian chief.
قلعة الشونة	Pg	*Kŭlât esh Shûneh.* The castle of the granary.
قلعة السودآ	Ph	*Kŭlât es Sôda.* The castle of black (basalt).
قرن حطين	Ph	*Kŭrn Hattîn.* The peak of Hattin.
قصر بنت الملك	Qh	*Kŭsr Bint el Melek.* The palace of the king's daughter.
لوبية	Ph	*Lûbieh.* Lûbieh. p.n.; لوبيا, French beans.
معذر	Pi	*Mâdher.* Mâdher. p.n.
مخاضة ام جونية	Qi	*Makhâdet Umm Jûnieh.* The ford of the grouse.
مخاضة ام المعزة	Qi	*Makhâdet Umm el Mâza.* The ford of the goats.
مخاضة ام سدرة	Rg	*Makhâdet Umm Sidreh.* The ford of the lotus tree.
المنصورة	Og	*El Mansûrah.* El Mansûrah; see p. 9.
معصرة عيسى	Qg	*M'âsret 'Aîsa.* The wine-press of Jesus.
المعترضة	Ph	*El Mâterideh.* The opposing, or hindrance.

المطمومة Ni *El Mathûmeh.* El Mathûmeh. p.n.; 'full,' 'complete.'

مزار شيخ محمد Oi *Mazâr Sheikh Muhammed.* The shrine of Sheikh Mohammed.

مدينة الايكة Ph *Medinet el Aikeh.* The City of the Grove. This is the name of the city to whose inhabitants the 'prophet Shuaib,' or Jethro, was sent, according to the Kor'ân; *see* my translation, vol. i, p. 249.

المجدل Qh *El Mejdel.* The watch-tower.

المنارة Qh *El Menârah.* The light-house.

المرحان Oh *El Merhân.* The folds.

مرج الباب Qg *Merj el Bâb.* The meadow of the gate.

مرج الذهب Oh *Merj edh Dhaheb.* The meadow of gold.

مرج حطين Ph *Merj Hattîn.* The meadow of Hattin. p.n.

مرج السنبل Oh *Merj es Sûnbûl.* The meadow of the ears of corn.

مرج الثبات Qg *Merj eth Thebât.* The plain of stability.

مسيل الحسماس Ng *Meseil el Hosmâs.* The stream of el Hosmâs. p.n.

مسحة Pi *Mes-hah.* The place of unction.

المشهد Ni *El Mesh-hed.* The shrine or place of martyrdom.

مشرة Pi *Meshârah.* Meshârah. p.n.; signifying covered with brush-wood or thick foliage; it also means a monument.

مغاير القرود Qg *Mûghâir el Kurûd.* The cave of the apes (or goblins).

المغار Og *El Mûghâr.* The caves. The Mearah of Josephus.

مغارة الافرنجي Pg *Mûgharet el 'Afranjy.* The cave of the Franks.

مغارة الاميرة Qg *Mûghâret el Emîreh.* The cave of the princess.

مغارة معطى Ph *Mûghâret Mâty.* The cave of Mâty. p.n.; meaning 'given.'

مغارة الزطية Qg *Mûghâret ez Zûttiyeh.* The cave of the gypsies; *see* 'Aila ez Zutt, p. 17.

المهر Qg *El Muhr.* The foal.

المنطار Pg *El Mûntâr.* The watch-tower.

ناصر الدين Qh *Nâsr ed Din.* Nasir ed Din. p.n.

النبى يشوع بن   Ph   *Neby Eshûa Ibn Amîn.* The prophet Elisha (Joshua), the
امين   son of Amin (the faithful one).

النبى حبقوق   Pg   *Neby Habkûk.* The prophet Habbakuk.

النبى شعيب   Ph   *Neby Shâib.* The prophet Shâib (Jethro); *see Medinat el
                                 Aikeh,* p. 131.*

النبى يونس   Ni   *Neby Yûnis.* The prophet Jonah.

نمرين   Ph   *Nimrîn.* Nimrin. p.n.; meaning well-watered. Heb.
                                 Nimrim.

رب عكيبا   Qh   *Rabbi 'Akîba.* Rabbi Akibah ; he was the great supporter
                                 of Barcochebas in his rebellion. *See* Besant and Palmer's
                                 "Jerusalem," p. 51.

رب كهنا   Qh   *Rabbi Kahna.* Rabbi Kahna. (Cohen, priest.)

رب ماهير   Qh   *Rabbi Maîr.* Rabbi Meyer.

رب موشى بن   Qh   *Rabbi Mûsha Ben Maimûn.* Rabbi Mûsa, the son of
ماعون   Maimûn (the celebrated Maimonides).

رأس البرج   Qi   *Râs el Burj.* The top of the tower.

رأس حزوة   Og   *Râs Hazweh.* The headland or hill-top of Hasweh. p.n.

رأس الكبيرة   Qi   *Râs el Kebîreh.* The large headland or hill-top.

رأس كرومان   Og   *Râs Krûmân.* The hill-top or headland of the vine-
                                 yards.

رأس المرماة   Qi   *Râs el Mermâleh.* The headland or hill-top of the sandy
                                 place.

رأس النخلة الصغير Qi   *Râs en Nukhleh es Saghîr.* The small headland or hill-top
                                 of the palm.

رأس الطاقية   Pg   *Râs et Tâkiyeh.* The hill-top or headland of the cornice
                                 (ledge or projecting ridge).

رأس التين   Oh   *Râs et Tîn.* The hill-top or headland of the fig.

روض غزال   Pi   *Raud Ghûzâl.* The meads of gazelles.

روض تفاح   Pi   *Raud Tuffâh.* The meads of apple.

الرينة   Ni   *Er Reineh.* Er Reineh. p.n. (dirt).

رجم العجمى   Ni   *Rûjm el 'Ajamy.* The cairn of the Persian

رجم شمر   Oh   *Rûjm Shummer.* The cairn of wild fennel.

رقبة الخان | *Qg* | *Rukbet el Khân.* The esplanade of the Caravanserai.

رمانة | Nh | *Rummânch.* The pomegranate (Rimmon).

سهل الاحما | { Ph
{ Pi }
(Qi) | *Sahel el Ahma.* The level plain (unapproachable plain).

سهل عرابة | Nq | *Sahel 'Arrâbeh.* The plain of the 'Arrâbeh (chariot).

سهل العطل | Oh | *Sahel el 'Atl.* The vacant plain.

سهل البطوف | { Nh }
{ Oh } | *Sahel el Buttauf.* The plain of el Buttauf.	p.n.

سهل الحيطا | *Qg* | *Sahel el Heit.* The plain of the wall.

سهل الطبوية | Pg | *Sahel et Tahâwîyeh.* The plain of the cooks.

سهل تورعان | *Oh* | *Sahel Tôr'ân.* The plain of Tôr'ân; *see* p. 123.

سهل ودعان | Pi | *Sahel Wad'ân.* The plain of deposit.

سارونا | Pi | *Sárôna.* Sarônâ. p.n.; Lashuron.

سمخ | Qi | *Semakh.* Semakh; *see Khûrbet Semakh,* p. 50.

الشعبة | *Pi* | *Esh Shâbeh.* The spur.

الشاغور | Og | *Esh Shâghûr.* The Shâghûr; *see* p. 28.

شفا حريدون | *Qi* | *Shefa Hureidânch.* The plain of the lizards.

شفا ام العلق | Pi | *Shefa Umm el 'Alak.* The edge or margin with the leeches.

شفا ام زعيطا | Qi | *Shefa Umm Zâît.* The margin or edge of the stream.

شيخ ابى شالة | *Qg* | *Sheikh Abu Shâleh.* Sheikh Abu Shâleh; the man with the shawl.

شيخ ابى زعرورة | Qi | *Sheikh Abu Zârûrah.* Sheikh Abu Zârûrah; father of hawthorn. Sheikh is here, like Weli, probably used for 'a saint's tomb.'

شيخ على ذناب | Qg | *Sheikh 'Aly Diâb.* Sheikh Ali Diâb. p.n.; 'wolves.' A common Arabic name.

شيخ على الصياد | Qg | *Sheikh 'Aly es Seiyâd.* Sheikh Ali, the fisherman. p.n.

شيخ حمدان | Oi | *Sheikh Hamdân.* Sheikh Hamdân. p.n.

شيخ حسن | *Qg* | *Sheikh Hasan.* Sheikh Hasan. p.n.

شيخ قدوم | Qh | *Sheikh Kaddûm.* Sheikh Kaddûm; *cf.* Cadmus.

شيخ المجدوب　Nc　*Sheikh el Mejdûb.* Sheikh el Mejdûb. p.n.; meaning parched or barren land.

شيخ المزغيت　Qi　*Sheikh el Mezeighît.* Sheikh el Mezeighit. p.n.; perhaps مزغند 'angry.'

شيخ محمد العجمى　　*Sheikh Muhammed el 'Ajamy.* Sheikh Mohammed, the Persian.

شيخ محمد القلاعى　Rg　*Sheikh Muhammed el Kŭlâây.* Sheikh Mohammed, of the castle.

شيخ محمد الويبدانى　Qg　*Sheikh Muhammed el Weibdânĕh.* Sheikh Mohammed el Weibdâni. p.n.

شيخ ناشى　Pg　*Sheikh Nâshy.* Sheikh Nâshy. p.n.; growing.

شيخ الرسلان　Qh　*Sheikh er Rŭslân.* Sheikh er Rŭslân. p.n.; from ارسل 'to send'; *cf.* رسول 'apostle.'

شيخ شباب　Qi　*Sheikh Shehâb.* Sheikh Shehâb. p.n.; meaning a shooting star, a common Arabic name; the full form is *Shehâb ed dín,* the shooting star or firebrand of religion. It alludes to the Mohammedan legend that shooting stars are brands with which the angels pelt the devils when they are found eavesdropping at the gates of heaven; *see* my Kor'ân, vol. i, p. 51, note 2.

الشجيرة　Oh　*Esh Shejerah.* The tree.

شجرة الكلب　Qi　*Shejeret el Kelb.* The tree of the dog.

شجرات المباركات　Qg　*Shejerât el Mubârakât.* The blessed trees.

شجرة المستراح　Ph　*Shejeret el Musterâh.* The tree of the resting-place.

شجرة الصالحان　Qh　*Shejeret es Sâlchân.* The tree of Sâlchân. p.n.

سن النبرة　Qi　*Sinn en Nâbrah.* The tooth of the eminence (Sennabris).

الصيرة　Oh　*Es Sîreh.* The sheep-fold.

ست سكينة　Qh　*Sitti Sekîneh.* Lady Sekineh. p.n.; the word means 'tranquillity,' but is used in the Kor'ân for the Hebrew 'Shechina.' *See* my translation of the Kor'ân, vol. i, p. 38, note 2.

طاحونة ابى شوشة　Qg　*Tâhûnet Abu Shûsheh.* The mill of the father of the top-knot.

طاحونة الآغا　Qi　*Tâhûnet el 'Âghâ.* The mill of the 'Âghâ; a Turkish title.

طاحونة العرب	Qi	*Tâhûnet el 'Arab.* The mill of the Arab.
طاحونة العرابية	Og	*Tâhûnet el 'Arrâbîyeh.* The mill of the carriage.
طاحونة البستان	Pg	*Tâhûnet el Bestân.* The mill of the garden.
طاحونة الدولة	Qi	*Tâhûnet ed Daulch.* The mill of the Government or soldiery.
طاحونة الفولية	Qh	*Tâhûnet el Fûlîyeh.* The mill of bean-sellers.
طاحونة الجديدة	Pg	*Tâhûnet el Judeiyidch.* The mill of the dykes ; *see Wâdy el Judeiyidch,* p. 11.
طاحونة القبو	Qg	*Tâhûnet el Kabû.* The mill of the vault.
طاحونة الليمونة	Pg	*Tâhûnet el Leimûnât.* The mill of the lemon-trees.
طاحونة المكيسة	Og	*Tâhûnet el Mukeiseh.* The mill of the drainage channels.
طاحونة النبي شعيب	Qg	*Tâhûnet en Neby Shâib.* The mill of the prophet Shâib (properly Shuâib), *i.e.*, Jethro.
طاحونة نجمة الصبح	Qg	*Tâhûnet Nejmet es Subh.* The mill of the morning star.
طاحونة السلامية	Og	*Tâhûnet es Sellâmîyeh.* The mill of the Selâmîyeh. p.n. ; applied to the natives of Baghdad, which is called مدينة السلام *Medinat es Selâm,* the city of places.
طاحونة الشرار	Oi	*Tâhûnet esh Sherrâr.* The mill of the sparks ' of the wicked man.'
طاحونة السكر	Qi	*Tâhûnet es Sukker.* The sugar-mill.
طاحونة الثوم	Pg	*Tâhûnet et Tôm.* The mill of garlic.
الطاطورة	Qh	*Et Tâtûrah.* Et Tâtûrah. p.n.
طواحين الفجاس	Qi	*Tawâhîn el Fejjâs.* The mills of the haughty oppressor.
الطابغة	Qg	*Et Tâbghah.* Et Tâbghah. p.n.
تل البطمة	Nh	*Tell el Butmeh.* The mound of the terebinth tree.
تل حوم	Qg	*Tell Hûm.* The mound of Hûm. p.n.
تل الهنود	Qg	*Tell el Henûd.* The mound of the Indians.
تل معون	Qh	*Tell Maûn.* The mound of Maûn. p.n. ; meaning ' provisions,' or perhaps ' the spring.'
تل مريبض	Qg	*Tell Mureibid.* The mound of the lairs.
تل الناعم	Qi	*Tell en Nâam.* The mound of soft soil.

تلول الجلمة	Oh	*Tellûl el Jalameh.* The mounds of the hill.
الطيرة	Ni	*Et Tîreh.* The fortress.
تورعان	Oh	*Tôr'ân.* Tor'ân. p.n.; *see 'Ain Tor'ân,* p. 123.
طبرية	Qh	*Tûbarîya.* Tiberias.
ام جونية	Qi	*Umm Jûnieh.* Umm Jûnieh. p.n.
ام القناطير	Qi	*Umm el Kanâtir.* The building with the arches.
وادى اغار	Oi	*Wâdy Aghâr.* The valley of caves or hollows.
وادى ابى فريوة	Pg	*Wâdy Abu Fureiweh.* The valley of Abu Fureiweh; *see 'Ain Furaiwiyât,* p. 1.
وادى ابى العميس	Ph	*Wâdy Abu 'l 'Amîs.* The valley of Abu 'l 'Amîs. p.n.
وادى عيلبون	Og	*Wâdy 'Ailbûn.* The valley of 'Ailbûn; *see 'Ailbûn,* p. 121.
وادى العين	Nq	*Wâdy el 'Ain.* The valley of the spring.
وادى عين الناطف	Og	*Wâdy 'Ain en Nâtef.* ⎫ The valley of the dropping; *see*
وادى عين ناطف	„	*Wâdy 'Ain Nâtef.* ⎬ *en Nâtef,* p. 115.
وادى عمود	Pg	*Wâdy 'Amûd.* The valley of the column.
وادى البركة	Oh	*Wâdy el Birkeh.* The valley of the pool.
وادى الدلافة	Nq	*Wâdy ed Dellâfeh.* The valley of dropping water.
وادى الذيب	Ph	*Wâdy edh Dhîb.* The valley of the wolf.
وادى فجاس	Qi	*Wâdy Fejjâs.* The valley of the haughty oppressor.
وادى الحمام	Ph	*Wâdy el Hamâm.* The valley of the pigeon; perhaps *hammâm,* warm bath.
وادى حرفيشة	Pg	*Wâdy Hurfîsheh.* The valley of Hurfîsheh. p.n.
وادى الجاموس	Qg	*Wâdy el Jâmûs.* The valley of the buffalo.
وادى الجرابان	Oh	*Wâdy el Jerâbân.* The valley of the plantations.
وادى الجزائر	Nq	*Wâdy el Jizâir.* The valley of the islands.
وادى قيسارية	Pg	*Wâdy Kaisâriyeh.* Cæsarea.
وادى قسطة	Oi	*Wâdy Kastah.* The valley of Kastah; *see Khûrbet Kastah,* p. 128.
وادى كفر كنا	Nh	*Wâdy Kefr Kenna.* The valley of the village of Kenna. p.n.

وادى كرازة Qg *Wády Kerázeh.* The valley of Kerázeh. p.n.; Chorazin.

وادى الخانوق Nh *Wády el Khánúk.* The valley of the gorge.

وادى الخشب Nq *Wády el Khasheb.* The valley of timber.

وادى الخرب Og *Wády el Khárb.* The valley of the ruins.

وادى الكروم Og *Wády el Kerúm.* The valley of vineyards.

وادى قصيب Qh *Wády Kúseib.* The valley of reeds.

وادى المدى Oi *Wády el Mady.* The valley of the mada tree. *See Ain el Mady,* p. 122.

وادى مامليا Pg *Wády Mámelia.* The valley of Mámelia. p.n.

وادى المعلقة Pi *Wády el Muállakah.* The valley of the overhanging rock.

وادى المدورة Qg *Wády el Mudauwerah.* The valley of the round (rocks).

وادى المغر Oh *Wády el Mughr.* The valley of caves; but perhaps from *mŭghr,* red ochre.

وادى المغرة Pi *Wády el Mughrah.* The valley of red ochre.

وادى محمد الخلف Ph *Wády Muhammed el Khalef.* The valley of Mohammed the successor. p.n.

وادى مزيبلة Pg *Wády Muzeibleh.* The valley of dung-heaps.

وادى الزكة Ph *Wády el Muzekka.* The valley of el Muzekka. *See* p. 129.

وادى الناشف Rg *Wády en Náshef.* The dry valley.

وادى النيلة Qg *Wády en Nileh.* The valley of indigo.

وادى الربضية Pg *Wády er Rŭbŭdiyeh.* The valley of the lairs.

وادى رمانة Nh *Wády Rummáneh.* The valley of the pomegranate trees.

وادى سعد Ph *Wády Sád.* The valley of Sád. p.n.

وادى سارونا Pi *Wády Sáróna.* The valley of Sárona. p.n.

وادى سلامة Og *Wády Sellámeh.* The valley of Sellámeh. p.n. (peace).

وادى شعارة Pi *Wády Sh'árah.* The valley of thick foliage. *See Khŭrbet Sh'árah,* p. 129.

وادى الشرق Og *Wády esh Sherk.* The eastern valley.

وادى شبابة Ph *Wády Shubbábeh.* The valley of young men.

وادى الشونة	Pg	*Wâdy esh Shûneh.* The valley of the barn.
وادى شومر	Pi	*Wâdy Shômer.* The valley of wild fennel.
وادى سيرين	Pg	*Wâdy Sîrin.* The valley of Sirin. p.n.
وادى التفاح	Pg	*Wâdy et Tuffâh.* The valley of apples.
وادى ام نيلة	Qg	*Wâdy Umm Nîleh.* The valley of indigo.
وادى الوبيدة	Qh	*Wâdy el Weibdeh.* The valley of the mountain ravine.
وادى الزيد	Oh	*Wâdy ez Zeid.* The valley of ez Zeid, a common Arabic name (signifying increase).
وادى زحلق	Rg	*Wâdy Zuhlûk.* The slippery valley.
وعر عتمة	Pg	*Wâr'Atmeh.* The rugged rocks of darkness.
وعرة الحوارنة	Ph	*Wâret el Hûârneh.* The rugged rocks of the cotton carders ; family name.
وعرة الصنوع	Pg	*Wâret es Senûâ.* The rugged rocks of Senûâ. p.n. ; handicraft.
وعرة السودآ	{Qg / Pq}	*Wâret es Sôda.* The black rugged rocks (basalt).
وطاة السروبة	Oh	*Watât es Serûby.* The low ground of flowing water.
ولى ابرهيم	Rg	*Weli Ibrahîm.* Saint (or the saint's tomb of) Abraham.
يافوق	Pg	*Yâkûk.* Yâkûk. p.n.
يما	Qi	*Yemma.* Yemma. p.n. ; Caphar Yama.
الزعرور	Ni	*Ez Zârûr.* The hawthorn.
الزاوية	Oh	*Ez Zâwieh.* The corner or hermitage.
زريبة الربضية	Qg	*Zeribet er Rŭbŭdîyeh.* The outlet of Rŭbŭdiyeh. q.v.

SHEET VII.

عبدون   Jk   *'Abdûn.* 'Abdûn. p.n.; connected with 'Abd, 'a slave.' Heb. עבדון 'a worshipper,' which occurs as a proper name several times in the Bible. *See Khŭrbet 'Abdeh,* p. 46.

عين العبال   Jj   *'Ain el 'Ajjâl.* The spring of the hastener, or 'the herd of cattle.'

عين العسل   Il   *'Ain el 'Asl.* The spring of honey.

عين القنطرة   Jj   *'Ain el Kantarah.* The spring of the arch.

عيون حيدرة   Jj   *'Ayûn Heiderah.* The springs of the declivity.

بئر ابي بازة   Jk   *Bîr Abu Bâzeh.* The well of Abu Bâzeh. p.n.; meaning 'of the hawks.'

بئر سليمان المرزوق   Il   *Bîr Suleimân el Marzûk.* The well of Suleimân el Marzûk. p.n. *El Marzûk* means 'provided for.'

بركة عين ام الفحمة   Il   *Birket 'Ain Umm el Fahmeh.* The pool of the spring where the charcoal is.

بركة عتا   Il   *Birket 'Ata.* The pool of 'Ata. p.n.

بركة البطيخ   Jl   *Birket el Battîkh.* The pool of the water-melon. *See Safed el Battîkh,* p. 32.

بركة بلاقيس   Il   *Birket Belâkîs.* The pool of Belâkîs. p.n.

بركة صفرا   Il   *Birket Sŭfra.* The yellow or empty pool.

البرج و قيل خربة طنطورة   Ij   *El Burj.* The tower; also called *Khŭrbet Tantûrah,* the ruin of the 'horn.' *See Tell et Tantûr,* p. 117.

دريمة (اوذريمة)   Jj   *Dreihemeh (or Dhreihemeh).* D'reihimeh. p.n.; meaning either 'small silver coins,' 'money,' or 'enclosed gardens.'

دولة النمرة   Il   *Dûkat en Nimreh.* The well-watered enclosure. *cf. Wâdy en Nimûr,* p. 37.

الحنانة   Ij   *El Hannâneh.* The trough. *See Bir el Hannâneh,* p. 107.

جزيرة العبال   Il   *Jeziret el 'Ajjâl; see* Sheet V, p. 110.

جزيرة الملاط   Ik   *Jeziret el Melât.* The island of Melât. q.v. below.

جزيرة المڤر   Jj   *Jeziret el Mŭkr.* The island of el Mŭkr; perhaps a mistake for معر 'red earth.' معڤر, as here written, means 'brine for salting fish.'

جزيرة النعمي   Jj   *Jeziret el Namy.* The island or promontory of en Namy. p.n.; perhaps ناعمة 'soft soil.'

جسر الزرقا   Ik   *Jisr ez Zerka.* The bridge of the River Zerkā.

الجليمة   Jj   *El Jleimeh.* The little hill. See *Khŭrbet el Jelameh*, p. 27.

قيصارية   Ik   *Kaisârieh.* Cæsarea.

كبارة   Jk   *Kebârah.* Kebâra. p.n.; from كبر 'to be great.' In vulgar Arabic it means 'capers.'

كفر لام   Jj   *Kefr Lâm.* The village of Lâm. p.n.

الكنيسة   Ik   *El Kenîseh.* The Church.

خروبة ابو شوشة   Jj   *Karrûbet Abu Shûsheh.* The locust tree (*Carob Ceratonia Siliqua*) 'with the tuft.'

الخشم   Jk   *El Khashm.* The Nose. A prominent precipice projecting over the plain.

خربة حيدرة   Jj   *Khŭrbet Heiderah.* The ruin of the declivity.

خربة ابريكتاس   Il   *Khŭrbet Ibreiktâs.* The ruin of Ibreiktâs. p.n.; perhaps from بريكةآس 'the myrtle pool.'

خربة منصور العتاب   Jk   *Khŭrbet Mansûr el 'Akâb.* The ruin of Mansûr (victory) of the rocks. Also called القبى el Kabâi, 'the domed,' and المعلقة el Mâlakah, 'the overhanging place,' being at the edge of the precipice.

القلعة   Ik   *El Kŭlâh.* The Castle.

الملاط   Ik   *El Melât.* El Melât. p.n.; meaning 'plastered.'

المزرعة   Jj   *El Mezrâh.* The sown land.

نهر الدفلة وقين   Ij   *Nahr ed Dufleh.* The River of the Oleander. Also called

نهر القراجة   Ik   *Nahr el Karâjeh.* The River of el Karâjeh. p.n.

نهر المفجير   Il   *Nahr el Mefjir.* The river of the gushing stream or outlet.

نهر الزرقا   Ik   *Nahr ez Zerka.* The azure or glittering river.

الصفرا   Ij   *Es Sŭfra.* The yellow; a marsh.

سرفند   Jj   *Sûrafend.* Sarafend. p.n. Heb. צרפן.

الشيخ حلو   Jl   *Sheikh Helu.* Sheikh Helu. p.n.; 'sweet.'

طاحونة ابو نور   Jk   *Tâhûnet Abu Nûr.* The mill of Abu Nûr. p.n.; meaning 'father of light.'

طنطورة   Ij   *Tantûrah.* The peak. *See* p. 117.

تل الاخضر   Il   *Tell el Akhdar.* The green mound.

تل بارك   Jk   *Tell Bârak.* The mound of Bârak. p.n.; meaning either 'blessing,' 'kneeling,' or 'the breast.' (بَرْك = بَارِك) 'breast.' LANE.)

وادى الدفلة   Jj   *Wâdy ed Dufleh.* The Valley of the Oleanders.

SHEET VIII.

ابو مدور	Kl	*Abu Madwar.* Properly *Mudawwar,* 'the rounded.'
ابو شوشة	Lj	*Abu Shûsheh.* The father of the tuft.
العفولة	Nj	*El 'Afûleh.* El 'Afûleh. p.n.; from Heb. עֹפֶל *ophel,* 'a hill or tower.' In Arabic the word means 'ruptured' (a woman).
عين ابو زريق	Lj	*'Ain Abu Zereik.* Spring of the magpie.
عين اتاروك	Mj	*'Ain Atârûk.* The spring of Atârûk. p.n.; *see* p. 105.
عين البيضاء	{Lj} {Mj}	*'Ain el Beida.* The white spring.
عين برتة	Ll	*'Ain Bertah.* The spring of Berta. q.v.
عين البئر	Lk	*'Ain el Bîr.* The well-spring.
عين الدالية	Lk	*'Ain el Dâlieh.* Spring of the trailed vine.
عين داود	Lk	*'Ain Dâûd.* The spring of David.
عين الدرامك	Mj	*'Ain ed Derâmik.* The spring of the soft ground; pl. of درمك
عين الفوار	Mj	*'Ain el Fûwâr.* The spring of the fountain.
عين غدران	Lk	*'Ain Ghadrân.* The spring of swamps.
عين غزال	Jj	*'Ain Ghûzâl.* The spring of the gazelle.
عين الحجة	Lk	*'Ain el Hajjeh.* The spring of the pilgrimage, or of the footpath.
عين الحسن	Ml	*'Ain el Hasn.* The spring of el Hasan. p.n.; the son of 'Ali.
عين ابرهيم	Lk	*'Ain Ibrahîm.* Abraham's spring.
عين اسمعين	Jk	*'Ain Ismâîn.* Ishmael's spring.
عين الخروبة	Ml	*'Ain el Kharrûbeh.* The spring of the locust-tree.
عين القنب	Lj	*'Ain el Kinnib.* The spring of hemp.

عين القُبَّة Mj 'Ain el Kubbeh. The spring of the dome.

عين القصب Lj 'Ain el Kăsab. The spring of the reeds.

عين القصيبة Lk 'Ain el Kăscibeh. The spring of the reeds.

عين المراح Kk 'Ain el Maráh. The spring of the sheep-fold.

عين المدينة Lj 'Ain el Medîneh. The spring of the city.

عين الميتة Jk 'Ain el Meiyiteh. The spring of the dead woman, or 'the dead spring.'

عين النبي Lk 'Ain en Neby. The Prophet's spring (near Sheikh Iskander). q.v.

عين الرز Mj 'Ain er Roz. The spring of rice.

عين السهلة Ll 'Ain es Sahleh. The spring of the plain.

عين الصرافة Lj 'Ain es Seráfeh. The spring of the exchange or banking-place.

عين الشعرة Lk 'Ain esh Sha'reh. The spring of the thick foliage.

عين السكر Lj 'Ain es Sikr. The spring of the dyke.

عين الست Lj 'Ain es Sitt. The lady's spring.

عين السنطات Lj 'Ain es Sŭmtát. Vulgar for es Sŭntát. The spring of the thorny acacia trees.

عين السوس Lj 'Ain es Sûs. The spring of liquorice.

عين التينة Lj 'Ain et Tineh. The spring of the fig-tree.

عين التريمة Lj 'Ain et Tureimeh. The spring of Tureimeh. p.n.

عين أم الغزلن Lk 'Ain Umm el Ghŭzlán. The spring with the gazelles.

عين أم الصمّن Kl 'Ain Umm es Summit. The spring with the hard earth.

عين الزيتونة Lk 'Ain ez Zeitûneh. The spring of the olive-tree.

عقد العزرية Jl 'Akid el 'Aziriyeh. The arch of el 'Aziriyeh. p.n.

عقادة Lk 'Akkádeh. The building with a coping stone. See Bestán el 'Akkád, p. 3.

العلتيان Jl El 'Aleiyân. The high (spring).

عانين Lk 'Ânîn. 'Ânin. p.n.; perhaps from Heb. עינים 'fountains.'

عراق المواشت Jk 'Arák el Mŭáshet. Perhaps an error for مواشط The cliffs of the tire-women.

عراق الذائف   Kj   *'Arâk en Nâtef.* The cliff of the dripping water.

عرعرة   Ll   *'Arârah.* The juniper tree.

ارض الرميلي   Jj   *Ard er Rumeily.* The sandy land.

العريش   Lk   *El 'Arîsh.* The hut.

العراقة   Ml   *El 'Arrâkah.* The cliffs.

عشيرة التواثة   Lj   *Ashîret et Tawâthah.* The clan of the Tawatha ; name of a tribe of Turkomans.

اشلول الخواجة   Mj   *Ashlûl el Khawâjah.* The waterfalls of the merchant. The word *Khawâjah*, 'merchant,' is Persian, and is applied to European travellers in the Levant. In Morocco, *Tâjir*, also meaning merchant, is used in the same sense.

اشلول المجحية   Aj   *Ashlûl el Mujahîyeh.* Waterfalls of el Mujahiyeh. *Majah* means to upset a bucket in a well. The name is applied to the spring-head of the Kishon.

عيون البنات   Jl   *'Ayûn el Benât.* The springs of the girls.

عيون الفرث   Mj   *'Ayûn el Farth.* The springs of the filth.

عيون حضيضون   Jl   *'Ayûn Hadeidûn.* The springs of the low ground.

عيون حماءة   Mj   *'Ayûn Hûmmâmeh.* The springs of the 'hot bath,' or of the 'pigeon.'

عيون الرز   Mk   *'Ayûn er Roz.* The springs of rice.

عز الدين   Nl   *'Azz ed Dîn.* 'Azz ed Din. p.n.; signifying Glory of the Faith.

باقا الغربية   Kl   *Bâka el Gharbîyeh.* The western Bâka. p.n.

البليان   Lk   *El Baliân. cf. Neby Baliân,* p. 152.

البارد   Ml   *El Bârid.* The cool.

باطن السما   Ml   *Bâtn es Sema.* The inside or belly of the sky. *Bâtn* also means 'a watercourse on rugged ground.'

باطن الطويل   Lk   *Bâtn et Tawîl.* The long channel.

بلاد الروحة   Kj   *Belâd er Rûhah.* The country of er Rûhab. *See Dahet er Rûhah,* p. 146.

برتة   Ll   Bertah. p.n.; signifying 'cutting.'

بسمة   Nl   *Besmeh.* Besmeh. p.n.; perhaps connected with בשם. 'balsam.'

بستان علي يعقوب   *Ll*   *Bestân '-Aly Yâkûb.* The garden of 'Aly Yâkûb.

بئر عارة   *Kk*   *Bir 'Árah.* The well of 'Árah. p.n.

بئر ابو شقرا   *Lk*   *Bir Abu Shakra.* The well with the red colour.

بئر البطنية   *Ml*   *Bir el Batnîyeh.* The well of the low ground or channel.

بئر بلعمة   *Nl*   *Bir Belâmeh.* The well of Belâmeh. p.n.; perhaps Heb. יִבְלָעָם Ibleam.

بئر بورين   *Kl*   *Bir Bûrîn.* The well of Bûrin. p.n.; perhaps from بُور, 'fallow land.'

بئر الحجة   *Ml*   *Bir el Hajjeh.* The well of the footpath, or of the pilgrimage.

بئر حراميس   *Kj*   *Bir Harâmis.* The well of barren seasons.

بئر اسير   *Kl*   *Bir Isîr.* The well of Isîr. p.n.; perhaps Asîr, 'a captive.'

بئر كفر قارع   *Kj*   *Bir Kefr Kârâ.* The well of Kefr Kârâ. q.v.

بئر المالح   *Ml*   *Bir el Málch.* The salt well.

بئر المويلح   *Nj*   *Bir el Muweilih.* The brackish well.

بئر الناطف   *Kj*   *Bir en Nâtef.* The dripping well.

بئر النبي بليان   *Kk*   *Bir en Neby Baliân.* The well of Neby Baliân. q.v., p. 152.

بئر رابا   *Lj*   *Bir Râba.* The well of Râba. p.n.

بئر الشلاف   *Nj*   *Bir esh Shellâf.* The well of the water which descends from a height.

بئر الشرب   *Nj*   *Bir esh Sherb.* The well of drinking.

بئر السنجب   *Nl*   *Bir es Sinjib.* The well of the squirrel.

بئر الصفصاف   *Jk*   *Bir es Sûfsâf.* The well of the osier.

بئر ام الكديش   *Kk*   *Bir Umm el Kedîsh.* The well with the pack-horse.

بئر الزرقآ   *Kk*   *Bir ez Zerka.* The azure well.

بركة البطيخ   *Jl*   *Birket el Battîkh.* The pool of the water-melon.

بركة الفولة   *Nj*   *Birket el Fûleh.* The pool of el Fûleh. q.v.

البريج   *Jk*   *El Bureij.* The little tower.

بريج القراني   *Kk*   *Bureij el Kûrâny.* The annexed little tower.

U

المبريكة   *Jk*   *El Bureikeh.* The little pool; a village.

برج الخيل   *Jk*   *Burj el Kheil.* The tower of cavalry.

برقين   *Nl*   *Bŭrkîn.* Bŭrkin. p.n.; connected with برقة, which means 'sandy soil covered with flints.'

بصيلة   *Lj*   *Băseileh.* The little onion.

البويشات   *Lk*   *El Bŭweishât.* El Bûweishât. p.n.

دالية الروحة   *Kj*   *Dâliet er Rûhah.* The trailing vine of er Rûhah. p.n.; either from روحة 'repose,' or روحآ 'shallow.'

دردارا   *Jl*   *Dardâra.* The elm-tree.

دبة الزعتر   *Kl*   *Debbet ez Zâter.* The plot of ground (covered with) thyme.

ظهر المصطبة   *Lk*   *Dhahr el Mastabah.* The ridge of the bench.

ظهرة الصافي   *Kk*   *Dhahret es Sâfi.* The ridge of smooth or bright rock.

دير الهوا   *Lk*   *Deir el Hawa.* The monastery of the wind.

ازبوبا   *Mk*   *Ezbûba.* Ezbûba. p.n.; either connected with زبيب 'raisins'; or, it may be, from ذبوب 'dried up.'

الفخيرة   *Mj*   *El Fakhîreh.* The glorious.

فراسين   *Ll*   *Ferâsîn.* Ferâsîn. p.n.

الفريديس   *jj*   *El Fureidîs.* The little paradise.

الغنام وقيل كفر يعروب   *Nl*   *El Ghannâm.* The plunderers or shepherds. Also called *Kefr Yârûb.* The village of Yârûb. p.n.; perhaps Yârab, son of Cahtân, the progenitor of the Arabs.

الغاطي   *Jl*   *El Ghâty.* The shed. There is a wicker hut at the place which has also the name خربة دردارة *Khŭrbet Dardâra,* the ruin of the elm-tree.

الحربوشية   *jj*   *El Harbûshîyeh.* The place invested by vipers.

إجزم وقيل اثزم   *Jj*   *Ijzim,* or *Ikzim,* as variously spelt by natives. The first means 'cut off,' the second 'worthless.'

خنافس   *Mj*   *Ikhueifis.* Properly *Khanâfis.* 'Beetles'; a ruined tower.

جعارا   *Lj*   *J'ârah.* The hyæna. The word also means a rope which is tied round the waist to prevent the water-drawer from falling into the well.

الجَبْحَمة   *Lk*   *El Jahmeh.* The flare. (A small ruin on a hill-top, perhaps formerly a beacon.)

جبع   *Jj*   *Jebá.* Heb. גבע 'a hill.'

جنين   *Nl*   *Jenin.* Jenin. p.n.; (*cf.* Heb. עֵין גַּנִּים *en Ganim,* 'the spring of gardens.')

جنجار   *Nj*   *Junjár.* Junjár, name of a herb.

جويديرة   *Nk*   *Jüweidireh.* The stone sheep pens.

قضا حيفا   *Jk*   *Kada Haifa.* The district of Haifa.

قضا ناصرة   *Nj*   *Kada Násirah.* The district of Nazareth.

كنير   *Kk*   *Kannir.* Kannir. p.n.

كوكب   *Ml*   *Kaukab.* Kaukab. p.n. *See* p. 110.

كفر اذان   *Ml*   *Kefr Adán.* The village of Adhán. p.n.; meaning in Arabic, 'the call to prayer.' Perhaps the כפר עותני Capher Outheni of the Talmud.

كفر قارع   *Kk*   *Kefr Kárá.* The village of the gourd.

كفر تود   *Ml*   *Kefr Kúd.* The village of Kúd. p.n. The ancient Capercotia.

كفرين   *Lk*   *Kefrein.* The two villages.

كفريرة   *Ml*   *Kefreireh.* Kefreireh. p.n.

كركور   *Kl*   *Kerkúr.* Kerkúr. p.n.

خلة الفول   *Lk*   *Khallet el Fúl.* The dell of beans.

خلة النجاس   *Lk*   *Khallet en Nejás.* The dell of pears.

خلة الزوق   *Lj*   *Khallet ez Zúk.* The dell of the village.

خروبة   *Nl*   *Kharrúbeh.* The locust-tree (*Ceratonia siliqua*).

الخطمية   *Kk*   *El Khatmîyeh.* The marsh-mallows.

الخيخيرة   *Mj*   *El Kheikheirah.* El Kheikheireh. p.n. (Snoring).

خربة العبهرية   *Kk*   *Khürbet el 'Abhariyeh.* The ruin of the mock-orange (*Styrax officinalis*).

خربة ابى عامر   *Ml*   *Khürbet Abu 'Amir.* The ruin of Abu 'Amir. p.n.

خربة ابى رجمان   *Kl*   *Khürbet Abu Rájmán.* The ruin with the cairns.

خربة على قوتا *Ll* Khůrbet 'Aly Kôka. The ruin of 'Aly Kôka. p.n.

خربة عنين *Ml* Khůrbet 'Anîn. The ruin of 'Anîn. q.v., p. 143.

خربة عارة *Kk* Khůrbet 'Árah. The ruin of 'Árah. p.n.

خربة ببلون *Jk* Khůrbet Bablûn. The ruin of Bablûn. p.n.

خربة باسيلا *Kl* Khůrbet Bâsîla. The ruin of Bâsîla. p.n. Basilius.

خربة بيدوس *Kl* Khůrbet Beidûs. The ruin of Beidûs. p.n.

خربة بيت راس *Lj* Khůrbet Beit Râs. The ruin of the house of the hill-top.

خربة البيار *Lk* Khůrbet el Biâr. The ruin of the wells.

خربة بئر اسير *Kl* Khůrbet Bîr Isîr. The ruin of Bîr Isîr. q.v.

خربة البرك *Kk* Khůrbet el Burak. The ruin of the pools.

خربة بصيلة *Lj* Khůrbet Buscileh. The ruin of Buscilah. q.v., p. 146.

خربة ظهرة حماد *Lk* Khůrbet Dhahret Hammâd. The ruin of the ridge of Hammâd. p.n.

خربة الدفيس *Jk* Khůrbet ed Dufeis. The ruin of ed Dufeis. p.n. (glittering).

خربة القربة *Lj* Khůrbet el Farrîyeh. The ruin of the quails.

خربة الفنيطر *Kk* Khůrbet el Funeiter. The ruin of el Funeiter. p.n.

خربة احديثية *Jk* Khůrbet el Hadeithîyeh. The ruin of el Hadeithîyeh. p.n.; from حداثة a 'novelty' or 'accident.'

خربة حنا *Jj* Khůrbet Hanna. John's ruin.

خربة حنانة *Jj* Khůrbet Hannâneh. The ruin of the cistern. See Bîr el Hannâny, p. 107.

خربة الإخرين *Ll* Khůrbet el Ikhrein. The ruin of the excrement.

خربة جبجب *Ml* Khůrbet Jebjeb. The ruin of the flat ground.

خربة جرار *Lk* Khůrbet Jerrâr. The ruin of Jerrâr; family name, meaning 'waterpot-sellers.'

خربة جنزار *Ml* Khůrbet Jinzâr. The ruin of Jinzâr. p.n.

خربة كفر باسا *Kl* Khůrbet Kefr Bâsa. The ruin of Kefr Bâsa. q.v.

خربة الكلبي *Kk* Khůrbet el Kelbi. The ruin of the Kelbi; i.e., member of the Beni Kilâb tribe.

خربة قزازة *Jl*   *Khūrbet Kezāzeh.* The ruin of glass.

خربة الخنيزيرة *Jk*   *Khūrbet el Khaneizîreh.* The ruin of the swine.

خربة الخضيرة *Kk*   *Khūrbet el Khūdeirah.* The ruin of the verdure.

خربة الخزنة *Mj*   *Khūrbet el Khūzneh.* The ruin of the treasury.

خربة فيرة *Lj*   *Khūrbet Kirch.* The ruin of pitch.

خربة فتينة *Kj*   *Khūrbet Koteineh.* The ruin of Koteineh. p.n. This may mean either 'inhabitants,' or 'a pot'; or it may be connected with قطن *Kotn,* cotton. (*cf.* Talmudic קבנית.)

خربة القصب *Lk*   *Khūrbet el Kūsab.* The ruin of reeds.

خربة كوسية *Kl*   *Khūrbet Kūsieh.* The ruin of Kūsieh. p.n.

خربة منصورة *Kj*   *Khūrbet Mansûrah.* The ruin of Mansûrah ; *see* p. 9.

خربة الماوية *Lk*   *Khūrbet el Māwiyeh.* The ruin of the place of shelter (a ruined Caravanserai).

خربة المدكاكين *Ml*   *Khūrbet el Medekākîn.* This cannot be right. It is probably *Khūrbet em Dekkākin,* 'the ruin of the shops,' the article *el* often being corrupted into *em.*

خربة المدينة *Lj*   *Khūrbet el Medîneh.* The ruin of the city.

خربة المزرعة *Nj*   *Khūrbet el Mezrâh.* The ruin of the sown corn.

خربة المنطار *Ll*   *Khūrbet el Mūntâr.* The ruin of the watch-tower.

خربة الناظر *Mk*   *Khūrbet en Nâdhr.* The ruin of the overseer.

خربة ناسوس *Jj*   *Khūrbet Nâsûs.* The ruin of Nâsûs. p.n.

خربة نحلين *Ll*   *Khūrbet Nehalîn.* The ruin of Nehalin. *See Khūrbet Juzwar en Nūkhl,* p. 17.

خربة النزلة ${Ml \brace Jj}$ *Khūrbet en Nūzleh.* Ruin of the place of alighting.

خربة رصيصة *Jk*   *Khūrbet Raseiseh.* The ruin of rubble ; also called خربة السمرا *Khūrbet es Sumra,* 'the brown ruin'

خربة الريحانة *Kj*   *Khūrbet er Rîhâneh.* The ruin of sweet basil.

خربة الصابر *Kk*   *Khūrbet es Sâbir.* The ruin of the prickly-pear.

خربة سمارة *Ll*   *Khūrbet Samârah.* The ruin of nocturnal entertainments.

خربة شمسين *Kl*   *Khūrbet Shemsîn.* The ruin of Shemsin. p.n. *Shems,* 'the sun.'

خربة الشيخ   *Jj*   *Khûrbet esh Shîh.* The ruin of the Shih. *See Merj Sidr es Shîh,* p. 90.

خربة الست ليلى   *Jk*   *Khûrbet Sitt Leila.* The ruin of Lady Leila.

خربة السليمانية   *Jj*   *Khûrbet es Suleimânîyeh.* The ruin of the Solomon family.

خربة السمرا   *Kl*   *Khûrbet es Sumra.* The brown ruin.

خربة سروج   *Ml*   *Khûrbet Surûj.* Ruin of saddles; or perhaps 'lamps,' from *seraj.*

خربة الطرم   *Ml*   *Khûrbet et Turm.* The ruin of Turm, meaning 'honey,' or a kind of tree so called.

خربة ام البصل   *Kl*   *Khûrbet Umm el Bûsl.* The ruin with the onions.

خربة ام البطم   *Nl*   *Khûrbet Umm el Butm.* The ruin with the terebinth trees.

خربة ام الحفة   *Ml*   *Khûrbet Umm el Haffeh.* The ruin on the edge (or extremity), being at the end of a mountain spur.

خربة ام الجمال   *Jj*   *Khûrbet Umm el Jemâl.* The ruin of the camels. Also called *Khûrbet el Harbushîyeh. See el Harbûshîyeh,* p. 146.

خربة ام الكديش   *Kk*   *Khûrbet Umm el Kedîsh.* The ruin of the packhorse.

خربة ام القتف   *Kl*   *Khûrbet Umm el Kûtûf.* The ruin with the St. John's wort.

خربة ام الريحان   *Ll*   *Khûrbet Umm er Rîhân.* The ruin with the sweet basil.

خربة الزبادنة وقيل وادى عارة   *Kl*   *Khûrbet ez Zebâdneh.* The ruin of the people of Zebdah; *see Khûrbet Zebad,* p. 51. Also called *Wâdy 'Arah,* from the valley of that name.

خبيزة   *Kk*   *Khobbeizeh.* Marsh-mallow.

قفين   *Kl*   *Kuffîn.* Kuffin; from قف 'a rugged hill.'

الكفيرة   *Mj*   *El Kufeireh.* The little village.

قمبازة   *Kj*   *Kumbâzeh.* Kumbâzeh; perhaps the Persian گنبذ 'gun-bazeh,' 'a dome' or 'cupola.'

القصبية   *Kk*   *El Kûsabîyeh.* The reedy place.

قصر عين الشريعة   *Ll*   *Kûsr 'Ain esh Sheriâh.* The palace of the spring of the watering-place.

قصر فقس   *Kk*   *Kûsr Fûkkis.* The palace of the unripe melon.

اللجّون	*Mk*	*El Lejjûn.* The Latin Legio.
لدّ	*Mj*	*Ludd.* Ludd. p.n.
المعصر	*Kk*	*El Máäser.* The wine-press.
المغازين	*Nl*	*El Magházin.* The warriors. (A sacred place.)
المأكورة	*Jj*	*El Mâkûrah.* El Mâkûrah. p.n.
المعلّكة	*Ll*	*El Málakah.* El Málakah; perhaps معلّكة 'overhanging.'
ماعاس	*Jk*	*Má-más.* Mámas. p.n.; apparently Majumas.
المراح	*Jk*	*El Maráh.* The resting-place by night-time.
المرجحة	*Ll*	*El Marcihah.* The night resting-places.
الذئاب وقيل ام الذئاب	*Kl*	*El Medhtâb.* The place of wolves; also called *Umm edh Dhiâb,* with the same meaning.
المنسي	*Mj*	*El Mensî.* The forgotten.
مرج ابن عامر	*Mj*	*Merj Ibn 'Âmir.* Meadow of the son of 'Âmir. p.n.
المسعدي	*Jl*	*El Mesâdy.* The place of the gladioli (*su'd*). This flower grows all over the plain.
المشيرفة	*Lk*	*El Mesheirfeh.* The high places.
مشرع الصفا	*Lj*	*Mishrá es Sûfa.* The clear watering-place.
المستقي	*Kk*	*Miska.* The drinking-place, or place of the *Sâkia* or water-wheel.
مغارة الغول	*Kk*	*Mûghâret el Ghûleh.* The cave of the ghoul.
المجيعية		*El Mujahiyeh.* The place of springing (of water).
متيبلة	*Nk*	*Mukeibileh.* The front place.
المناطر	*Mj*	*El Mânâtir.* The watch-towers.
المنطار	*Ll*	*El Mântâr.* The watch-tower.
مرتفعة	*Mk*	*Murtefeh.* Murtefa. p.n.; perhaps مرتفع high place.
مشمش	*Lk*	*Mûsmûs.* Mûsmûs. p.n.; meaning compactly built. Robinson gives *Mushmush* (مشمش), i.e., 'apricot.'
نغر ام الخروص	*Jk*	*Naghr Umm el Kharûs.* The brackish spring with the reservoir.

ناحية جنين *Mk* *Nâhiet Jenîn.* The commune of Jenîn ; also called بلاد الحارثة *Belad el Hâritheh.* The land of tilth; the part on this sheet is called also شفعة الغربي *Shefât el Gharby,* 'the western federation.' Compare *Shefât el Kibly,* Sheet IX.

نهر المقطع *Mj* *Nahr el Mukûttâ.* The River el Mukûttâ. *See* p. 114.

النبي بليان *Kk* *Neby Baliân.* The Prophet Baliân. Perhaps *Billiyan,* 'one who has gone to an unknown place.'

النفاخة *Jk* *En Nefâkhah.* 'The blower.' *See en Nefakhîyeh,* p. 31.

نزلة المعاصي *Kl* *Nûzlet el M'âsfy.* The place of alighting of the side road.

الرضيضة *Kl* *Er Raseiseh.* The rubble.

رمانة *Mk* *Rummâneh.* 'Pomegranates' (Rimmon).

سهل عرابة *Ml* *Sahel* (or *Merj*) *'Arrâbeh.* The plain (or meadow) of 'Arrâbeh. p.n. ; meaning 'steppe.'

سهل إجزم *Jj* *Sahel Ijzim.* Plain of Ijzim. q.v.

سالم *Mk* *Sâlim.* Salim. p.n. (Salem).

شعروية الشرقية *Ll* *Shârawîyet esh Sherkîyeh.* The eastern woodland.

شفيا *Jj* *Shefeia.* The edge or margin.

شجرة فرهود *Ll* *Shejeret Farhûd.* The tree of the kid.

شيخ عمير *Jj* *Sheikh 'Ameir.* Sheikh 'Ameir. p.n.

شيخ حمدان *Mk* *Sheikh Hamdân.* Sheikh Hamdân. p.n.

شيخ حسن العروري Mk *Sheikh Hasan el 'Arûry.* Sheikh Hasan, 'the possessed.'

شيخ اسكندر *Lk* *Sheikh Iskander.* Sheikh Alexander. p.n.

شيخ خليف *Ll* *Sheikh Khalîf.* Sheikh Khalif. p.n. ; 'the successor.'

شيخ ماذي *Jj* *Sheikh Mâdhy.* Sheikh Madhi. p.n. ; meaning 'pure honey.'

شيخ منصور *Nl* *Sheikh Mansûr.* Sheikh Mansûr. p.n. (Victor). The building beside it is sometimes called *Khûrbet Belâmeh. See Bîr Belâmeh,* p. 145.

شيخ ميسر *Kl* *Sheikh Meisir.* Sheikh Meisir. p.n. ; meaning a certain gambling game with arrows.

شيخ محمد *Kl* *Sheikh Muhammed.* Sheikh Mohammed.

شيخ محمد التلولي *Lk*   *Sheikh Muhammed et Tellûlî*. Sheikh Mohammed of the mounds.

شيخ مصلي *Ml*   *Sheikh Musally*. Sheikh Musally. p.n.; meaning one who prays.

شيخ سعادة *Nl*   *Sheikh S'âdeh*. Sheikh Saadeh. p.n. (fortunate).

شيخ سنداحاوي *Kl*   *Sheikh Sandâhâwy*. Sheikh Sandâhâwy. p.n.; perhaps Saint Eva.

شيخ شبلة *Ml*   *Sheikh Shibleh*. Sheikh Shibleh. p.n. *See* "Memoirs."

شيخ الاباق *Jl*   *Sheikh el Utâk*. The Sheikh of the apartment.

شيخ زيد *Ml*   *Sheikh Zeid*. Sheikh Zeid. p.n.

سيلي *Mk*   *Sily*. Sily. p.n.

السنابل *Kj*   *Es Sinâsîl*. Vulgar for *es Silâsil*, the concatenation (applied to small valleys).

سنديانة *Jk*   *Sindiâneh*. Oaks (*Quercus pseudo coccifera*).

السوامر *Jj*   *Sââmir*. Places of nocturnal entertainment; pl. of سامر.

صبارين *Kk*   *Sŭbbârîn*. Sŭbbârin, from صبارة 'rough ground.' cf. Heb. צְבָרִים 'heaps.' In the official lists it appears as *Khŭdeirât Subbârin*, خضيرات صبارين cf. *Khŭrbet Khŭdeirah*, p. 149.

طاحونة الاساور *Kl*   *Tâhûnet el Asâwir*. The mill of the bracelets.

تعنك *Mk*   *Tânnuk*. Taânnuk. p.n. Heb. תַּעֲנָךְ Taanach, 'sandy.'

طربنه *Mj*   *Tarbanah*. Tarbanah. p.n.

تل ابو حماد *Jk*   *Tell Abu Hammâd*. The mound of Abu Hammâd. p.n. (A second name for *Khŭrbet Bablûn*.)

تل ابو قديس *Mk*   *Tell Abu Kŭdeis*. The mound of Abu Kŭdeis. p.n. (dim. of *Kuds*, 'holy').

تل افرين *Jl*   *Tell Afrein*. The mound of the oven or reservoir.

تل الاغبارية *Lj*   *Tell el Âghbâriyeh*. The mound of Âghbâriyeh, an Arab tribe. The name means 'dust-coloured,' hence 'wolves.'

تل الساور *Kl*   *Tell el Asâwir*. The mound of the bracelets.

تل الذهب *Mk*   *Tell edh Dhaheb*. The mound of gold.

X

تل الذرور   *Jl*   *Tell edh Dhrûr.* The mound of the Dhrûr (a plant, *Calamus aromaticus*).

تل الدودحان   *Jk*   *Tell el Dôdchân.* The mound of Dôdchân. p.n. (دودح) 'short or corpulent.')

تل قيمون   *Lj*   *Tell Keimûn.* The mound of Keimûn. p.n.

تل المتسلّم   *Mj*   *Tell el Mutasellim.* Mound of the governor.

تل شدود   *Mj*   *Tell Shadûd.* Tell Shadûd. p.n. ; 'strong.'

تل ثورة   *Lj*   *Tell Thôrah.* The mound of the bull.

تلول الجحاش   *Lj*   *Tellûl el Jehâsh.* The mounds of the asses.

الطيبة   *Mk*   *Et Taiyibeh.* The goodly.

أم العبير   *Lk*   *Umm el 'Abhar.* Growing the mock orange (*Styrax officinalis*).

أم العلق   *Jk*   *Umm el 'Alak.* Producing leeches.

أم البنادق   *Kj*   *Umm el Benâdik.* The place of hazel-nuts, or 'guns.'

أم البطيمات   *Lk*   *Umm el Buteimât.* The place with the terebinths.

أم الدرجة   *Kj*   *Umm ed Derajeh.* The place with the steps.

أم الفحم   *Lk*   *Umm el Fahm.* Producing charcoal.

أم الحارثة   *Lk*   *Umm el Hârithch.* The place with the ploughland.

أم الحشورة } المحشرة   *Lj*   *Umm el Hashûreh.* For *el Mehasherah*, the place for collecting water ; a rock-cut tank.

أم القلايد   *Mj*   *Umm el Kelâid.* Literally 'the mother of necklaces' ; but it may mean the place with a system of irrigating.

أم الخروص   *Kk*   *Umm el Kharûs.* The place with the reservoirs.

أم الخطاف   *Ll*   *Umm el Khatâf.* The place with the hooks.

أم القبة   *Kj*   *Umm el Kubbeh.* The place with the dome.

أم الشوف   *Kk*   *Umm esh Shûf.* The place with the harrow.

أم الطوس   *Jj*   *Umm et Tôs.* The place with the cups ; but perhaps from the Aramaic טוש 'to cover with cement.'

أم التوت   *Jj*   *Umm et Tût.* Producing mulberries.

أم الزينات *Kj* *Umm ez Zeinât.* The place of ornamentation or of festivals.

الورقاني *Mj* *El Warakâny.* The leafy.

وادي ابو لحية *Lj* *Wâdy Abu Lehieh.* The valley of the man with the beard.

وادي عارة *Kl* *Wâdy 'Ârah.* The valley of 'Ârah. p.n. (This is a village so called.)

وادي عارة *Kl* *Wâdy 'Ârah.* The valley of 'Ârah. p.n.

وادي العريان *Lk* *Wâdy el 'Erriân.* The naked valley.

وادي العصل *Ll* *Wâdy el 'Asl.* The valley of the rhododendron. It may also mean 'crooked.'

وادي بلعمة *Nl* *Wâdy Belâmeh.* The valley of Belâmeh. *See Bir Belâmeh.* p. 145.

وادي بير اسير *Kl* *Wâdy Bir Isir.* The valley of Bir Isir. q.v., p. 145.

وادي الدفلة *{Jk Lj}* *Wâdy ed Dufleh.* The valley of the oleander.

وادي الفوار *Kj* *Wâdy el Fuwâr.* The valley of the fountain.

وادي حنو *Jj* *Wâdy Henû.* The valley of the curve.

وادي الخالد *Lk* *Wâdy el Khâlid.* The valley of Khâlid. p.n.

وادي الخضيرة *Jl* *Wâdy el Khudeirah.* The valley of verdure.

وادى قدران *Kk* *Wâdy Kudrân.* The valley of the pots; perhaps an error for *Ghadrân,* 'swamps,' as the valley is very swampy.

وادي القصب *{Lj Lk}* *Wâdy el Kûsab.* The valley of the reeds.

وادي ماذي *Jj* *Wâdy Mâdhy.* The valley of Mâdhy. *See Sheikh Mâdhy,* p. 152.

وادي مالح *Jl* *Wâdy el Mâleh.* The salt valley.

وادي مصيدي *Jj* *Wâdy Mâsidy.* The valley of Mâsidy. p.n.

وادي المطابن *Kj* *Wâdy el Matâbin.* The valley of the pits for keeping a fire in.

وادي المتلة *Kj* *Wâdy el Mikteleh.* The valley of slaughter.

وادي الملح *Kj* *Wâdy el Milh.* The valley of salt.

وادي المويلى *Nj* *Wâdy el Muweily.* The valley of Muweily. p.n.

وادي النذيف Nl *Wâdy en Nîf.* The valley of Nîf. p.n.

وادي نغل Jk *Wâdy Nughl.* The valley of the penfold or den.

وادي راسين Kl *Wâdy Râsein.* The valley of two heads or hill-tops.

وادي سلهب Ml *Wâdy Selhab.* The long valley.

وادي صمنطار Kl *Wâdy Samantâr.* The valley of Samantâr. p.n.

وادي شمّة Nl *Wâdy Shemmah.* The valley of odour.

وادي الشقاق Kj *Wâdy esh Shûkâk.* The valley of clefts.

وادي السناجق Kj *Wâdy es Sinâjik.* Valley of the provinces.

وادي الست Lk *Wâdy es Sitt.* The valley of the lady.

وادي ام البنادق Kj *Wâdy Umm el Benâdik.* The valley with the hazel-nuts.

وادي الحمور Ll *Wâdy el Yahmûr.* The valley of the wild ass or of the roebuck.

وادي زبرية Jj *Wâdy Zibriyeh.* The valley of Zibriyeh; from *Zibr,* books, writings, inscriptions.

يعبد Ll *Yâbid.* Yâbid. p.n.

الياسون Ml *El Yâmôn.* El Yâmôn. p.n.

زبدة Ll *Zebdah.* Zebdah. p.n. *See Khûrbet Zebed,* p. 51.

زبد Mk *Zebed.* Zebed. p.n. *See Khûrbet Zebed,* p. 51.

زلفة Mk *Zelefeh.* The cisterns.

الزرغانية Jk *Ez Zerghânîyeh.* Ez Zerghânîyeh. p.n.

زمارين وقيل طابلين Jk *Zimmârîn.* Also called *Tubbalîn.* The first means flute-players, the second drummers.

SHEET IX.

عين العاصي Pk 'Ain el 'Asy. The spring of el 'Asy. p.n., meaning ' rebellious.'

عين بالا Ql 'Ain Bâla. The spring of Bâla. p.n.

عين الباز Oj 'Ain el Bâz. The spring of the hawk.

عين البيضآ Pj 'Ain el Beida. The white spring.

عين البستان Qj 'Ain el Bestân. The spring of the garden.

عين البيرة Pj 'Ain el Bireh. The spring of the fortress. Heb. בירה.

عين دابو Oj 'Ain Dâbû. The spring of Dâbû. p.n.

عين الدغيم Ql 'Ain ed Daghcim. The whitish spring.

عين دنا Pj 'Ain Denna. The spring of the amphora. See Denna, p. 160.

عين حيّة Qj 'Ain Haiyeh. The spring of the serpent.

عين الحلو وقيل المزوقة Qj 'Ain el Helu or 'Ain el Muzôkah. The sweet spring, or the spring of the painted or plated (stone).

عين جالود Ok 'Ain Jâlûd. Goliath's spring (for جالوت).

عين الجمعين Pk 'Ain el Jemâin. The spring of the two assemblies.

عين الجوسق Pl 'Ain el Jôsak. The spring of the lofty building.

عين الخنازير Ql 'Ain el Khaneizir. The spring of the swine.

عين القصب Pj 'Ain el Kâsab. Spring of reeds.

عين لسد Pj 'Ain Lesed. The spring of sucking (as a kid does).

عين المدوع Pl 'Ain el Madûâ. The spring of sprinkling.

عين المعيبرة Ql 'Ain el Majerah. The spring of the fillet.

عين مكبوز Ql 'Ain Mak-hûz. The spring of Makhûz. p.n.

عين المالحة {Qj / Ql} 'Ain el Mâllah. The salt spring.

عين الميتة {Nk / Ql} 'Ain el Meiyiteh. The dead spring. It had ceased to flow, and was re-opened by 'Abd el Hâdy, governor of Jenin, whence its name. There is another of the same name.

عين الملعب Pk 'Ain el Melâb. The spring of the playhouse (close to the Beisân Theatre).

عين المغربة Ql 'Ain el Magharrebah. The spring of the cross-breeds between Jinns (genii or demons) and Men; but perhaps it should be Mughâribeh, 'Arabs of the West.'

عين المخزقة Pk 'Ain el Mukhuzakah. The spring of the impaling stake.

عين النخلة Ql 'Ain en Nukhleh. The spring of the palm-tree.

عين نصيرة Pl 'Ain Nŭseirah. The spring of Nŭseirah. p.n.

عين نصرة Pl 'Ain Nŭsrah. The spring of Nŭsrah. p.n. Both this and the last are so named from the slaughter of the Abu Nuseir Arabs at the place.

عين العليق Qj 'Ain el 'Olleik. The spring of the brambles; or عين ام حجاير 'Ain Umm Hajeir. The spring with the small stone.

عين العليقة Qj 'Ain el 'Olleikah. The spring of the bramble-bush.

عين الريحانية Ok 'Ain er Rîhâniyeh. The spring of sweet basil.

عين السودآ Qk 'Ain es Sôda. The black spring.

عين السخنة Pk 'Ain es Sokhneh. The warm spring.

عين الصفصافة Ql 'Ain es Sŭfsâfeh. The spring of the osier-willow.

عين التينة Qk 'Ain et Tîneh. The spring of the fig-tree.

عين طبعون Ok 'Ain Tŭbâûn. The spring of Tŭbâûn. See p. 117.

عين طمرة Oj 'Ain Tŭmrah. The spring of Tŭmrah. q.v., p. 117.

عين ام عمود Pl 'Ain Umm 'Amûd. The spring with the columns.

عين ام الفلوس Ql 'Ain Umm el Flûs. The spring with the coins.

عين ام الغزلن Ok 'Ain Umm el Ghŭzlân. The spring with the gazelles.

عين ام حية Ql 'Ain Umm Haiyeh. The spring of the serpent.

عين ام السواليل Pj 'Ain Umm es Sawâlîl. The spring of little streams.

عين ام سدرة {Ql}{Qj} 'Ain Umm Sidreh. The spring with the lotus-tree (Zizyphus lotus).

عين ام صابون Qj 'Ain Umm Sâbôn. The spring where the soap is made, at Khûrbet Umm Sâbôny. q.v.

عين الزهرة   Pk   *'Ain ez Zahrah.* The spring of the flower.

عين الزاوِيان   Qj   *'Ain ez Zâwiyân.* The spring of tares.

العُقود   Pk   *El 'Akûd.* The arches. The Beisân Theatre.

علم الدين   Pk   *'Alam ed Dîn.* The banner of the faith. A Sheikh's tomb at Beisân.

عرب البيتاوي   Oj   *'Arab el Beitâwy.* The Beitâwy Arabs.

عرب البلدِية   Qj   *'Arab el Belediyeh.* The Belidiyeh (of the country) Arabs.

عرب بنى فهيد   *'Arab Beni Fuheid.* The Beni Fuheid (sons of the panther) Arabs.

عرب البشتوي   Qj   *'Arab el Beshtâwy.* The Beshtawy Arabs.

عرب الغزية   Ql   *'Arab el Ghuzzawiyeh.* The Ghuzzawiyeh (raiding) Arabs.

عرب السقر   {Pl / Ql}   *'Arab es Sûkr.* The Sukr (falcon) Arabs.

عراق الرجوة   Nj   *'Arâk er Rahweh.* The cliff of the knoll.

عراق الشوح   Pk   *'Arâk es Shûh.* The cliff of the kite.

عرانة   Nl   *'Arrâneh.* Arrâneh. p.n.; perhaps from عران 'a hyæna's den.'

عربونة   Ok   *'Arrûbôneh.* Arrûbôneh. p.n.; meaning in Arabic 'earnest money.'

عولم   Pj   *'Aûlam.* 'Aûlam. p.n.; perhaps عيلم, 'a full well'; or from علام the stone of the *nebuk*, or fruit of the lote-tree.

عيون الخروة   Nj   *'Ayûn el Khirwah.* The springs of the castor-oil plant.

عيون المخير   Pj   *'Ayûn el Mukheiyir.* The springs of the charitable person.

عيون السدر   Oj   *'Ayûn es Sidr.* The springs of the lotus-tree.

عيون الثعلب   Nk   *'Ayûn eth Thâleb.* The springs of the fox.

عيون التراب   Nj   *'Ayûn et Trâb.* The springs of earth.

بصّة ابو الاجاج   Pk   *Basset Abu el Ajâj.* The marsh with the pungent-tasting water.

بصّة دبّة القتّاف   Pl   *Basset Debbet el Kûttâff.* The marsh of the tract of St. John's wort.

بصّة الديوان   Pl   *Basset ed Diwân.* The marsh of the bench (from an old sarcophagus overturned in it).

بتة حيط الشيد   Pl   *Basset Heit esh Shíd.* The marsh of the mortar-wall.

بتة الخيسة   Ql   *Basset el Khísah.* The marsh of the reservoir.

بتة المندسي   Pk   *Basset el Mandesi.* The marsh of el Mandesi. This is evidently a mistake, and may either be مندسى *man-dessí*, 'buried'; or مهندسى *m'hendasi*, 'of the architect.'

بتة الرشيد   Pk   *Basset er Rashíd.* The marsh of er Rashid. p.n.; meaning 'orthodox.' *cf.* Haroun Alraschid.

بيسان   Ql   *Beisân.* Beisân. p.n. The Hebrew בֵּית שְׁאָן Bethshean. Talmud ביסן.

بيت قاد   Ol   *Beit Kâd.* The house of Kâd. p.n.

بئر الخنير   Nj   *Bîr el Hâfiyîr.* The well of the excavations.

بئر الشيبان   Ok   *Bîr esh Sheibân.* The well of Sheibân. p.n. Probably from the well-known tribe of the Beni Sheibân. The word means 'a cold frosty day,' and the root from which it is derived means to be 'hoary.'

بئر السويد   Nk   *Bîr es Sûweid.* The blackish well; but *see Jubb Suweid,* p. 5.

البيرة   Pj   *El Bîreh.* The fortress. *See 'Ain el Bîreh,* p. 157.

بركة ام الغزلان   Oj   *Birket Umm el Ghûzlân.* The pool with the gazelles.

دابو   Oj   *Dâbû.* Dâbû. p.n.

ديرابى نعيف   Ol   *Deir Abu Dâîf.* The convent of Abu Dâîf. p.n. = father of the weak or lean one.

دير غزالة   Ol   *Deir Ghûzâleh.* The convent of the gazelles.

دير السدان   Ol   *Deir es Sudân.* The convent of the veil, or the temple-keeper. *cf. 'Ain Abu Sudân,* p. 13.

الدلهمية   Qj   *Ed Delhemiyeh.* Ed Delhemiyeh, family name, from *Delham,* meaning 'a wolf.'

دنة   Pj   *Denna.* The Amphora. It may however be from Hebrew דַּנָּה 'lying low.'

الضحاك   Qj   *Edh Dhahâk.* The wide road. Applies to the junction of *Wâdy el Bîreh* with the Jordan.

الدوابة   Qk   *Ed Dûwâyeh.* The scum. Applied to the course of the stream at Beisân.

اندور   Oj   *Endôr.* Endor. Hebrew עֵין דֹּר.

اسدود   Pl   *Esdûd.* Dams (a ruined dam to the stream).

فتوع   Ol   *Fukûû.* Toadstools.

الفولة   Nj   *El Fûleh.* The beans.

الحمام   Qk   *El Hûmmâm.* The warm bath (an artificial reservoir).

جبول   Qk   *Jabbûl.* Jabbûl. p.n. *cf.* Heb. גְּבוּל boundary.

جامع الاربعين غزاوي   Pk   *Jamiâ el Arbûin Ghardwi.* The mosque of the Forty Warriors (at Beisân).

الجار   Pl   *El Jâr.* Abounding in herbage.

جبل ابي مدور   Ol   *Jebel Abu Madwar.* The mountain with the round rock.

جبل التلالى   Ok   *Jebel el Kaleily.* The mountain with the peaks.

جلمة   Nk   *Jelameh.* Properly جليمة 'the hill.'

جلبون   Ol   *Jelbôn.* Jelbûn. p.n.

جليل   Pl   *Jeljel.* Jeljel. p.n. *cf.* Heb. גִּלְגָּל.

جلتموس   Ol   *Jelkamûs.* Jelkamûs. p.n. Perhaps גַּל קִמּוֹס The mound of thistles.

جسر الخان   Pk   *Jisr el Khân.* The bridge of the Caravanserai.

جسر المقطوع   Qk   *Jisr el Maktûâ.* The cut off bridge.

جسر المجامع   Qj   *Jisr el Mujâmiâ.* The bridge of the place of assembling. *See* "Memoirs."

الجيزل   Ql   *El Jizil.* 'Cut off.'

تبر الذيابة   Pj   *Kabr edh Dhiâbeh.* The grave of the Dhiâbeh (men of the Dhiâb family).

قنات الحكيمية   Qk   *Kanât el Hakeimîyeh.* The canals of the Hakeimîyeh; family name.

قنات السخنة   Pk   *Kanât es Sokhny.* The canals of warm water.

قنات الطعنة   Qk   *Kanât et Tâneh.* The canals of the spear-thrust, or of the 'taunt.'

قنات ام حيل   Qk   *Kanât Umm Heil.* The canals with the alluvial deposit.

قنات الوقف   Qk   *Kanât el Wôkif.* The canal of the property belonging to a religious endowment.

القنطرة   Pk   *El Kantarah.* The arch.

كوكب الهوا   Qj   *Kaukab el Hawa.* Kaukab, 'of the wind.' Kaukab, meaning a star, is a name of frequent occurrence.

كفر مصر   Pj   *Kefr Misr.* The village of the town (or of Egypt).

كفرة   Pj   *Kefrah.* The village.

خان الاحمار   Pk   *Khân el Ahmâr.* The Caravanserai of the asses.

الخضر   Nl   *El Khŭdr.* El Khŭdr. *See* p. 28.

خربة عباة   Nl   *Khŭrbet 'Aba.* The ruin of the '*Aba*, a cloak worn by the Arabs.

خربة ادمآ   Qj   *Khŭrbet Admah.* Properly *admâ*, the ruin of the brown soil (Heb. אֲדָמָה 'red earth'), from the purple basaltic soil. *See* Section A.

خربة عين الحية   Qj   *Khŭrbet 'Ain el Haiyeh.* The ruin of the spring of the serpent.

خربة برغشة   Ol   *Khŭrbet Barghashah.* The ruin of the Barghash family. The name, meaning 'a gnat,' is borne by several Arab families, amongst others by the Sultan of Zanzibar.

خربة بدرية   Qj   *Khŭrbet Bedrîyeh.* Probably for *Beidariyeh*, ruin of the threshing-floor.

خربة بيت إلفا   Pk   *Khŭrbet Beit Ilfa.* The ruin.

خربة بقع   Qj   *Khŭrbet Beka.* The ruin of the speckled ground, or 'of the valley.'

خربة بير طيس   Nj   *Khŭrbet Bir Tibas.* The ruin of the plastered well. طيس also means 'abundant water,' and 'black.'

خربة فتيقيعة   Ok   *Khŭrbet Fukeikiâh.* The ruin of toadstools.

خربة العشة   Qk   *Khŭrbet el 'Esh-sheh.* The ruin of the nest.

خربة الحداد   Pj   *Khŭrbet el Haddâd.* The ruin of the blacksmith.

خربة الحكيمية   Qk   *Khŭrbet el Hakeimîyeh.* The ruin of el Hakeimîyeh. *See* Kanât el Hakeimîyeh, p. 161.

خربة الحمرآ   Pl   *Khŭrbet el Hŭmra.* The red ruin (from the colour of the soil).

خربة المجديدة   Ok   *Khŭrbet el Jûdeideh.* The ruin of the dykes (*i.e.*, coloured streaks in the rocks).

خربة قارا Oj *Khŭrbet Kâra.* The ruin of Kâra; probably قارة signifying a detached hill, or large stone.

خربة قمل Pj *Khŭrbet Kummil.* The ruin of lice.

خربة مألوف Oj *Khŭrbet Mâlûf.* The ruin of the familiar friend.

خربة المغاير Ol *Khŭrbet el Mŭghair.* The ruin of the caves.

خربة المجدع Pl *Khŭrbet el Mujeddá.* Ruin of the cropped or cut-off place. *See* "Memoirs," under 'Megiddo.'

خربة النجار Nl *Khŭrbet en Nejjâr.* The ruin of the carpenter.

خربة صابر Pk *Khŭrbet Sâbir.* The ruin of the patient one.

خربة السامرية Pl *Khŭrbet es Sâmrîyeh.* Ruin of the Samaritans.

خربة سيرة Oj *Khŭrbet Sîreh.* The ruin of the fold.

خربة الصفصافة Oj *Khŭrbet es Sûfsâfeh.* The ruin of the osier-willow.

خربة الطاقة Qj *Khŭrbet et Tâkah.* The ruin of the arch or shelf.

خربة طبعون Ok *Khŭrbet Tubâûn.* The ruin of Tubâûn. *See Tubâûn,* p. 117.

خربة طونس Pk *Khŭrbet Tûnis.* The ruin of Tûnis. p.n.; the root signifies 'very dark.'

خربة ام العلق Qj *Khŭrbet Umm el 'Alak.* The ruin with the leeches.

خربة ام غوادي Nj *Khŭrbet Umm Ghawâdy.* The ruin of those who sally forth in the morning; also called خربة ام الرايات *Umm el Râiât,* the mother of standards.

خربة ام صابوني Qj *Khŭrbet Umm Sabôuy.* The ruin of the mother of the soap-maker.

خربة يبلا Pk *Khŭrbet Yebla.* The ruin of Yebla. p.n. *cf.* Arabic وبل 'a shower.'

خربة الزاويان Qj *Khŭrbet ez Zâiwiyân.* The ruin of tares (darnel).

الخزنة Pk *El Khâznch.* The treasury.

القلعة Oj *El Kŭlâh.* The castle (a rock).

قومية Ok *Kŭmieh.* Stature, support.

قرية دحي Oj *Kuryet Dûhy.* The village of Dûhy. *See Neby Dûhy.*

مخاضة عبارة Qk *Makhâdet 'Abârah.* The ford of the crossing. *See* "Memoirs," under 'Bethabara.'

مخاضة عبد الله Qk *Makhâdet 'Abdallah.* The ford of 'Abdallah. p.n.

مخاضة ابى ناج Ql *Makhâdet Abu Nâj.* The ford of Abu Nâj. p.n.

مخاضة اقلة Ql *Makhâdet Aklch.* The ford of Akleh (an Arab woman who was drowned there).

مخاضة عورتة Ql *Makhâdet 'Awertah.* The ford of 'Awertah. p.n.

مخاضة البرتقانى Qk *Makhâdet el Bortukâni.* The ford of the orange.

مخاضة الدحمية Qj *Makhâdet ed Delhemîyeh.* The ford of ed Delhemiyeh. q.v.

مخاضة الذيابة Ql *Makhâdet edh Dhiâbeh.* The ford of the Dhiâbeh (family name, meaning 'wolves').

مخاضة الكركار Ql *Makhâdet el Kerkâr.* The ford of the grinding or driving along.

مخاضة كرم الحاج صالح Ql *Makhâdet Kurm Hâj Sâleh.* The ford of the vineyard of el Hâjj Sâleh. El Hâjj is the title given to anyone who has made the pilgrimage to Mecca.

مخاضة القطاف Qj *Makhâdet el Kûtâf.* The ford of the vine-dresser.

مخاضة النسوان Qk *Makhâdet el Niswân.* The women's ford.

مخاضة نقب الباشا Qk *Makhâdet Nŭkb el Bâsha.* The ford of the pass of the Pacha.

مخاضة نقب فارس Ql *Makhâdet Nŭkb Fâris.* The ford of the pass of the horseman.

مخاضة نقب الشايبات Ql *Makhâdet Nŭkb esh Shâyebât.* Ford of the pass of the old men.

مخاضة رزينة Qk *Makhâdet Razîneh.* The ford of stagnant water.

مخاضة الرمانة Ql *Makhâdet er Rummâneh.* The ford of the pomegranate tree.

مخاضة الشيخ داود Ql *Makhâdet esh Sheikh Dâûd.* The ford of Sheikh David.

مخاضة الشيخ حسين Qk *Makhâdet esh Sheikh Husein.* The ford of Sheikh Husein.

مخاضة الشيخ قاسم Qj *Makhâdet esh Sheikh Kâsim.* The ford of Sheikh Kâsim.

مخاضة السفرة Ql *Makhâdet es Sŭfrah.* Ford of the journey (or 'of the table').

مخاضة الصغير Ql *Makhâdet es Sughaiyir.* The little ford.

مخاضة تل حمود Qk *Makhâdet Tell Hammûd.* The ford of Hammud's mound.

مخاضة ام الحجار Ql *Makhâdet Umm el Hajâr.* The ford with the stones.

مخاضة ام القرانيس Qk *Makhâdet Umm el Kerânis.* The ford of the promontories.

مخاضة ام توتة Qk *Makhâdet Umm Tûteh.* The ford of the mulberry-tree.

مخاضة ام الوحل Qj *Makhâdet Umm el Wahl.* The ford with the mud.

مخاضة الزاعبي Qk *Makhâdet ez Z'âbi.* The ford of ez Z'âbi. p.n.; meaning a 'traveller'; but the root زعب also means 'filling with water.'

مخاضة زور الشومر Ql *Makhâdet Zôr esh Shômar.* Ford of the hollow of fennel.

مخاضة زور السمسم Ql *Makhâdet Zôr es Simsim.* The ford of the curve of sesame.

المسكنية Pk *El Maskaniyeh.* The habitations.

المزار وقيل الوزر Ok *El Mazâr.* The shrine; called also *el Wezr*, 'the burden.'

المفرق Pl *El Mefrak.* The parting (of roads).

المراغة Pj *El Merâghah.* The place where cattle roll themselves about.

المشرع Qk *El Mishrâ.* (Beisân Special Survey.) The watering-place.

الموبرة Qk *El Môbarah.* The quarry.

المغاير Ol *El Mŭghair.* The caves.

مغارة ابي ياغي Qk *Mŭghâret Abu Yâghî.* The cave of Abu Yâghi. p.n.

مغارة الجوفة Pl *Mŭghâret el Jûfeh.* The hollow cave.

مغارة الشامي Qj *Mŭghâret esh Shâmy.* The cave of the Syrian.

مغارة التل *Mŭghâret et Tell (Beisân).* The cave of the mound.

مغارة ام الحلس Nj *Mŭghâret Umm el Halas.* The cave of the verdure.

مغارة الزعيدة Pk *Mŭghâret ez Zâideh.* The cave of Zâideh. p.n.

مقام سيدنا عيسى Oj *Mukâm Sidna 'Aisa.* The station of our Lord Jesus.

المنطار {Pl / Qj} *El Mŭntâr.* The watch-tower.

المنطار الابيض Pl *Mŭntâr el Abeid,* properly *Abyadh.* The white watch-tower.

المنطار الازرق Pl *Mŭntâr el Azrak.* The blue watch-tower.

المرتص‎   Pk   *El Murŭssus.* The place of the rubble.

المطلة‎   Qk   *El Mutelly.* The look-out.

نهر جالود‎   Pk   *Nahr Jâlûd.* Goliath's river. *See Ain Jâlûd,* p. 157.

نواحي جنين‎   *Nawâhy Jenîn.* The communes of Jenîn; also called بلد الحارثة‎ *Belâd el Hârîtheh,* or the land of tilth.

الناعورة‎   Oj   *En Nâûrah.* The water-wheel.

النبي دانيان‎   Pj   *Neby Dânîân.* The Prophet Daniel.

النبي دحي‎   Oj   *Neby Dŭhy.* The Prophet Dŭhy. p.n.

نين‎   Oj   *Nein.* Nain. p.n. **Naïn.**

نخلة السيبة‎   Ql   *Nukhlet es Sîbeh.* The palm of the stream.

نورس‎   Ok   *Nûris.* Nûris. p.n.

راس الشيبان‎   Ok   *Râs esh Sheibân.* The hill-top of Sheibân. *See Bîr esh Sheibân,* p. 160.

الريحانية‎   Ok   *Er Rîhânîyeh.* The place of sweet basil; perhaps a family name.

السبعين‎   Nl   *Es Sebâîn.* The Seventy.

السراي‎   Pk   *Es Serâi.* The court-house (at Beisân).

شفعة القبلي‎   Ol   *Shefât el Kibly.* The southern federation; part of the *Nâhiet Jenîn,* or commune of Jenin.

الشيخ العجمي‎   Ol   *Sheikh el 'Ajamy.* The Persian Sheikh.

الشيخ الرحاب‎   Pl   *Sheikh Arehâb.* The Sheikh of er Rehâb ('the broads').

الشيخ ازعة‎   Pj   *Sheikh Azâbeh* or *ez Zâbeh. See Makhâdet ez Z'âbi,* p. 165. Sheikh means 'elder,' 'chief,' 'saint,' &c. Saint's tomb.

الشيخ برقان‎   Ok   *Sheikh Barkân.* Sheikh Barkân. p.n. *cf. Bŭrktn,* p. 146.

الشيخ ذئب‎   Ol   *Sheikh Dîb.* Sheikh Dib ('wolf,' a common Arab name).

الشيخ الحلبي‎   Pk   *Sheikh el Halebi.* The Aleppine Sheikh.

الشيخ حسن‎   Ok   *Sheikh Hasan.* Sheikh Hasan. p.n.

الشيخ ميماس‎   Ok   *Sheikh Mîmâs.* p.n. *See Deir Mîmâs,* p. 20.

الشيخ محمد‎   Qk   *Sheikh Muhammed.* Sheikh Mohammed. p.n.

الشيخ محمد القابو Ql *Sheikh Muhammed el Kábu.* Sheikh Mohammed of the Dome.

الشيخ المخيشيق Ql *Sheikh el Mukheishik.* Sheikh el Mukheishik. p.n.; meaning either 'galloping about,' or 'large spoons'; it is applied to large trees.

الشيخ صالح Xl *Sheikh Sáleh.* Sheikh Sáleh. p.n.; meaning 'righteous.'

الشيخ الصماد Ql *Sheikh es Semád.* Sheikh es Semád. p.n.; meaning 'fighting.'

اشلاليف Oj *Esh Shellálif.* The descending waters.

الشريعة Qk *Esh Sheríah.* The watering-place. This is the modern title of Jordan. The root means also to 'flow straight on or through.'

شونة طمرة Oj *Shúnet Tũmrah.* The barn of Tũmrah. q.v., p. 117.

شتا Pk *Shátta.* Shũtta; probably شـ 'a river bank.'

سيدنا عبد الرحمن
ابو العوف Ol *Sídna 'Abd er Rahman Abu el 'Aúf.* 'Abd er Rahman, 'our lord' (the servant of the Merciful), 'father of 'Aúf'; p.n.

سيرين Pj *Sírin.* Folds (a village).

سولم Nj *Sólam.* Sólam. p.n. Heb. שׁוּנֵם Shunem.

سندلة Nk *Sũndela.* p.n.; meaning 'thick-headed.' The word also means 'sandal wood,' and 'a stool.'

الطيبة Pj *Et Taiyibeh.* The goodly.

طاحونة الاشدف Pk *Táhũnet el Ashdaf.* The mill of el Ashdaf. Probably from *Shádũf,* a contrivance for raising water for irrigation purposes.

طاحونة الرجة Pk *Táhũnet el Arjah.* The mill of the counterpoise.

طاحونة ارسان المطلق Ok *Táhũnet Arsán el Matlak.* The mill of the rugged ground of the open pasture-land.

طاحونة العصي Pk *Táhũnet el 'Asy.* The mill of el 'Asy. *See Ain el 'Asy,* p. 157.

طاحونة البقر Qj *Táhũnet el Bakr.* The mill of the cow.

طاحونة احمرآ Pl *Táhũnet el Hũmra.* The red mill.

طاحونة جلجل Ql *Táhũnet Jeljel.* The mill of Jeljel. q.v., p. 161.

طاحونة الجَرِم Pk *Tâhûnet el Jerm.* The mill of the sheep-shearing, or of the crime.

طاحونة الجِسر Ok *Tâhûnet el Jisr.* The mill of the bridge.

طاحونة الجوسَق Pl *Tâhûnet el Jôsak.* The mill of the terrace.

طاحونة الخراوش Pk *Tâhûnet el Khrâush.* The mill of the old household stuff; it may also mean ' of foam.'

طاحونة الخربة Ok *Tâhûnet el Khûrbeh.* The mill of the ruin.

طاحونة القوسي Pk *Tâhûnet el Kôsi.* The mill of the bowman.

طاحونة المالحة Pk *Tâhûnet el Mâlhah.* The salt mill.

طاحونة المرتفعة Ql *Tâhûnet el Mirtefâh.* The elevated mill.

طاحونة الرأس Ok *Tâhûnet er Râs.* The mill of the (spring) head, near the spring-head of 'Ain Jâlûd.

طاحونة الساخني Pk *Tâhûnet es Sâkhny.* The mill of es Sâkhiny. p.n. *See* '*Ain es Sokhneh,* p. 158.

طاحونة الشيخ Ok *Tâhûnet esh Sheikh.* The mill of the Sheikh. *See Sheikh Azâbeh,* p. 166.

طاحونة الشيخة Ql *Tâhûnet esh Sheikhah.* The mill of the female Sheikh.

طاحونة الشيخ امين Qj *Tâhûnet esh Sheikh Amîn.* The mill of Sheikh Amin. p.n. ; ' trusty.'

طاحونة الشيخ ابرهيم Ok *Tâhûnet esh Sheikh Ibrahîm.* The mill of Sheikh Abraham, p.n.

طاحونة الطوم Pk *Tâhûnet et Tôm.* The mill of et Tôm. p.n.

طاحونة الويبدة {Ql } {Ok } *Tâhûnet el Weibedeh.* Mill of the mountain pass. There are two of this name.

طواحين الحسنية Ql *Tawâhîn el Hasaniyeh.* Mills of the Hasan family or sect.

طواحين الرشيد Pk *Tawâhîn er Rashîd.* The mill of er Rashid.

طواحين الوادي Pk *Tawâhîn el Wâdy.* The mills of the valley.

تل ابي فرج Ql *Tell Abu Faraj.* The mound of Abu Faraj. p n.

تل ابي الجمل Qj *Tell Abu el Jemel.* The mound of the man with the camel.

تل العجول وقيل تل العبد Oj *Tell el 'Ajjûl.* The mound of black mud. Or *el 'Ajjâl,* of ' the hastener,' or ' the herd ' (a volcanic crater).

تلّ الباشا	Ql	*Tell el Básha.* The mound of the Pasha.
تلّ الذّنابة	Ql	*Tell edh Dhiábeh.* The mound of the Dhiábeh. p.n.; (wolves).
تلّ الفار	Nj	*Tell el Fár.* The mound of the mouse.
تلّ الفرّ	Ok	*Tell el Ferr.* The mound of flight.
تلّ الحرفة	Oj	*Tell el Harfeh.* The mound of the brink or border; but perhaps from حَرف, a sort of grain like mustard.
تلّ الحزنبل	Nj	*Tell el Hazambal.* The mound of the 'hazambal,' a kind of sweet herb.
تلّ الحصن	Pk	*Tell el Hosn.* The mound of the fortress.
تلّ الجسر	Pk	*Tell el Jisr.* The mound of the bridge.
تلّ الجزل	Ql	*Tell el Jizil.* The mound of el Jizil. q.v., p. 161.
تلّ الخنزير	Ql	*Tell el Khanciztr.* The mound of the swine.
تلّ المالحة	Ql	*Tell el Málhah.* The salt mound.
تلّ المصطبة	Qk	*Tell el Mastabah.* The mound of the bench.
تلّ المقرقش	Pj	*Tell el Makurkush.* The mound of el Makurkush. p.n.; meaning in vulgar Arabic any hard food, like chick pease.
تلّ المنشية	Ql	*Tell el Menshíyeh.* The mound of Menshiyeh. p.n.; 'grown.'
تلّ نمرود	Ql	*Tell Nimrúd.* The mound of Nimrod.
تلّ الرّعيان	Ql	*Tell er Ráîan.* The mound of the shepherds.
تلّ الصارم	Pl	*Tell es Sárem.* The mound of the sharp scimitar.
تلّ الشيخ داود	Ql	*Tell esh Sheikh Dáúd.* The mound of Sheikh David. q.v.
تلّ الشيخ حسن	Ok	*Tell esh Sheikh Hasan.* The mound of Sheikh Hasan. q.v.
تلّ الشيخ قاسم	Qj	*Tell esh Sheikh Kásim.* The mound of Sheikh Kâsim. q.v.
تلّ الشمدين	{Pl / Qj}	*Tell esh Shemdin.* The mound of Shemdin. p.n.
تلّ الشوك	{Qj / Pl}	*Tell esh Shôk.* The mound of thorns.
تلّ الشقف	Ql	*Tell esh Shúkf.* The mound of the cleft.

z

تل الزنبقية Qj   *Tell ez Zanbakîyeh.* The mound of white lilies.

تلول فروانة Pl   *Tellûl Farwanah.* The mounds of Farwanah. p.n.

تلول الثوم Pl   *Tellûl eth Thûm.* The mounds of garlic.

تلول الزهرة Pk   *Tellûl ez Zahrah.* The mounds of the flower.

طيرة ابى عمران Pj   *Tîret Abu 'Amrân.* The fort of Abu 'Amrân. p.n.

طيرة الخاربة Pj   *Tîret el Khârbeh.* The ruined fort.

طمرة Oj   *Tûmrah.* Tûmrah. p.n. *See* p. 117.

الطوال Qk   *Et Tuwâl.* The long; a long ridge.

ام العمدان Ql   *Umm el 'Amdân.* The place with the pillars.

ام سريسة Pl   *Umm Sarrîseh.* The place with the lentisk (*Pistachio lentiscus*), which grows here.

ام الشرسية Qk   *Umm esh Sherashîyeh.* Mother of shrubs; a tree.

ام التوت Ol   *Umm et Tût.* The place with the mulberries.

وادي ابى حديدة Oj   *Wâdy Abu Hadîdeh.* The valley with the piece of iron.

وادي العين Nj   *Wâdy el 'Ain.* The valley of the spring.

وادي العتي Oj   *Wâdy el 'Asy.* The valley of el 'Asy. *See* '*Ain el 'Asy*, p. 157.

وادي الاسمر Oj   *Wâdy el Asmar.* The brown valley.

وادي عيون الكحل Pj   *Wâdy 'Ayûn el Kahel.* The valley of the springs of manganese.

وادي بليق Oj   *Wâdy Beleik.* The valley of Beleik. p.n. The word means 'piebald,' and was the name of a horse which though thought little of used to run well. There is a proverb in Arabic, يجرى بليق ويذم بليق 'Buleik runs, but Buleik is despised.' It may be a diminutive form connected with بلقا Belkâ, a common geographical name in Arabia, meaning either 'variegated soil,' or 'a fertile open valley.'

وادي بعلى وقيل وادي ام توتة Qj   *Wâdy Bâly,* or *Wâdy Umm Tûteh.* Valley of the high lands; or the valley with the mulberries.

وادي البيرة Qj   *Wâdy el Bîreh.* The valley of el Bîreh. *See el Bîreh*, p. 160.

وادي دابو Pj   *Wâdy Dâbû.* The valley of Dâbû. *See Dâbû.*

وادي العشة Qk *Wâdy el 'Esh-sheh.* The valley of the nest.

وادي حمود Pj *Wâdy Hammûd.* The valley of Hammûd. p.n.

وادي الحريّة Ok *Wâdy el Harriyeh.* The valley of soft sandy soil.

وادي المختلية Pj *Wâdy el Hoktiyeh.* The valley of the Hoktiyeh; family name, meaning 'dwarfish,' or 'agile.'

وادي الحفير Nk *Wâdy el Hûfiyir.* The valley of the excavation.

وادي الحمرآ Pl *Wâdy el Hûmra.* The red valley.

وادي الجمل Oj *Wâdy el Jemel.* The valley of the camel.

وادي الجديد Ok *Wâdy el Jûdîd.* The new valley; but *see Khûrbet el Jûdeideh,* p. 162.

وادي الكليلى Ok *Wâdy el Kaleily.* The valley of el Kaleily. *See Jebel el Kaleily,* p. 161.

وادي التنطرة Pk *Wâdy el Kantarah.* The valley of the arch.

وادي الخنزير Qk *Wâdy el Khaneizîr.* The valley of the swine.

وادي الخرّار Pk *Wâdy el Kharrâr.* The valley of the murmuring waters.

وادي الخزنة Pk *Wâdy el Khûzneh.* The valley of the treasures.

وادي المخزقة Pk *Wâdy el Mukhuzakah.* The valley of el Mukhuzakah. *See 'Ain el Mukhuzakah,* p. 158.

وادي النصف Ol *Wâdy en Nûsf.* The half-way valley.

وادي نصورة Oj *Wâdy Nûsûrah.* The valley of Nûsûrah. *See 'Ain Nuseirah,* p. 158.

وادي الرصيف Ol *Wâdy er Rasîf.* The valley of the firm or solid (rock.)

وادي ريضان Ok *Wâdy Rîdân.* The valley of meadows or gardens; pl. of روضة.

وادي الرمل Ok *Wâdy er Rûml.* The valley of sand (in this case basaltic *débris*).

وادي صابر Pk *Wâdy Sâbir.* The valley of the patient one.

وادي شليطا Pk *Wâdy Shelleit.* The valley of the ploughshare.

وادي السد Pl *Wâdy es Sidd.* The valley of the dam.

وادي السدر Oj *Wâdy es Sidr.* The valley of the lotus tree (*Zyziphus lotus*).

Z 2

وادي شرار Pj *Wády esh Sherrár.* The valley of the wicked one, or of 'sparks.'

وادي شوباش Pl *Wády Shúbásh.* Probably for شوباشى, in vulgar Arabic 'the agent or bailiff of an estate'; from the Turkish صوباشى ; or perhaps it is for شوبص 'intricate brambles,' 'a spinny.' ROBINSON spells it كبوش *Kubósh.*

وادي صبحة Oj *Wády Subhah.* The reddish-white valley.

وادي ام الضبع Pl *Wády Umm ed Dubá.* The valley with the hyænas.

وادي ام ربيع Pj *Wády Umm Rabíá.* The valley of the spring pasturage.

وادي الشعلة Nj *Wády esh Shuálah.* The valley of the flame; but perhaps from Heb. שׁעֹל 'a hollow.' *See Bír Abu Shuálch,* Sheet XVIII.

وادي ام طنطور Nj *Wády Umm Tantúr.* The valley with the peak. *See* p. 117.

وادي ام ولهان Qj *Wády Umm Walhán.* The valley of the mother of the distracted one.

وادي الواويات Oj *Wády el Wáwiyát.* The valley of jackals.

وادي يبلا Pj *Wády Yebla.* The valley of Yebla. *See Khúrbet Yebla,* p. 163.

وادي زريق Pj *Wády Zereik.* The bluish valley.

وطا الجالود Pk *Watá el Jálúd.* The low ground of the Nahr Jálúd. *See* p. 166.

زعترة Nl *Záterah.* Thyme.

زبع (و مكيت) Pk *Zebá (wa Mekeit).* Zebá, also called Mekeit. Both proper names. The first meaning 'snorting with rage,' and *cf.* زبوعة 'a whirlwind.'

زرعين Nk *Zer'ín.* Zer'in. p.n. It exactly corresponds with the Heb. יִזְרְעֶאל Jezreel.

SHEET X.

ابو زبورا Im *Abu Zabûra.* Abu Zabûra. p.n.; the most probable meaning is, 'cased with stones.'

عين توبة In *'Ain Tûbeh.* The spring of repentance. Probably an error for عين طوبى the spring of good water.

عين الزرقية Io *'Ain ez Zerkîyeh.* The blue spring.

عرب الحوارث In *'Arab el Hawârith.* Arabs of the Hawârith tribe. p.n.; signifying 'tillers of the soil,' 'ploughmen.'

عرب النفعية In *'Arab en Nefeiât.* Arabs of the Nefeiát tribe. p.n.; signifying 'a staff.'

ارسوف Ho *Arsûf.* Arsûf. p.n.

عيون ابو شهوان Im *'Ayûn Abu Shehwân.* The springs of Abu Shehwân. p.n. (The father of the sensual one.)

عيون البقة Io *'Ayûn el Bassch.* The springs of the marsh.

عيون الغزلن In *'Ayûn el Ghûzlân.* The springs of the gazelles.

عيون حسن Jm *'Ayûn Hasan.* The springs of Hasan. p.n.

عيون القصب Jm *'Ayûn el Kŭsab.* The springs of the reeds.

عيون المصعير وقيل عيون الزرانيق Jm *'Ayûn Mesáïyîr.* The springs of Mesáïyir. p.n.; meaning either 'distorted,' or 'journeying hard by night in search of water.' Also called *'Ayûn ez Zerânîk,* 'springs of the channels.'

عيون الوردات Jn *'Ayûn el Werdât.* Springs of going down to water.

بحرة تطورآ Ho *Bahret Katûrah.* The lake of Katûrâ; name of a plant; but perhaps from تطور 'a rain cloud.'

بص الهندي In *Bass el Hindi.* The marsh of the Indian.

بصة ابن علق Io *Basset Ibn 'Alak.* The marsh of the son of leeches.

بئر الابابشة Io *Bîr el Abâbsheh.* The well of pot-herbs.

بثر البليقة   Hn   *Bîr el Beleikeh.* The well of Beleikeh. p.n. *See Wâdy Beleik*, p. 170.

بثر غنيم   In   *Bîr Ghaneim.* The well of Ghaneim. p.n. ; 'plundering'.

بثر الرشرش   Im   *Bîr er Rishrish.* The well of the willow (*Agnus castus*).

بثر اليزك   In   *Bîr el Yezek.* The well of the sentinel.

بثر زيد   Ho   *Bîr Zeid.* Zeid's well. A common Arabic name, meaning 'increase.'

بركة عبدالله   Im   *Birket 'Abdallah.* The pool of 'Abdallah. p.n.

بركة عتا   Im   *Birket 'Ata.* The pool of 'Ata. p.n.

بركة عطيفة   Ho   *Birket 'Atifeh.* The pool of 'Atifeh. p.n. ; 'mild,' 'kind.'

بركة الحصو   Im   *Birket el Hasu.* The pebbly pool.

بركة المزورقة   Hn   *Birket el Muzôrakah.* The pool of el Muzôrakah. p.n. ; either from زورق 'a boat,' or تزورق 'to excrete.'

بركة النجار   Im   *Birket en Nejjâr.* The pool of the carpenter. In Hebrew נגר means 'flowing water.'

بركة شيشان   Ho   *Birket Shîshân.* The pool of Shishân. p.n.

بركة سبيلة   Ho   *Birket Subeily.* The pool of the little path.

بركة زيزان   Ho   *Birket Zeizân.* The pool of squills.

بيارة كاورك   In   *Biyâret Kâwirk.* The wells of Kâwirk. The word is a Persian one, meaning a sort of cucumber.

الدسوقية   In   *Ed Dusûkiyeh.* The full or clear water.

حرم علي ابن عليم   *Haram 'Aly Ibn 'Aleim.* The sanctuary of 'Aly Ibn 'Aleim. p.n. ; called for short الحرم el Haram.

هراب المسكة   Io   *Hurâb el Miskeh.* The tanks of the water-holes.

الحمام   Jo   *El Hûmmâm.* The bath.

القنطور   Ho   *El Kantûr.* The arch.

قربونة   Jo   *Karbôneh.* Karbôneh. p.n. ; 'offering,' 'sacrifice.'

كفر سابا   Jo   *Kefr Sâba.* The village of Sâba. p.n. ; the Aramaic כבר סבא.

الخشاش   Jm   *El Khashâsh.* The entrance.

خربة الجزيرة   In   *Khûrbet el Jezîreh.* The ruin of the island.

خربة الجيوش	Io	*Khûrbet el Jiyûsch.* The ruin of Jiyûsch. p.n.; from the village of *Jiyûs.* It is also called البرابة *el Harâbeh,* the tanks.
خربة مد الدير	Im	*Khûrbet Madd ed Deir.* The ruin of the glebe of the monastery.
خربة مليكة	Io	*Khûrbet Maleika.* The ruin of Maleika. p.n.; meaning perhaps 'level ground.'
خربة المنطار	Ho	*Khûrbet el Mântâr.* The ruin of the watch-tower.
خربة سابية	Io	*Khûrbet Sâbieh.* The ruin of Sâbieh. *cf. Kefr Sâba,* near which it lies, p. 174.
خربة الشيخ محمد	Im	*Khûrbet esh Sheikh Muhammed.* The ruin of Sheikh Muhammed. p.n.
خربة الزبابدة	In	*Khûrbet ez Zebâbdeh.* The ruin of the Zebâbdeh; family name. *cf. Khûrbet Zebed,* p. 51.
خربة الزرقية	Io	*Khûrbet er Zerkiyeh.* The ruin of ez Zerkiyeh. *See 'Ain ez Zerkiyeh,* p. 173.
مد الاكراد	Io	*Madd el Akrâd.* The property of the Kurds.
مجاهد شيخة	Im	*Mejâhed Sheikhah.* The conflict or industry of the (female) elder or saint.
المسرف	Ho	*El Mesraf.* The place of flooding. The name applies to a tunnel for letting out the waters of the marsh.
مينة ابو زبرا	Im	*Minet Abu Zabûra.* The harbour cased with stones.
مسكة	Jo	*Miskeh.* The water-hole.
المواحة	Im	*El Muâlha.* The salt soil.
مغارة ابو سماحا	Im	*Mughâret Abu Semâhâ.* The cave of Abu Semâhâ. p.n.
المغاير	In	*El Mughâir.* The caves.
مغار الشرف	Im	*Mughâr esh Sherif.* The chief or imminent caves.
مغر الابابشة	Ho	*Mûghr el Abâbsheh.* The caves of the pot-herbs.
مخالد وقيل مدينة ابواب	In	*Mukhâlid.* Mukhâlid. p.n. It is also called *Medinet Abûâb,* the city of the gates.
نهر الفالق	Hu	*Nahr el Fâlik.* The splitting river, from the rock cutting.
نهر اسكندرونة	Im	*Nahr Iskanderûneh.* The river of Alexandroskene.

راس البلوط	Im	*Râs el Mallûl.* The hill-top of the oaks.
شيخ حسين	Im	*Sheikh Husein.* Sheikh Husein. p.n.
شيخ محمد	Im	*Sheikh Muhammed.* Sheikh Muhammed. p.n.
تبسور	Io	*Tabsôr.* Tabsôr. p.n.
تل العرف	Im	*Tell el 'Arf.* The mound of the crest.
تل الحتف	Im	*Tell el Hekaf.* The mound of the sand-dune.
تل الإنشار	Im	*Tell el Ifshâr.* The mound of the Ifshâr. p.n. (It is the name of a Turkoman tribe.)
تل مسعود	Im	*Tell Mas'ûd.* The mound of Mas'ûd. p.n. ; ' happy.'
ام خالد	In	*Umm Khâlid.* The mother of Khâlid.
ام صور	Jn	*Umm Sûr.* The mother of rock.

SHEET XI.

عبد المغيث Ln *'Abd el Mughith.* 'Abd el Mughith. p.n.; meaning 'servant of the Helper' (a name of God).

ابو نار Km *Abu Nâr.* Abu Nâr. p.n.; meaning 'father of fire.'

ابو شعر Mm *Abu Shâr.* Abu Shâr. p.n.; meaning 'father of hair' or thick foliage.

عين ابى عتيدة Mo *'Ain Abu 'Akeideh.* The spring with the coping-stone.

عين ابى حمور No *'Ain Abu Hammûr.* The spring of asses.

عين عسكر No *'Ain 'Askar.* The spring of 'the army' (at 'Askar).

عين العسل No *'Ain el 'Asl.* The spring of honey.

عين البطة Mn *'Ain el Basseh.* The spring of the marsh.

عين بيت إلما Mo *'Ain Beit Ilma.* The spring of Beit Ilma. p.n. *Beit* means 'house of.'

عين البيادر Mo *'Ain el Beiâdir.* The spring of the threshing-floors.

عين دفنة No *'Ain Dufna.* The spring of the oleander.

عين الفاتورة No *'Ain el Fâkûrah.* The spring of Fâkûrah. p.n.; from فتر 'to dig a well.'

عين الفؤاد Mo *'Ain Fûâd.* The spring of the heart.

عين الفوار Mo *'Ain el Fûwâr.* The spring of the fountain.

عين الغربية Mo *'Ain el Gharbîyeh.* The western spring.

عين هارون Mn *'Ain Hârûn.* Aaron's spring.

عين حوض الفراديس Mn *'Ain Haud el Fûrâdîs.* The spring of the cistern of the 'paradises.'

عين الحفيرة Mm *'Ain el Hûfîreh.* The spring of the excavation.

عين كاكوب No *'Ain Kâkûb.* The spring of blossoms.

عين قريون *'Ain Kariûn.* The spring of Kariûn. p.n.; perhaps Caraites. In Nablus.

عين كفر روما   Mn *'Ain Kefr Rûma.* The spring of the village of Rûmah. p.n.; or 'of the Greeks.'

عين الخضر   Mn *'Ain el Khûdr.* The spring of el Khudhr. *See p. 28.*

عين الكسية   Mo *'Ain el Kisieh.* The spring of el Kisieh. p.n.; from كسا 'to clothe.'

عين كفرات   Mn *'Ain Kuferât.* The spring of the villages.

عين القصب   Mo *'Ain el Kûsab.* The spring of reeds.

عين المزرب   Mo *'Ain el Mezrab.* The spring of the water-spout.

عين المخنة   No *'Ain el Mûkhnah.* The spring of el Mûkhneh. Probably from مخنة the embouchure of a stream.

عين ساريس   No *'Ain Sârîn.* 'Ain Sârin. p.n.

عين السيف   Mn *'Ain es Seif.* The spring of the sword.

عين السفلا   Mn *'Ain es Sifla.* The lower spring, west of Burka.

عين سلامة   Mo *'Ain Selâmeh.* The spring of Selâmeh. p.n.

عين الشبطبط   Km *'Ain esh Shabâtbât.* The spring of the herb called 'shepherd's staff,' or 'dill' (*Galmin aparinoides*).

عين الشرق   Mo *'Ain esh Sherk.* The eastern spring.

عين الشريش   Mo *'Ain esh Sherîsh.* The spring of Sherish (shrubs).

عين الصبيان   Mo *'Ain es Subiân.* The spring of the boys.

عين الصور   Mo *'Ain es Sûr.* The spring of the rock.

عين الثعلب   Mo *'Ain eth Thâleb.* The spring of the fox.

عين ام القناطر   Ln *'Ain Umm el Kanâtir.* The spring with the arches; between *'Ain el Hâfireh* and *el Bizârieh.*

عين زكريا   Mn *'Ain Zakarîya.* The spring of Zachariah.

عجة   Mm *'Ajjeh.* Ajjeh. p.n.

عماد الدين   No/Km *'Amâd ed Dîn.* 'Amâd ed Din. p.n.; meaning 'the support of the Faith.'

اماتين/ماتين   Lo *Amâtîn* or *Mâtein.* Amâtin or Mâtein. p.n.

عنبتا   Ln *'Anebta.* Anebta. p.n. Grapes.

عنزا   Mm *'Anza.* The goats.

عرابة   Mm   *'Arrâbeh.* A steppe.

العراق   Mo   *El 'Arâk.* The cliff or bank.

عراق الديّة   No   *'Arâk et Tiych.* The cliff or bank of wandering.

عسيرة الحطب   No   *'Asîret el Hatab.* The difficult place of timber.

عسيرة القبليّة   Mo   *'Asîret el Kibliych.* The southern difficulty.

عسكر   No   *'Askar.* Army.

عطّارا   Ln   *'Attâra.* 'Attàra. p.n. *cf.* Heb. עברות.

عتيل   Km   *'Attîl.* 'Attil. p.n. ; meaning 'severe.'

عيون الدالي   Jm   *'Ayûn ed Dâly.* The springs of the bucket.

عيون الفرز   Mo   *'Ayûn el Firz.* The springs of the road through the hills.

عيون الحوض   Mn   *'Ayûn el Haud.* The springs of the cistern.

عيون الحفير   Jn   *'Ayûn el Hâfyir.* The springs of the excavations.

عيون الجحاش   Jm   *'Ayûn el Jehâsh.* The springs of the ass.

عيون المجنبات   Km   *'Ayûn el Jennahât.* The springs of the Indian-canes.

عيون الكفي   Jn   *'Ayûn el Kufy.* The springs of el Kufy. If it come from كفّ *Kaff,* it may mean 'a hill' or 'stone-heap'; but if it come from كفو *Kafu,* it means 'pursuing.'

عيون وادي خراج   Mo   *'Ayûn Wâdy Kharâj.* The springs of Wâdy Kharâj. q.v.

عيون الزوطيّه   Jm   *'Ayûn ez Zûtiych.* The springs of the Zûtiych ; perhaps زطيّة of the Zutts (Jâts) gipsies. *See 'Aitâ ez Zutt,* p. 17.

عزبة (حمانة) ابي حرفيل   Ko   *'Azbet (Hammâdeh) Abu Harfil.* The summer pasturage of Abu Harfil. p.n.

عزون   Ko   *'Azzûn.* 'Azzùn. p.n.

باقا بني صعب   Lo   *Bâka Beni Sâb.* Bâka of the Beni Sâb (Arabs). p.n.

باقا الغربيّة   Km   *Bâka el Gharbiych.* The western Bâka.

باقا الشرقيّة   Km   *Bâka esh Sherkiych.* The eastern Bâka.

بلاطا   No   *Balâta.* Flagstones; but *see* "Memoirs," under the name. Also "Memoirs," Sheet III, *Khûrbet Belât.*

باطن النوري   Mm   *Bâtn en Nûry.* The inside or watercourse of the gipsy.

بيت بزين Lo   *Beit Bezzin.* The house of Bezzin. p.n. Samaritan
בית בזין.

بيت إيبا Mo   *Beit Îba.* The house of Îba. p.n.

بيت إمرين Mn   *Beit Imrîn.* The house of Imrin. p.n.

بيت جنا Jo   *Beit Jiffa.* The house of Jiffa. p.n.

بيت ليد In   *Beit Lîd.* The house of Lid. p.n.

بيت أوذن Mo   *Beit Udhen.* The house of Udhen. p.n. ('permission').

بلعة Lm   *Beláh.* Beláh. p.n.; meaning either 'swallowing,' or 'the hole in a millstone.'

بنات يعقوب Kn   *Benât Yâkûb.* Jacob's daughters.

بيارت ابي زريق Ko   *Biâret Abu Zereik.* The wells of the magpie.

بئر العبد Jm   *Bîr el 'Abd.* The well of the slave.

بئر العقاربة Km   *Bîr el 'Akâribeh.* The well of scorpions.

بئر العاصم Km   *Bîr el 'Âsem.* The well of 'Âsem. p.n. ('the innocent').

بئر عسيرة Mo   *Bîr 'Asîreh.* The well of 'Asireh. p.n., meaning 'difficult'; it is also written in the notes عسير, which is connected with the root عصر 'to press.'

بئر عصور Ln   *Bîr 'Asûr.* The well of 'Asûr. *See* last entry.

بئر البشم Mm   *Bîr el Bushm.* The well of the balsam.

بئر الدير Km   *Bîr ed Deir.* The well of the monastery.

بئر الفوارة Mm   *Bîr el Fuwârah.* The well of the fountain.

بئر غزالة Mo   *Bîr Ghûzâleh.* The gazelle's well.

بئر الحنوتا Jo   *Bîr el Hanûtah.* The well of the tavern or booth.

بئر احمرا Jo   *Bîr el Hûmra.* The well of red earth.

بئر الخارجة Mm   *Bîr el Khârjeh.* The outer well.

بئر النبي شمعون Jo   *Bîr en Neby Shem'ôn.* The well of the prophet Simeon.

بئر المنقب Nm   *Bîr en Nŭkb.* The well of the pass.

بئر الرجوم Ko   *Bîr er Rujûm.* The well of the cairns.

بئر الساق Mm   *Bîr es Sâk.* The well of the leg or stalk.

بئر الشيخ Jo *Bir esh Sheikh.* The well of the Sheikh.

بئر شيخ علي Lo *Bir Sheikh 'Aly.* The well of Sheikh 'Aly.

بئر الصفا Mm *Bir es Sûfa.* The pure or clear well.

بئر الصفصاف Jn *Bir es Sûfsâf.* The well of osier-willows.

بئر يعقوب No *Bir Yâkûb.* Jacob's well; called by the native Christians بئر السامرية *Bir es Sâmrîyeh,* 'well of the Samaritan woman.'

بئر زينا Mm *Bir Zeita.* The well of Zeita. q.v.

بركة البط Jo *Birket el Butt.* The pool of the ducks.

بركة دير اصليا Ln *Birket Deir Estia.* The pool of the convent of Estia.

بركة الخراب Lm *Birket el Khûrâb.* The pool of the ruins.

بركة النقعة Kn *Birket en Nukâh.* The pool of the swamp.

البزارية Ln *El Bizâriah.* The well of seeds or pot-herbs.

بورين {Jn Mo} *Bûrîn.* Bûrin. p.n. Perhaps connected with بئر 'a pit.'

البرج Mm *El Burj.* The tower.

برج العطوط Jn *Burj el 'Atôt.* The overthrown tower.

برقا Mn *Burka.* Sandy soil covered with dark stones, especially flints.

البطمة Kn *El Butmeh.* The terebinth (*Pistachio terebinthus*).

دورتا No *Dawerta.* Also called خربة الدوارة *Khûrbet ed Dawârah,* ruin of the circular enclosure or tumulus (of which word it seems to be a corruption).

دير ابان Kn *Deir Abân.* The convent of Abân. p.n.

دير عسفين Jn *Deir 'Asfîn.* The convent of 'Asfîn. p.n. ('deviating').

دير البندق Mo *Deir el Bunduk.* The monastery of the hazel-nut.

دير الغصون Km *Deir el Ghûsûn.* The convent of the branches.

دير سرور Ln *Deir Serûr.* The convent of joy. Also called خربة الناصرة *Khûrbet en Nâsirah,* 'the ruin of Nazareth': خربة السفانة *Khûrbet Sefâneh,* 'the ruin of Sefâna,' p.n.; خربة المسقوفي *Khûrbet el Muskûfy,* the roofed-in ruin.

ديرشراف   Mn   *Deir Sheráf.* The monastery of the nobles.

الدكاك   No   *Ed Dekák.* Dekâk. The word means 'level sands'; but it may be a corruption of *dekákín,* 'shops.' this being a cave, and the word is commonly applied to Jewish tombs by the peasants.

ذنابة   Kn   *Dennábeh* or *Dhennábeh.* The lower part of a valley through which water runs.

ظهر حريق الصوفي   Mm   *Dhahr Harîk es Sûfy.* The ridge of the burning of the Sûfy. The *Sûfis* are a mystic Dervish sect, chiefly found in Persia.

ظهر ام الخرامية   Mm   *Dhahr Umm el Kharrâmîyeh.* The ridge with the holes in the rock. خرامية, vulgar for خارية. The word also applies to a sect of heretics, and means 'Epicureans,' being derived from the Persian.

علّر   Lm   *'Ellâr.* 'Ellar. p.n.

فحمة   Mm   *Fahmeh.* The charcoal.

فلامية   Ko   *Felâmieh.* Felâmieh. The marsh. *cf.* Heb. פֵּילוֹבִיא.

فندقومية   Mn   *Fendakûmieh.* Πεντακόμια.

فرتا   Lo   *Ferâta.* Ferâta. p.n. Perhaps the Samaritan עברה.

فرعون   Kn   *Fer'ôn.* Fer'ôn. p.n. Pharaoh is so called in the Kor'ân.

الفندق   Lo   *El Funduk.* The inn.

فردیسیا   Kn   *Fûrdîsia.* Paradise.

الفوار   *El Fûwâr.* The fountain.

حزن يعقوب   No   *Hizn Yâkûb.* Mourning of Jacob. *See* "Memoirs." It is the supposed place where Joseph's coat was brought to Jacob. It is also called جامع الخضر *Jámiä el Khúdr,* 'mosque of el Khŭdr.' *See* p. 28.

اكتابا   Kn   *Iktâba.* Inscription.

الامام علي بن ابي طالب   *Imâm 'Aly Ibn Abu Tâleb.* The Imâm 'Ali, son of Abu Tâleb, the cousin and son-in-law of the Prophet.

ارتاح   Kn   *Irtâh.* Rest.

جامع العمود   No   *Jámiä el 'Amûd.* The mosque of the pillar.

الجامع الكبير   No   *Jámiä el Kebîr.* The great mosque. For the names of other mosques in Nâblus. *See* "Memoirs."

جبع   Mn   *Jebá.* Jebá. Heb. נֶבִי 'a hill.'

جبل بئر عصور   Ln   *Jebel Bir 'Asûr.* The mountain of Bir 'Asûr. q.v.

جبل اسلامية   No   *Jebel Eslâmîyeh.* The mountain of (Sitt) Eslamiyeh. q.v. The mountain is also called very often Jebel et Tôr, as is also Gerizim.

جبل الطور   No   *Jebel et Tôr.* The mountain of 'the mount' *par excellence.* Ebal, Gerizim, Olivet, and Tabor are all so named.

جلمة   Km   *Jelameh.* The heap. *See* Sheet IX.

جنسنيا   Mn   *Jennesinia.* Jennesinia. p.n.

جت   Km   *Jett.* Jett. Heb. גַת 'wine-press.'

جنيد   Mo   *Jineid.* Jineid; diminutive of جند, which may either mean 'an army,' 'rough ground,' or 'a city.' *See* also *'Ain el Jenadiyeh*, p. 63.

جنفافوط   Lo   *Jinsâfût.* Jinsâfût. p.n.

جسر دار سواد   Kn   *Jisr Dâr Suâd.* The bridge of the house of Suâd. p.n.

جسر المكتبة   Jm   *Jisr el Maktabah.* The inscribed bridge. Also called جسر قاقون *Jisr Kâkôn*, the bridge of Kâkôn. p.n.

جيوس   Ko   *Jiyûs.* Jiyûs. p.n.

الجديدة   Nm   *El Judeideh.* The dyke.

جورة عمرآ   Mo   *Jûrat 'Amrâ.* The lowland of 'Amrâ. p.n.

جورة الحوطي   Km   *Jûrat el Hûtî.* The walled-in hollow.

الجبرية   Nm   *El Jûrbah.* The plantation.

قبر يوسف   No   *Kabr Yûsef.* Joseph's tomb.

قاقون   Jm   *Kâkôn.* Kâkôn. p.n.

قلقلية   Jo   *Kalkilieh.* Kalkiliyeh. p.n.; meaning 'a kind of pomegranate'; also 'gurgling of water.'

كنا   Kn   *Keffa.* Keffa; connected with كَف 'the palm of the hand.'

كفر عبوش   Ko   *Kefr 'Abbûsh.* The village of 'Abbûsh. p.n. (decent).

كفر جمال   Ko   *Kefr Jemmâl.* The village of the camel-driver.

كفر قدوه   Lo   *Kefr Kaddûm.* The village of Kaddûm; from قدم 'front,' 'eastern.' Heb. קָדְמָה.

كفر قلين ﴿
قليل ﴾   No   *Kefr Kullin*, or *Kullil.* The village of Kullin, or Killil. p.n. Samaritan כליאל.

كفر لاتف   Ko   *Kefr Lâkif.* The village of Lâkif, meaning a ruinous structure, especially a well or cistern.

كفر اللبد   Ln   *Kefr el Lebad.* The village of the felt-cloth.

كفر راعي   Lm   *Kefr Râŷ.* The village of the shepherd.

كفر رمان   Ln   *Kefr Rummân.* The village of pomegranates (Rimmon).

كفر سا وقيس ﴿
خربة الزرعة ﴾   Jo   *Kefr Sâ.* The village of Sâ. p.n. It is also called *Khûrbet el Mezrâh.* 'The ruin of the sown land.'

كفر سيب   Km   *Kefr Sîb.* The village of the stream.

كفر صور   Ko   *Kefr Sûr.* The village of the rock.

كفر زيباد   Ko   *Kefr Zîbâd.* The village of Zibâd. *See Zebed,* p. 51.

خلة العتبة   Lo   *Khallet el 'Akabeh.* The dell of the ascent.

خلة قطانية   Lu   *Khallet Katânieh.* The dell of millet.

الخراب   Lm   *El Khûrâb.* The ruins.

خور الحمام   Jm   *Khaur el Hummâm.* The hollow of the bath.

خربة ابي كميش   Kn   *Khûrbet Abu Kemeish.* The ruin of Abu Kemeish. p.n. (father of canes).

خربة العتل   Mn   *Khûrbet el 'Akil.* The ruin of the fortification, or of the fetter.

خربة العتود   No   *Khûrbet 'Akûd.* The ruin of the coping-stones.

خربة عسافة   Lo   *Khûrbet el 'Asâfeh.* The ruin of deviation.

خربة عسكر   Lo   *Khûrbet 'Askar.* The ruin of the army.

خربة اوفار   Mo   *Khûrbet Aufâr.* The ruin of ample spaces.

خربة بيت ساما   Km   *Khûrbet Beit Sâma.* The ruin of the house of Sâma. p.n. (lofty).

خربة بيت سلوم   Mo   *Khûrbet Beit Sellûm.* The ruin of the house of Sellûm. p.n. A local form connected with سلم whence come *salâm,* 'peace,' and the common names Suleimân, Selâmeh, &c.

خربة البشم   Mm   *Khûrbet el Bushm.* The ruin of the balsam (*Amyris opobalsamum*).

خربة ديدبان   Mm *Khŭrbet Deidebân.* The ruin of the watch-tower. *See Jebel ed Deidebeh,* p. 109.

خربة الدير   Ln *Khŭrbet ed Deir.* The ruin of the monastery.

خربة فاحص   Ko *Khŭrbet Fâhas.* The ruin of the searcher.

خربة الحاج رحال   Mm *Khŭrbet el Hâj Rahhâl.* The ruin of el Hâj (Pilgrim) Rahhâl. p.n.; meaning 'traveller.'

خربة حمارة   Km *Khŭrbet Hamârah.* The ruin of asses.

خربة الهوا   No *Khŭrbet el Hawa.* The ruin of the wind.

خربة حسين   Km *Khŭrbet Husein.* The ruin of Husein. p.n.

خربة ابن الحاج حماد   Kn *Khŭrbet Ibn Hâj Hammâd.* The ruin of the son of the Hâjj (Pilgrim) Hammâd.

خربة ابريكة   Jo *Khŭrbet Ibreikeh.* The ruin of the little pool.

خربة ابثان   Km *Khŭrbet Ibthân.* The ruin of gardens, or of soft soil.

خربة اسكندر   Ln *Khŭrbet Iskander.* The ruin of Alexander. p.n.

خربة جافا   Mm *Khŭrbet Jâfa.* The ruin of Jâfa. p.n.

خربة جفرون   Mo *Khŭrbet Jafrûn.* The ruin of Jafrûn. Perhaps from جفرة 'a partially filled-up well.'

خربة الجلمة   Jn *Khŭrbet el Jelameh.* The ruin of the hill.

خربة جريبان   Mm *Khŭrbet Jureibân.* The ruin of Jureibân. p.n. *cf. el Jûrbah,* p. 183.

خربة قابوبة   Ln *Khŭrbet Kâbŭbah.* The vaulted ruin.

خربة القيصومة   Jm *Khŭrbet el Keisûmeh.* The ruin of the southern-wood (a plant).

خربة كفر قوس   No *Khŭrbet Kefr Kûs.* The ruin of the village of the cloister.

خربة الكروم   Mn *Khŭrbet el Kerûm.* The ruin of the vineyards.

خربة الخارجة   Mm *Khŭrbet el Khârjeh.* The outer ruin.

خربة الخريجة   Jo *Khŭrbet el Khareijeh.* The outside ruin.

خربة كفرات   Mn *Khŭrbet Kuferât.* The ruin of villages.

خربة قنيسا قنيسا   No *Khŭrbet Kuleisa* or *Kuneisa.* The ruin of the conical cap or mitre.

خربة القمقم Ln *Khŭrbet el Kumkum.* The ruin of the decanter or water-bottle.

خربة قرقف Lm *Khŭrbet Kŭrkŭf.* The ruin of the Kŭrkŭf, a kind of small bird.

خربة قوسين Mn *Khŭrbet Kûsein.* The ruin of the two cloisters.

خربة قوسين السهل Ln *Khŭrbet Kûsein es Sahel.* The ruin of the two cloisters on the plain.

خربة الكويب Mn *Khŭrbet el Kuweib.* The ruin of the drinking-cup.

خربة لوزة No *Khŭrbet Lôzeh.* The ruin of the almond-tree.

خربة المغازون Jo *Khŭrbet el Maghazûn.* The ruin of the warriors.

خربة المعلكة Kn *Khŭrbet el Mâlakah.* The ruin of the place of the mastic plant.

خربة مسين Lm *Khŭrbet Massîn.* The ruin of Massîn. p.n.

خربة المدحدرة Ko *Khŭrbet el Mudahderah.* The ruin of the roller.

خربة مخنة الفوقآ No *Khŭrbet Mukhnah el Fôka.* The ruin of upper Mukhnah. See *'Ain el Mûkhnah,* p. 178.

خربة مخنة التحتآ No *Khŭrbet Mukhnah et Tahta.* The ruin of lower Mukhnah. q.v.

خربة المنطرة No *Khŭrbet el Mûntarah.* The ruin of the watch-tower.

خربة النيرة Ln *Khŭrbet en Neirabeh.* The ruin of the furrows (scored in the sand by wind).

خربة نشا Jo *Khŭrbet Nesha.* The ruin of tender herbs.

خربة نيب Nn *Khŭrbet Nîb.* The ruin of Nîb. p.n. (The word means 'canine tooth.')

خربة راشين Ln *Khŭrbet Râshîn.* The ruin of Râshîn. p.n.

خربة ارزازة Jo *Khŭrbet er Ruzzâzeh.* The ruin of the rice-growers.

خربة سباتا Nn *Khŭrbet Sebâta.* The ruin of Sebâta (a herb so called).

خربة صياد Xn *Khŭrbet Seiyâd.* The ruin of hunters.

خربة الشريم Mn *Khŭrbet esh Shureim.* The ruin of the cracks.

خربة صير Ko *Khŭrbet Sîr.* The ruin of the fold.

خربة طفسة Mo *Khŭrbet Tafsah.* The ruin of filth.

خربة ياتياح Kn *Khûrbet Teiyâh.* The ruin of the plough-tail.

خربة ام غنمة Mn *Khûrbet Umm Ghanmeh.* The ruin of Umm Ghanmeh. p.n.; 'mother of spoil.'

خربة وصيل Lm *Khûrbet Wuseil.* The ruin of the arrival.

خربة يوبك Ko *Khûrbet Yaubek.* The ruin of Yobek. p.n.

خربة يودا Nn *Khûrbet Yehûda.* The ruin of Judah.

خربة زهران Kn *Khûrbet Zahrân.* The ruin of flowers.

خربة زيتا Mm *Khûrbet Zeita.* The ruin of Zeita. q.v.

الخسفى No *El Khûsfey.* The perennial well, close to *'Ain el Kûsab,* west of Nâblus.

قباطية Nm *Kûbâtieh.* The Copts' place.

قبيبة الظهور Mn *Kubeibet edh Dhahûr.* The little dome of the ridges.

القلعة No *El Kŭlâh.* The castle.

قلنسوة Jn *Kûlûnsaweh.* The conical cap or mitre.

كور Lo *Kûr.* Digging. Perhaps *Kuwar,* pl. of كورة a town.

قرية حجا Lo *Kuryet Hajja.* The town of Hajja. p.n.; 'pathway.' Samaritan קרית הגה.

قرية جيت Lo *Kuryet Jît.* The town of Jit. p.n.

القرنين Lo *El Kŭrnein.* The two peaks.

قوسين Mo *Kûsein.* The two cloisters.

قصر الفحيص Mm *Kasr el Fuheis.* The house of the even or scratched-up ground.

مادبا Mo *Mâdema.* Mâdema. p.n.

المحرونة Mm *El Mahrûneh.* The place of cotton carding.

المجدل {Jm Ko} *El Mejdel.* The watch-tower.

ميثلون / ميتلون Nm *Meithalûn* or *Meitalûn.* Meithalûn or Meitalûn. *See* *Khûrbet Mithelieh,* p. 112.

مرج الغرق Nm *Merj el Ghŭrûk.* The meadow of drowning; flooded in winter.

مركبي Km *Merkeby.* Merkeby. p.n.; from كب, 'to ride,' 'to place one thing on another.'

مركي   Mm   *Merkeh.* Merkeh. p.n. *cf. Neby Murâkin.*

مسلية } مثلية {   Nm   *Meselieh* or *Methelia.* Meselieh or Methelia. p.n. *See Khŭrbet Mithilia,* p. 112.

مشارق الجرار   Mm   *Meshârik el Jerrâr.* The eastern district of the Jerrâr (family).

المسرب   Lo   *El Mesrab.* The place of pasture, or 'the channel.'

المدحدل   Kn   *El Mudahdal.* The stone roller.

مغارة   Mm   *Mŭghârah.* The cave; 'a ruin.'

مغارة الحاج خليل   Ko   *Mŭghâret el Hâj Khŭlîl.* The cave of el Hâjj (Pilgrim) Khŭlil. p.n. This is a common name, and is derived from Abraham, who is called *Khŭlîl Allah,* the friend of God.

نابلس   No   *Nâblus.* Neapolis.

نخيل   Mm   *Nakheil.* The little palm.

الناقورة   Mn   *En Nâkûrah.* En Nâkûrah (the horn or trumpet). p.n.; the root نقر means 'to peck.' *See* p. 53.

النبي عجاج   Mm   *Neby 'Ajâj.* The Prophet 'Ajjâj. p.n.; 'dusty.' The shrine is in 'Ajjeh, whence the name in all probability.

النبي عرابين   Mm   *Neby 'Arâbin.* The Prophet 'Arâbin, in 'Arrâbeh. q.v.

النبي الياس   Ko   *Neby Elyâs.* The Prophet Elias.

النبي هارون   Nn   *Neby Harûn.* The Prophet Aaron.

النبي حزقين   Lm   *Neby Hazkîn.* The Prophet Ezekiel.

النبي لوين   Mn   *Neby Lawin.* The Prophet Levi.

النبي مراكين   Mm   *Neby Murâkin.* The Prophet Murâkin. p.n.

النبي رابي   Lo   *Neby Râby.* The Prophet Râby, 'tutor.'

النبي ساءة   Nm   *Neby Sâmeh.* The Prophet Sâmeh; perhaps Shem, in Kŭbâtieh.

النبي سراقة   Jo   *Neby Serâkah.* The Prophet Serâkah. p.n.; 'thieving.'

النبي شمعون   Jo   *Neby Shem'ôn.* The Prophet Simeon.

النبي سيلان   Mn   *Neby Sîlân.* The Prophet Sîlân. p.n.; near Sileh, on the south.

النبي يعرود بن يعقوب	Mm	*Neby Yārūd Ibn Yâkûb.* The Prophet Jared, the son of Jacob.
النبي يحيى بن زكريا	Mm	*Neby Yahya Ibn Zakariya.* The Prophet John, the son of Zachariah ; the church in Samaria.
النبي يمين	Jo	*Neby Yemin.* The Prophet Yemin (right hand). *cf. Benjamin.*
نصف جبيل	{Kn}{Mn}	*Nusf Jebil.* The watershed. (Vulgarly pronounced *Nusijbin.*)
نزلة الشرقية	Lm	*Nâzlet esh Sherkîyeh.* The eastern settlement. The word is applied to small suburbs of a village.
نزلة التينات	Km	*Nâzlet et Tînât.* The settlement of the fig-trees.
نزلة الوسطى	Lm	*Nâzlet el Wusta.* The middle settlement. Also called الفقرا *el Fukra,* 'the poor.'
اولاد يعقوب العشرة		*Oulâd Yâkûb el 'Asherah.* The ten sons of Jacob.
رفيضة	Mo	*Rafidia.* The infidels.
الرامة	Lm	*Er Râmeh.* The height. Heb. Ramah.
رامين	Ln	*Râmîn.* Râmin. p.n.
الرأس	Kn	*Er Râs.* The hill-top.
راس العين	No	*Râs el 'Ain.* The spring-head. It is also called المرصرصة *el Merûsrûsa,* which means either 'strongly-built,' or 'a spring with hard soil and stones surrounding it.'
راس البرج	Ko	*Râs el Burj.* The hill-top of the tower.
راس القاضي	Mo	*Râs el Kâdy.* The hill-top of the judge.
راس قليلة	Kn	*Râs Kuleileh.* The top of the little hill (قلة).
راس الشويفات	Ln	*Râs esh Shûweifât.* Hill-top of esh Shûweifât. Probably from شوف 'a harrow.'
رجال الاربعين	Mo	*Rijâl el Arbaîn.* The forty men.
رجال الغارة	Ko	*Rijâl el Ghârah.* The men of the plundering. A name applied to *Mûgharet el Hajj Khûlil.*
روجيب	No	*Rûjib.* Rûjib. p.n. Perhaps from رجبة 'a prop for a tree.'
سهل عسكر	No	*Sahel 'Askar.* The plain of 'Askar, q.v.

سهل المخنة   No   *Sahel el Mŭkhnah.* The plain of el Mŭkhnah, q.v., p. 178.

سهل روجيب   No   *Sahel Rûjîb.* The plain of Rûjib. q.v., p. 189.

صيدا   Lm   *Saida.* Saida. p.n.; from صيد 'hunting.'

صانور   Mm   *Sânûr.* Sânûr. p.n.; it may be the Hebrew צנור, 'an aqueduct.'

سبسطية   Mn   *Sebŭstieh.* Sebaste (Samaria).

سفارين   Ln   *Sefârîn.* Sefârin. p.n.

شعروية الشرقية   Lm   *Shârawîyet esh Sherkîyeh.* The Eastern Shârawîyeh. The word means 'woodlands.'

شعروية الغربية   Jm   *Shârawîyet el Gharbîyeh.* The Western Shârawîyeh.

الشيخ عبدالله   Ln   *Sheikh 'Abdallah.* Sheikh 'Abdallah. p.n.; 'servant of God.'

الشيخ ابو إسمعين   No   *Sheikh Abu Ismaîn.* Sheikh Ismael's father. p.n.

الشيخ ابو سعد   Lo   *Sheikh Abu Sâd.* Sheikh Abu Sâd. p.n.

الشيخ ابو السخا   No   *Sheikh Abu es Sakha.* Sheikh Abu's Sakha. p.n.; meaning 'liberal.'

الشيخ ابو الوفا   Mm   *Sheikh Abu el Wafa.* Sheikh Abu'l Wafa. p.n.; meaning 'faithful to his promise' in *Zâwieh.*

الشيخ احمد   {Lo / Mo}   *Sheikh Ahmed.* Sheikh Ahmed. p.n.

الشيخ احمد الثعبان   Mo   *Sheikh Ahmed eth Thâbân.* Sheikh Ahmed eth Thâbân ('the snake').

الشيخ احميد   Mn   *Sheikh Ahmeid.* Sheikh Ahmeid. p.n.; diminutive of Ahmed.

الشيخ عيسى   Lu   *Sheikh 'Aîsa.* Sheikh Jesus. p.n.

الشيخ علي   {Lo / No}   *Sheikh 'Aly.* Sheikh 'Aly. p.n.

الشيخ علي النوباتي   Mo   *Sheikh 'Aly en Nôbâty.* In Bûrin, Sheikh 'Aly the musician.

الشيخ عمر   Km   *Sheikh 'Amr.* Sheikh 'Amr. p.n.

الشيخ عم ي   Mn   *Sheikh 'Amry.* Sheikh 'Amry. p.n.

الشيخ عطا   Lo   *Sheikh 'Ata.* Sheikh 'Ata. p.n.; meaning 'gift.'

الشيخ العروري	Mm	*Sheikh el 'Arûry.* The possessed or mad Sheikh, or ·the Sheikh from 'Arûrah.'
الشيخ بيت الزاكي	No	*Sheikh Beit ez Zâky.* The Sheikh of the Beit ez Zâky (' pure house or family ').
الشيخ بيازيد	Mn	*Sheikh Beiyâzîd.* Sheikh Bayâzid. p.n. ; name of a well-known saint and mystic.
الشيخ غانم	No	*Sheikh Ghânim.* Sheikh Ghânim. p.n. (' plunderer ' or ' shepherd ').
الشيخ حميد		*Sheikh Hameid.* Sheikh Hameid. p.n. ; same as Ahmeid. q.v.
الشيخ حدان	Km	*Sheikh Hamdân.* p.n.
الشيخ حسن	Ln	*Sheikh Hasan.* Sheikh Hasan. p.n. ; in B. Lid.
الشيخ ابن بدرية	Mm	*Sheikh Ibn Bedrîyeh.* Sheikh Ibn Bedriyeh. p.n.
الشيخ الحريش	Mn	*Sheikh el Hureish.* Sheikh el Hureish. p.n. (' scabby ').
الشيخ ابرهيم الادهم	Mn	*Sheikh Ibrahîm el Ad-hem.* Sheikh Ibrahim el Adhem, a well known Moslem worthy.
الشيخ اسمعيل	Mn	*Sheikh Ismâîl.* Sheikh Ishmâil. p.n. ; in Sebûstieh.
الشيخ كساب	Mm	*Sheikh Kessâb.* Sheikh Kessâb. p.n. ; meaning ' one who earns his living.'
الشيخ الحرس	Lo	*Sheikh el Mahras.* Sheikh el Mahras (' the guard ').
الشيخ مسعود	Jn Ln	*Sheikh Mas'ûd.* Sheikh Mas'ûd ('fortunate').
الشيخ مشرف	Ko	*Sheikh Mesherraf.* Sheikh Mesherraf (' honoured ').
الشيخ محمد	Ln	*Sheikh Muhammed.* Sheikh Mohammed. p.n. ; at Kûsein es Sahel.
الشيخ محمد البغدادي	Xm	*Sheikh Muhammed el Baghdâdy.* Sheikh Mohammed of Bagdad, in Jûrba.
الشيخ نافع	Mo	*Sheikh Nâfiâ.* Sheikh Nâfiâ. p.n. (' useful ').
الشيخ ناصر	Ln	*Sheikh Nâsir.* Sheikh Nâsir. p.n. (' victor ' or ' helper ').
الشيخ سعد	Lo	*Sheikh Sâd.* Sheikh Sâd. p.n. (' fortunate ').
الشيخ صلاح	Jm	*Sheikh Salâh.* Sheikh Salâh. p.n. (' righteous') ; a sacred tree.
الشيخ سلمان الفارسي	Mo	*Sheikh Selmân el Fârsi.* Sheikh Selmân el Fârsi. p.n. ; a companion of the Prophet.

الشيخ شعلا   Mn   *Sheikh Shálá.* Sheikh Shálá ('flame'); and one of the objects of worship of the Nuseiriyeh.

الشيخ السيرة   Mo   *Sheikh es Sîreh.* The Sheikh of the fold.

الشيخ التبّان   Lo   *Sheikh el Tebbân.* Sheikh Tebbân. p.n. ('the straw-dealer').

الشيخ ثلبي   Nm   *Sheikh Theljy.* Sheikh Theljy ('the snowy').

الشيخ بحيى   Kn   *Sheikh Yahyah.* St. John the Baptist; also called *Rijál ez Zâwich*, 'the men of the hermitage.'

شلاليف   Jm   *Shellâlif.* Waters descending from a height; fallow lands.

شوفة   Kn   *Shûfeh.* Shûfeh. p.n. شوف 'a harrow,' also 'polishing.' In vulgar Arabic it means 'seeing.'

شويكة   Km   *Shuweikeh.* Shuweikeh. p.n. Thorns.

سدر حسن السبحة   Jm   *Sidr Hasan es Subhah.* The lotus-tree of Hasan es Subha. Perhaps حسن الصباح Hasan es Sebbâh, founder of the sect of the Assassins.

سيلة الظهر   Mn   *Sîlet edh Dhahr.* Sileh of the ridge.

سريس وقيل سيريس   Nn   *Sirîs* or *Sîrîs.* Siris. p.n.

الست اسلامية   No   *Sitt Eslâmîyeh.* The lady of Islâm. p.n.

الست القامة   Ko   *Sitt el Kâmeh.* The lady of el Kâmeh. p.n.; meaning 'upright,' or 'a statue.'

صوفين   Jo   *Sûfîn.* p.n. Probably the Heb. צופים Zophim.

صور   Mo   *Sûr.* Rock.

الصورتين   Mo   *Es Sûratein.* The two pictures or statues.

سرا   Mo   *Sûrra.* Sarra. p.n.

طاحونة ابو طافش   Kn   *Tâhûnet Abu Tâfish.* The mill of Abu Tâfish. p.n.

طاحونة العزيزية   Mn   *Tâhûnet el 'Azîzîyeh.* The mill of 'Azîziyeh. Family name. This is north-west of *Deir Sherâf.*

طاحونة البلدية   Mn   *Tâhûnet el Beledîyeh.* The village mill, west of *Deir Sherâf.*

طاحونة القصبية   Mn   *Tâhûnet el Kûsabîyeh.* The mill of the people of 'Ain el Kûsab.

طاحونة مسعود   Mn   *Tâhûnet Mas'ûd.* The mill of Mas'ûd, west of Sebûstieh. See *Sheikh Mas'ûd.*

طاحونة ام الحرادين   Mn   *Tâhûnet Umm el Harâdîn.*   The mill with the lizards.

طاحونة ام راسين   Ln   *Tâhûnet Umm Râsein.*   The mill with the two heads (of water).

طاحونة الواويات   Ln   *Tâhûnet el Wâwîyât.*   The mill of the jackals.

الطيبة   Kn   *Et Taiyibeh.*   'The goodly.'

طواحين راعين   Ln   *Tawâhîn Râmîn.*   The mills of Râmîn.   South of that village are ruined mills marked 'Mills.'

طواحين وادي شعير   Mo   *Tawâhîn Wâdy Shâir.*   The mills of the valley of barley, including the following from east to west (marked 'mill' on plan) :—

طاحونة العتبة   *Tâhûnet el 'Akabeh.*   The mill of the steep, or mountain road.

طاحونة رفيضية   *Tâhûnet Rafîdia.*   The mill of Rafîdia (near that village).

طاحونة الشيخ   *Tâhûnet esh Sheikh.*   The mill of the Sheikh.

طاحونة الخدشنة   *Tâhûnet el Khashneh.*   The mill of the rough ground.

طاحونة الصروية   *Tâhûnet es Sûrrawîyeh.*   The mill of the people of Sûrra (from the village).

طاحونة النوفلية   *Tâhûnet en Nûfelîyeh.*   The mill of the Naufel family ; a common Syrian name.

طاحونة الطالبية   *Tâhûnet et Tâlebîyeh.*   The mill of the Tâlebiyeh family.

طاحونة التوسيدية   *Tâhûnet el Kûseinîyeh.*   The mill of the Kuseimeh people (from the village).

طاحونة القرتية   *Tâhûnet el Kerâiyeh.*   The mill of the Kerâiyeh (family name).

طاحونة البلدية   *Tâhûnet el Beledîyeh.*   The mill of the Beledîyeh, country people.

طاحونة العزيزية   *Tâhûnet el 'Azîzîyeh.*   The mill of the 'Azîziyeh ; family name.

[These occur along the valley, but are not written on the plan.]

تل   Mo   *Till.* Mound.

تل دوثان   Mm   *Tell Dôthân.* The mound of Dôthân. p.n.; Heb. דֹּתָן Dothan, 'two wells.'

تل اشثاف   Jm   *Tell Ishkâf.* The mound of the clefts.

تل القزاعي   Nm   *Tell el Kezâiy.* The mound of el Kezâiy. p.n.

تل خيبر   Nm   *Tell Kheibar.* Mound of Kheibar. Kheibar is the name of a famous Jewish-Arabic tribe, and of several places, notably one near Mecca.

تل منصيف   Ko   *Tell Manasîf.* The mound of the watershed.

تل صديع   Jn   *Tell Subîh.* The reddish coloured mound.

الطيرة   Jo   *Et Tîreh.* The fort.

تبراس   Lm   *Tubrâs.* Tubrâs. p.n.

طول كرم   Kn   *Tûl Keram.* The long (place) of the vineyard.

وادي ابو قصلان   Mn   *Wâdy Abu Kaslân.* The valley of Abu Kaslân; from قصل 'treading corn,' 'cutting unripe corn for horses,' or 'flowers of the *sellam* plant.'

وادي ابو نار   Km   *Wâdy Abu Nâr.* The valley with the fire.

وادي العيون   Ko   *Wâdy el 'Ayûn.* The valley of the springs.

وادي عزون   Ko   *Wâdy 'Azzûn.* The valley of 'Azzûn. q.v.

وادي باجورة   Km   *Wâdy Bajûrah.* The valley of Bâjûra. Perhaps from بجر 'an excrescence.'

وادي البرودة   Nn   *Wâdy el Barûdeh.* The cool valley. The word also means 'a gun' in vulgar Arabic.

وادي البستاني   *Wâdy el Bestâny.* The valley of the gardener.

وادي البرك   {Kn Mo}   *Wâdy el Burak.* The valley of the pools.

وادي البصل   Lm   *Wâdy el Bûsl.* The valley of the onions.

وادي دعوك   Mm   *Wâdy Dâûk.* The valley of Dâûk ('rubbing,' or 'rolling on the ground').

وادي الامير   Mn   *Wâdy el Emîr.* The valley of the prince.

وادي الفول   Km   *Wâdy el Fûl.* The valley of beans.

وادي العميق   Lm   *Wâdy el Ghamîk.* The deep valley.

وادي الغول   Nm   *Wâdy el Ghûl.* The valley of the ghoul (demon).

وادي الحاج موسى   Lo   *Wâdy el Hâj Mûsa.* The valley of el Hâjj (pilgrim) Moses.

وادي الحوض   Ko   *Wâdy el Haud.* The valley of the cistern.

وادي الحزيزة   *Wâdy el Huzeizeh.* The valley of the scab.

وادي اربد   Nn   *Wâdy Irbid.* The valley of Irbid. p.n.

وادي الجاموس   Km   *Wâdy el Jâmûs.* The valley of the buffalo.

وادي جرّا   Mo   *Wâdy Jerr'â.* The valley of the sand-hill.

وادي جنيد   Mo   *Wâdy Jineid.* The valley of Jineid. q.v.

وادي كار   Ko   *Wâdy Kâr.* The valley of Kâr (large stores of corn).

وادي الكلبة   Ln   *Wâdy el Kelbeh.* The valley of the bitch.

وادي القرام   Ko   *Wâdy el Kerrâm.* The valley of stumps.

وادي الخراج   Mo   *Wâdy el Kharâj.* The valley of the exit, or 'of the tax.'

وادي اللزام   Mm   *Wâdy el Lezzâm.* The valley of el Lezzâm. p.n.; meaning 'necessary.'

وادي المالح   Jm   *Wâdy el Mâleh.* The salt valley.

وادي مسين   Lm   *Wâdy Massîn.* The valley of Massîn. p.n.

وادي الملك   Nm   *Wâdy el Melek.* The king's valley.

وادي المسرب   Ko   *Wâdy el Mesrab.* The valley of the pasturage or channel.

وادي المسبق   Mn   *Wâdy el Misbak.* The valley of the race.

وادي النمل   Ko   *Wâdy en Naml.* The valley of the ants.

وادي النمر   Kn   *Wâdy en Nimr.* The valley of the leopard, or of 'abundant water.'

وادي النقعة   Ln   *Wâdy en Nukâh.* The valley of the swamp.

وادي النصراني   Mm   *Wâdy en Nusrâny.* The valley of the Christian.

وادي الرز   Km   *Wâdy er Roz (ruzz).* The valley of rice.

وادي سريس   Nn   *Wâdy Sarrîs.* The valley of the lentisk.

وادي السبين   Ko   *Wâdy es Sebîn (for Sebîl).* The valley of the highway.

2 C 2

وادي الشعير   Ln   *Wâdy esh Shâîr.* The valley of barley.

وادي الشامي   Km   *Wâdy esh Shâmy.* The Syrian or Damascene valley.

وادي الشجور   No   *Wâdy esh Shejûr.* The valley of trees.

وادي الشرق   Mo   *Wâdy esh Sherk.* The eastern valley, near 'Ain esh Sherk.

وادي صير   Ko   *Wâdy Sîr.* The valley of the fold.

وادي الصميلي   Lo   *Wâdy es Summeily.* The valley of the Summeily. p.n. صمل 'to be hard and dry.'

وادي التين   Kn   *Wâdy et Tîn.* The valley of figs.

وادي يهودا   Nn   *Wâdy Yehûda.* The valley of Judah.

وادي زيمر   Kn   *Wâdy Zeimer.* The valley of Zeimer. p.n. Perhaps Zâmir, 'a piper.'

وادي زيبار   *Wâdy Zîbâr.* The valley of Zibâr. *See Wâdy Zibrîyeh,* p. 156.

وادي زبل   Mo   *Wâdy Zibl.* The valley of manure.

ولي ابي جود   Mo   *Wely Abu Jûd.* The Saint (or saint's tomb of) Abu Jûd. p.n.; meaning 'liberal.'

ولي الزهراوي   Mm   *Wely ez Zihrâwy.* The Saint (or saint's tomb of) ez Zihrâwy. p.n.

الوريا   Ln   *El Wiria.* The touchwood.

ياصيد   Nn   *Yâsîd.* Yâsid. p.n.

يما   Km   *Yemma.* Yemma. p.n.

زواتا   Mo   *Zawâta.* Zawâta. p.n.

الزاوية   Mm   *Ez Zâwieh.* The corner or hermitage.

زيتا   Km   *Zeita.* The olive.

زلفة   Jm   *Zelefeh.* The cistern.

SHEET XII.

ابو قندول   Pn   *Abu Kandôl.* Producing the thorn tree (*Spartium aspara-thoides*); also called طلعة الوحوش *Talât el Wahûsh,* 'the wild beasts' ravine.'

عينون   On   *'Ainûn.* Springs. (Ænon. *See* "Memoirs.")

عين البيضآ   {Pm / Nn}   *'Ain el Beida.* The white spring.

عين الدبور   Oo   *'Ain ed Dabbûr.* The spring of bees or wasps.

عين الدليب   On   *'Ain ed Duleib.* The spring of Duleib; perhaps from دولاب *dûlâb,* 'a water-wheel.'

عين الفارة   On   *'Ain el Fârah.* The spring of the open valley.

عين الغزال   Qm   *'Ain el Ghûzâl.* The spring of the gazelle.

عين حابوس   Qm   *'Ain Hâbûs.* The imprisoned spring.

عين الحلوة   Qm   *'Ain el Helweh.* The sweet spring.

عين الحلوة   Pn   *'Ain el Helweh.* The sweet spring, as distinct from *'Ain el Mâleh* ('the salt spring') near it.

عين الكبيرة   No   *'Ain el Kebîreh.* The great spring.

عين الحمة   I'm   *'Ain el Hûmmeh.* The hot spring (thermal).

عين القديرة   *'Ain el Kudeirah.* The spring of the little pots.

عين قرعان   Qm   *'Ain Kûrân.* The spring of gourds.

عين المالح   Pn   *'Ain el Mâleh.* The salt spring.

عين الميتة   {Pn / Oo}   *'Ain el Meiyiteh.* The dead spring.

عين مسقى   Oo   *'Ain Miska.* The spring of irrigation.

عين الردغة   Qm   *'Ain er Ridghah.* The spring of mud or clay.

عين الصغيرة   No   *'Ain es Saghireh.* The small spring.

عين الساقوطا   Qm   *'Ain es Sâkût.* The dropping spring (a small cascade).

عين الشمسية   Qm   'Ain esh Shemsîyeh. The sunny spring.

عين شبلة   Oo   'Ain Shibleh. The spring of Shibleh. Perhaps Hebrew
שִׁבֹּלָה 'stream.'

عين شعب البئر   Oo   'Ain Shâb el Bîr. The spring of the spur of the well.

عين الشق   Pm   'Ain esh Shukk. The spring of the cleft. A cliff with
a spring beneath, and a small precipice below the spring.

عين الصبيان   'Ain es Subiân. The spring of the boys.

عين التبان   Nn   'Ain et Tebbân. The spring of the straw-dealer.

عين أم طيونة   Pn   'Ain Umm Teiyûneh. The spring with the clay.

عقابة   Om   'Akâbeh. The steep or mountain road.

عقبة البطمة   Po   'Akabet el Butmeh. The ascent of the terebinth.

عرب البلاوني   'Arab el Belâuny. The Belâuny Arabs. p.n.

عرب الفحيلات   'Arab el Faheilât. The Faheilât Arabs. p.n.

عرب السردية   'Arab es Sardîyeh. The Sardiyeh Arabs. p.n.

عراق ابي الحشيش   Qo   'Arâk Abu el Hashîsh. The cliff or bank with the grass.

عراق ابي زتى   Po   'Arâk Abu Zâk. Cliff or bank of Abu Zâk. The last
word means 'crying out,' or 'driving cattle.'

عراق العقبة   On   'Arâk el 'Akabeh. The cliff or bank of the mountain
road.

عراق الاصبح   Qo   'Arâk el Asbâh. The brown cliff or bank.

عراق الجزري   Om   'Arâk el Bizry. The cliff or bank of el Bizry. Name
of a man of Tûbâs, who keeps his cattle in this cavern.

عراق الدغيلي   Po   'Arâk ed Dugheily. Cliff or bank of the thicket.

عراق الحمام   Pm   'Arâk el Hamâm. The cliff of the doves.

عراق الجناب   Pm   'Arâk el Jenâb. The cliff or bank of the Jenâb. A kind
of terebinth is so called.

عراق خليل المسلح   On   'Arâk Khaleil el Meslah. The cliff or bank of the little
dell of the arsenal.

عراق الخبي   Om   'Arâk el Khubby. The cliff or bank of hiding.

عراق المردوم   Pm   'Arâk el Mardûm. The closed up cliff bank or cavern.

عراق المحيرة I'm 'Arâk el Môbarah. The cliff of the quarry.

عراق السهيلي On 'Arâk es Saheily. The cliff or bank of the plain.

عرقان النمر Qo 'Arkân en Nimr. The cliffs or bank of the leopard.

عيون الأساور Qn 'Ayûn el Asâwir. The springs of the bracelets. Compare *Tâhûnet el Asâwir.* Also called عيون ام الجمال 'Ayûn Umm el Jemâl, the springs with the camels.

عزموطا No 'Azmût. 'Aznût. p.n.

البصة Qm El Basseh. The marsh.

بيت دجن Oo Beit Dejan. The house of Dagon.

بيت فوريك No Beit Fûrîk. The house of Fûrik. p.n.

برذلة Pm Berdeleh. Berdeleh. p.n.

البقيع Po El Bukeiâ. The plain, or open valley.

برج الفارعة On Burj el Fârâh. Tower of Fârâh. q.v.; from the valley.

برج المالح Pn Burj el Mâleh. Tower of Mâleh. q.v.; from the valley.

بصيلية Po Buseiliyeh. The place of onions (or any bulbous plant).

دبة الساقوط Qm Debbet es Sâkût. The plot of ground of Sâkût. *See 'Ain es Sâkût,* p. 197.

الصعيجة Po Ed Deijah. The incline.

الدير {Qm / On} Ed Deir. The monastery.

دير الحطب No Deir el Hatab. The convent of timber.

ظهرة حمصة Po Dhahret Homsah. The ridge of the chick-pea.

ظهرة الميدان Qo Dhahret el Meidân. Ridge of the plain or exercise ground (*Meidân el 'Abd*).

ظهرة موفيا Qo Dhahret Môfia. The ridge of Môfia. p.n.; meaning 'fulfilling.'

ظهرة ام الكبيش On Dhahret Umm el Kubeish. The ridge of the little ram.

الفاطور Qm El Fâtûr. The fissures, rocks and a spring.

الفرش Po El Fersh. The open field; literally, 'carpet,' or 'bed.' *See Fersh Iskander,* p. 109.

حبس قطوي Pn *Habs Katwy.* The ecclesiastical property, where the sand-grouse is found.

حجر الاصبح Po *Hajr el Asbâh.* The reddish-brown stone. *See* "Memoirs."

الحمرآ‎ Po *El Hamrah.* The red.

الحمّام Pn *El Hŭmmâm.* The bath.

حوطة سادونة Po *Hûtet Sâdûneh.* The enclosure of Sâdûneh. p.n. *See* '*Ain Abu Sudân,* p. 13.

إستال احلوة Qn *Iskâl el Helweh.* The stone-heaps of Helweh (the sweet). Heb. סקל 'to stone.' Also called *Abu Kehakfr.*

جبل الكبير Oo *Jebel el Kebîr.* The great mountain.

جبل طمون On *Jebel Tammûn.* The mountain of Tammûn. q.v.; from the village.

جرف ام رمانة Po *Jorf Umm Rummâneh.* The precipitous bank with the pomegranates.

جوار بيت فار Oo *Jôwâr Beit Fâr.* The hollow places of Beit Fâr ('the mouse house').

قبور الدلاحمة Qo *Kabûr ed Delâhmeh.* Graves of the Delâhmeh folk (said to be certain derwishes). *cf. Delhemîyeh,* p. 160.

قطايا منتش On *Katât Mantash.* The crags of the place of tender herbage.

خلة ابي سياج Pn *Khallet Abu Siâj.* The dell with the thorn fence.

خلة ابي زيادد I'm *Khallet Abu Ziâdeh.* The dell of Abu Ziâdeh; a common proper name, meaning 'father of increase.' *cf. Abu Zeid.*

خلة البد Pm *Khallet el Bedd.* The dell of the idol; بذ an 'idol' or 'temple.'

خلة اليَد Pn *Khallet el Hadd.* Dell of the boundary.

خلة الحمام Pn *Khallet el Hŭmmâm.* Dell of el Hŭmmâm ('the bath').

خلة قاسم On *Khallet Kâsim.* Kâsim's dell.

خلة الربيع Po *Khallet er Robîa.* The dell of spring pasturage.

خلة الشقيف Pn *Khallet esh Shŭkîf.* The dell of the cleft.

خلة السوق *Khallet es Sûk.* The dell of the market.

خلة ام خبيزة Pm *Khallet Umm Khobbeizeh.* The dell with the marsh mallows.

خُلَّة اُمّ رِكاب   Po   *Khallet Umm Rukâb.* The dell with the hill-tops. *See Merkeby,* p. 187.

النَخوانِيق   On   *El Khawânîk.* The gorges.

النَخضر   Pm   *El Khŭdr.* El Khŭdhr. *See* p. 28.

خربة ابى على   Nn   *Khŭrbet Abu 'Aly.* Abu 'Aly's ruin. p.n.; ruined house.

خربة العقبة   Pn   *Khŭrbet el 'Akabeh.* The ruin of the steep or mountain road.

خربة عرقان الصخور   Oo   *Khŭrbet 'Arkân es Sakhûr.* The ruin of the rocky cliffs.

خربة عطوف   Pn   *Khŭrbet 'Atûf.* The ruin of 'Atûf. p n.; meaning 'kind.'

خربة بيت فار   Oo   *Khŭrbet Beit Fâr.* The ruin of the home of the mouse.

خربة بئر الشويهة   Oo   *Khŭrbet Bîr esh Shŭeiheh.* The ruin of the well of the little sheep.

خربة فروة   Nn   *Khŭrbet Ferweh.* The ruin of the waste land.

خربة حية   No   *Khŭrbet Haiyeh.* The ruin of the serpent.

خربة حندوس   On   *Khŭrbet Handûs.* The ruin of the beggar's dole.

خربة إبزيق   Om   *Khŭrbet Ibzîk.* The ruin of Ibzik. p.n. بزق signifies 'to spit,' and 'to sow the ground.' *See* also "Memoirs," under the name 'Bezek.'

خربة جبريش   Pm   *Khŭrbet Jebrîsh.* The ruin of Jebrish. p.n.

خربة جليجيل   No   *Khŭrbet Juleijil.* The ruin of Juleijil; diminutive or plural of جلجل *Jeljel,* q.v., p. 161.

خربة قاعون   Pm   *Khŭrbet Kâaûn.* The ruin of Kâaûn. p.n. *See* "Memoirs," under name *Kaiua.*

خربة القارور   Qo   *Khŭrbet el Kârûr.* The ruin of the glass bottle (urinometer).

خربة كشدة   On   *Khŭrbet Kashdeh.* The ruin of the Kashdeh, a kind of vegetable.

خربة كفر بيتا   No   *Khŭrbet Kefr Beita.* The ruin of the village of Beita. p.n. (house).

خربة كفر دك   Om   *Khŭrbet Kefr Dukk.* The ruin of the village of the little hill.

خربة المالح   Pn   *Khŭrbet el Mâleh.* The ruin of el Mâleh. *See* '*Ain el Mâleh.*

2 D

خربة موفيا   Qo   *Khŭrbet Môfia.* The ruin of el Môfia. p.n. *See Dharet Môfia,* p. 199.

خربة النهم   Nm   *Khŭrbet en Nahm.* The ruin of en Nahm. نهم means 'to pelt with stones'; نهم *nahm* is the name of an ancient Arab idol.

خربة عفر   On   *Khŭrbet 'Odhfer.* The ruin of 'Odhfer. p.n.

خربة ربرابة   Om   *Khŭrbet Rabrâbah.* The ruin of Rabrâbah. p.n. ربرب means 'herd of wild oxen.' A ruin close to Râba, and called generally *el Kûsr,* 'the castle.'

خربة رؤس الديار   Oo   *Khŭrbet Rûs ed Diâr.* The ruin of the hill-top of the houses.

خربة صفرية   Nm   *Khŭrbet Safiriych.* The ruin of Safiriyeh. p.n. *cf. Khŭrbet Safeirai* (next entry).

خربة الصفيرد   Pn   *Khŭrbet es Sefeirah.* The ruin of the withered herbage (صفار), or 'of the empty house' (diminutive of صفر).

خربة السلهب   Om   *Khŭrbet es Selhab.* The ruin of the tall man.

خربة السرب   Nn   *Khŭrbet es Serb.* The ruin of the pasturage or channel.

خربة الشيخ نصر الله   No   *Khŭrbet Sheikh Nasr Allah.* The ruin of Sheikh Nasr Allah. p.n. (God's help).

خربة الشرابة   No   *Khŭrbet esh Sherârbeh.* The ruin of the drinkers.

خربة السميط   On   *Khŭrbet es Smeit.* The ruin of the little acacia. Also called *Khŭrbet Sumtah,* with the same meaning.

خربة السمرآ   Pn   *Khŭrbet es Sumra.* The brown ruin.

خربة السويدة   Qn   *Khŭrbet es Sûweideh.* The blackish ruin. It is also the name of the blue thrush (*Pterocincla cyaneus*). *See* also *Jubb Suweid,* p. 5.

خربة تلفيت   Om   *Khŭrbet Telfit.* The ruin of Telfit. p.n.

خربة تل الفخار   Oo   *Khŭrbet Tell el Fokhâr.* The ruin of the mound of the potters.

خربة ثعلة   Po   *Khŭrbet Thâlah.* The ruin of Thâlah (name of an Arab tribe).

خربة ام حراز   Qo   *Khŭrbet Umm Harrâz.* The ruin of Umm Harrâz. p.n., (the mother of the preserver).

خربة ام الحسن   Pn   *Khŭrbet Umm el Hasn.* The ruin of Umm el Hasan. p.n.

خربة ام الحصر | Pn | *Khŭrbet Umm el Hosr.* The ruin of 'the mother of confined space.' Perhaps حصر may be a corruption of حجار, 'a fortress.'

خربة ام الاقبا | Pn | *Khŭrbet Umm el Ikba.* The ruin with vaults.

خربة ام الجرين | Po | *Khŭrbet Umm el Jurein.* The ruin with the troughs.

خربة ام القاسيم | | *Khŭrbet Umm el Kâsim.* The ruin of Umm el Kâsim. p.n. (Kasim's mother).

خربة ام قيسمة | Po | *Khŭrbet Umm Keismeh.* The ruin of Umm Keismeh. p.n.

خربة ام القطن | Pn | *Khŭrbet Umm el Kotn.* The ruin with the cotton.

خربة ام الكبيش | On | *Khŭrbet Umm el Kubeish.* The ruin of Umm el Kubeish. p.n. (meaning 'mother of the little ram').

خربة ام الشيبك | Pn | *Khŭrbet Umm esh Sheibik.* The ruin of the intricacies. *See Wâdy esh Sheibik.*

خربة يرزة | Pn | *Khŭrbet Yerzeh.* The ruin of Yerzeh. p.n.

كتف العوسيا | Po | *Kitf el 'Aûsia.* The shoulder of el 'Aûsia. This may either mean 'a sort of sheep,' or cavities in rough ground holding water (عاسية).

الكفير | Om | *El Kufeir.* The little village.

القصر | Om | *El Kŭsr.* The palace.

حنف ابو لوزة | Qo | *Lahf Abu Lôzeh.* The foot of the hill with the almond tree.

حنف جادر | On | *Lahf Jâdir.* Foot of the hill of Jâdir. q.v.

مخاضة ابي الشرع | Qn | *Makhâdet Abu 'l Ashert.* The ford with the cracks (in the banks of the river).

مخاضة ابي سهاسيل | Qn | *Makhâdet Abu Sahâsîl.* The ford of Abu Sahâsil. p.n.

مخاضة ابي سدرة | Qo | *Makhâdet Abu Sidreh.* The ford with the lotus tree (*Zizyphus lotus*).

مخاضة ابي سوس | Qm | *Makhâdet Abu Sûs.* The ford with the liquorice plant.

مخاضة دخيني | Qm | *Makhâdet Dukheini.* The ford of millet.

مخاضة فتح الله | Qm | *Makhâdet Fatah Allah.* The ford of Fath-Allah, a common proper name, meaning 'God's victory.' The expression *Fatah - Allâh*, or more commonly in vulgar Arabic *Yeftah-Allah*, 'may God open,' is used when looking forward to some piece of luck—as the first money taken by a tradesman, &c.

مخاضة فتال الصفاح   Qm   *Makhâdet Fettâl es Sûfâh.* The ford of windings of the flat rock.

مخاضة الجرو   Qn   *Makhâdet el Jerro.* The ford of el Jerro. p.n.; 'the whelp'; name of a traditional hero.

مخاضة المسعودي   Qm   *Makhâdet el Mas'ûdî.* The ford of el Mas'ûdî; Arab tribal name.

مخاضة السعيدية   Qn   *Makhâdet es Sâidîyeh.* The ford of the Sâidîyeh Arabs.

مخاضة الشرار   Qm   *Makhâdet esh Sherâr.* The ford of esh Sherâr. p.n. See *Wâdy esh Sherâr,* p. 171.

مخاضة التركمانية   Qo   *Makhâdet et Turkomânîyeh.* The ford of the Turcomans.

مخاضة ام الحمر   Qm   *Makhâdet Umm el Imghar.* The ford with the red earth.

مخاضة ام سدرة   Qo   *Makhâdet Umm Sidreh.* The ford with the lotus tree (*Zizyphus lotus*).

مخاضة الوحادن   Qn   *Makhâdet el Wahâdneh.* The ford of the hollows or declivities.

مخاضة الزقومة   Qn   *Makhâdet ez Zakkûmeh.* The ford of the zakkûm, a tree found at Jericho, and used for inlaying in ornamental wood objects.

مكسر الحسان   Po   *Maksar el Hisân.* Place where the horse was killed.

مراح ارار   Oo   *Marâh Arrâr.* The nightly resting-place (penfold) of dung; or of calling cattle.

مرما فياض   Qm   *Marma Fiâd.* The heap of Fiyâd (properly Faiyâdh, a common Arab name, meaning 'liberal'). (A soldier having here been killed by the Arabs, and a cairn erected over him.)

ميدان العبد   Qo   *Meidân el 'Abd.* The plain or exercise ground of the slave.

المراقة   Oo   *El Merâkah.* The place of flowing; also called مدب ابن العراق *Medebb Ibn el 'Arâk,* the watercourse, son of (*i.e.*, proceeding from) the cliff; a cliff with water streaming down.

مرجة الخراف   Qo   *Merjet el Akhrâf.* The meadow of the lambs.

مشارق الجرار   {Om}{Nn}   *Meshârik el Jerrâr.* The eastern estates of the Jerrâr (family).

مشارق نابلس   {Oo}{Po}   *Meshârik Nâblus.* The eastern estates of Nâblus. q.v.

المحبرة   Pn   *El Mobarah.* The quarry.

مغارة ام العمود   Po   *Mughâret Umm el 'Amûd.* The cave with the pillars.

مغارة ام الجرين   Po   *Mughâret Umm el Jurein.* The cave with the troughs.

مغارة ام الزروب   Oo   *Mughâret Umm ez Zerûb.* The cave with the water-pipes.

المجيلبات   Po   *El Mujelbât.* The patches of pasture.

المخبي   Om   *El Mukhubby.* The hiding-place. *See* "Memoirs," under name 'Choba.'

الملسا   Oo   *El Mulassa.* The smooth spot.

منطار موبيا   Qo   *Mantâr Môfia.* The watch-tower of el Môfia. q.v.

منطار الشق   Qm   *Mantâr esh Shukk.* The watch-tower of the cleft.

النبي بلان   No   *Neby Belân.* The Prophet Belân. p.n.

نبي رابين   Om   *Neby Râbîn.* The Prophet Râbin. p.n.

النبي طوبا   Om   *Neby Tôba.* The Prophet Tôbâ. p.n.

نقب العرايس   Po   *Nukb el 'Arais.* The pass of the brides.

نقب الاسفر   Po   *Nukb el Asfar.* The pass of the journeys, or books.

نقب السكار   Po   *Nukb es Sakâr.* The pass of dykes.

اولادات الجوادرة   On   *Ouladât el Jawâdireh.* The children (clans) of the Jâdir people.

رابا   Om   *Râba.* Râba. p.n.

راس ابي شوشة   Pm   *Râs Abu Shûsheh.* The hill-top with tuft.

راس العقري   Om   *Râs el 'Akra.* The hill-top of the estate or real property.

راس البد   Pm   *Râs el Bedd.* The hill-top of the idol.

راس حمادة   Pn   *Râs Hammâdeh.* The hill-top of Hammâdeh. p.n.; probably حامية 'pasture.'

راس حمصة   Po   *Râs Homsah.* The hill-top of the chick-pea.

راس الحمير   Qm   *Râs el Humeiyir.* The hill-top of the asses.

راس ابزيك   Om   *Râs Ibzik.* The hill-top of Ibzik. *See Khûrbet Ibzîk,* p. 201.

راس جادر   On   *Râs Jâdir.* The hill-top of the wall (a very steep mountain).

راس الـجـبـسـيـن   Qm   *Râs el Jibsîn.* The hill-top of gypsum.

راس مراح الـواويات   Pn   *Râs Marâh el Wâwiyât.* The hill-top of the fold of the jackals.

راس نقب الـبـقـر   Qo   *Râs Nâkb el Bakr.* The hill-top of the pass of the cows.

راس الـرابـي   Pm   *Râs er Râby.* The hill-top of er Râbi. p.n.; inhabitant of Râba. q.v.

راس الـرمـالـي   Pm   *Râs er Rummâly.* The hill-top of the sands. This is an outbreak of basalt called *Rûml;* singular رَمْلَة, which means in the classical language a black streak, whence the name.

راس ام الـخـروبـة   Po   *Râs Umm el Kharrûbeh.* The hill-top with the carob tree (*Ceratonia siliqua*).

راس ام الـخـبـيـزة   Qo   *Râs Umm el Khobbeizeh.* The hill-top with the marsh-mallow.

راس ام زوفـة   Qr   *Râs Umm Zôkah.* The hill-top with the quicksilver, the colour being very white.

رجم الـيـقـلـوم   *Rujm el Yaklûm.* The cairn of Yaklûm. p.n.

سـادة الـنـحـلـة   Po   *Sâdet en Nahleh.* The cliffs of the bee (probably the Hebrew נחל 'a torrent'). *Sâdeh,* properly *Sâddeh,* means acting as a *sidd* or *sadd, i.e.,* 'rampart,' applied, in the Kor'ân, ch. XVIII, v. 92, to 'a mountain' which forms an inaccessible barrier. In common parlance it applies to a vertical cliff.

سـادة الـتـالـب   Qn   *Sâdet et Tâleb.* The vertical cliff, or the cliff of the mountain goat.

سـهـل ام الـقـب   Pn   *Sahel Umm el Ikba.* The plain of Umm el Ikba. q.v.

سـهـل الـطـيـره   Xm   *Sahel et Teireh.* The plain of the flight. Perhaps טירה 'fortress.'

سـالـم   No   *Sâlim.* Sâlim. p.n. Salem.

شـعـب الـشـنـار   On   *Shâb esh Shinnâr.* The spur of the Greek partridge.

شـيـخ غـانـم   On   *Sheikh Ghânim.* Sheikh Ghânim. p.n.

شـيـخ حـمـود   No   *Sheikh Hammûd.* Sheikh Hammûd. p.n., in 'Azmût.

شـيـخ حـزقـيـن   Om   *Sheikh Hazkîn.* Sheikh Ezekiel.

شـيـخ كـامـل   Oo   *Sheikh Kâmil.* Sheikh Kâmil. p.n. (perfect).

شـيـخ مـحـمـد   { Pn / No }   *Sheikh Muhammed.* Sheikh Mohammed. p.n.

شيخ السُمَيِط   On   *Sheikh es Smeit.* The Sheikh of es Smeit. *See Khûrbet es Smeit*, p. 202.

شجرة الشيخ كامل   Oo   *Shejeret esh Sheikh Kâmil.* The tree of Sheikh Kâmil. The dedication of a tree to an idol was as ancient Arab practice, and the custom survives in the desert to the present day. *See* my translation of the Kor'ân, Vol. I, p. 132, n. 2.

الشِرار   Qm   *Esh Sherâr.* Esh Sherâr. p.n. *See* p. 171.

سد البلتاوي   Qo   *Sidd el Belkâwy.* The cliff of the open space.

سير   Nm   *Sîr.* The fold.

سوفي الخريبات   Pn   *Sûfy el Khŭreibât.* The grounds of the ruins.

طواحين وادي النارية   *Tawâhîn Wâdy el Fârâh.* The mills in Wâdy Fârâh. q.v.

[*The following are not written on the Plan.*]

طاحونة المحمودية   *Tâhûnet el Mahmûdiyeh.* The mill of the Mahmûd family or sect.

طاحونة البشتاوية   *Tâhûnet el Bushtawîyeh.* The mill of the Bushtawiyeh family.

طاحونة المجدوبة   *Tâhûnet el Mejdûbeh.* The mill of the barren land.

طاحونة العزرية   *Tâhûnet el 'Azirîyeh.* The mill of the 'Aziriyeh family.

طاحونة الشبيبية   *Tâhûnet el Shubeibîyeh.* The mill of the Shubeibiyeh family.

طاحونة التيسيات   *Tâhûnet el Keisiyât.* The mill of the Keisiyât Arabs.

[*The above are all near* 'Ain el Beida *and the south springs of* Fârâh. (Nn).]

طاحونة المشاقية   *Tâhûnet el Mashâkîyeh.* The mill of the Mashâka family.

طاحونة الدبورية   *Tâhûnet ed Dŭbbârîyeh.* The mill of the mountain pass (also a family name).

طاحونة الوبيدة   *Tâhûnet el Weibdeh.* The mill of the mountain pass.

طاحونة الدكيبية   *Tâhûnet ed Dŭkeibîyeh.* The mill of the people of Dŭkeib. *See* 'Ain Dŭkeib, p. 197.

[*The above are by the northern springs of* Fârâh (On), *near* 'Ain Dŭkeib.]

طاحونة العسافية On *Tâhûnet el 'Asâfîyeh.* The mill of the 'Asâfîyeh. p.n.

طاحونة التلماوية Po *Tâhûnet el Atmawîyeh.* Mill of the high or well-filled channel.

طاحونة الاطم Po *Tâhûnet el Attûm.* The mills of el Attûm. p.n. *See* p. 168.

طاحونة البريكة Oo *Tâhûnet el Bureikeh.* The mill of the little pool.

طاحونة الفاروقية Oo *Tâhûnet el Fârûkîyeh.* The mill of the Fârûkîyeh family or sect. The word فاروق means 'discriminator,' the Caliph Omar was so called.

طاحونة الفرش Po *Tâhûnet el Fersh.* The mill of el Fersh. q.v., p. 199.

طاحونة الخزام Po *Tâhûnet el Khazâm.* The mill of el Khazâm. Probably خزامي *Khuzâma,* a fragrant herb with a red flower. *See Wâdy Umm el Khazâm. Khezâm* means 'a nose-ring.'

طاحونة الخزية Po *Tâhûnet el Khazîyeh.* The mill of the Khazîyeh. p.n. (meaning 'disgrace').

طاحونة السلطانيات On *Tâhûnet es Sultaniyât.* The mill of the king's highway.

طاحونة التمونية Oo *Tâhûnet et Tammûnîyeh.* The mill of Tammûn. q.v.

طاحونة ام سفاح Po *Tâhûnet Umm Sefâh.* The mill with the pouring water. It should perhaps be صفاح 'flat rock.'

طاحونة الترة Po *Tâhûnet et Turreh.* The mill of the river bank.

طلعة ابو عيد Po *Talât Abu 'Aid.* The ascent or ravine of Abu 'Aid (a common p.n.).

طلعة الطويته Po *Talât et Tûlîyeh.* The longer ascent.

طمون On *Tammûn.* p.n. طمّ 'to overflow.'

تعنا Oo *Tâna.* Tâna; perhaps the Hebrew הַתַּאֲנָה. *See* "Memoirs," under 'Tanath Shiloh.'

تنين Om *Tannîn.* Tannîn. p.n.; in Arabic it means 'a dragon,' but it may be connected with the Hebrew תנה, 'habitation.'

تياصير Om *Teiâsîr.* Teiâsîr. p.n.

تل ابي رمح Po *Tell Abu Rumh.* The mound of Abu Rumh (the man with the lance).

تل ابي سدرة Qo *Tell Abu Sidreh.* The mound with the lotus tree.

تل ابي سفرة Pn *Tell Abu Sifry.* The mound with the round table.

تل ابي سوس	Qm	*Tell Abu Sûs.* The mound with the liquorice.
تل دبلقة	Pm	*Tell Dablakah.* The mound of Dablakah. p.n.
تل الفارعة	On	*Tell el Fârah.* The mound of el Fârah. *See* p. 197.
تل فص الجمل	Pn	*Tell Fass el Jemel.* The mound of the camel's joint.
تل الحمة	Pm	*Tell el Hûmmeh.* The mound of the thermal spring. *See* '*Ain el Hûmmeh,* p. 197.
تل القبور	Qm	*Tell el Kabûr.* The mound of graves; an Arab cemetery.
تل التذية	Oo	*Tell el Kadhiyeh.* The mound of powdered earth.
تل الردغة	Qm	*Tell er Ridghah.* The mound of mud or clay.
تل الزقومة	Qm	*Tell er Zakkûmeh.* The mound of the Zakkûm tree. *See Makhâdet ez Zakkûmeh,* p. 204.
نغرة القبور	Qm	*Thoghret el Kabûr.* The pass of the graves.
طوباس	On	*Tûbâs.* Tûbâs. p.n. *cf.* Heb. תֵּבֵץ Thebez.
طبقة الحلوة	Pn	*Tûbkat el Helweh.* The terrace of Helweh. *See '*Ain el Helweh,* p. 197.
التوانيك	Oo	*Et Tuwânik.* Tuwânik. p.n. *cf.* Heb. הַבֵּן 'extremity.'
طويل الذياب	Pn	*Tuweil edh Dhiâb.* The tall rock, or long ridge of the Dhiâb family. p.n. (wolves); also called طويل ابو هلال *Abu Helâl,* 'father of the crescent,' from its shape.
ام العمدان	Qm	*Umm el 'Amdân.* The mother of, *i.e.,* the place with columns.
ام لدرج	Qo	*Umm ed Deraj.* The mother of, *i.e.,* the place with steps. The district is so called from the nature of the ground, which is broken up in steps and terraces.
ام الجرين	Po	*Umm el Jurein.* The mother of the little troughs.
ام القطفة	Po	*Umm el Kûtfeh.* The place with the St. John's wort.
ام الرجمان	Oo	*Umm er Rujmân.* The place with the cairns.
ام الزروب	Po	*Umm ez Zerûb.* The place with the water-pipes.
وادي ابي لوز	Qo	*Wâdy Abu Lôz.* The valley with the almond.
وادي ابي سدرة	Qo	*Wâdy Abu Sidreh.* The valley with the lotus tree.
وادي اذرا	Pn	*Wâdy Adhra.* The grey valley. Also called

2 E

وادي عرقان الغرة		*Wâdy 'Arkân et Turreh.* Valley of the cliffs of the river bank; being rocky near the Ghor.
وادي بيدان	No	*Wâdy Beidân.* The valley of deserts.
وادي البقيع	Qo	*Wâdy el Bukeia.* The valley of the plateau, or open valley.
وادي الضبع	Pm	*Wâdy ed Dubâ.* The valley of the hyenas.
وادي فارعة	On	*Wâdy Fârâh.* The valley of the open valley. *See* p. 197.
وادي النو	Pn	*Wâdy el Fûu.* The valley of the madder plant (which grows in it).
وادي فص الجمل	Qn	*Wâdy Fass el Jemel.* The valley of the camel's joint.
وادي الحلوة	Pn	*Wâdy el Helweh.* The valley of el Helweh (the sweet). *See 'Ain Helweh,* p. 197.
وادي حتيم	Pn	*Wâdy Heteim.* The blackish valley; but it may be from حتمة the outlet of a water-dam.
وادي البرة	Pm	*Wâdy el Hirreh.* Valley of the wild cat.
وادي حمتة	Po	*Wâdy Homsah.* The valley of chick-peas.
وادي الحمير	Pm	*Wâdy el Humeiyir.* The valley of asses.
وادي الحمة	On	*Wâdy el Hûmmeh.* The valley of the thermal bath.
وادي جمع	Pm	*Wâdy Jemiâ.* The valley of assembly.
وادي جورة القطفي	Qn	*Wâdy Jûrat el Kûtûfy.* The valley of the hollow of St. John's wort.
وادي كشده	On	*Wâdy Kashdeh.* The valley of the Kashdeh. *See Khûrbet Kashdeh,* p. 201.
وادي كو ابو ديه.	Pn	*Wâdy Kaû Abu Deiyeh.* The valley of the fissure or outlet of Abu Daiyeh. p.n.
وادي الكلخ	Pn	*Wâdy el Kelkh.* The valley of Kelkh, a plant resembling fennel.
وادي كردة	Pm	*Wâdy Kerdeleh.* The valley of Kerdela. p.n.
وادي الخشنة	Pm	*Wâdy el Khashneh.* The rough valley.
وادي الخضر	Pm	*Wâdy el Khûdr.* The valley of el Khûdr. q.v., p. 28.
وادي القمح	Nn	*Wâdy el Kûmh.* The valley of wheat.
وادي المالح	Qm	*Wâdy el Mâleh.* The valley of salt water.

وادي المَرَاش   Nn   *Wâdy el Marâsh*. The valley of Merâsh, a herb so called.

وادي مرما فياض   Qm   *Wâdy Marma Fiâd*. The valley of Marma Fiâd. q.v., p. 204.

وادي الميتة   Pn   *Wâdy el Meiyiteh*. The valley of the dead spring. *See 'Ain el Meiyiteh.*

وادي المحجرة   Pn   *Wâdy el Mêbarah*. The valley of the quarry.

وادي المداخل   Pn   *Wâdy el Mudâhel*. The valley of the narrow-mouthed caves.

وادي المجلبات   Po   *Wâdy el Mujelbât*. The valley of Mujelbât. q.v.

وادي مخناوي   Om   *Wâdy Mukhnâwy*. The valley of the Mukhnâwy. *See 'Ain el Mukhnah*, p. 178.

وادي الرصيف   On   *Wâdy er Resîf*. The valley of the solid (rock).

وادي السريس   Nn   *Wâdy es Sarrîs*. The valley of the lentisk (*Pistachio lentiscus*).

وادي السلهب   Om   *Wâdy es Selhab*. The valley of the tall man.

وادي سلمان   Pm   *Wâdy Selmân*. The valley of Selmân. p.n.

وادي شعيب   Qn   *Wâdy Shâîb*. The valley of the little spur; or it may be ' of Jethro.' *See Neby Shâaib*, p. 132.

وادي الشجور   No   *Wâdy esh Shejûr*. The valley of trees.

وادي شوباش   Om   *Wâdy Shûbâsh*. The valley of Shûbâsh. *See* p. 172.

وادي الشق   Qm   *Wâdy esh Shûkk*. The valley of the cleft.

وادي سوفي الخربيات   Pn   *Wâdy Sûfy el Khûrcibât*. The valley of the grounds of the ruins. *See Sûfy el Khûrcibât*, p. 207.

وادي السمرا   Pn   *Wâdy es Sumra*. The brown valley.

وادي طوباس   Om   *Wâdy Tûbâs*. The valley of Tûbâs. q.v.

وادي الطبقة   Qn   *Wâdy et Tûbkah*. The valley of the terrace.

وادي ام درج الزقوم   Qn   *Wâdy Umm Deraj ez Zakkûm*. The valley with the steps of the Zakkûm tree. *See Tell ez Zakkûmeh*, p. 204, and *Umm ed Deraj*, p. 209.

وادي ام الحسن   Pn   *Wâdy Umm el Hasn*. The valley of Umm el Hasn. p.n.

وادي ام الاتبا   Pn   *Wâdy Umm el Ikba*. The valley of the vaulted ruin.

وادي ام الجبرين   Po   *Wâdy Umm Jurein.* The valley of the ruin with the troughs.

وادي ام الخنروبة   {On} {Po}   *Wâdy Umm el Kharrûbeh.* The valley of the ruin with the locust tree.

وادي ام الخزام   Po   *Wâdy Umm el Khazâm.* The valley of the Khuzama. *See Tâhûnet el Khazâm,* p. 208.

وادي ام الشيبك   Pn   *Wâdy Umm esh Sheibik.* The valley with the intricate irrigation.

وادي يرزة   Pn   *Wâdy Yerzeh.* The valley of Yerzeh. *See Khûrbet Yerzeh,* p. 203.

وادي الوحيش   Pn   *Wâdy el Waheish.* The valley of the wild beasts.

زبابدة   Nm   *Zebâbdeh.* Zebabdeh. p.n. *See Khûrbet ez Zebed,* p. 51.

SHEET XIII.

العنزية Ir *El 'Aneizīyeh.* El 'Aneiziyeh. p.n.; it would mean the family of 'Oneizah (an ancient Arabic female name), or 'the men from 'Oneizah' (a place in Arabia) ; but it is probably connected with the great Arab tribe of the 'Anazeh. 'Oneizah also means 'a she goat.'

الاربعين مغازي Ir *El 'Arbaîn Maghâzi.* The Forty Champions. Also called جامع الابيض *Jâmiâ el Abeid,* 'the white mosque.'

عرب ابي كشك Hp *'Arab Abu Kishk.* Arabs of the Abu Kishk family. *Kishk* is a Persian word, meaning 'cheese.'

عرب العريشية Hp *'Arab el 'Areishîyeh.* The Arabs of the huts (living in cane huts). They are also called *es Sûârki. See* Sheet XVI.

عرب السوطرية {Gr / Hr} *'Arab es Sûterîyeh.* The Sûteriyeh Arabs. p.n.

عرب الطيورية {Gr / Hr} *'Arab et Tiûrîyeh.* The Tiûriyeh Arabs ; p.n.; collective plural of طير 'a bird.'

عيون قارا Hr *'Ayûn Kâra.* The springs of Câra. *See Khûrbet Kâra,* p. 163.

بيت دجن Hq *Beit Dejan.* The house of Dagon, בֵּית דָּגוֹן.

بيارة عبود Hr *Biârct 'Abbûd.* The wells of 'Abbûd. p.n.

بيارة حيدرة Hq *Biâret Heiderah.* The wells of the declivity.

بيارة القاضي Hr *Biârct el Kady.* The wells of the judge (cadi). (Near the last.)

بيارة المفتي Hr *Biârct el Mufti.* The wells of the mufti ; the law-officer superior to the cadi. (These are close to the last.)

بيارة محمد علي Hr *Biârct Muhammed 'Aly.* The wells of Mohammed 'Ali. (Also close to the last. All these wells are named from their founders.)

بيارة السقا Hr *Biârct es Sukka.* The wells of irrigation.

بير عدس | Ip | *Bîr 'Adas.* The well of lentiles.

بير الزيبا | Ir | *Bîr ez Zeiba.* The plentiful well.

بركة بنت الكَافر | Ir | *Birket Bint el Kâfir.* Pool of the daughter of the infidel or pagan (near the *Jâmiâ el Abiad*), at Ramleh. The *Kanât Bint el Kâfir* leads to it, and an aqueduct from near Abu Shûsheh flows into it.

بركة الفواخير | Ir | *Birket el Fawâkhîr.* The pools of precious things ; or ' of pottery.'

بركة الجاموس | Ip | *Birket el Jâmûs.* The pool of the buffalo.

بركة الجلوس | Hq | *Birket el Jallûs.* The pool of el Jallûs. p.n. جلس means 'a pond'; جلوس 'sitting,' 'ascending a throne.'

بركة القمر | Gq | *Birket el Kŭmr.* The pool of the moon. The old port of Jaffa, now silted up.

البرك | Hr | *El Burak.* The pools.

برك ليل | Hp | *Burak Leil.* The pools of Leil. p.n. (night).

الدكاكين | Gr | *Ed Dekâkîn.* The shops. Here applied to *Kokim*, rock-cut tombs, as in many other instances, from their resemblance to an Arab shop.

الذهبيَّات | Ir | *Ed Dhahabiyât.* The golden things. It also means 'a well-known company or sect of doctors who used to report the sayings of Mohammed, which, under the name of حديث *hadîth* (traditions), form the سنّة *sunneh,* or supplementary law of Islâm.' In the Egyptian dialect the word would mean ' Nile pleasure boats.'

ظهر سلمة | Hq | *Dhahr Selmeh.* Ridge of Selmeh. p.n.; from the village.

فجا | Ip | *Fijja.* Fijja. p.n.; meaning a broadway, especially between two mountains.

فرخيّة | Ip | *Ferrikhîyeh.* Ferrikhîyeh. p.n. (فرخ 'a chicken,' 'a sprouting plant.')

ابن إبراك | Hq | *Ibn Ibrâk.* The son of Ibrâk. p.n.

الإمام علي | Hq | *Imâm 'Aly.* The Imâm 'Aly. *See* p. 182. The name applies to the *kubbeh* or ' shrine' at *Bâret Heiderah.*

الجليل | Hp | *El Jelîl.* El Jelîl. p.n. The word in Arabic means 'illustrious,' 'grand'; but it is a survival of the Hebrew גָּלִיל, signifying 'a district' or ' circuit.'

جريشة‎   Hp   *Jerîsheh.* Jerisheh. p.n.; from جرش‎ 'to pound' or 'grind.'

جميزة الجوز‎   Hr   *Jimmeizet el Jôz.* The twin sycamore.

جميزة الطيرة‎   Hr   *Jimmeizet et Tîreh.* The sycamore of the fort.

جنداس‎   Ir   *Jindâs.* Jindâs. p.n.

جسر العبابنة‎   Jq   *Jisr el 'Abâbneh.* The bridge of the 'Abâbneh; family name.

جسر جنداس‎   Ir   *Jisr Jindâs.* The bridge of Jindâs. *See* above.

جسر السودة‎   Iq   *Jisr es Sûdah.* The bridge of the blackish stony ground, (or 'of the black water').

قتا يافا‎   {Hq Iq}   *Kada Yâfa.* The district of Jaffa. *See* "Memoirs," p. 44.

قناة بنت الكافر‎   Ir   *Kanât Bint el Kâfir.* The canal of the daughter of the pagan.

كفر عانا‎   Iq   *Kefr 'Âna.* The village of 'Âna. p.n.

كفر جنس‎   Ir   *Kefr Jinnis.* The village of Jinnis. p.n.

كيوتا‎   Ip   *Keibûta.* Also pronounced *Geibûta.* p.n.

الكنيسة‎   Ir   *El Kenîseh.* The church.

الخراب‎   Jq   *El Khŭrâb.* The ruins.

خربة ابيار الليمون‎   Hr   *Khŭrbet Abiâr el Leimûn.* The ruin of the wells of lemons.

خربة الدبة‎   Gr   *Khŭrbet ed Dŭbbeh.* The ruin of the plot of ground.

خربة الظهرية‎   Jr   *Khŭrbet edh Dhâheriyeh.* The ruin of the ridge.

خربة الغرن‎   Hr   *Khŭrbet el Fum.* The ruin of the oven or reservoir.

خربة حدرة‎   Hp   *Khŭrbet Hadrak.* The ruin of the declivity.

خربة الحية‎   Ip   *Khŭrbet el Haiyeh.* The ruin of the serpent.

خربة لولية‎   Hr   *Khŭrbet Lûlieh.* The ruin of Lûlieh. p.n.

خربة الراس‎   Jr   *Khŭrbet er Râs.* Ruin of the top.

خربة شعيرة‎   Iq   *Khŭrbet Shâireh.* The ruin of barley.

خربة السوالمية‎   Ip   *Khŭrbet es Sŭâlimîyeh.* The ruin of the Sâlem family.

خربة صرفند‎   Hr   *Khŭrbet Sûrafend.* *See Sûrafend.*

خربة وبصة   Ip   *Khŭrbet Wabsah.* The ruin of abundant herbage beside a stream.

قلعة راس العين   Jp   *Kŭlit Râs el '.lin.* The castle of the spring-head.

لد   Ir   *Ludd.* Ludd. p.n. Heb. לד Lod.

المقاصر   Ir   *El Makâsir.* The villas ; a piece of ground cast of Birket cl Jânûs.

المخرس   IIp   *El Makhras.* The place where there is no echo.

المقتلة   Hr   *El Maktalah.* The place of slaughter ; once said to have been infested with robbers.

مينة روبين   Gr   *Minet Rûbin.* The harbour of Rûbin. *See Neby Rûbin.*

المرقبل المحمودية   Ip   *El Mirr.* The passage. Also called *el Mahmûdiyeh,* ' the property of Mahmûd.'

مقام الامام علي   Ir   *Mukâm cl Imâm 'Aly (Ibn Ali Tâleb).* The shrine of the Imâm Ali ibn Ali Tâleb, the cousin, son-in-law, and successor of Mohammed. *See p. 182.*

مقام النبي دان   Jr   *Mukâm en Neby Dân.* The shrine of the Prophet Dân.

مقام النبي هودا   Jq   *Mukâm en Neby Hûda.* The shrine of the Prophet Sudah.

ملبس   Ip   *Mulebbis.* Obscure.

نهر العوجا   IIp   *Nahr cl 'Aujd.* The meandering river.

نهر الباردة   Hp   *Nahr el Bârideh.* The cold river.

نهر روبين   Gr   *Nahr Rûbin.* The river of Rûbin. *See Neby Rûbin.*

النبي دانيال   Jr   *Neby Dâniâl or Dâniân.* The Prophet Daniel.

النبي دنون   Ir   *Neby Dannûn.* The Prophet Dannûn. p.n.

النبي كفل   Jq   *Neby Kifil.* The Prophet Kifil. *See* Kor'ân, ch. XXI, v. 85, "And Ishmael and Idris, and Dhu 'l Kifl,* all these were of the patient, and we made them enter into our mercy : verily, they were among the righteous."

---

* " That is, Elias, or, as some say, Joshua, and some say Zachariah, so called because he had a ' portion ' from God most high, and ' guaranteed ' his people ; or because he had ' double ' the work of the prophets of his time, and their reward ; the word *Kifl* being used in the various senses of ' portion,' ' sponsorship,' and ' double.'—*Beidhâwi's Arabic Commentary,* quoted in Palmer's translation of the Kor'ân, Vol. II, p. 53.

النبي مقدام Ir *Neby Mukdâm.* The prophet Mukdâm. p.n.; (foremost).

النبي روبين Gr *Neby Rûbîn.* The prophet Reuben.

النبي صالح Ir *Neby Sâleh.* The prophet Sâleh. p.n.; (righteous). A Mukâm in the *Jamiâ el Abiad* at Ramleh.

النبي سلمي Hq *Neby Selmy.* The prophet Selmy. p.n.; (in Selmeh).

النبي ثاري Jq *Neby Târi.* The prophet Thâri (blood avenger).

العمريات Ir *El 'Omriyât.* The lands of the 'Omar family. Ground east of *Birket el Jâmûs.*

الرملة Ir *Er Ramleh.* Sandy.

رنتية Iq *Rantieh.* Rantieh. Also spelt رنتي, *Ranti.* p.n.

الرسيم Ip *Er Reseim.* The little vestige of buildings.

سافرية Ir *Sâfiriyeh.* Sâfiriyeh. p.n.

ساكيا Iq *Sâkia.* Sâkia. p.n. It is probably a mistake for ساقية.

الساقية Hp *Es Sâkiah.* The water-wheel.

ساكنة ابي درويش Hq *Sâknet Abu Derwîsh.* The settlement of the father of the derwish.

ساكنة ابي كبير Hq *Sâknet Abu Kebîr.* The settlement of Abu Kebîr. p.n.; (great father).

ساكنة حماد Hq *Sâknet Hammâd* or ساكنة الدنايتي *Sâknet ed Denâity,* the settlement of Hammâd. p.n. Also called Säknet ed Danâity. p.n.

ساكنة العبيد Hq *Sâknet el 'Abîd.* The settlement of the slaves.

ساكنة العراينة Hq *Sâknet el 'Arâineh.* The settlement of the 'Arâineh. p.n.

ساكنة الجبايتة Gq *Sâknet el Jebâliyeh.* The settlement of the hill people. The official list has four others. (*See* "Memoirs," under ' Yâfa.')

ساكنة المسلخ Hq *Sâknet el Maslakh.* The settlement of the skinner, or ' place of flaying.'

ساكنة الشيخ ابرهيم Gq *Sâknet esh Sheikh Ibrahîm.* The settlement of the Sheikh Ibrahim, named from Sheikh Ibrahim el 'Ajamy.

ساكنة السبيل Hr *Sâknet es Sebîl.* The settlement of the wayside fountain. Applies to the buildings near Sebil Abu Nabbût.

2 F

ساكنة المصرية Hq *Sâknet el Músríyeh.* The Egyptian settlement just north of the town. Also called ساكنة الرشيد *Sâknet er Rushîd.*

ساكنة البرسيان Hq *Sâknet el Brusiân.* The settlement of the Prussians (a German colony).

سارونا Hq *Sârôna.* Sârôna. p.n. The name given by the Germans to the farm and settlement round it.

السبيل Ir *Es Sebîl.* The wayside fountain.

سبيل ابو نبوط Hq *Sebîl Abu Nabbût.* The wayside fountain of Abu Nabbût. p.n. (He was Governor of Jaffa at the commencement of the century.)

سبيل شاهين اغا Hr *Sebîl Shâhîn Âgha.* The fountain of Shâhîn Âgha. p.n.

سلمة Hq *Selmeh.* Selmeh. p.n. Peaceful.

الشيخ عباس Iq *Sheikh 'Abbâs.* Sheikh 'Abbâs. *See Abbasîyah,* p. 1. Also called *Sheikh 'Abd er Rahmân.*

الشيخ عبد النبي Hp *Sheikh 'Abd en Neby.* Sheikh 'Abd en Neby. p.n.; meaning 'the Prophet's servant.'

الشيخ عبدالرحمن Lr *Sheikh 'Abd er Rahmân.* Sheikh 'Abd er Rahmân. p.n.; meaning 'servant of the merciful.' North of Ludd.

الشيخ عبدالله Ir *Sheikh 'Abdallah.* Sheikh 'Abdallah. p.n. Servant of God. Just south of the houses in Ramleh.

الشيخ ابو عدس Ip *Sheikh Abu 'Adas.* The Sheikh Abu 'Adas. 'Father of lentils,' at Bir 'Adas.

الشيخ العجمي Ir *Sheikh el 'Ajamy.* The Persian Sheikh. East of Ludd.

الشيخ علي Hq *Sheikh 'Aly.* Sheikh 'Aly. p.n.

الشيخ مراد Hq *Sheikh Amrâd.* Sheikh Amrad, properly Murâd. p.n.; meaning 'desired.'

الشيخ دربات Hq *Sheikh Durbâth.* Sheikh Durbâth. p.n.

شيخ البلوطة Ip *Sheikh el Balhitah.* 'The Sheikh of the oaks.' From the trees near.

الشيخ ابرهيم العجمي Gq *Sheikh Ibrahîm el 'Ajamî.* Sheikh Abraham, the Persian.

الشيخ القطاناني Hq *Sheikh el Katânâny.* The Sheikh el Katânâny. *See Katanneh,* Sheet XVII.

الشيخ معنس Hp *Sheikh Muánnis.* Sheikh Muánnis. p.n. Perhaps from عنس 'a rock.'

الشيخ محمد السطاري   Ir   *Sheikh Muhammed es Settâri.* Sheikh Mohammed es Settâri. p.n.; سطار means 'a ruler of lines,' or a 'story-teller.'

الشيخ طاها   Ir   *Sheikh Tâha.* Sheikh Tâha (cook). In the same enclosure with *Neby Dannûn.*

الصير   Ip   *Es Sîr.* The fold (a ruined cattle-house). The name occurs twice.

الست نفسة   IIr   *Sitt Nefîsah.* The lady Nefisah. A common female name, meaning 'precious.'

سميل   IIp   *Summeil.* Summeil. p.n.; (hard or withered).

صرفند   Ir   *Sûrafend.* Sûrafend. p.n. Aram. צריפין, 'huts' (Buxtorf).

طاحونة الوسطى   Ip   *Tâhûnet el Wusta.* The central mill.

تل ابي زيتون   IIp   *Tell Abu Zeitûn.* The mound with the olive.

تل الذهب   Jp   *Tell edh Dhaheb.* The mound of gold.

تل العجول   Jp   *Tell el 'Ajjûl.* The mound of 'Ajjul. *See* p. 224.

تل الحلو   Ip   *Tell el Helu.* The sweet mound.

تل الملول   Jp   *Tell el Mallûl.* The mound of the oaks.

تل الخمر   Ip   *Tell el Mukhmar.* The mound of the wine-bibber.

تل النورية   Ip   *Tell en Nûriyeh.* The mound of the gypsies.

تل الرقيت ونيل احدبة   IIp   *Tell er Rekkeit.* The mound of the shore; also called *el Hadabeh,* 'the sand dune.'

تل السلطان   Gr   *Tell es Sultân.* The mound of the Sultan.

وادي ابو لجة   Ip   *Wâdy Abu-Lejja.* The valley with the abyss.

وادي عيون ضو   Gr   *Wâdy 'Ayûn Dô.* The valley of the springs of light; but compare Heb. צוא 'filth.' The word is commonly used in connection with water.

وادي الفرش   IIr   *Wâdy el Fersh.* The valley of the carpet (open ground).

وادي حنين   IIr   *Wâdy Hanein.* The valley of Hanein (properly Honein), a common geographical name in Arabia; the word means the cry of a she-camel to her colt.

وادي الاشكر   Ip   *Wâdy el Ishkar.* El Ishkar. p.n. *Ashkar* would mean 'grateful,' and *ushkur* might be translated herbage, especially 'hemlock.'

2 F 2

وادي جميلة	Ip	*Wâdy Jamîleh.* The valley of Jamileh; feminine p.n. ('handsome').
وادي قربة	Iq	*Wâdy Kereikah.* The valley of level ground.
وادي لد	Ir	*Wâdy Ludd.* The valley of Ludd. q.v.
وادي نصرة	IIq	*Wâdy Nusrah.* The valley of victory or help.
وادي الرزيقات	Gr	*Wâdy er Rezeikât.* The valley of sustenance.
وادي ريزيا	Ir	*Wâdy Rîzia.* The valley of Rizia. p.n.
وادي الشقارق	Ir	*Wâdy esh Shekârik.* The valley of the parrakeets or green magpies.
وادي الشلال	Iq	*Wâdy esh Shellâl.* The valley of the cascade.
وادي السرار	Ir	*Wâdy es Sûrâr.* The valley of the fertile spot; perhaps an error for الصرار 'pebbles.'
يافا	Gq	*Yâfa.* Jaffa. Heb. יָפֶי 'beauty.'
يازور	IIq	*Yâzûr.* Yâzûr. p.n. The original name is said by the natives to have been *Adâha.* *See* the "Quarterly Statement," 1874, p. 5.
اليهودية	Iq	*El Yehûdîyeh.* The Jewish place, family, tribe, or female.

ابو العوف   Nq   *Abu 'l 'Auf.* Abu 'l 'Aûf. The first word means father, *i.e.*, 'possessed of.' Aûf means 'flocking' (as birds round a well, &c.), and also 'pasturage.'

ابو العون   Jp   *Abu 'l 'Aûn.* The father of the flock.

ابو تش   Mr   *Abu Kûsh.* The father of Kûsh (a bucket).

ابو رجيلة   Mr   *Abu Rujeileh.* The father of the runlet. The hill on which is '*Ain Jelazûn.*

ابو شخيدم   Lr   *Abu Shukheidim.* The father of Shukheidim. p.n. Perhaps شخيدب 'earth-worms.'

عابود   Kq   *'Abûd.* 'Âbûd. p.n.; from عبد 'to worship.'

عبوين   Mq   *'Abwein.* 'Abwein. p.n.

عين عبرود - يبرود   Mr   *'Ain 'Abrûd* or *'Ain Yebrûd.* The spring of 'Abrûd or Yabrûd.

عين ابوس   Mp   *'Ain Abûs.* The spring of Abûs. p.n.

عين عجول   Mq   *'Ain 'Ajjûl.* The spring of Ajjûl. q.v., p. 219.

عين العليا   Mr   *'Ain el 'Alia.* The upper spring.

عين الوا   Lp   *'Ain Alûa.* The spring of the meanderings.

عين ارطبة   Lr   *'Ain Artabbeh.* The spring of Artabbeh. Perhaps from رطب 'moist, fresh.'

عين عروث   Nr   *'Ain 'Arûth.* The glittering spring, or the spring of 'denudation.'

عين بدران   Lq   *'Ain Badrân.* The spring of Badrân (a stinking herb).

عين البقار   Mp   *'Ain el Bakkâr.* The spring of the cowherds, below '*Ain Abûs.*

عين بكوم   Lq   *'Ain Bakkûm.* The spring of Bakkûm. p.n.; 'dumb.'

عين البص   Lr   *'Ain el Bass.* The spring of the marsh.

عين بيت اللو   Lr   *'Ain Beit Ello.* The spring of Beit Ello. q.v.

عين بنيّاك   Kq   *'Ain Beniák.* The spring of Beniák. p.n.

عين بريت   Np   *'Ain Berkit.* The spring of Berkit. p.n. *See* "Memoirs."

عين دارة   Mr   *'Ain Dárah.* The spring of Dárah (house, enclosure, or circle).

عين دير العقبان   Mr   *'Ain Deir el 'Okbán.* The spring of Deir 'Okbán. q.v.

عين ظهرة   Mr   *'Ain Dhahrah.* The spring of the ridge.

عين الدلبة   Mp   *'Ain ed Dilbeh.* The spring of the plane-tree (at Yásúf).

عين ايوب   Lr   *'Ain Eyúb.* Joab's (or Job's) well.

عين فوّارة   Kq   *'Ain Fúwárah.* The spring of the fountain.

عين الفوقا   Lq   *'Ain el Fóka.* The upper spring.

عين الغربيت   Nr   *'Ain el Gharbiyeh.* The western spring, on the west of Selwád.

عين الحجر   Kq   *'Ain el Hajr.* The spring of the rock.

عين الحلفا   Mr   *'Ain el Halfa.* The spring of the bulrushes.

عين احمم   Mr   *'Ain el Hamam.* The spring of the pigeons.

عين الحرميّة   Mr   *'Ain el Haramiyeh.* The robber's spring. p.n.

عين الحراشة   Lr   *'Ain el Harásheh.* The spring of rough ground.

عين اشعار   Mq   *'Ain Ish'ár.* The spring of thick foliage.

عين جنتا   Lr   *'Ain Jannata.* The spring of Jannata. p.n. ; 'garden.'

عين جرب   Mr   *'Ain Jereb.* The spring of the plantations.

عين جزون   Mr   *'Ain Jelazún.* The spring of Jelazún. p.n.

عين جلجيا   Mq   *'Ain Jiljilia.* The spring of Jiljilia. q.v.

عين الجوز   {Mp} {Lr}   *'Ain el Józ.* The spring of the walnut.

عين الجوزة   Lp   *'Ain el Józeh.* The spring of the walnut-tree.

عين جورة الحشة   Nr   *'Ain Júrat el Hishsheh.* The spring of the water of the cupola.

عين قانت   Lr   *'Ain Kánieh.* The crimson spring.

عين الكلبة   Lq   *'Ain el Kelbeh.* The spring of the bitch.

عين "التنة" Lr '*Ain el Kubbeh*. The spring of the dome.

عين كفريخ Mr '*Ain Kufriyeh*. The spring of the infidels.

عين قريوت Np '*Ain Kuriyût*. The spring of Kuriyût. q.v.

عين القسيس Mr '*Ain el Kussis*. The spring of the Christian) priest.

عين الليمون Lq '*Ain el Leimûn*. The spring of the lemon.

عين اللوزة Mr '*Ain el Lôzeh*. The spring of the almond tree.

عين المقورة Mq '*Ain el Mâkûrah*. The spring of the cropped herbage, or of the felled timber, or of the ham-strung beast; also called '*Ain 'Abwein*, being at the village of '*Abwein*.

عين المعصر Lp '*Ain el Mâaser*. The spring of the winepress.

عين مترون Lq '*Ain Matrûn*. The defiled spring.

عين المرج Mr '*Ain el Merj*. The spring of the meadow on the northeast side of the village of 'Ain Sinia).

عين المزراب Mr '*Ain el Mezrâb*. The spring of spouts.

عين المغاسل Lr '*Ain el Mûghâsil*. The spring of washings.

عين المكر Kq '*Ain el Mukr*. The spring of the hole.

عين الخوي Lp '*Ain el Mûtwey*. The walled spring.

عين الموس Mr '*Ain el Mûs*. The spring of the razor.

عين نافع Lq '*Ain Nâfâ*. The spring of the profitable one.

عين نعلين Kr '*Ain Nâlin*. The spring of Nâlin. q.v.

عين النويتف Lp '*Ain en Nuêitif*. The spring of little droppings.

عين الرجا Mp '*Ain er Raja*. The spring of hope.

عين روحان Np '*Ain Rûjân*. The spring of Rûjân. *See Khûrbet Rûjân*, Sheet XV.

عين الصعب Lq '*Ain es Sâbeh*. The spring of difficulty.

عين السكا Lp '*Ain es Sakka*. The spring of the irrigation, or of the pelican.

عين الساقي Mr '*Ain es Sâky*. The irrigating spring.

عين سرينا Kq '*Ain Sarina*. The spring of Sarina. p.n.

عين سيلون Np '*Ain Seilûn*. The spring of Seilûn. q.v.

عين الشلال　Lp　*'Ain esh Shellâl.*　The spring of the cascades.

عين الشرقية　Nr　*'Ain esh Sherkiyeh.*　The eastern spring.

عين سينيا　Mr　*'Ain Sinia.*　The spring of Sinia.　p.n.; probably Heb. יְשָׁנָה Jeshanah.

عين السلطان　Mr　*'Ain es Sultân.*　The sultan's spring.

عين الصرار　Nr　*'Ain es Surâr.*　The spring of pebbles.　*See Wâdy Surâr,* Sheets XVI–XVII.

عين صردد　Mr　*'Ain Surdah.*　The spring of Surdah.　q.v.

عين التنور　Lp　*'Ain et Tannûr.*　The spring of the oven, or reservoir.

عين التحتا　Lr　*'Ain et Tahta.*　The lower spring.

عين تلفيت　Np　*'Ain Telfît.*　The spring of Telfit.　q.v.

عين تبنة　Lq　*'Ain Tibneh.*　The spring of Tibneh.　q.v.

عين الطايور　Mq　*'Ain et Tiyûr.*　The spring of the birds.

عين ترمس عيّا　Nq　*'Ain Turmus 'Aya.*　The spring of Turmus 'Aya.　q.v.; at the village so named.

عين ام القصب　Np　*'Ain Umm el Kŭsab.*　The spring with the reeds.

عين ام سراج　Lr　*'Ain Umm Serâj.*　The spring with the lamps.

عين ام طوخ　Lr　*'Ain Umm Tûkh.*　The spring of Umm Tûkh.　The word has no meaning in Arabic, but is a very common topographical name in Egypt.

عين ينبوع　Lq　*'Ain Yanbûâ.*　The perennial spring.

عين ينبوع اليابس　Lq　*'Ain Yanbûâ et Yâbis.*　The dry perennial spring.

عين اليسيرة　Lr　*'Ain el Yasîreh.*　The spring of el Yasireh.　q.v.

عين يبرود　Mr　*'Ain Yebrûd.*　The spring of Yebrûd.　q.v.

عين وادي ريّا　Lq　*'Ain Wâdy Reiya.*　The spring of Wâdy Reiya.　q.v.

عين الزرقآ　{Kq / Lr}　*'Ain ez Zerka.*　The azure spring.

عين الزبية　Mr　*'Ain ez Zubeîyeh.*　The spring of the pit.　Also called عين الجبية *'Ain el Jubbeiyeh,* which means the same.

عجول وقيل عجرول　Mq　*'Ajûl or 'Ajjûl.*　The first means 'a calf,' and also 'black mud'; the last 'hasty.'

عَتَبَة جغامة Mr *'Akabet Jaghâmeh.* The steep or mountain road of Jaghâmeh. p.n.

عالِم الهدا Mp *'Alim el Hada.* ' Knowing guidance.' A wely, or saint's tomb.

العمري Mr *El 'Amry.* El 'Amry. p.n. A wely's tomb at Abu Kûsh. There is another at Surdah. *See* Sheet XVII.

عمورية Mq *'Amûrieh.* 'Amûrieh. p.n. (Ammorites).

عراق الاحمر Kq *'Arâk el Ahmar.* The red cliff or bank.

عراق فريج Kq *'Arâk Freij.* The cliff with the crevasse.

عراق الناطوف Jr *'Arâk en Nâtûf.* The cliff of drippings.

عراق النمري Kp *'Arak en Nimry.* The cliff of en Nimry ; *i.e.,* Arab of the Nimr (leopard) tribe.

عرارا Kp *'Arâra.* 'Arâra. p.n.

الاربعين Nq *El Arbâin.* The Forty. (Probably the Forty Martyrs of Cappadocia, whose shrines are common throughout the East.)

ارنوطية Mr *Arnûtieh.* The Albanians (Arnouts).

عرورا Lq *'Arûra.* 'Arûra. p.n.

اشقاف داود Nr *Ashkâf Dâûd.* David's cliffs.

اشقاف جلجال Nr *Ashkâf Jiljâl.* The cliffs of Gilgal.

عطارة Mq *'Attâra.* 'Attarah (Ataroth).

اودلة Np *Audelah.* Audelah. p.n.

عورتة Np *'Awertah.* 'Awerta. p.n.

عيون ابي زبعة Lr *'Ayûn Abu Zemâh.* The springs of Abu Zemâh. p.n.

عيون السقي Mr *'Ayûn es Saky.* The irrigating springs.

العزير Np *El 'Azeir.* Eleazar. The shrine of Eleazar.

العزيرات Np *El 'Azeirât.* The tomb of Phinehas.

عزون ابن عتمة Kp *'Azzûn Ibn 'Atmeh.* The wild olive of Ibn 'Atmeh (the son of darkness).

بهرة كوفة Ir *Bahret Kûfah.* The lake of Kûfa. p.n.

البلوع Mr *El Balûd.* The gulley-hole.

بربارا Kq *Barbára.* Saint Barbara.

باطن حراشة Lr *Bâtn Harásheh.* The rugged knoll.

باطن السنام Kq *Bâtn es Sinám.* The knoll of the camel's hump.

بلاد الجمعين الاول والثاني {Kp Lp Mp} *Belâd el Jemâin (el Awal wa 'ttâni).* 'The land of the two assemblies,' the first and the second. It is also called جورة مردا *Jûrat Merda;* the vicinity of Merda, from the village so called.

بلعين Kr *Belâin.* Belâin. p.n.

بياض عابود Lq *Beiâd 'Âbûd.* The white places of 'Âbûd. p.n.

بيتا Np *Beita.* Beita. p.n. *Beit,* a house.

بيت الو Lr *Beit Ello.* The house of Ello. p.n. Heb. אֵילוֹ Elou.

بيت كوفة Jr *Beit Kûfah.* The house of Kûfa. p.n.

بيت نبالا Jr *Beit Nebâla.* The house of archery.

بيت ريما Lq *Beit Rima.* The house of Rima. The Hebrew בית רימא.

بيتين Mr *Beitîn.* Beitîn; the Hebrew בֵּית אֵל Bethel.

بنات بري Kq *Benât Bŭrry.* Daughters of the desert-dweller—a cave in the side of a precipice.

بنو حارة القبلية Lr *Beni Hârith el Kibliyeh.* Southern district of the Ben Hârith. p.n.; (Aretas).

بنو حارة الشمالية Lr *Beni Hârith esh Shemâliyeh.* The northern Beni Hârith.

بنو حمار Kr *Beni Hŭmâr.* The Beni Hŭmâr (ass).⎫

بنو مرة {Mq Nq} *Beni Murreh.* The Beni Murreh. ⎬ All well-known

بنو سالم Nr *Beni Sâlim.* The Beni Sâlim. ⎪ Arab tribes.

بنو زيد Lq *Beni Zeid.* The Beni Zeid. ⎭

بروكين Lq *Berûkîn.* Berûkin. p.n.

بدية Kp *Bidich.* Bidich. p.n.

بئر ابى عمار Kp *Bîr Abu 'Ammâr.* The well of the father of the builder.

بئر ابى قدح Kq *Bîr Abu Kadah.* The well with the bowl.

بئر ابى خشبة   Mr   *Bir Abu Khashabeh.* The well with the timber.

بئر عين يوسف   Lq   *Bir 'Ain Yûsef.* The well of Joseph's spring.

بئر الاشقر   Kp   *Bir el Ashkar.* The reddish-grey well.

بئر الدرج   Mq   *Bir ed Deraj.* The well of the steps.

بئر حزة   Lq   *Bir Hamzeh.* The well of Hamzeh (the Prophet's uncle), at Fürkhah.

بئر حارس   Lp   *Bir Hâris.* The well of Hâris. q.v.

بئر الكوف   Nr   *Bir el Kûf.* The well of el Kûf. p.n.

بئر كوزة   Np   *Bir Kûzah.* The well of Kûzah. q.v.

بئر المرج   Mq   *Bir el Merj.* The well of the meadow.

بئر مسعف   Kq   *Bir Mesâf.* The well with the fittings (bucket, rope, &c.

بئر المستا   Np   *Bir el Miska.* The well of irrigation.

بئر محدد   Jp   *Bir Muhaddad.* The circumscribed or bounded well, or the well with the ironwork.

بئر الرمانة   Mr   *Bir er Rummâneh.* The well of pomegranates.

بئر الصهريج   Lr   *Bir es Sahrij.* The well of es Sahrij. q.v.

بئر السبيل   Kp   *Bir es Sebîl.* The well of the wayside drinking fountain.

بئر الشما   Kq   *Bir esh Shemma.* The well of the gourd.

بئر ام الابريق   Lr   *Bir Umm el Ibrîk.* The well with the pitcher.

بئر الوطاوط   Jp   *Bir el Watâwit.* The well of the bats.

بئر الزيت   Mr   *Bir ez Zeit.* The well of oil.

بركة الربا   Jq   *Birket er Ribba.* The well of er Ribba; perhaps ربا, *ribâ*, 'the hillocks.'

بركة الوقع   Jq   *Birket el Wakâ.* The pool of falling.

{ بدروس / بدرس }   Jr   *Budrus.* Budrus. p.n. *See* "Memoirs."

البرج   Mq   *El Burj.* The tower.

برج بردويل   Mr   *Burj Bardawîl.* Baldwin's tower.

برج بيتين   Mr   *Burj Beitin.* The tower of Bethel. Also called
برج مخرون   *Burj Makhrûn,* the tower of Makhrûn; perhaps a corruption of مخور a bank cut away by a stream.

برج الحنية   Jq   *Burj el Ḥaniyeh.* The leaning tower.

برج اللسانة   Nr   *Burj el Lisâneh.* The tower of the tongue or promontory.

برج الياخور   Lp   *Burj el Yakhûr.* The tower of the stable. *Kurâwa Ibn Hassan. See "Memoirs."*

برجمس   Nr   *Bûrjmûs.* Perhaps the Greek Περγαμος; it is a knoll of rock—a very remarkable feature.

الدواقير   Mr   *Ed Dawâkîr.* The hollows or furrows.

دير ابو مشعل   Kr   *Deir Abu Meshâl.* The monastery with the cresset (a beacon).

دير ابي سلامة   Jr   *Deir Abu Selâmeh.* The monastery of Abu Selâmeh. p.n.

دير علا   Jq   *Deir 'Alla.* The monastery of 'Alla. p.n.

دير تمار   Lr   *Deir 'Ammâr.* The monastery of the builder.

دير عرابي   Kq   *Deir 'Arâby.* The monastery of the Arab.

دير بلوطا   Kq   *Deir Ballût.* The monastery of the oak.

دير دقلة   Kq   *Deir Dakleh.* The monastery of Dakleh. The word means 'long-shanked sheep.'

دير الدرب   Lp   *Deir ed Derb.* The monastery of the road. This is a large tomb.

دير استيا   Lp   *Deir Estia.* The monastery of Estia. p.n.

دير غسانة   Lq   *Deir Ghüssâneh.* The monastery of the Ghassâneh (an old Christian Arab tribe).

دير ابزيع   Lr   *Deir Ibziâ.* The monastery of Ibziâ. p.n.; ('nimble').

دير الجبالي   Mp   *Deir el Jâly.* The monastery with the wall or parapet.

دير جريردارجرد   Nr   *Deir Jerîr, or Dâr Jerîr.* The monastery or house of Jerir, a celebrated Arab poet.

دير القديس   Kr   *Deir el Ķâddîs.* The monastery of the saint.

دير قلعة   Kq   *Deir Ķûlâh.* The monastery of the castle.

دير النظام   Lq   *Deir en Nidhâm.* The monastery of the marshal.

دير سعيدة   Lr   *Deir Sáideh.* The monastery of Sáideh. p.n. ; ' happy.'

دير سمعن   Kq   *Deir Simán.* The monastery of St. Simeon.

دير السودان   Lq   *Deir es Sudân.* The monastery of the negroes.

دير طريف   Jr   *Deir Tureif.* The monastery of Tureif ('the end'). *See* "Memoirs," under ' Bethariph.'

الديورة   Kq   *Ed Diûrah.* The monasteries.

دورة   Mr   *Dûrah.* Dûrah. p.n. ; from دور ' a circle.'

فرخة   Lq   *Fŭrkhah.* Fŭrkhah. p.n.

غرس الجندي   Kq   *Ghars el Jindy.* The plantation of the soldier.

حبلة   Jp   *Hableh.* Hableh. p.n. (' pregnant ').

الحبس   Jr   *El Habs.* The religious endowment (the same as وقف *wakf*, ' mosque property ').

حديثة   Jr   *Hadîthih.* Hadithih. p.n. ; ' new.'

حرم النبي شيت   Lq   *Haram en Neby Shît.* The sanctuary of the prophet Seth, in Fŭrkhah.

حارس   Lp   *Hâris.* The watch.

حوارة   Np   *Huwârah.* White marl.

اسقاقا   Mp   *Iskâka.* Roads situate at a junction.

جانية   Lr   *Jânieh.* Jânieh. p.n. ; جنى to gather fruit.

جامع الشيخ علي   Nr   *Jâmiä Sheikh 'Aly.* The mosque of Sheikh 'Aly, at Dâr Jerir.

جامع اليتيم   Nq   *Jâmiä el Yeteim.* The mosque of the orphan, or ' the unique one,' an epithet of Mohammed.

جبل البرد   Lp   *Jebel el Barad.* The cool mountain.

جبل قليلة   Np   *Jebel Kuleileh.* The mountain of the peaks.

جبل الرخوات   Nq   *Jebel er Rakhwât.* The mountain of the soft stones.

جبل صبيح   Np   *Jebel Sûbîh.* The reddish-white mountain.

جمالة   Lr   *Jemmâlah.* Jemmâla. *cf.* Heb. Gamala.

جمعين   Mp   *Jemmâin.* Jemmâin. p.n. ; from *Jemáäh*, ' a company.'

جيبيا   Lq   *Jíbia.* Jibia. p.n. (Geba).

جلجلية   Mq   *Jiljilia.* Jiljiliyeh. *See* "Memoirs," under 'Gilgal.'

جلجولية   Jp   *Jiljúlieh. See* last paragraph.

جمزو   Jr   *Jimzu.* The sycamores. Heb. גִּמְזוֹ.

جفنا   Mr   *Jufna.* Gophna. p.n. *See* "Memoirs."

جورة الحشة   Nr   *Júrat el Hishsheh.* The hollow of the cupola.

جورة مردة   Mp   *Júrat Merda.* The district of Merda (a village so-called).

جردة   Kr   *Jurdeh.* Bare ground.

قبر الملك فردوس   Lp   *Kabr el Melek Ferdús.* The tomb of King Ferdûs (Paradise).

قبور اليهود   Jr   *Kábur el Yehúd.* The tomb of the Jews (at Khurbet Midieh).

الكفر   Kq   *El Kefr.* The village.

كفر عين   Lq   *Kefr 'Ain.* The village of the spring ('Ain Bakkâm).

كفر حارس   Lp   *Kefr 'Háris.* The village of the guard.

كفر إنشآ   Jq   *Kefr Insha.* The village of production, or of letter-writing.

كفر قاسم   Jp   *Kefr Kásim.* The village of Kásim. p.n. *See Neby Kasim,* p. 10.

كفر مالك   Nr   *Kefr Málik.* The village of the landlord.

كفر ناعمة   Lr   *Kefr Námah.* The village of Námah. p.n.

خلة عواد   Kq   *Khallet 'Auwád.* The dale of 'Auwâd. p.n.

خلة البلوع   Lr   *Khallet el Balúa.* The dell of the gulley-hole.

خلة اللوزة   Nr   *Khallet el Lôzeh.* The dell of the almond-tree.

خلة الرمان   Lq   *Khallet er Rummân.* The dell of the pomegranates.

خلة الصليب   Kq   *Khallet es Salíb.* The dell of the cross (near 'Âbûd, which is a Christian village).

خلة شوريا   Nq   *Khallet Shúria.* The dell of Shûria (a certain plant which grows by the seashore).

خلة السلطان   Nr   *Khallet es Sultân.* The Sultan's dell.

خلّة الزيغانية   Jq   *Khallet ez Zeighâniyeh.* The dell of ez Zeighâniyeh (the Royston rooks).

خان ابى الحاج فارس   Mr   *Khân Abu Hâj Fâris.* The Caravanserai of the father of the pilgrim knight.

خان لبن   Mq   *Khân Lubban.* The Caravanserai of Lebonah.

خان الساوية   Np   *Khân es Sâwieh.* The Caravanserai of Sâwieh, 'the level.' (a village so-called).

خروبة البرج   Jq   *Kharrûbet el Burj.* The locust-tree of the tower.

خروبة مرج عبيد   Jq   *Kharrûbet Merj 'Obeid.* The locust-tree of the meadow of 'Obeid. p.n.; meaning 'little slave.'

خربة عبد المهدي   Mr   *Khûrbet 'Abd el Mahdy.* The ruin of 'Abd el Mahdy. p.n.; servant of the guide (the last Imam, a sort of second advent of Mohammed, the Moslem Messiah).

خربة عبد النبي   Kr   *Khûrbet 'Abd en Neby.* The ruin of the Prophet's servant: also called خربة دبلجة *'Khûrbet Deblejeh.'* p.n.

خربة ابيني   Jr   *Khûrbet Abeiny.* The ruin of Abeiny. *cf.* Heb. אֶבֶן 'a stone.'

خربة ابو فلاح   Nq   *Khûrbet Abu Felâh.* The ruin of the father of agriculture (perhaps *fellah,* 'peasant').

خربة ابو حامد   Jq   *Khûrbet Abu Hâmid.* The ruin of Abu Hâmid. p.n.

خربة عفريتة   Np   *Khûrbet 'Afritch.* The ruin of the female demon.

خربة عين الحرامية   Np   *Khûrbet 'Ain el Haramîyeh.* The ruin of the robbers' spring.

خربة عين القصر   Mr   *Khûrbet 'Ain el Kûsr.* The ruin of the spring of the castle.

خربة عين اللوزة   Mr   *Khûrbet 'Ain el Lôzeh.* The ruin of the spring of the almond-tree.

خربة عين المحيمة   Lr   *Khûrbet 'Ain el Muheimeh.* The ruin of the spring of el Muheimeh. p.n.; ('Ain Eyûb).

خربة الاقرع   Lp   *Khûrbet el Akrâ.* The bald ruin.

خربة علة   Kq   *Khûrbet 'Aleh.* The ruin of the high place.

خربة العليآ   Nr   *Khûrbet el 'Alia.* The upper ruin.

خربة الياطا   Mq   *Khûrbet 'Aliâta.* The ruin of mortar or cement.

خربة علي مالكها   Jq   *Khŭrbet 'Aly Mâlkǐna.*   The ruin of 'Aly, the inhabitant of Kefr Mâlik.

خربة عامر   Kp   *Khŭrbet 'Âmir.*   The ruin of 'Âmir (a common Arab proper name).

خربة عمورية   Nq   *Khŭrbet 'Amûrieh.*   The ruin of 'Amûrieh.   q.v.

خربة عنير   Lr   *Khŭrbet 'Annîr.*   The ruin of 'Annir.   p.n.

خربة العراق   Kq   *Khŭrbet el 'Arâk.*   The ruin of the cliff or bank.

خربة أرطبة   Lr   *Khŭrbet Artabbeh.*   The ruin of Artabbeh.   *See 'Ain Artabbeh,* p. 221.

خربة نزار   Jq   *Khŭrbet 'Azzâr.*   The ruin of 'Azzâr.   p.n. ; (the repeller).

خربة عزون   Mp   *Khŭrbet 'Azzûn.*   The ruin of the wild olive-tree ; also called خربة الرويسون *Khŭrbet er Ruweisûn* (رويس, a little hill-top).

خربة بلاطة   Kq   *Khŭrbet Balâtah.*   The ruin of flagstones.

خربة براعيش   Kq   *Khŭrbet Barââîsh.*   Also called برغش the ruin of gnats. *See Khŭrbet Barghasha,* p. 162.

خربة البد   Lr   *Khŭrbet el Bedd.*   The ruin of el Bedd (an idol temple).

خربة البيضآء   {Jq}
{Jr}   *Khŭrbet el Beida.*   The white ruin.

خربة بيت الحبس   Lp   *Khŭrbet Beit el Habs.*   The ruin of the religious endowment.

خربة بيت يمين   Kp   *Khŭrbet Beit Yemin.*   The ruin of the house of Yemin.   p.n.

خربة برنيقية   Jp   *Khŭrbet Bernîkieh.*   The ruin of Bernikieh (a kind of root so-called).

خربة برقيت   Np   *Khŭrbet Berkît.*   The ruin of Berkit.   p.n. *See* " Memoirs," under ' Borceos.'

خربة البساتين   Kp   *Khŭrbet el Besâtîn.*   The ruin of gardens.

خربة بئر الزيت   Mr   *Khŭrbet Bîr ez Zeit.*   The ruin of the well of oil.

خربة البيرة   Jq   *Khŭrbet el Bîreh.*   The ruin of the well; but *see Bîreh,* p. 157.

خربة البرناط   Jq   *Khŭrbet el Bornât.*   The ruin of the hat.

خربة بدرس   Jr   *Khŭrbet Budrus.*   The ruin of Budrus.   *See* p. 227.

خربة بور الجان   Kp   *Khŭrbet Bûr el Jân.*   The ruin of the hole of the genii or demons.

خربة البرج Mr *Khŭrbet el Burj.* The ruin of the tower.

خربة البريج Lp *Khŭrbet el Bureij.* The ruin of the little tower.

خربة برهام Mr *Khŭrbet Bûrhâm.* The ruin of Bûrhâm. p.n.

خربة دار حية Nr *Khŭrbet Dâr Haiyeh.* The ruin of the house of the serpent.

خربة دار ابرهيم Kq *Khŭrbet Dâr Ibrahim.* The ruin of Ibrahim's house. (It is modern.)

خربة دثرة وقيل دسرة Jr *Khŭrbet Dathrah,* or *Dasrah.* The ruin of Dathrah, or Dasrah; the first meaning 'caulking.' the second 'abundance.'

خربة الدير Lq *Khŭrbet ed Deir.* The ruin of the monastery.

خربة دير الفقيا Nq *Khŭrbet Deir el Fikia.* The ruin of the convent of Fikia ; perhaps فقي 'the doctor.'

خربة دير القسيس Kp *Khŭrbet Deir el Kŭssîs.* The ruin of the monastery of the (Christian) priest.

خربة دير العقبان Mr *Khŭrbet Deir el 'Okbân.* The ruin of the monastery of the eagles.

خربة دير شباب Mr *Khŭrbet Deir Shebâb.* The ruin of the monastery of the youths.

خربة دكرين Jq *Khŭrbet Dikerin.* The ruin of Dakrin. p.n.; from ذكر remembrance.

خربة الديس Nr *Khŭrbet ed Dîs.* The ruin of the forest.

خربة الدوير Jp Ip (Kq) *Khŭrbet ed Duweir.* The ruin of the little monastery.

خربة الامير Kq *Khŭrbet el Emîr.* The ruin of the prince.

خربة اشقارا Nr *Khŭrbet Eshkâra.* The reddish-gray ruin.

خربة الفخاخير Lp *Khŭrbet el Fakhâkhîr.* The ruin of potsherds.

خربة فارة Lp *Khŭrbet Fârah.* The ruin of Fârah. p.n.

خربة غرابة Mq *Khŭrbet Ghŭrâbeh.* The ruin of the raven.

خربة حامد Kr *Khŭrbet Hâmid.* The ruin of Hâmid. p.n.

خربة حماد Kp *Khŭrbet Hammâd.* The ruin of Hammâd. p.n.

خربة حنونة Kr *Khŭrbet Hanûneh.* The ruin of the privet-blossom.

2 II

خربة الحركة Lr *Khŭrbet el Harakah.* The ruin of motion.

خربة حراشة Lr *Khŭrbet Harâsheh.* The ruin of rough ground.

خربة حرموش Jr *Khŭrbet Harmûsh.* The ruin of Harmûsh. p.n.; perhaps from *harmâs*, 'hard ground.'

خربة حزيمة Kp *Khŭrbet Hazîmah.* The ruin of the breast or girth.

خربة حوش Mp *Khŭrbet Hôsh.* The ruin of the court-yard.

خربة الحمام Jr *Khŭrbet el Hŭmmâm.* The ruin of the bath.

خربة ابعنة Jr *Khŭrbet Ibânneh.* The ruin of Ibânneh. p.n.

خربة افتاز Mp *Khŭrbet Ifkâz.* The ruin of Ifkâz. p.n.; the word means 'breaking an eggshell.'

خربة جرعا Mp *Khŭrbet Jerr'â.* The ruin of the sandhill on which vegetation thrives.

خربة كمونة Mp *Khŭrbet Kammûneh.* The ruin of cummin seed.

خربة كفر عانا Mr *Khŭrbet Kefr 'Anâ.* The ruin of the village of 'Anâ. *See* "Memoirs," under ענת Anath.

خربة كفر بارة Jp *Khŭrbet Kefr Bârah.* The ruin of the village of Bâra. p.n.; perhaps from بأر 'to dig a well.'

خربة كفر فيديا Lr *Khŭrbet Kefr Fidia.* The ruin of the village of Fidia ('ransom').

خربة كفر حتا Jp *Khŭrbet Kefr Hatta.* The ruin of the village of Hatta. p.n.

خربة كفر ثلث Kp *Khŭrbet Kefr Thilth.* The ruin of the village of the third part. *See* "Memoirs."

خربة كفر توت Lq *Khŭrbet Kefr Tût.* The ruin of the village of the mulberry. *See* "Memoirs," under 'Tuta.'

خربة قيس Mq *Khŭrbet Keis.* The ruin of the Keis tribe, whose feuds with the Yemani faction have disturbed Syria since the time of the Caliphate.

خربة الكلخ Jr *Khŭrbet el Kelkh.* The ruin of hemlock.

خربة كسفة Jp *Khŭrbet Kesfa.* The ruin of the segment.

خربة الخريش Jp *Khŭrbet el Khôreish.* The ruin of the scratch.

خربة كفرية Lr *Khŭrbet Kufrîyeh.* The ruin of the pagans or infidels.

خربة كريسنة Kr — *Khŭrbet Kurcisinneh.* The ruin of vetches; diminutive of كرسنة *karsanah.*

خربة كركش Lp — *Khŭrbet Kŭrkŭsh.* The ruin of Kŭrkŭsh. p.n.; perhaps كركس *karkas,* 'tying up a beast.'

خربة كرم عيسى Kp — *Khŭrbet Kurm 'Aïsa.* The ruin of the vineyard of Jesus. ('Aïsa is the modern Arabic equivalent for Emmanuel; the place is a modern ruin.)

خربة المايات Mr — *Khŭrbet el Massâyât.* The ruin of el Massâyât. The word means 'birds' crops.'

خربة الميدان Lr — *Khŭrbet el Meidân.* The ruin of the plain or exercise ground.

خربة ميدروس Lr — *Khŭrbet Meiderûs.* The ruin of Meiderûs p.n.

خربة مزارعة Nr — *Khŭrbet Mezâráh.* The ruin of the sown lands.

خربة المزرعة Kp — *Khŭrbet el Mezráh.* The ruin of the sown land.

خربة مدية Jr — *Khŭrbet Midieh.* The ruin of Midieh. p.n. *See* "Memoirs," under 'Modin' or 'Modaith.'

خربة المقاطر Mr — *Khŭrbet el Mukâtir.* The ruin of the censers (a ruined church).

خربة المنطار Jq — *Khŭrbet el Muntâr.* The ruin of the watch-tower.

خربة المرابعة Nr — *Khŭrbet el Murâbáh.* The ruin of el Murâbâh, which means 'helping to load a camel,' &c.; it is probably *murabbadh,* 'square.'

خربة مرارة وقيل كفر مر Mr — *Khŭrbet Murârah,* or *Kefr Murr.* The village of bitterness; it is in the district of the Beni Murreh, from which it was probably named.

خربة المتيين Kp — *Khŭrbet el Muteiyin.* The plastered ruin.

خربة المطوي Lq — *Khŭrbet el Mŭtwy.* The ruin of the well lined with stones.

خربة نعلان Lr — *Khŭrbet Nâlân.* The ruin of Nâlân (near Nâlin). Perhaps from *nâl,* 'a sandal.'

خربة نصر Nr — *Khŭrbet Nasr.* The ruin of Nasr. p.n.; ('help').

خربة نجارا Kp — *Khŭrbet Nejjâra.* The ruin of the carpenters.

خربة الرافض Nq — *Khŭrbet er Râfid.* The ruin of the heretics.

خربة الراس {Qq} {Mr} — *Khŭrbet er Râs.* The ruin of the hill-top.

خربة راس الطيرة   Kp   *Khŭrbet Râs et Tîreh.* The ruin of the hill-top of the fort of Tireh. Perhaps the village Tireh (Sheet XI), visible from this top.

خربة راشانية   Lr   *Khŭrbet Râshânîyeh.* The ruin of Râshâniyeh. The ruin of the Râshâniyeh (persons who come uninvited to a feast). It may be from *rashan,* the mouth of a river.

خربة روبين   Lr   *Khŭrbet Rûbîn.* 'Reuben's ruin.'

خربة سامكية   Nr   *Khŭrbet Sâmkîyeh.* The ruin of the Sâmkiyeh. Family name ('high,' 'lofty').

خربة سطي وقيل شطي   Mr   *Khŭrbet Satty,* or *Shatty.* The ruin of the coast.

خربة السهلات   Mq   *Khŭrbet es Sahlât.* The ruin of the plains; situated on a flat plateau.

خربة سلمية   Mr   *Khŭrbet Selemîyeh.* The ruin of the Selemiyeh. Family name.

خربة السمانة   Kp   *Khŭrbet es Semâneh.* The ruin of the quail.

خربة شبتين   Kr   *Khŭrbet Shebtîn.* The ruin of Shebtin. q.v.

خربة شهادة   Kp   *Khŭrbet Shehâdeh.* The ruin of the martyrdom or testimony.

خربة الشجيرة   Mp   *Khŭrbet esh Shejerah.* The ruin of the tree.

خربة الشلال   Lp   *Khŭrbet esh Shellâl.* The ruin of the cascade, so-called from 'Ain esh Shellâl. q.v.

خربة الشقف   Lq   *Khŭrbet esh Shŭkf.* The ruin of the cleft.

خربة الشونة   Nq   *Khŭrbet esh Shûneh.* The ruin of the barn.

خربة سيح   Lr   *Khŭrbet Sîh.* The ruin of the surface water.

خربة سريسيا   Kp   *Khŭrbet Sirisia.* The ruin of Sirisia. p.n. *See* "Memoirs," under ' Beth Sarisa.'

خربة صوم ــ كفرصوم   Lr   *Khŭrbet Sôm,* or *Khŭrbet Kefr Sôm.* The ruin of fasting ; also called 'the ruin of the village of fasting.'

خربة السمرآ   Kp   *Khŭrbet es Sumra.* The brown ruin.

خربة سوسية   Kq   *Khŭrbet Sûsieh.* The ruin of the liquorice-plant.

خربة تنورة   Kp   *Khŭrbet Tannûrah.* The ruin of the oven or reservoir.

خربة طنطورة   Nr   *Khŭrbet Tantûrah.* The ruin of the peak. *See* p. 117.

خربة الحظيرة Np *Khŭrbet et Tireh.* The ruin of the fort.

خربة ام البريد Jp *Khŭrbet Umm el Bureid.* The ruin from which the post starts. p.n.

خربة ام الاكبا Jq *Khŭrbet Umm el Ikba.* The ruin of Umm Ikba. p.n.

خربة ام الطواق Jq *Khŭrbet Umm et Tawáky.* The ruin with the arches or ledges.

خربة وادي عبّاس Mr *Khŭrbet Wâdy 'Abbás.* The ruin of the valley of 'Abbás. p.n.

خربة وادي العسس Nr *Khŭrbet Wâdy el 'Asas.* The ruin of the valley of the patrol.

خربة وادي السرت Mr *Khŭrbet Wâdy es Seráh.* The ruin of the valley of Seráh. p.n.; (swift).

خربة زكريّا Jr *Khŭrbet Zakariya.* The ruin of Zakariya ; near Neby Zakariya. q.v.

خربة زبدة Jr *Khŭrbet Zebdah.* The ruin of butter. *See Khŭrbet Zebed,* p. 51.

خربة زيفزفيّة Jr *Khŭrbet Zeifizfiyeh.* The ruin of the lotus or red jujube tree زيزف.

خربثا ابن حارث Kr *Khŭrbetha Ibn Hárith.* The Khŭrbetha (probably synonymous with *Khurbeh,* 'ruin') of the son of Hárith ; named from the district of Beni Hárith.

الخضر $\left\{ \begin{array}{l} \text{Nr} \\ \text{I.q} \end{array} \right\}$ *El Khŭdr.* El Khŭdr. *See* p. 28.

قبّية Kr *Kibbieh.* Domed.

قيرة Mp *Kireh.* Pitch ; though it may be from the Heb. קיר 'a wall.'

قبالن Np *Kŭbalän.* Fronting. (The chief sheikh of the 'Adwän Arabs is called *Kublân ;* in his case it is probably a corruption of the Turkish *kuplân,* 'a lion.')

كوبر Lr *Kûbar.* Kûbar. p.n.

القلعة Mp *El Kŭláh.* The castle—a plot of ground where a castle is said to have stood.

قلعة فردوس Lp *Kŭlât Ferdûs.* The castle of (King) Ferdûs.

قلعة القمقمة Lp *Kŭlât el Kumakmeh.* The castle of the pots.

قولة Jq *Kûleh.* Kûleh. p.n.

قرايا إبن حسن Lp   *Kŭráwa Ibn Ḥasan.* The towns of Ibn Ḥasan. The old name according to the inhabitants is *Shâm et Tawîl.* q.v.

قرايا إبن زيد Lq   *Kŭráwa Ibn Zeid.* The towns of Ibn Zeid, in the Beni Zeid district.

قربيوت Nq   *Kuriyât.* The towns.

قرن الزوان Kq   *Kŭrn ez Zawân.* The peak of tares.

قرنة بزغ Kq   *Kŭrnet Bezâ.* The peak of Bezâ ('growing').

قرنة بئر التل Lp   *Kŭrnet Bìr et Tell.* The peak of the well of the mound.

قرنة الحرامية Jp   *Kŭrnet el Ḥaramiyeh.* The robbers' peak.

قرنة الصنوبر Kq   *Kŭrnet es Sinóbar.* The peak of the cypress tree.

قصر علي سليمان Kq   *Kŭsr 'Aly Suleimân.* The palace (house) of 'Aly Suleimân. p.n.

قصر جيزة Mp   *Kŭsr Jîzeh.* The palace on the coast or valley side.

قصر الكنز Lp   *Kŭsr el Kinz.* The palace of the treasure.

قصر مغارة البنيان Lp   *Kŭsr Mŭghâret el Bunián.* The palace of the cave of the foundation.

قصر منصورة Lp   *Kŭsr Mansûrah.* The palace of Mansûrah. *See* p. 9.

قصر عبيد Jq   *Kŭsr 'Obeid.* The palace of 'Obeid. p.n. ; by Kharrûbet Merj 'Obeid. q.v.

قصر الصباح Lp   *Kŭsr es Subâh.* The reddish-white palace.

قصر السنامة Lp   *Kŭsr es Senâmeh.* The palace of the camel's hump.

قوزة Mp   *Kûzah.* Samaritan קוזה and קרית נקתה. *See* "Memoirs." The word in Arabic means a vegetable like the egg-plant.

اللبن Mq   *El Lubban.* The milk (white). Heb. לבונה Lebonah ; from the white cliff beyond the village.

لبن رنتيس Kq   *Lubban Rentîs.* The milk-white spot of Rentîs. p.n.

المحجرة Kq   *El Maḥjerah.* The stony place.

مزار عبد الحق Mp   *Mazâr 'Abd el Ḥakk.* The shrine of 'Abd el Ḥakk. p.n. ; meaning 'servant of the true God.'

مجدل يابا Jq   *Mejdel Yâba.* The watch-tower of Yâba. Also called مجدل الصديق *Mejdel es Saddik,* 'the watch-tower of the truth-teller,' from a Sheikh Saddik el Jemâiny (*i.e.,* of Jemâin village), whose seat (*Kursy*) was there.

مردا	Mp	*Merda*. Merda. p.n. Samaritan מירדה. *cf. Khŭrbet Mird*, Sheet XVIII.
مرج عيد	Nq	*Merj 'Aid*. The meadow of the feast.
مرج القصاصى		*Merj el Kasásy*. The meadow of stones or gypsum.
مرج كسفة	Ip	*Merj Kesfa*. The meadow of Kesfa. p.n.
مرج الصوان	Lq	*Merj es Suwân*. The meadow of flints.
مرج الينبوت	Nq	*Merj el Yambût*. The meadow of thistles or of fenugreek.
المربعة	Kq	*El Merŭbah*. Probably *el Merŭbbah*, 'the square.'
مسآء	Kp	*Mesha*. Gravelly soil.
مشارق البيتاوي	Np	*Meshârik el Beitâwy*. The eastern district of Beita (a village).
المزيرعة	Jq	*El Mezeirah*. The sown lands.
مزيرعة القبلية	Lr	*Mezeirât el Kiblîyeh*. The southern sown land.
مزرعة	Lq	*Mezrâh*. The sown land.
مزرعة الشرقية	Nq	*Mezrât esh Sherkîyeh*. The eastern sown land.
مدية	Jr	*Midieh*. Midieh. p.n. *See Khŭrbet Midieh*, p. 235.
مسمار	Kq	*Mismâr*. The nail.
مقاطع عابود	Kq	*Mokâtâ 'Âbûd*. The quarries of 'Âbûd. q.v.
المدعدر	Kp	*El Mudâdir*. El Mudâdir. p.n.; perhaps from دعر, which in vulgar Arabic means a dirty grey colour.
مغاير اكمة	Kq	*Mŭghâir Akmeh*. The caves of the hillock.
مغاير خنفس	Kp	*Mŭghâir Khunfis*. The caves of the beetle.
مغارة ابى يمين	Kp	*Mŭghâret Abu Yemîn*. Cave of Abu Yemin. p.n.
مغارة قولية	Mr	*Mŭghâret Kûlich*. The cave of Kûlich. q.v.
مغارة السنطة	Np	*Mŭghâret es Sumtah*. The cave of the acacia.
مغارة ام البزوز	Kr	*Mŭghâret Umm el Buzûz*. The cave with the furniture or clothes. It is marked *Cave* in *Wâdy Malâkeh*.
مغارة الزوان	Kq	*Mŭghâret ez Zawân*. The cave of tares.
مغر ابى شعار	Jr	*Mŭghr Abu Sh'ar*. The caves with the thick overgrowth.

مغر الحمام   Kp   *Mŭghr el Hamám.* The caves of the pigeons.

مغر جنادس   Kq   *Mŭghr Jinâdis.* The caves of Jinâdis. The word is the plural of Jindâs. q.v.

مقام الامام علي   Jr   *Mukám el Imâm 'Aly.* The shrine of Imâm 'Aly (Ibn Abu Tâleb).

مقام النبي يحيى   Jq   *Mukám eh Neby Yahyah.* The shrine of the prophet Yahya. *i.e.,* St. John the Baptist. The place is also called مار حنا *Mâr Hanna,* St. John (the Christian appellation).

المثر   Np   *El Mukr.* The water-pits.

نعالين   Kr   *N'âlîn.* Nâlin. p.n. ; from *na'l,* a sandal.

النبي عنير   Lr   *Neby 'Annîr.* The prophet 'Annir. p.n.

النبي اقدام   Mq   *Neby Ekdâm.* The prophet Ekdâm ('foremost').

النبي ايوب   Lr   *Neby Eyûb.* The prophet Job.

النبي كفل   Lp   *Neby Kifl.* The prophet Kifl. *See* p. 216.

النبي لوشع   Lp   *Neby Lûshà.* The prophet Lûshà, apparently a corruption of יהושׁע Joshua, with the Arabic article *al.* Joshua's tomb has been shown here at least since 1283 A.D.

النبي منصورب   Np   *Neby Mansûry.* The prophet Mansûry ('victor').

النبي نون   Lp   *Neby Nûn.* The prophet Nûn, Joshua's father, whose tomb has also been shown here since 1283 A.D.

النبي عالح   Lq   *Neby Sâleh.* The prophet Sâleh ('righteous'). p.n.

النبي سائر   Lq   *Neby Sâ-îr.* The wandering, or journeying prophet.

النبي يوسف   Mr   *Neby Yûsef.* The prophet Joseph.

النبي زكريآ   Jr   *Neby Zakarîya.* The prophet Zachariah.

رافات   Kq   *Râfât.* Râfât. p.n. ; meaning 'acts of kindness.'

الراس   Lp   *Er Râs.* The hill-top—a village.

راس عيش   Lp   *Râs 'Aîsh.* The hill-top of life or bread.

راس الاقرع   Jr   *Râs el Akrà.* The bald hill-top.

راس عمار   Kq   *Râs 'Ammâr.* The hill-top of the builder.

راس عطية   Jp   *Râs 'Atîyeh.* The hill-top of 'Ata. p.n. *See Sheikh 'Ata,* below.

راس الدار   Mp   *Râs ed Dâr.* The hill-top of the house.

راس كركر   Lr   *Râs Kerker.* The hill-top of Kerker. p.n.

راس المنطار   Np   *Râs el Mantâr.* The hill-top of the watch-tower. A small round watch-tower exists on it.

راس الشيخ   Nq   *Râs esh Sheikh.* The hill-top of the Sheikh.

راس الطرفينة   Mr   *Râs et Turfineh.* The hill-top of the extremity; also called راس الساقي, *Râs es Saky,* the hill-top of the water-works, from the springs *'Ayûn es Saky.*

راس الزيمرة   Nr   *Râs ez Zeimerah.* The hill-top of the pipers.

رنتيس or رنطيس   Kq   *Rentis.* Rentis. p.n.

رجال صوفة   Kq   *Rijâl Sûfah.* The men of wool, *i.e.,* Sûfi, dervishes.

رمون   Nr   *Rummôn.* Rummôn. The Heb. כֵּלֶע הָרִמּוֹן Rock Rimmon.

سهل كفر إسطونا   Np   *Sahel Kefr Istûna.* The plain of Kefr Istûna. q.v. It is the eastern part of Merj el Kasâsy.

الصهريج   Lr   *Es Sahrîj.* The cistern—a sacred building in Jânieh.

الساوية   Np   *Es Sâwieh.* The level place.

سلفيت   Mp   *Selfit.* Selfit. p.n.; perhaps سلفة a levelled sown field.

سليتا   Kp   *Selita.* p.n.; perhaps سليتا the name of an Arab tribe.

سلواد   Nr   *Selwâd.* Selwâd. p.n.

سيلون   Nq   *Seilûn.* Heb. שִׁילֹה Shiloh.

سنيرية   Kp   *Senirieh.* Senirieh. p.n.

عرتا   Kp   *Serta.* Serta. p.n.

شعب الذياب   Jr   *Shâb edh Dhiâb.* The hill-spur of the Dhiâb (wolves); family name.

شعب الحريق   Mp   *Shâb el Harîk.* The hill-spur of the conflagration.

شابوني   Lr   *Shâbûny.* Shâbûny. p.n. (*shâbin,* soft, delicate); a sacred place.

شبتين   Kr   *Shebtîn.* p.n.

2 I

الشيخ عبدالله (Lp) (Mr) (Lr) *Sheikh 'Abdallah.* Sheik 'Abdallah. p.n.; meaning 'servant of God.' It is a tree in Khŭrbet Satty (Kr); the name occurs four times on the Sheet. The practice of dedicating trees to saints is a very ancient Arab custom. *See* p. 207.

الشيخ عبد الرحمن Nr *Sheikh 'Abd er Rahman.* Sheikh 'Abd der Rahmân ('the servant of the Merciful One'); in Rŭmmôn.

الشيخ ابو عزير Mp *Sheikh Abu 'Ozeir.* Sheikh Abu 'Ozeir, *i.e.*, the father of Ezra, *i.e.*, Eleazar.

الشيخ ابو شوشة Mp *Sheikh Abu Shûsheh.* The sheikh with the top-knot. South of Merda, west of Sheikh Muhammed.

الشيخ ابو يوسف Mr *Sheikh Abu Yûsef.* The Sheikh Abu Yûsef (father of Joseph).

الشيخ ابو زرد Mp *Sheikh Abu Zarad.* The sheikh with the mail-coat.

الشيخ احمد (Lq) (Nq) (Mr) (Lr) *Sheikh Ahmed.* Sheikh Ahmed. p.n.; occurs four times.

الشيخ احمد العبمي Mr *Sheikh Ahmed el 'Ajamy.* Sheikh Ahmed, the Persian, in 'Ain Sinia.

الشيخ احمد الاربعين Kr *Sheikh Ahmed el 'Arbâîn.* Sheikh Ahmed of the Forty.

الشيخ احمد الفراديس Mp *Sheikh Ahmed el Furâdis.* Sheikh Ahmed of the paradises.

الشيخ احمد المقيصب Mp *Sheikh Ahmed el Mukeisib.* Sheikh Ahmed of el Mukeisab. p.n. South of Merda.

الشيخ احمد النوباني Kr *Sheikh Ahmed el Nôbâny.* Sheikh Ahmed of Beit Nûba, in Shebtîn.

الشيخ احمد الطويل Kq *Sheikh Ahmed et Tawîl.* Sheikh Ahmed the long.

الشيخ احمد اليمني Jr *Sheikh Ahmed el Yemeny.* Sheikh Ahmed of Yemen.

الشيخ عيسى Lr *Sheikh 'Aîsa.* Sheikh 'Aîsa. p.n.; (Jesus).

الشيخ علي Mp *Sheikh 'Aly.* Sheikh 'Aly. p.n. In Yâsûf (Kp), a second also occurs called جامع البطمة *Jâmia el Butmeh*, the mosque of the terebinth.

الشيخ علي الامانات Lp *Sheikh 'Aly el Amânât.* Sheikh 'Aly of the trusts.

الشيخ علي الاربعين Mp *Sheikh 'Aly el Arbâîn.* Sheikh 'Aly of the Forty. It is in 'Ain Abús.

الشيخ عمر   Nq   *Sheikh 'Amr.* Sheikh 'Amr. p.n.

الشيخ عطا   Lp   *Sheikh 'Ata.* Sheikh 'Ata. p.n.; meaning 'gift.'

الشيخ براز الدين   Jp   *Sheikh Baráz ed Dín.* Sheikh Baráz ed Din. p.n.; meaning 'the champion of the Faith.'

الشيخ بدر   Mp   *Sheikh Bedr.* Sheikh Bedr. p.n.; 'full-moon.'

الشيخ بشارة   Mp   *Sheikh Bishárah.* Sheikh Beshára. p.n.; 'good news' (in village of Huwárah).

الشيخ الغرباوي   Jr   *Sheikh el Gharbáwy.* The western (Moorish) Sheikh. q.v. Also called ابو صبحة *Abu Subhah*, the reddish-white, or ابو سبحة 'the man with the rosary.'

الشيخ غيث   Lr   *Sheikh Gheith.* Sheikh Ghaith. p.n.; 'succour.'

الشيخ حمدان   Kr   *Sheikh Hamdán.* Sheikh Hamdán. p.n.

الشيخ حسين   Lp Lq Mp Nq   *Sheikh Husein.* Sheikh Husein. p.n.

الشيخ اسليم (ابو عمر)   Jr   *Sheikh Islím (Abu 'Amr).* Sheikh Selim. p.n.

الشيخ جميل   Lr   *Sheikh Jemíl.* Sheikh Jemil. p.n.; ('beautiful').

الشيخ كامل   Nq   *Sheikh Kámil.* Sheikh Kámil. p.n.; ('perfect').

الشيخ قطرواني   Mr   *Sheikh Katrawány.* The Sheikh from Katrah.

الشيخ كواش   Kq   *Sheikh Kauwásh.* The Sheikh Kauwásh. p.n.

الشيخ خضر   Lq   *Sheikh Khŭdr.* Sheikh Khŭdr. *See* p. 28. (Mq in 'Abwein; Nr in Rámmón.)

الشيخ خليل   Mq   *Sheikh Khŭlíl.* Sheikh Khŭlil. p.n. ('friend'). Abraham is called *Khŭlíl Allah*, 'the friend of God.' East side of 'Ajŭl.

الشيخ قمرة   Lr   *Sheikh Kumrah.* Sheikh Kumrah. p.n.; 'dusty white.'

الشيخ مسعود   { Lq Lr }   *Sheikh Mas'úd.* Sheikh Mas'úd. p.n.; 'fortunate.'

الشيخ المحوطا   Mp   *Sheikh el Mehaûwat.* The Sheikh of the walled place. *See* Merda, towards the north.

الشيخ مبارك   Nq   *Sheikh Mubárak.* Sheikh Mubárak. p.n.; 'blessed Sheikh A sacred tree.

الشيخ محمد   Mp   *Sheikh Muhammed.* Sheikh Mohammed. In south-east corner of Merda.

الشيخ محمد العجمي   Nq   *Sheikh Muhammed el 'Ajamy.* Sheikh Mohammed the Persian.

الشيخ محمد العروري   Lq   *Sheikh Muhammed el 'Arûry.* Sheikh Mohammed of 'Arûra.

الشيخ محمد الشلتاوي   Kr   *Sheikh Muhammed esh Shiltâwy.* Sheikh Mohammed of Shilta. On west side of Shilta.

الشيخ المفضل   Np   *Sheikh el Mufŭddel.* Sheikh el Mufŭddel. p.n. ('the preferred.')

الشيخ موسى   Lr   *Sheikh Mûsa.* Sheikh Moses.

الشيخ نافوخ   Kq   *Sheikh Nâfûkh.* Sheikh Nâfûkh. *Nâfûkh* means 'liquorice-root.'

الشيخ عبيد راحيل   Jr   *Sheikh 'Obeid Râhîl.* Sheikh 'Obeid Râhil. p.n. 'Obeid the traveller.'

الشيخ عثمان   Lp   *Sheikh 'Othmân.* Sheikh 'Othmân. p.n.

الشيخ الرفاعي   Kq   *Sheikh er Rafâây.* The Sheikh er Rafâây; *i.e.*, member of the order of dervishes founded by Ahmed ibn er Rifâ'i, who pretend to have power over fire, snakes, &c.

الشيخ الرافاني   Mr   *Sheikh er Râ-fâty.* The Sheikh from Râ-fât.

الشيخ رضوان   Lq   *Sheikh Redwân.* Sheikh Redwân. p.n. *Ridhwân,* name of the angel who guards paradise.

الشيخ صالح   {Lr Mr Nr Nq}   *Sheikh Sâleh.* Sheikh Sâleh. p.n.; 'righteous.' The last (Nq) is at Sinjil.

الشيخ سرايا   Mp   *Sheikh Sarâya.* Sheikh Sarâya. p.n.; signifying 'night journeying.'

الشيخ سليم-زيد   Nq   *Sheikh Selîm, or Sheikh Zeid.* p.n. Sheikh Selim, or Sheikh Zeid; both common Arabic names.

الشيخ صبيح   Kp   *Sheikh Subîh.* Sheikh Subih. p.n.; 'bright.'

الشيخ تقي الدين   Mp   *Sheikh Takî ed Dîn.* Sheikh Taki ed Din. p.n.; 'the pious one of the Faith.'

الشيخ طاروتية   Mq   *Sheikh Târûtieh.* Sheikh Târûtâyeh. p.n.

الشيخ التيم Lq *Sheikh et Teim.* The Sheikh of et Teim—name of an old Arab tribe. 'Teim' means servant, and was used in pagan times, like *'Abd*, in combination with the name Allah, *Teim Allah*, 'the servant of Allâh'; but it never occurs in Moslem cognomens.

الشيخ يوسف {Mr / Mq} *Sheikh Yûsef.* In 'Ajûb. Joseph. p.n.

الشيخ زهد Mp *Sheikh Zehed.* Sheikh Zehed. p.n.; 'abstinence.'

الشيخ الزيتاوي Mp *Sheikh ez Zeitâwy.* The Sheikh from Zeita, in the village of Jemmâin.

شمس الدين Jp *Shems ed Dîn.* Shems ed Dîn, a common Arabic name, meaning 'sun of the Faith.'

شلتا Kr *Shilta.* Shilta. p.n.

شقبة Kp *Shukbah.* The crevasse, cleft, or narrow pass.

شقيف الذبان Kp *Shŭkîf edh Dhibbân.* The cleft of flies.

شقيف الرين Kq *Shŭkîf er Rîn.* The cleft of dirt.

شقيف التينة Kq *Shŭkîf et Tîneh.* The cleft of the fig-tree.

سنجل Xq *Sinjil.* Saint Gilles. *See* "Memoirs."

الست مريم Mp *Sitt Miriam.* The Lady Mary (Virgin).

الصير Kq *Es Siyer.* The sheep-cotes.

سلطان عبد القادر Jr *Sultân 'Abd el Kâdir.* Sultân 'Abd el Kâdir. In Beit Nebâla.

صردة Mr *Surdah.* Surdah; probably Heb. הצְּרֵדָה Zereda.

صوانة راس سليم Lq *Sûwânet Ras Salim.* The flints of the hill-top of Selim. p.n.

التبلية Kq *Et Tabalieh.* The place where sweet herbs for seasoning grow. The name applies here to a valley.

طاحونة الدلب Lr *Tâhûnet ed Dilb.* 'Mill of Dilbeh.' q.v. In Wâdy Dilbeh.

الطايبة Nr *Et Taiyibeh.* 'The goodly.'

الطارود Mp *Et Târûd.* 'The projecting'; a prominent peak.

تلفيت Xp *Telfit.* Telfit. p.n.

التل {Nq Nr} *Et Tell.* The mound.

تل عصور Nr *Tell 'Asûr.* The mound of 'Asûr. p.n.; in Arabic it means 'ages.'

تبنة Lq *Tibneh.* Straw; probably the Heb. תִּמְנַת Timnath.

التناني Kq *Et Tinâny.* The Tinâny. p.n.

الطيرة Jq *Et Tireh.* The fort.

ترمس عايا Nq *Turmus 'Âya.* Turmus 'Âya. p.n. Probably תורמיסיא Thormasia.

ام اللبد Jq *Umm el Lebed.* The mother of felt-cloth.

ام صفة Lq *Umm Suffah.* The mother of flat ground. Probably Heb. מִצְפָּה Mizpeh. It is also called كفر اشوح *Kefr Ishwâh,* or كفر اشوا *Kefr Ishûa,* Joshua's village.

عوريف Mp *'Ûrif.* 'Ûrif. p.n.

وادي عبد النبي Nr *Wâdy 'Abd en Neby.* The valley of the prophet's servant. North of Dâr Jerir,

وادي عباس Lq *Wâdy 'Abbâs.* The valley of 'Abbâs. p.n.

وادي ابي غار Lp *Wâdy Abu 'Ammâr.* The valley of Abu 'Ammâr. *See Bîr 'Ammâr.*

وادي ابو الحيات Nr *Wâdy Abu 'l Haiyât.* The valley with the snakes; perhaps حيوة *haiyât,* 'life.'

وادي ابي كنديس Kp *Wâdy Abu Kindis.* The valley with the helebore.

وادي عابود Kq *Wâdy 'Âbûd.* The valley of 'Âbûd. q.v.

وادي عدسية Jq *Wâdy 'Adasîyeh.* The valley of lentils.

وادي العين Nr *Wâdy el 'Ain.* The valley of the spring.

وادي عين دارة Lq *Wâdy 'Ain Dârah.* The valley of 'Ain Dâra. q.v.

وادي علي Mq *Wâdy 'Aly.* The high valley.

وادي غار Nq *Wâdy 'Ammâr.* The valley 'Ammâr. *See Wâdy Abu 'Ammâr.*

وادي غورية Nq *Wâdy 'Amûrieh.* The valley of 'Amuriyeh. *See Khŭrbet 'Amûrieh.*

وادي عرصَة   Mr   *Wâdy 'Arsah.* The valley of the quadrangle.

وادي أرطبَة   Lr   *Wâdy Artabbeh.* The valley of Artabbeh. *See Khûrbet Artabbeh,* p. 232.

وادي عودة   Kq   *Wâdy 'Aûdeh.* The valley of 'Aûdeh. p.n.

وادي عواد   Lp   *Wâdy 'Auwâd.* The valley of 'Auwâd; both this and the last are common Arab names.

وادي العيون   Kp   *Wâdy el 'Ayûn.* The valley of springs.

وادي عزون   Kp   *Wâdy 'Azzûn.* The valley of the wild olive.

وادي البيوتة   Kp   *Wâdy el Bahûteh.* The valley of the liars. Perhaps for وادي البحونة 'the valley of mines.'

وادي بربارا   Kq   *Wâdy Barbâra.* The valley of St. Barbara.

وادي البلاط   Mq   *Wâdy el Belât.* The valley of flagstones.

وادي البشارات   Mp   *Wâdy el Bushârât.* The valley of good news, or of the Beshâra family.

وادي دير بلوطا   Kq   *Wâdy Deir Ballût.* The valley of the convent of the oaks.

وادي دير غسانة   Kq   *Wâdy Deir Ghŭssâneh.* The valley of Deir Ghŭssâneh. q.v.

وادي الدلبة   Lr   *Wâdy ed Dilbeh.* The valley of the plane-tree.

وادي درمة   Kp   *Wâdy Dirmeh.* The valley of the soft ground.

وادي غريب   Mp   *Wâdy Gharîb.* The valley of the stranger, near *Khûrbet Ghŭrâbeh.*

وادي غياش   Jr   *Wâdy Ghiâsh.* The valley of Ghiâsh; either from غوش 'birch tree,' or from غوش, which in vulgar Arabic means 'to attack treacherously.'

وادي دميس   Mr   *Wâdy Hamîs.* The valley of the tramping of camels' feet. Also called—

وادي كفرية   *Wâdy Kufrîyeh* (from *Khûrbet Kufrîyeh*). q.v.

وادي الهياج   Mp   *Wâdy Heiyaj.* The valley of the fray (near Shâb el Harîk. q.v.

وادي إشعار   Mq   *Wâdy Ishâr.* See *'Ain Ishâr,* p. 222.

وادي جنتا   Kr   *Wâdy Jannata.* The valley of Jannata. p.n.; from Heb. גנּה 'garden.'

وادي جروفة   Nq   *Wâdy Jarûfah.* The valley of steep banks cut out by the stream.

وادي الجناب   Np   *Wâdy el Jenâb.* The side valley, or the valley of the court.

وادي جمعين   Mp   *Wâdy Jemmâin.* The valley of Jemâin. q.v.

وادي القنابس   Nr   *Wâdy el Kanâbis.* The valley of women قنبس; perhaps from تنافس 'snakes.'

وادي كانة   Jp   *Wâdy Kânah.* Probably an error of the native scribe for قانة. Heb. קָנָה Kanah. *See* "Memoirs."

وادي القتيل   Lq   *Wâdy el Katîl.* The valley of the slain (west of Kefr 'Ain).

وادي الكلب   Lr   *Wâdy el Kelb.* The valley of the dog.

وادي قرىقة   Jr   *Wâdy Kereikah.* The valley of level ground.

وادي الخروب   Kq   *Wâdy el Kharrûb.* The valley of the locust-tree.

وادي الخليل   Nq   *Wâdy el Khâlîl.* The valley of Khûlil. *See* p. 243.

وادي الخرب   Mr   *Wâdy el Khûrb.* The valley of ruins.

وادي قيرة   Lp   *Wâdy Kirch.* The valley of Kirch. q.v., p. 237.

وادي الكوة   Nr   *Wâdy el Kôla.* The valley of the rising ground.

وادي الكوب   Mq   *Wâdy el Kûb.* The valley of the cup.

وادي القصر   Kq   *Wâdy el Kûsr.* The valley of the palace.

وادي اللحام   Kp   *Wâdy el Lehhâm.* The valley of the butcher.

وادي المعاصر   Lp   *Wâdy el Mâaser.* The valley of the wine-presses.

وادي الماجور   I r   *Wâdy el Mâjûr.* The valley of the flower-pot ; another name for *Wâdy el Kelb.*

وادي ملاكة   Kr   *Wâdy Malâkeh.* The valley of the smooth stone, or of the trowel.

وادي مرج عرزب   Mq   *Wâdy Merj 'Erzy.* The valley of the meadow of 'Erzy. p.n.

وادي محيصن   Nr   *Wâdy Muheisin.* The valley of fortifications.

وادي موسى   Nq   *Wâdy Mûsa.* The valley of Moses.

وادي المصلب   Kr   *Wâdy el Muslib.* The valley of the Crucifixion.

وادي المطوي   Lp   *Wâdy el Mûtwy.* The valley of el Mûtwy. *See 'Ain el Mûtwy,* p. 223.

وادي الناطوف | Kr | *Wâdy en Nâtûf.* The valley of drippings.

وادي الندى | Lr | *Wâdy en Neda.* The valley of dew.

وادي النمر | Nr | *Wâdy en Nimr.* The well-watered valley; it contains several springs.

وادي نقيب | Mr | *Wâdy Nâkîb.* The valley of the warden.

وادي نصير | Nr | *Wâdy Nuseir.* The valley of Nuseir. p.n.; meaning 'help.'

وادي عمير | Lp | *Wâdy 'Omeir.* The valley of 'Omeir. p.n.; diminutive of 'Omar. South of Deir Estia.

وادي رباح | Jp | *Wâdy Rabâh.* The valley of gain or profit. It is also called
وادي ريحا | | *Wâdy Riha,* either from *Rîh,* the wind, or *Rîhân,* sweet herbs, especially basil.

وادي راشد | Lq | *Wâdy Râshid.* The valley of Râshid. p.n.; meaning 'orthodox.'

وادي ريّا | Lq | *Wâdy Reiya.* The valley of sweet fragrance, or of quenching thirst.

وادي الرمح | {Mq} {Np} | *Wâdy er Rumh.* The valley of the lance.

وادي شاهين | Jr | *Wâdy Shâhîn.* The valley of the falcon.

وادي سعد | Lr | *Wâdy Sâd.* The valley of Sâd. p.n.

وادي صحوري | Jq | *Wâdy Sahûry.* The desert valley.

وادي سيلون | Nq | *Wâdy Seilûn.* The valley of Seilûn. q.v.

وادي الشامي | Kr | *Wâdy esh Shâmy.* The Damascus or Syrian valley.

وادي صباح | Kq | *Wâdy Subâh.* The reddish-white valley.

وادي الصرار | Lt | *Wâdy es Surâr.* The valley of pebbles.

وادي صرا | Nq | *Wâdy Sûrra.* The valley of Sûrra. q.v., Sheet XV.

وادي الطنطورة | Nr | *Wâdy et Tantûrah.* The valley of the peak. *See* p. 117.

وادي الطيور | Mq | *Wâdy et Tiyûr.* The valley of birds.

وادي ياسوف | Mp | *Wâdy Yâsûf.* The valley of Yâsûf. q.v.

وادي زعورة | Nr | *Wâdy Zaûrah.* The bare valley; north of Khûrbet Mezârâh.

وادي الزرقا | {Kq} {Lr} | *Wâdy ez Zerka.* The valley of the azure stream.

2 K

وادي الزيتون   Mr   *Wâdy ez Zeitûn.*   The valley of olives.

الياسيرة   Lr   *El Yâsîreh.*   El Yâsireh ('case'); a sacred place.

ياسوف   Mp   *Yâsûf.* Yâsûf. p.n.   Samaritan יסבה. *See* "Memoirs."

يبرود   Mr   *Yebrûd.* Yebrûd. p.n.; from *Bârid* ('cold').

يتما   Np   *Yetma.* Yetma. p.n.

الزاقور   Jp   *Ez Zâkûr.* Ez Zâkûr. p.n.   (From زقر = صقر a hawk).

زاوية   Kp   *Zâwieh.* Corner, hermitage.

زيتا   Mp   *Zeita.* Olive.

زيرة السلع   Kq   *Zîret es Selâ.* The segment of the cleft rock.

SHEET XV.

عبد القادر   Pp   *'Abd el Kâdir.* 'Abd el Kâdir. p.n.; 'servant of the Almighty.'

ابو سيف   Op   *Abu Seif.* The man with the sword.

ابو تلول   Pr   *Abu Tellûl.* The place with the mounds.

الاحما   Pp   *El Ahma.* The inaccessible ; a hill top.

عين عولم   Np   *'Ain 'Aûlam.* The spring of 'Aûlam. p.n.; from علم a sign-post, or mountain.

عين العوجة   Or   *'Ain el Aûjah.* The crooked spring.

عين بضة الريح   Qp   *'Ain Basset er Rîh.* The spring of the marsh of the wind.

عين فضايل   Oq   *'Ain Fûsâil.* The spring of the breastwork ; but see the " Memoirs," under ' Phasaelis.'

عين الجهينة   Oq   *'Ain el Jeheineh.* The spring of the Joheineh, a well-known Arab tribe.

عين جوزة   Qp   *'Ain Jôzeleh.* The spring of the wild dove.

عين الكرزلية   Pp   *'Ain el Kurzeleiyeh.* The spring of the shepherds. Aram. בַרְזֵל.

عين سامية   Nr   *'Ain Sâmieh.* The lofty spring.

عين السخن   Oq   *'Ain es Sokhn.* The warm spring.

عين ام عمير   Oq   *'Ain Umm 'Omeir.* The spring of Umm 'Omeir ; fem. p.n.

عقربة   Op   *'Akrabeh.* Scorpion. ('Ακραββείν, Acrabi.)

عليط النجمة   Or   *'Allit en ' Nejmeh.'* The 'Allit tree of en Nejmeh, on *Jebel en Nejmeh* (' the star ').

عرب الكعبنة   Pr   *'Arab el Kâbneh.* The Kâbneh Arabs. p.n.

عرب المساعيد   Qp   *'Arab el Mesâîd.* The Arabs of the Mas'ûd tribe. (Pl. of *Mas'ûd.*)

عراق العلية   Pp   *'Arâk el 'Alieh.* The upper cliff.

2 K 2

عراق دار شعله   Nq   *'Arâk Dâr Shuâleh.* The cliff of the house (or circuit) of the flame, or in vulgar parlance, 'of the cavern.'

عراق الرصيفة   Pp   *'Arâk er Rasîfeh.* The solid cliff.

الاراكة   Pq   *el Arâkah.* The thorn tree (*Palinous aculeatus*).

عراقيب المريغات   Or   *'Arakîb el Mereighât.* The passes, or winding tracks of the pasture lands.

عرقان العوري   Op   *'Arkân el 'Aûrî.* The cliffs, banks, or buttresses of el 'Aûry. p.n.

عرقان البقر   Op   *'Arkân el Bakkar.* The cliffs, banks, or buttresses of el Bakkâr. p.n. ; 'cattle-drover.'

عرقان الغالية   Pq   *'Arkân el Ghâlieh.* The cliffs, banks, or buttresses of the civet.

عرقان الرب   Pq   *'Arkân er Rubb.* The cliffs, banks, or buttresses of Syrop.

عرقان السبع   Op   *'Arkân es Sebâ.* The cliffs, banks, or buttresses of the lion.

عرقان الصوانة   Pq   *'Arkân es Shwâneh.* The cliffs, banks, or buttresses of flint.

عرقوب مقصر اليانم   Pp   *'Arkûb Maksar en Naim.* The pass, or winding track of the watering-place of the sleeper ; مقصر water around which cattle lie.

العوجة   Qq   *El 'Aûjah.* The crook or bend.

اوسرين   Np   *Aûsarîn.* High ground.

البغلات   Pr   *El Baghalât.* The mules ; a mound.

بئر ابو درج   Pp   *Bîr Abu Deraj.* The well with the steps.

بئر الدوا   Op   *Bîr ed Dôwa.* The well of medicine.

بئر الحوار   Nq   *Bîr el Hûwâr.* The well of white marl.

بئر ترتلة   Np   *Bîr Tirteleh.* The well of humming or singing.

بركة فسايل   Pq   *Birket Fûsâil.* See *Ain Fûsâil,* p. 251.

دلوك   Op   *Dalûk.* Dalûk. *cf.* Heb. דלק 'to burn'; it is a high peak with a cairn ; beneath is Wâdy en Nâr, 'valley of fire.' It is on the Kurn Surbabeh range, and was probably an ancient beacon.

الدامية   {Op / Qp}   *Ed Dâmieh.* The bloody spot.

دير ابو سكوب   Nq   *Deir Abu Sekûb.* The monastery with the flowing (water).

ظَهْرة البَلْقَا Pr *Dhahret el Belka.* The ridge of the speckled or open ground.

الخَرس وقيل الخُتيان Pp *Edh Dhirs.* 'The rough hill.' A mound where a fight occurred between the *Mesáïd* and *Hejeikah* Arabs thirty years ago; also called *el Hatiàn.* p.n.

دُومَة Oq *Dômeh.* 'The lotus' (*Zizyphus lotus*).

دَوْنَة المَصْنع Pp *Dûkat el Masná.* The enclosure of the cistern.

ارْنَات البُومَة Pp *Erdâk el Bûmeh.* The winkings of the owl; a set of caves on a cliff.

الغُور Qr *El Ghôr.* The hollow or low land.

الحَبِج Qq *El Habej.* The trysting place of the tribe; remains of an old camp close by.

حَتِج الزير Pq *Habej ez Zîr.* The tryst of ez Zîr, a famous chief. *See* "Memoirs."

الحَبْس Nq *El Habs.* The religious endowment.

حانوت العُلَيْن Or *Hânût el 'Alein.* The booth of 'Alein. p.n.

حيشة اجْدُوع Op *Hîshet el Jeddûá.* The copse of Jeddûá. q.v.

الحِصْن Op *El Hosn.* The fort (at 'Akrabeh).

الحُفِيرة Pp *El Hûfîreh.* The excavation.

جالُود Nq *Jâlûd.* Jâlûd. p.n.; Goliath.

جَبَل البِتَّة Oq *Jebel el Büttah.* The isolated mountain.

جَبَل صَفْحَة الروس Nq *Jebel Safhat er Rûs.* The mountain of the flat rocks on the tops; south-east of *Jalûd.*

جَبَل التَّى Op *Jebel et Teyi.* Probably التائي the flat-topped mountain.

اجْدُوع Op *El Jeddûá.* Cut off (a hill).

جِفا نون Op *Jefa Nûn.* Nûn's rubbish (a hill near *Neby Nûn*).

جُحير عَقْرَبَة Op *Jehîr 'Akrabeh.* The den of 'Akrabeh. q.v.

جِسْر الدامِيَة Qp *Jisr ed Dâmieh.* The bridge of ed Dâmieh. q.v.

جُورة الوَحْرُبة Pp *Jûrat el Wahrubbeh.* Hollow of the tanks; near *el Masná.* q.v.

جُورِيش Np *Jûrîsh.* Jûrîsh. p.n.

قبور عيال حرب   Pr   *Kabûr 'Aiâl Harb.* The graves of the family of Harb (one of the Heteimât Arabs).

قبر الهلالي   Pr   *Kabr el Helâly.* The grave of the man of the Beni Helâl Arabs.

كمّونية   Op   *Kammûnieh.* The place of cummin seed.

قنة القراوا   Pp   *Kanat el Kûrâwa.* The canal of Karâwa. (The towns.)

قنة المنيل   Or   *Kanat el Manîl.* The canal of el Manil. p.n.

قنة موسى   Or   *Kanat Mûsa.* The canal of Moses.

قنة ام الهريري   Pp   *Kanat Umm el Hureiry.* The canal with much water.

الكبابير   Pp   *El Kebbabîr.* The capers or 'the drums.' Limekilns in which the mortar used on Kŭrn Sŭrtŭbeh was made.

كفر عتيا   Np   *Kefr 'Atya.* The village of 'Atya. *See* "Memoirs," under ' Caphar Atân.'

خلّة عسيم   Op   *Khallet 'Assim.* The dell of 'Aseim. p.n.

خلّة الفوة   Pp   *Khallet el Fûleh.* The dell of the beans.

خلّة زريق   Nq   *Khallet Zereik.* The blue dell.

الخضر   Oq   *El Khŭdr.* See p. 28.

خراب ابو غريب   Op   *Khŭrâb Abu Gharîb.* The ruins of the father of the stranger.

خربة ابي ملول   Nq   *Khŭrbet Abu Malûl.* The ruin with the oaks.

خربة ابى راشد   Nr   *Khŭrbet Abu Râshid.* The ruin of Abu Râshid. p.n.

خربة ابى ريسة   Op   *Khŭrbet Abu Rîsah.* The ruin of Abu Risa. p.n.

خربة عين عينة   Np   *Khŭrbet 'Ain 'Ainah.* The ruin of the spring of 'Aina. *See* "Memoirs," under ' Annath.'

خربة الاراكا   Pq   *Khŭrbet el Arâkah.* The ruin of el Arâkah. q.v. Also called خربة الشجرة *Khŭrbet esh Shejerah,* 'ruin of the tree.'

خربة العوجة الفوقا   Or   *Khŭrbet el 'Aûjah el Fôka.* The upper ruin of el Aûjah (the crook or bend).

خربة العوجة التحتاني   Pr   *Khŭrbet el A'ûjah et Tahtâni.* The lower ruin of el Aûjah.

خربة عوليم   Np   *Khŭrbet 'Aûlim.* The ruin of Aûlim. *See* 'Ain 'Aûlam, p. 251.

خربة البيوضات   Pr   *Khŭrbet el Beiyûdât.* The ruin of the white places.

خربة بني فاضل   Oq   *Khŭrbet Beni Fâdl.* The ruin of the Beni Fâdl (an Arab tribe).

خربة بركة القصر   Nq   *Khŭrbet Birket es Kŭsr.* The ruin of the pond of the castle.

خربة فصايل   Pq   *Khŭrbet Fŭsâil.* The ruin of Fŭsâil. *See 'Ain Fŭsâil*, p. 251, and " Memoirs," under ' Phasaelis.'

خربة جبيط   Oq   *Khŭrbet Jibeit.* The ruin of the hollow thing, or of the idol (an old Syrian word).

خربة قَتَوَل   Nr   *Khŭrbet Kaswal.* The ruin of treading corn.

خربة كفر استونا   Nq   *Khŭrbet Kefr Istûna.* The ruin of Istûna. p.n. ; perhaps ستون (Persian) a column.

خربة الكرك   Nq   *Khŭrbet el Kerek.* The ruin of the fortress.

خربة الكروم   Op   *Khŭrbet el Kerûm.* The ruin of the vineyards.

خربة قرقفة   Np   *Khŭrbet Kŭrkŭfah.* The ruin of the Kŭrkŭfah (a small bird so-called).

خربة المراجم   Oq   *Khŭrbet el Merâjem.* The ruin of the cairns.

خربة مراس الدين   Op   *Khŭrbet Merâs ed Dîn.* The ruin of Merâs ed Din, a sheikh's name.

خربة المنطار   Oq   *Khŭrbet el Mŭntâr.* The ruin of the watch-tower.

خربة النجمة   Op   *Khŭrbet en Nejmeh.* The ruin of en Nejmeh. q.v.

خربة راس الطويل   Nq   *Khŭrbet Râs et Tawîl.* The ruin of the tall hill-top.

خربة ردين   Nr   *Khŭrbet Rudein.* The ruin of the little sleeve.

خربة روجان   Np   *Khŭrbet Rûjân.* The ruin of Rûjân. p.n.

خربة صبوبة   Op   *Khŭrbet Subbûbeh.* The ruin of the slope.

خربة سامية   Nr   *Khŭrbet Sâmuh.* The lofty ruin.

خربة صرا   Nq   *Khŭrbet Sarra.* The ruin of the solid rock.

خربة سيا   Nq   *Khŭrbet Sia.* The ruin of Sia. *See Merj Sia.*

خربة السمرآ   Pr   *Khŭrbet es Sŭmrah.* The brown ruin.

خربة الطويل   Op   *Khŭrbet et Tuweiyil.* The ruin of the tall peak.

خربة وادي ناصر   Op   *Khŭrbet Wâdy Nâsir.* The ruin of Wâdy Nâsir. q.v.

خربة يانون   *Khŭrbet Yânûn.* The ruin of Yânûn. q.v.

الكفل ابو عمار   Np   *El Kifl Abu 'Ammâr.* El Kifl, the father of 'Ammâr. See p. 216.

كيليا   Or   *Kîlia.* Kilia; perhaps from اكول hilly ground.

كتار الشريعة   Qp   *Kitâr esh Sheriâh.* The hummocks of the Jordan; applies to the isolated mud hills near the river.

قبة النجمة   Or   *Kubbet en Nejmeh.* Dome of en Nejmeh. q.v.

قلاسون   Oq   *Kulâsôn.* Kulâsôn. p.n. قلاص water coming up in a well.

قراوا المسعودي   Pp   *Kŭrâwa el Mas'ûdy.* Kŭrâwa. p.n. (towns) of the men of the *Mesâîd* tribe; the Arabs who live here.

قرن صرطبة   Pp   *Kŭrn Sŭrtŭbeh.* The peak of Sŭrtŭbeh. Talmudic סרטבא.

قرنة الجوالدة   Oq   *Kurnet el Jawâlideh.* The peak of the men of *Jâlûd.*

قرنة المثيلي   Oq   *Kurnet el Mitheily.* The peak of Mitheily. See *Khŭrbet Mithelia,* p. 112; but it may be from نيل 'dogs' grass.'

قرنة ام عمير   Oq   *Kurnet Umm 'Omeir.* The peak of the mother of Omeir. p.n.

القصر   Or   *El Kŭsr.* The palace.

قصرة   Np   *Kŭsrah.* One palace.

حف اللوزة   Pp   *Lahf el Lôzeh.* The hill-foot of the almond tree.

حف سالم   Op   *Lahf Salim.* The hill-foot of Salim. p.n.

المعذبة   Pr   *El Mâdhbeh.* Either the place of '*Adhab* (a kind of alkali plant), or place of sweet water.

مفيض جوزلة   Qq   *Mafîd Jôzeleh.* The debouchure of Jôzeleh (the place where the stream joins the Jordan). See *'Ain Jôzeleh,* p. 251.

مخاضة الصايدة   Qp   *Makhâdet es Sâideh.* The ford of the fishermen.

المخروق   Qp   *El Makhrûk.* The pierced (a cutting in the rock through which the road passes).

منفصلة الجوزة   Qq   *Mankattat el Jôzeleh.* The separated ground off of el Jôzeleh (see above). It applies to the very broken ground near the river, separated from the rest of the *Ghôr.*

مَنْقَطَّة المَلّاحة   Qq   *Mankattat el Mellâhah.* The separated ground of the salt marsh.

المَسْكَرَة   Pr   *El Maskerah.* The sugar works, or place of growing sugar canes.

المَزَار   Oq   *El Mazâr.* The shrine.

الدَّكَاكِين   Pp   *Medekâkin.* Em Dekkâkin, for el Dekkâkin, 'the shops,' *i.e.*, 'caves.'

الحَجِيرَة   Pp   *El Mehajerah.* The stony place.

مَجِدَل بَنِي فَاضَل   Op   *Mejdel Beni Fâdl.* The watch-tower of the Beni Fâdhil (an Arab clan).

مَلّاحة امْ عَفِين   Pq   *Mellâhet Umm 'Afein.* The salt marsh with the stinking or rotten soil.

مَرْحَان ابِي صَلِيجَة   Pp   *Merhân Abu Salijeh.* The folds of Abu Salija. p.n.; meaning 'the father of the silver ingot.'

مَرْج سِيا   Nq   *Merj Sîa.* The level meadow; it is a sink, and has no outlet.

المَرْمَالة   Pq   *El Mermâleh.* The sandy place.

مَطِيل الذِيب   Pq   *Meteil edh Dhîb.* The peak of the wolf.

مَغَارَة العَتِي   Pp   *Mûghâret el 'Asy.* The cave of the rebellious one.

مَغَارَة بَرِكَة ابِي عَلُول   Or   *Mûghâret Birket Abu 'Alûl.* The cave of the pool of Abu 'Alûl. p.n.

مَغَارَة الدَّكَانَة   Oq   *Mûghâret ed Dekkâny.* The cave of the shop; the word commonly applied to ancient tombs.

مَغَارَة الحَبْلَى   Oq   *Mûghâret el Hably.* The cave of the terrace, or of the pregnant woman.

مَغَارَة الحَتِينَة   Op   *Mûghâret Heteiny.* The cave of the mountain peaks (another name for Mûghâret Jerfa).

مَغَارَة جَرْفَا   Op   *Mûghâret Jerfa.* The cave of the jorf (or bank).

مَغَارَة المَطْعَم   Op   *Mûghâret el Mutâm.* The cave of the feeding.

مَغَارَة الرَاس   Nq   *Mûghâret er Râs.* The cave of the hill-top.

مَغَارَة الرِدْقَة   Oq   *Mûghâret er Ridkah.* The cave of er Ridkah. p.n. *See Tell er Ridghah, p.* 209.

مَغَارَة الصَحْصِيلِي   Pp   *Mûghâret es Sahseily.* The cave of es Sahseily. p.n. perhaps from صحل to be hoarse.

2 L.

مغارة ام عمر   Nr   *Mughâret Umm 'Amir.* The cave of the hyæna.

مغارة ام خرما   Oq   *Mughâret Umm Khorma.* The cave with the prominence (near Mughâret er Ridkah).

مغارة وادي الحبيس   Nr   *Mughâret Wâdy el Habîs.* The cave of the valley of the religious endowment.

مغارة وادي جابر   Pp   *Mughâret Wâdy Jâbr.* The cave of Wâdy Jâbr. q.v.

المغائر   Oq   *El Mûgheir* (properly *mûghâïr*). The caves.

مغر المغربي   Pp   *Mûghr el Mughrabîn.* The caves of the Moors.

المناطر وقيل خربة   Nr   *El Munâtir.* The watch-towers. Called also
راس الدير   *Khûrbet Ras ed Deir.* The ruin of the convent hill-top.

منظار البنيك   Pp   *Mûntâr el Benîk.* The watch-tower of el Benik. *cf. 'Ain Benâk*, Sheet XIV.

المصلبة   Pr   *El Musellabeh.* The place of the cross.

المسطرة   Pp   *El Musetterah.* The inscribed place.

النبي نون   Op   *Neby Nûn.* The prophet Nûn. *See* p. 240.

النجمة   Or   *En Nejmeh.* The star; a high and very conspicuous mountain.

نبي شمايل   Nr   *Neby Shâmaïl.* The prophet Samuel; near Kefr Mâlik.

نقب حربة   Pp   *Nûkh Harbeh.* The pass of the lance.

العروة   Np   *El 'Ormeh.* The dam.

الرفاعى   Op   *Er Rafâi.* The Refâ'iy (dervish). *cf. Sheikh er Rafâây,* p. 244.

راس فاعلي   Oq   *Râs Fââly.* The hill-top of the pederast.

راس احفيرة   Pp   *Râs el Hûfîreh.* The hill-top of the excavation.

راس قنيطرة   Pp   *Râs Kaneiterah.* The hill-top of the arch.

راس المريغات   Pp   *Râs el Mereighât.* The hill-top of pastures.

راس الراهب   Op   *Râs er Râhib.* The hill-top of the monk.

راس الطويل   Nq   *Râs et Tawîl.* The tall hill-top.

راس وادي علي   Oq   *Râs Wâdy 'Aly.* The hill-top of the valley of 'Aly. p.n.

الرشاش   Oq   *Er Rishâsh.* The sprinkling (a spring).

| رجم ابي ﻋﺤﻴﺮ | Oq | *Rujm Abu Meheir.* The cairn of Abu Meheir. p.n. |

رجم ابي ﻋﺤﻴﺮ — Oq — *Rujm Abu Meheir.* The cairn of Abu Meheir. p.n.

رجم الرفيف — Nr — *Rujm er Refeif.* The glittering cairn.

رجم الﺼﺎﻳﻎ — Pp — *Rujm es Sáigh.* The cairn of the goldsmith.

ﺳﺎدة الﻔﻘﻴﻪ — Pp — *Sádet el Fikiah.* The cliffs of the jurisprudist.

ﺳﻬﻞ داليﺔ — Or — *Sahel Dálich.* The plain of the trailing vine.

ﺳﻬﻞ اﻓﺠﻢ — Op — *Sahel Ifjim.* The plain of Ifjim. p.n. ; Heb. צבם.

ﺳﻬﻞ ﺣﺪﺣﻮد — Or — *Sahlet Hûdhûd.* The plain of the short one ; perhaps of the 'lapwing,' ﺣﺪﺣﺪ, from the legend of this bird carrying Solomon's message to the Queen of Sheba.

ﺷﻌﺐ ابي ﺟﻤﻴﺔ — Oq — *Sháb Abu Jamíych.* The hill-spur with the reservoir of water.

ﺷﻌﺐ الﻐﻮرانيﺔ — Qp — *Sháb el Ghôrániych.* The hill-spur of the Arab of the Lowlands (singular of *Ghawárineh*, the name applied to the Arabs of the Ghor).

ﺷﻌﺐ ﻃﻌﻴﻤﺔ — Nq — *Sháb Taîmeh.* The hill-spur of food.

الﺸﻴﺦ العزير — Oq — *Sheikh el 'Azeir.* Sheikh Eleazar or Ezra.

الﺸﻴﺦ ﺣﺎﺗﻢ — Np — *Sheikh Hátim.* Sheikh Hátim. p.n. Hátim of the Taiy tribe was celebrated for his liberality ; he was a contemporary of Mohammed.

الﺸﻴﺦ ابرﺣﻴﻢ — Pr — *Sheikh Ibrahím.* Sheikh Abraham.

الﺸﻴﺦ ﺟﺮاح — Oq — *Sheikh Jerráh.* Sheikh Jerráh. p.n. Surgeon (at el Mûgheir).

الﺸﻴﺦ ﻣﺤﻤﺪ — Nq — *Sheikh Muhammed.* Sheikh Mohammed.

الﺸﻴﺦ زيﺪ — Nr — *Sheikh Zeid.* Sheikh Zeid. p.n.

ﺷﺠﺮة الﺤﻠﻮة — Nq — *Shejeret el Helweh.* The sweet tree.

ﺷﻮنﺔ الﻤﺼﻨﻊ — Pp — *Shûnet el Masná.* The barn of the cistern.

ﺳﻴﺎح النﻘﺐ — Pp — *Siáh en Nûkb.* The surface water of the pass.

ﺳﺪ ﺣﺮيز — Pq — *Sidd Haríz.* The fortified cliff.

ﻃﺎﺣﻮنﺔ البﻌﻠﺔ — Pp — *Táhûnet el Bálah.* The mill of the hill ; but probably connected with the name Baal.

ﻃﺎﺣﻮنﺔ الﻘﺪريﺔ — Pp — *Táhûnet el Kadriych.* The mill of the Kadriych (family name) ; close to *'Abd el Kádir.*

طلعة ابو عيد   Pp   *Talât Abu 'Âid.* The ascent of Abu 'Âid (a common Arab name).

طلعة عمرة   Pq   *Talât 'Amrah.* The ascent of 'Amrah. p.n.

طلعة العزيات   Or   *Talât el 'Azbât.* The ascent of the unmarried girls.

طلعة الغرين   Pp   *Talât el Kûrein.* The ascent of the little peak.

تل الابيض   Pp   *Tell el Abeid.* The white mound.

تل المزار   Pp   *Tell el Mazâr.* The mound of the shrine (of 'Abd el Kâdir).

تل المسطرة   Or   *Tell el Musetterah.* The mound of the inscribed place.

تل الريشة   Or   *Tell er Rîsheh.* The mound of the feather.

تل شيخ الذياب   Pq   *Tell Sheikh edh Dhiâb.* The mound of *Sheikh 'Aly Dhiâb* of the Adwân tribe. (An artificial mound.)

تل السويد   Op   *Tell es Sûweid.* The blackish mound.

تل الطروني   Or   *Tell et Trûny.* The mound of et Trûny. p.n. ; *târûny,* a sort of silk.

تلول البصة   Qp   *Tellûl el Basseh.* The mounds of the marsh.

ام حلال   Pp   *Umm Hallal.* The place where the rain falls.

ام المخالي   Pr   *Umm el Mekhâly.* The place where the nose-bags are kept.

ام الشرطا   Qq   *Umm esh Shert.* The place with the stripes or cracks.

وادي العباد   Nr   *Wâdy el 'Abbad.* The valley of the devotee.

وادي الابيض   {Op / Or}   *Wâdy el Abeid.* The white valley.

وادي ابو احتيات   Pr   *Wâdy Abu el Haiyât.* The valley with the snakes.

وادي ابو حمام   Op   *Wâdy Abu Hûmmâm.* The valley with the bath. There is a warm spring near it.

وادي ابى محمود   Oq   *Wâdy Abu Mahmûd.* Abu Mahmûd's valley.

وادي ابى عبيدة   Pr   *Wâdy Abu 'Obeideh.* Abu 'Obeideh's valley.

وادي ابى الرحم   Nq   *Wâdy Abu 'r Rahm.* The valley of Abu 'r Rahm. p.n. ; meaning 'the father of mercy.'

وادي ابى طلوان   Pq   *Wâdy Abu Tulwân.* The valley of Abu Tulwân. p.n. ; meaning ' father of expectation.'

وادي ابى زرقآ·	Oq	*Wâdy Abu Zerka.* The valley with the blue water.
وادي الاخراف	Nr	*Wâdy el Akhrâf.* The valley of the sheep.
وادي العلّلي	Oq	*Wâdy el 'Alâly.* The valley of the upper chamber, a name given to caves above a valley. *See el 'Aleiliât,* Sheet XVII. It is near *Rujm Abu Mcheir.*
وادي علي	Pq	*Wâdy 'Aly.* The valley of 'Aly. p.n.
وادي العمري	Oq	*Wâdy el 'Amry.* The valley of the 'Amry; a tribal name.
وادي عراق حبّاج	Oq	*Wâdy 'Arâk Hajâj.* The valley of the cliff (or rock) of the pilgrims.
وادي عراق الشبآ·	Oq	*Wâdy 'Arâk esh Shaheba.* The valley of the dun-coloured cliff.
وادي العسى	Nr	*Wâdy el 'Asa.* The valley of the rebellious one. *See* p. 257.
وادي العوجة	Pr	*Wâdy el Aujah.* The valley of the bend. *See el 'Aujah,* p. 252; which is a great bend just by the junction with Jordan.
وادي باب الخارجة	Oq	*Wâdy Bâb el Khârjeh.* The valley of the outer door.
وادي بقر	Pr	*Wâdy Bakr.* The valley of the cow.
وادي برشة	Oq	*Wâdy Barsheh.* The valley of variegated herbage; (from the flowers in it).
وادي الدالية	Or	*Wâdy ed Dâlieh.* The valley of the trailing vine.
وادي دار الجرير	Nr	*Wâdy Dâr el Jerîr.* The valley of the house of el Jerir; (name of a great Arabic poet).
وادي الدوا	Op	*Wâdy ed Dôwa.* The valley of medicine.
وادي الضبع	{Pq Oq}	*Wâdy ed Dubâ.* The valley of the hyena.
وادي فصايل	Pq	*Wâdy Fûsâil.* The valley of Fûsâil. q.v.
وادي الحمر	Pq	*Wâdy el Humr.* The red valley.
وادي الاجيم	Pp	*Wâdy el Ijjim.* The valley of Ifjim. *See Sahel Ifjim,* p. 259.
وادي جابر	Pp	*Wâdy Jâbr.* The valley of Jâbir. p.n.; meaning 'helper,' 'repairer,' or 'bone-setter.'
وادي جبيتا	Oq	*Wâdy Jibeit.* The valley of Jibeit. q.v.
وادي الجوزة	Qp	*Wâdy el Jôzeleh.* The valley of the wild dove.

وادي قطونيّة   Oq   *Wâdy Katûnîyeh.* The valley of vegetables.

وادي كيّس   Or   *Wâdy Keiyis.* The pretty valley.

وادي الكراد   Op   *Wâdy el Kerâd.* The valley of the Kurds.

وادي تريثر   Oq   *Wâdy Kercikir.* The valley of soft flat ground.

وادي كرزلية   Pp   *Wâdy el Kurzeleiyeh.* The valley of the shepherds. See *'Ain el Kurzeliyeh*, p. 251.

وادي مقثاية   Pq   *Wâdy Makthâyeh.* The valley of cucumber beds.

وادي مقور (مغور) الذيب   Pr   *Wâdy Mekûr (or Meghûr) edh Dhîb.* The valley of the water-holes of the wolf. Also called *Unkur edh 'Dhîb*, which has the same meaning.

وادي الملّاحة   Pr   *Wâdy el Mellâhah.* The valley of the salt marsh.

وادي المناخير   Op   *Wâdy el Menâkhîr.* The valley of the promontories.

وادي المرجم   Oq   *Wâdy el Merâjem.* The valley of the cairns.

وادي المراكب   Oq   *Wâdy el Merâkib.* The valley of the hillocks.

وادي المريغات   Pr   *Wâdy el Mereighât.* The valley of pasture lands.

وادي المدحدرة   Pr   *Wâdy el Mudahderah.* The valley of the rolling rocks.

وادي المسطّرة   Pq   *Wâdy el Musetterah.* The valley of the inscribed place.

وادي النار   Pp   *Wâdy en Nâr.* The valley of fire. *See Dalûk*, p. 252.

وادي ناصر   Oq   *Wâdy Nâsir.* The valley of Nâsir. p.n.; ('helper').

وادي النجمة   Or   *Wâdy en Nejmeh.* The valley of *en Nejmeh*. q.v.

وادي الرفيع   Or   *Wâdy er Rafâ.* The high valley.

وادي الرشاش   Oq   *Wâdy er Rishâsh.* The valley of er Rishâsh. q.v., p. 258.

وادي الريشة   Pr   *Wâdy er Rîsheh.* The valley of the feather.

وادي الرويحة   Oq   *Wâdy er Rûeihah.* The valley of repose.

وادي سدّة   Oq   *Wâdy Saddeh.* The valley of the cliff.

وادي سامية   {Oq / Nr}   *Wâdy Sâmieh.* The high valley.

وادي صرّا   Nq   *Wâdy Surra.* The valley of the solid rock.

وادي السبع	Op	*Wâdy es Sebâ.*	The valley of the lion.
وادي سباس	Nq	*Wâdy Sebbâs.*	The valley of Sebbâs. p.n.
وادي سباتا	Pr	*Wâdy Sebâta.*	The valley of Sebâta. p.n.
وادي شعب غصيب	Oq	*Wâdy Shâb Ghastb.*	The valley of the hill-spur of the confiscator.
وادي شعب الخارجة	Oq	*Wâdy Shâb el Khârjeh.*	The valley of the outer hill-spur.
وادي الشيخ عبيد	Oq	*Wâdy esh Sheikh 'Obeid.*	The valley of Sheikh 'Obeid. p.n.
وادي الصبحة	Op	*Wâdy es Subhah.*	The reddish (white) valley.
وادي التاح	Oq	*Wâdy et Tâh.*	The expanding valley.
وادي الطيبة	Nr	*Wâdy et Taiyibeh.*	The valley of et Taiyibeh ; (the goodly).
وادي طلعة الضبع	Pr	*Wâdy Talât ed Dubâ.*	The valley of the ascent of the hyena.
وادي نقور الذيب	Oq	*Wâdy Uukûr edh Dhîb.*	The valley of the water-holes of the wolf.
وادي ام كواك	Nq	*Wâdy Umm Kûâk.*	The valley of Umm (mother of) Kûâk ; perhaps an error for الكواخ booths.
وادي ام السويد	Pq	*Wâdy Umm es Suweid.*	The valley of the mother of black men. But *see Jubb Sûweid,* p. 5.
وادي الوعر	Pq	*Wâdy el Wâr.*	The valley of rugged rocks.
وادي ينبوع	Oq	*Wâdy Yanbûâ.*	The valley of the perennial spring.
وادي زقزقة	Pq	*Wâdy Zakaska* (properly *Zakazka*).	The valley of Zakaska. The name of a kind of ant; it also means the chirping of birds at dawn.
وادي زامور	Op	*Wâdy Zâmûr.*	The valley of the piper, or of the handsome youth.
وادي الزاوية	Oq	*Wâdy ez Zâwieh.*	The valley of the corner or hermitage.
وادي الزيت	Pp	*Wâdy ez Zeit.*	The valley of oil.

يانون Op   *Yânûn.* p.n.; perhaps יָנוֹחָה Janohah.

الزمارة Oq   *Ez Zemârah.* The valley of piping.   *cf. Wâdy Zâmûr,* p. 263.

زور ابو رفع Qq   *Zôr Abu Rafâ.* The raised low ground of the father (from its position between heights).

زور الحميري Qq   *Zôr el Hŭmeiry.* The low ground of el Hŭmeiry.   p.n.

زور الكلباني Qq   *Zôr el Kelbâny.* The low ground of the Kelbâny.   p.n.

SHEET XVI.

ابو شوشة Is *Abu Shisheh.* The father of the top-knot. *See* "Memoirs."

عين الخزنة IIt *'Ain el Khuzneh.* The spring of the treasury.

عين المقنّع It *'Ain el Mekenná.* The spring of the veiled one.

عين النينة It *'Ain en Nineh.* The spring of Nineh. p.n.

عين السجد It *'Ain es Sejed.* The spring of adoration.

عين ام سعدة It *'Ain Umm Sádeh.* The spring of Sádeh's mother. p.n.; but it may be connected with ساعد a channel.

عين يردة Js *'Ain Yerdeh.* The spring of Yerdeh; also called *'Ain Werdeh.* Both names have the same meaning of going down to fetch water, the first being the Hebrew form of the word.

عجور Ju *'Ajjûr.* Furrows on the sand made by the wind.

عاقر IIs *'Akir.* Barren. *See* "Memoirs."

عرب الملاحة Gs *'Arab el Melâllah.* The salt-making Arabs living on the shore.

عرب الرميلات Gs *'Arab er Rûmeilât.* The Arabs of the sand-dunes.

عرب السعادنة It *'Arab es S'ádneh.* The highland Arabs.

عرب السواركى Gs *'Arab es Suârki.* The Suârki Arabs.

عراق البقس Iu *'Arâk el Buks.* The cavern or buttress of the box-tree.

عراق الدير Iu *'Arâk ed Deir.* The cavern of the monastery; named after دير الذبان *Deir edh Dhibbán,* the monastery of flies. *See* "Memoirs."

عراق نعمان It *'Arâk Námán.* The cliff of Naaman.

بعليين IIu *Bálin.* Bálin. *See* "Memoirs."

برية Is *Barrîyeh.* The desert (of Ramleh) *See* "Memoirs."

2 M

البناية It *El Benâyeh.* The building.

برقوسية IIu *Berkûsieh.* For *Berkûshieh,* 'variegated.'

بشيت Gt *Beshshît.* For *Beit Shît,* House of Seth. The Mükâm of *Neby Shît* is beside it.

بيت أدراس (ادراس) Gu *Beit Durâs,* or *Beit Dârâs.* The house of the treading corn.

بيارة البدوي Eu *Biâret el Bedawi.* The wells of the Bedouin Arabs.

بيارة طاحة الصبحية Eu *Biâret Tâha es Subahîyeh.* The well of the reddish-white maze.

بئر ابي خشيبة Gs *Bir Abu Khasheibeh.* The well with easily-flooded ground.

بئر برقوسية Hu *Bîr Berkûsieh.* The well of the variegated ground.

بئر البرشين Hu *Bir el Burshein.* The well of variegated herbage.

بئر الداشر It *Bir ed Dâshir.* The abandoned well.

بئر الذكر Iu *Bir edh Dhekr.* The well of the male (or of remembrance).

بئر الغزلان Hs *Bir el Ghûzlan.* The well of the gazelles.

بئر الجمال Is *Bir el Jemâl.* The well of the camels.

بئر الجوخدار Gu *Bir el Jôkhadâr.* The well of the corpulent man.

بئر القشلة Eu *Bir el Kushleh.* The well of the barracks.

بئر المدوار Iu *Bir el Medwâr.* The circular well.

بئر النبه Gu *Bir en Nebah.* The notorious well.

بئر ربع المدوين Eu *Bir Reiâ el Mad-hûn.* The well of the slightly-wetted high ground.

بئر عائمة Hs *Bir Sâimeh.* The stagnant well.

بئر الشحم It *Bir esh Shahm.* The well of the fat.

بئر الشجرة Iu *Bir esh Shejerah.* The well of the tree.

بئر شقير Eu *Bir Shekeir.* The reddish-grey well.

بئر سميل Hs *Bir Summeil.* The well of the hard ground.

بئر الطياشة Is *Bir et Taiâsheh.* The inconstant well.

بئر الواد Iu *Bir el Wâd.* The well of the valley.

بئر زبلة Ju *Bir Ziblch.* The well of dung.

بوايك ابى صويلح Ft *Buâtk Abu Sûcileh.* The cattle-sheds of Abu Sûcileh.

البريج Ju *El Burcij.* The little tower.

برقة Gt *Burkah.* Sandy ground covered with flints.

البطاني الغربية Gt *El Butâni el Gharbîyeh.* The western Butâni.

البطاني الشرقية Gt *El Butâni esh Sherkîyeh.* The eastern Butâni.

دير العاشق It *Deir el 'Âshck.* Monastery of the lover.

دير الذبان Iu *Deir edh Dhibbân.* The monastery of flies.

دير المحسن Jt *Deir el Muhcisin.* The monastery of kind actions.

دير نعمان It *Deir Nâmân.* The monastery of Naaman. Also called *Khûrbet Hadâd,* 'the ruin of the blacksmith' (caves, foundations, and stones).

ظهيرة قريمدة Hs *Dhahret Kurcimdeh.* The ridge of the burnt bricks.

الذنبة Iu *Edh Dhenebbeh.* The lower part of a valley through which water flows.

عبدس Gu *'Ebdis.* 'Ebdis. p.n. In the official list and in Vande- velde it is written عدس *'Eddis.*

اسدود Ft *Esdûd.* Eshdûd. Heb. אַשְׁדּוֹד Ashdod, 'fortress.'

فخيت Is *Fekhît.* The crack (a quarry).

حجر البد Hs *Hajr el Bedd.* The stone of the idol temple.

الحاج سعيد Ht *El Hâj Saîd.* The pilgrim Saîd.

حمامة Fu *Hamâmeh.* The dove. (Perhaps حمّامة the warm bath).

حوض سكرير Gt *Haud Sukercir.* The cistern of Sukereir. p.n.

احمّام Gu *El Hûmmâm.* The bath.

جيليا It *Jîlia.* Jilia. p.n. *See* " Memoirs."

جميزة القاعة Fu *Jimmcizet el Kââh.* The sycamore of the open plain or courtyard.

جسر اسدود Ft *Jisr Esdûd.* The bridge of Ashdod.

جولس Fu *Jûlis.* Jûlis. p.n.

الجورة وقيل جورة عسقلون   Eu   *El Jûrah*, or *Jûrat 'Askelân*. The neighbourhood; or ' the neighbourhood of Ascalon'; probably יגור *Yagur* of the Talmud.

جورة الرملة   Is   *Jûrat er Rûmleh*. The neighbourhood of Ramleh, or the sandy place.

قضا يافا   {Hi/St}   *Kada Yâfa*. The district of Jaffa.

قطرة   Ht   *Katrah*. A drop. *See* " Memoirs."

الكنيسة   Gs   *El Kenîseh*. The Church. In Yebnah.

قرازة   It   *Kerâzeh*. Kerâzeh. p.n, Chorazin.

خلة عطوان   Iu   *Khallet 'Attûân*. The dell of 'Attûân; perhaps from عطون ' halting by a watering-place.'

خلة الذهب   It   *Khallet edh Dhaheb*. The dell of gold.

خلة المصاري   Ju   *Khallet el Masâri*. The dell of towns. In vulgar Arabic the word means small coins.

خلة الطويلة   Iu   *Khallet el Tuweileh*. The long dell. It is just north of *Bîr el Medwâr*.

خلة ام السيول   Iu   *Khallet Umm es Sûil*. The dell with the streams.

خان اسدود   Ft   *Khân (Esdûd)*. The Caravanserai of Ashdod.

الخيمة   Ht   *El Kheimeh*. The tent.

الخضر   Iu   *El Khŭdr*. El Khŭdr. *See* p. 28.

خلدة   It   *Khuldeh*. El Khuldeh. p.n. The perpetual.

خراب ابن زيد   Ju   *Khŭrâb Ibn Zeid*. The ruins of Zeid's son. p.n.

خربة العجوري   Hs   *Khŭrbet el 'Ajjûri*. The ruin of the man of 'Ajjûr. q.v.

خربة ابي عميرة   Iu   *Khŭrbet Abu 'Amîreh*. The ruin of the father of 'Amiry. p.n.

خربة عامر   Ju   *Khŭrbet 'Âmir*. The ruin of 'Âmir. p.n.

خربة عمورية   Iu   *Khŭrbet 'Ammûrieh*. The ruin of Amûrieh. p.n.

خربة عمر   Iu   *Khŭrbet 'Amr*. The ruin of 'Amr. p.n. All these are forms of the Arabic name 'Omar, and are identical with the ethnic name Amorite.

خربة العرب   Ju   *Khŭrbet el 'Arab*. The ruin of the Arabs.

خربة عصفورة   Jt   *Khŭrbet 'Asfûrah.* The ruin of the sparrow. Also called *Khŭrbet Umm el 'Aûsej,* 'the ruin with the thorn tree' (*Lycium Europeum*).

خربة عسقلون   Ju   *Khŭrbet 'Askalûn.* The ruin of Ascalon. *See "Memoirs."*

خربة الطريا   Iu   *Khŭrbet Atraba.* The ruin of fragrant herbs.

خربة البد   Hs   *Khŭrbet el Bedd.* The ruin of the idol, or idol temple.

خربة بيت فار   It   *Khŭrbet Beit Fâr.* The ruin of the house of the mouse (traces of ruins only).

خربة بلاس   Fu   *Khŭrbet Belâs.* The ruin of lentils.

خربة بردغة   Gu   *Khŭrbet Berdeghah.* The ruin of Berdegha ; perhaps برذعة soft ground.

خربة بشة   Fu   *Khŭrbet Beshshah.* The ruin of luxuriant vegetation.

خربة بزة   Fu   *Khŭrbet Bezzeh.* The ruin of cloth or furniture.

خربة بيار الكعبة   Gu   *Khŭrbet Biâr el Kâbeh.* The ruin of the wells of the cube : probably from the masonry cisterns.

خربة البيرة   {Fu} {Jt}   *Khŭrbet el Bîreh.* The ruin of the fortress. *See 'Ain el Bîreh, p. 157.*

خربة بير المدوار   Iu   *Khŭrbet Bîr el Medwâr.* The ruin of Bir Medwâr. q.v.

خربة دحيشة   Hs   *Khŭrbet Deheisheh.* The ruin of Deheisheh ; probably for دحيشة soft ground. This is the same as Khŭrbet el 'Ajjûri.

خربة ديران   Hs   *Khŭrbet Deirân.* The ruin of houses. Traces of ruins.

خربة دير البطم   Iu   *Khŭrbet Deir el Butm.* The ruin of the monastery of the Terebinth. *See "Memoirs."*

خربة ذكر   Iu   *Khŭrbet Dhekr.* The ruin of the male, or of the memorial.

خربة الذياب   Iu   *Khŭrbet edh Dhiâb.* The ruin of the Dhiâb clan (wolves).

خربة ذكرين   Iu   *Khŭrbet Dhikerîn.* The ruin of Dhikerin. p.n. ; from ذكر to remember. *See above, Khŭrbet Dhekr.*

خربة الفاتونة   Gs   *Khŭrbet el Fâtûneh.* The ruin of el Fâtûneh. p.n. ; from فتن to tempt.

خربة فرد   It   *Khŭrbet Fered.* The ruin of the unit.

خربة غيّانة   {Gs} {Fu}   *Khŭrbet Gheiyâdah.* The ruin of the forest.

خربة حبرآ Hs *Khŭrbet Hebra.* The ruin of abundant herbage; feminine of احبر

خربة هرماس Hs *Khŭrbet Hermás.* The ruin of the raging lion, or of the demon.

خربة اسطاس Hu *Khŭrbet Istás.* The ruin of Istás. p.n.

خربة جلدية Gu *Khŭrbet Jeledîyeh* The ruin of hard ground.

خربة البلخ Iu *Khŭrbet el Jelkh.* The ruin of the full valley.

خربة قلوس Iu *Khŭrbet Kallûs.* The ruin of the filled well.

خربة خسّة Fu *Khŭrbet Khasseh.* The ruin of lettuce.

خربة كفر عانا Il *Khŭrbet Kefr 'Ána.* The ruin of Kefr 'Ána. p.n.

خربة قرقفة Gu *Khŭrbet Kerkefeh.* The ruin of Kerkefeh, a kind of small bird.

خربة قطلانة Is *Khŭrbet Kutlâneh.* The ruin of Kutlâneh; from قطل 'to lop or amputate'; perhaps it should be قتلانة from قتل 'to slay.'

خربة لزقا *Khŭrbet Lezka.* The ruin of Lezka. p.n.

خربة اللوز Ju *Khŭrbet el Lôz.* The ruin of the almond.

خربة المعصبة Fu *Khŭrbet el Másebeh.* The ruin of el Másebeh; from عصب 'assembling round a cistern,' or from عصب 'ivy.'

خربة مكوس Fu *Khŭrbet Makkûs.* The ruin of the tolls or taxes.

خربة الميسية Is *Khŭrbet el Meisîyeh.* The ruin of mace-trees (*Cordia myxa*).

خربة المقنع It *Khŭrbet el Mekenná.* The ruin of the veiled one.

خربة ملات Is *Khŭrbet Melát.* The ruin of el Melát (plastered).

خربة المنسية Iu *Khŭrbet el Mensîyeh.* The ruin of the forgotten one.

خربة المخيزن IIt *Khŭrbet el Mukheizin.* The ruin of the storehouses.

خربة نينا IIt *Khŭrbet Nína.* The ruin of Nína. p.n.

خربة نويطح Ju *Khŭrbet Nuweitîh.* The ruin of those who butt with their horns. *See Sheikh Nattah,* p. 94.

خربة عكبر Ju *Khŭrbet 'Okbur.* The ruin of 'Okbur (the substance which bees carry on their legs to the hive).

خربة الرقدية Is *Khŭrbet er Rekediyeh.* The ruin of the recumbent ones.

خربة الرسم Iu *Khŭrbet er Resm.* The ruin of the traces of dwellings.

خربة رميلتا Gu *Khŭrbet Rumeiltah.* The ruin of the sandy place ; the soil here is partly sandy.

خربة الصافى Iu *Khŭrbet es Sâfi.* The ruin of clear water, or bright stones.

خربة سلوجة IIt *Khŭrbet Sallûjeh.* The ruin of the silkworm's cocoon.

خربة السجد It *Khŭrbet es Sejed.* The ruin of adoration.

خربة سلمة Is *Khŭrbet Selmeh.* The ruin of Selmeh. p.n.

خربة الشيخ خالد It *Khŭrbet esh Sheikh Khâlid.* The ruin of Sheikh Khalid. p.n.

خربة الشجرة Iu *Khŭrbet esh Shejerah.* The ruin of the tree. Its older name is خربة الشه *Khŭrbet esh Shah.* p.n.

خربة كرير Gt *Khŭrbet Sukereir.* The river of Sukereir. *See p. 267.*

خربة سميل Hs *Khŭrbet Summeil.* The ruin of hard ground.

خربة السمرآ Iu *Khŭrbet es Sumra.* The brown ruin.

خربة الستا Iu *Khŭrbet es Sŭtta.* The ruin of Sŭtta. p.n.

خربة طنيفة Iu *Khŭrbet Teneifseh.* The ruin of carpets. طنافـة plural of طنفـة .

خربة الطراطير Hu *Khŭrbet et Terâtîr.* The ruin of the conical caps.

خربة أم العقود Iu *Khŭrbet Umm el 'Akŭd.* The ruin with the vaults, or coping stones.

خربة أم الهمام Iu *Khŭrbet Umm el Hemâm.* The ruin of the mother of the hero.

خربة أم كلخة It *Khŭrbet Umm Kelkhah.* The ruin with the *kelkha* (a plant like fennel).

خربة أم الرياح Eu *Khŭrbet Umm er Riyâh.* The ruin with the fragrant odours, or 'where the winds blow.'

خربة أم زبيَّة Ju *Khŭrbet Umm Zebeileh.* The ruins with the dung heaps.

خربة ياسين Fu *Khŭrbet Yâsîn.* The ruin of Yâsin. p.n.

خربة يردة Js *Khŭrbet Yerdeh.* The ruin of Yerdeh. *See 'Ain Yerdeh,* p. 265.

الخرابة Gs *El Khŭrâbeh.* The ruined place. (Traces only.)

الغُبَيبة Hs *El Kubeibeh.* The little dome.

القُصطينة Hu *El Kustineh.* For *Kūstileh, i.e., Castellum.*

المَعلاوية Ht *El Māláwíyeh.* El Málâwiyeh. From علو 'to be high.'

المَنصورة Is *El Mansûrah.* El Mansûra. *See* p. 9.

المِجدل Eu *El Mejdel.* The watch-tower.

المِسمية Ht *El Mesmíyeh.* El Mesmiyeh. p.n.; from سمر 'to be lofty.'

مِينة القَلعة Ft *Mînet el Kŭláh.* The harbour of the castle.

المُستقى Fu *El Miska.* The place of irrigation.

مُستقى سليمان اغا Fu *Miska Suleimân.* The place of irrigation of Suleimân Agha (Turkish proper name).

مُستقا زُريق Fu *Miska Zereik.* The azure irrigation place.

المِغار Hs *El Mŭghâr.* The caves.

المِغارة Ht *El Mŭghârah.* The cave.

مغاير شيحة Is *Mŭghâir Shîhah.* The caves of Shihah ; but *see* p. 90.

مغاير تَميل Hs *Mŭghâir Summeil.* The caves of hard ground.

مغارة جائيحا Is *Mŭghâret Jâeihah.* The cave of Jâeihah. p.n.

مغارة خَلة قدوم Iu *Mŭghâret Khallet Kaddûm.* The cave of the dell of the front.

مغلس Iu *Mughullis.* Coming to water at the morning twilight, or making a raid at that time.

ناعة وقيل نيعة Is *Nâaneh* or *Nianeh.* The plant 'mint.' *See* " Memoirs."

ناحية المِجدل {Gh/Tu} *Nâhiet el Mejdel.* The commune of el Mejdel.

نهر سُكرير Ft *Nahr Sukereir.* The river Sukereir. *See* p. 267.

النبي برق Gt *Neby Bŭrk.* The prophet Bŭrk (lightning) ; in the village of Burkah. q.v.

النبي كندا Hs *Neby Kunda.* The prophet Kunda. p.n.

النبي شيت Gt *Neby Shît.* The prophet Seth.

النبي يونس Ft *Neby Yûnis.* The prophet Jonah.

الزيرات	Hs	*En Nezzeizât.* The exudations (applied to some pools).
رعنا	Iu	*Râna, vulgo Arâna.* The spur of a hill.
راس ابي حامد	Is	*Râs Abu Hâmid.* The hill-top of Abu Hâmid. p.n.
راس ديران	Hs	*Râs Deirân.* The hill-top of Deirân. *See Wâdy Deirân,* p. 276.
راس تبان	Gs	*Râs Kabbân.* The headland of the pair of scales.
رجم جيز	Jt	*Rujm Jîz.* The cairn on the valley side.
السبيل	Iu	*Es Sebîl.* The way-side fountain.
شحمة	Ht	*Shahmeh.* Shahmeh. p.n. ; 'fat.'
الشيخ عبدالله	Iu	*Sheikh 'Abdallah.* Sheikh 'Abdallah. p.n.
الشيخ ابو عمران	Iu	*Sheikh Abu 'Amrân.* Sheikh Abu Amran. p.n. *See Khûrbet Amr,* p. 268.
الشيخ ابو هريرة	Gs	*Sheikh Abu Hareireh.* Sheikh Abu Hareirah ; one of the Prophet's companions.
الشيخ ابو جهم	Fu	*Sheikh Abu Jahm.* Sheikh Abu Jahm. p.n.
الشيخ ابو جميزة	Is	*Sheikh Abu Jimmeizeh.* Sheikh Abu Jimmeizeh ; or it may be rendered 'the saint's tomb with the sycamore tree.'
الشيخ عوض	Ew	*Sheikh 'Awed.* Sheikh 'Awed ; a common Arabic name.
الشيخ داود	Iu	*Sheikh Dâûd.* Sheikh David.
الشيخ حامد	Fu	*Sheikh Hâmid.* Sheikh Hamid. p.n.
الشيخ ابرهيم المتبولي	Ft	*Sheikh Ibrahîm el Matbûli.* Sheikh Ibrahîm el Matbûli. *Matbûl* means 'lovesick,' but is used in a mystic sense.
الشيخ اسليم	Gs	*Sheikh Islim.* Sheikh Selim. p.n. ; west of Yebnah, in the village.
الشيخ جوعباس	Is	*Sheikh Jôâbâs.* Sheikh Jôâbâs. p.n.
الشيخ خير	Fu	*Sheikh Kheir.* Sheikh Kheir. p.n. ('good ').
الشيخ معروق الكلكي	Iu	*Sheikh Mârûf el Kelki.* Sheikh Mârûf, 'of the raft '; probably connected with the tradition of Hâj 'Aleiyân. q.v.
الشيخ محمد الاسمر	Iu	*Sheikh Muhammed el Asmar.* 'Sheikh Mohammed the brown.'
الشيخ محمد الجزاري	Is	*Sheikh Muhammed el Jezâri.* Sheikh Mohammed el Jezâri. p.n. ; named from the Tell.

الشيخ موسى طليع Js *Sheikh Músa Telliâ.* Sheikh Musâ, the picket or van-guard; from its lofty position.

الشيخ واهب Gs *Sheikh Wâheb.* Sheikh Wâheb. p.n. North-west of Yebnah.

الشيخة سعدة Gs *Sheikhah Sâdeh.* The (female) Sheikh or Saint Sâdeh p.n. Sister of the last. The name applies to a cave now closed.

شلال الغور Hs *Shellâl el Ghôr.* The waterfalls of the lowland.

سدرة الحريرية Fu *Sidret el Harîrîyeh.* The lotus-tree of the Harîrî family, or of the silk workers.

صيدون Is *Sîdûn.* Sidûn. p.n. Heb. Zidon.

السوافير الغربية Gu *Es Sûâfîr el Gharbîyeh.* The western nomads.

قومه سافرة اى ذووسفر لتد الخضر (ج سوافر)

السوافير الشمالية Gu *Es Sûâfîr esh Shemâlîyeh.* The northern nomads; also called *Sûâfîr Ibn 'Aûdeh,* the nomads of Ibn 'Aûdeh. p.n.; but this is said to be a modern name, from an inhabitant who died within the century.

السوافير الشرقية Gu *Es Sûâfîr esh Sherkîyeh.* The eastern nomads: also called *Sûâfîr Abu Huwâr;* also from an inhabitant who died within the century.

صوفية Iu *Sûfîeh.* The Súfis, a religious mystical order, so-called, it is said, from their dress صوف wool.

صميل Hu *Summeil.* Hard ground.

التنور Js *Et Tauwûr.* The oven or reservoir. See "Memoirs."

تل ابي حرازة Fu *Tell Abu Herâzeh.* The mound with the fortifications.

تل الاخضر Ft *Tell el Akhdar.* The green mound.

تل البطيخ Is *Tell el Batîkh.* The mound of the melon.

تل بكيش Is *Tell Bukkîsh.* The mound of Bukkîsh. p.n.

تل بطاشى It *Tell Butâshi.* The mound of the Battâshi. p.n. The root means to assault violently.

تل الفول Ht *Tell el Fûl.* The mound of the bean.

تل جزر Is *Tell Jezar.* The mound of Jezar. p.n. Heb. גֶּזֶר Gezer, 'cut off'; probably as being an outlier of the hills.

| تل الخروبة | Gs | *Tell el Kharrúbeh.* The mound of the locust-tree. |

Tell el Kharrúbeh. The mound of the locust-tree.

تل المرّة Ft *Tell el Murreh.* The mound of the brackish water, near the mouth of the river (but perhaps named from the Beni Murreh Arabs).

تل الصافى Iu *Tell es Sáfi.* The clear or bright mound, hence called Alba Specula. The name dates back to the 12th century, and originates from the white chalk cliff.

تل السلاقة Hs *Tell es Sellákah.* The mound of flowing water, having several pools near it.

تل الترمس Hu *Tell et Turmus.* The mound of the lupine.

تل زكريّا Ju *Tell Zakariya.* The mound of Zacharias.

تلول الفراني Fu *Tellúl el Feráni.* The mounds of el Feráni. p.n.

الثمدات Hs *Eth Themadát.* The puddles.

تبنة Ju *Tibnah.* Straw. *See* "Memoirs."

التينة Hu *Et Tíneh.* The fig-tree; (a village).

أم العبد Is *Umm el 'Abd.* The mother of the slave; (a tree).

أم العرايس Is *Umm el 'Aráis.* The mother of the brides; (a tree).

أم البناية Gt *Umm el Benáyeh.* The mother of the building.

أم المغر Iu *Umm el Mághr.* The mother of the cave.

وادي ابي خميس Iu *Wády Abu Khamís.* The valley of Abu Khamis. p.n. The father of the fifth, or of the den.

وادي ابي رجيلة Iu *Wády Abu Rejeileh.* The valley with the runlet.

وادي عمورية Iu *Wády 'Ammúrieh.* The valley of 'Ammuríyeh (Amorites).

وادي أطربا Iu *Wády Atraba.* The valley of Atraba. *See* Khúrbet Atraba, p. 269.

وادي البغل Jt *Wády el Baghl.* The valley of the mule.

وادي بحلس Hs *Wády Bahlas.* The valley of Bahlas. p.n.; (empty-handed).

وادي البيرة Fu *Wády el Bíreh.* The valley of the fortress.

وادي البرشين	Hu	*Wâdy el Burshein.* The valley of variegated herbage.
وادي ديران	Hs	*Wâdy Deirân.* The valley of houses.
وادي الظهر	Iu	*Wâdy edh Dhahr.* The valley of the ridge.
وادي الفار	IIu	*Wâdy el Fâr.* The valley of the mouse.
وادي فرحة	Gs	*Wâdy Ferhah.* The valley of joy.
وادي الغويط	Gu	*Wâdy el Ghûcit.* The valley of watered ground.
وادي الحماعس	Is	*Wâdy el Hamâmis.* The valley of the chick-peas.
وادي حد ابي عاقر	Is	*Wâdy Hamed Abu 'Akir.* The valley of Hammed, the father of the barren one. p.n.
وادي جاموس	IIs	*Wâdy Jâmûs.* The valley of the buffalo.
وادي قطرة	IIt	*Wâdy Katrah.* The valley of Katrah. q.v.
وادي الخب	Gt	*Wâdy el Khubb.* The valley of the Khubb (a thorny plant).
وادي الخليل	Jt	*Wâdy el Khalîl.* The valley of Khalil (*i.e., Abraham*).
وادي المالح	Is	*Wâdy el Mâleh.* The salt valley.
وادي المدوار	Iu	*Wâdy el Medwâr.* The valley of el Medwâr. *See Bîr el Medwâr,* p. 266.
وادي الميسية	It	*Wâdy el Meisîyeh.* The valley of the meis tree.
وادي المجمع	Gt	*Wâdy el Mejmâ.* The valley of the assembly.
وادي المقنع	IIt	*Wâdy el Mekennâ.* The valley of the veiled one.
وادي المناخ	It	*Wâdy el Menâkh.* The valley of the place where camels are made to halt and lie down.
وادي المربع	Ht	*Wâdy el Merubbâh.* The square valley.
وادي المغراقة	Fu	*Wâdy el Mûghrâkeh.* The flooded valley.
وادي المخيزن	IIt	*Wâdy el Mukheizin.* The valley of storehouses.
وادي النار	Iu	*Wâdy en Nâr.* The valley of fire.
وادي الندا	IIs	*Wâdy en Neda.* The valley of dew.
وادي الصافي	Iu	*Wâdy es Sâfi.* The clear or bright valley. *See Tell es Sâfi,* p. 275.

وادي الشحم	It	*Wâdy esh Shahm.* The valley of fat. *See Shahmeh.* p. 273.
وادي صندوقة	Fu	*Wâdy Sendûkah.* The valley of the box or chest.
وادي الصرار	It	*Wâdy es Sûrâr.* The valley of pebbles.
وادي الطحانات	Gs	*Wâdy et Tahhânât.* The valley of the millers.
وادي أم الهمام	Iu	*Wâdy Umm el Hemâm.* The valley of the mother of the hero or chief.
وادي الزيتون	Iu	*Wâdy ez Zeitûn.* The valley of the olive-trees.
ياسور	Gt	*Yâsûr.* Yâsûr. p.n. *See* " Memoirs."
يبنة	Gs	*Yebnah.* Yebnah. p.n. *See* "Memoirs."
زرنوقة	Hs	*Zernûkah.* The rivulet.

SHEET XVII.

ابو العينين   Ls   *Abu 'l 'Ainein.* The father of two springs (a sacred place) ; called also ابو الادحم *Abu 'l Edhem.* p.n.

ابو ديس   Nt   *Abu Dîs.* Abu Dis (a family name).

ابو صلاح   Ms   *Abu Saláh.* The father of righteousness. It is perhaps connected with the Aramaic צלח ' to cleave,' the place being a quarry.

عين ابي كرزم   Ms   *'Ain Abu Kerzem.* The spring of Abu Kerzem. p.n. ; ' the short-nosed man.'

عين ابي زيد   Lu   *'Ain Abu Zeid.* The spring of Abu Zeid. p.n.

عين ابي زياد   Mt   *'Ain Abu Zíâd.* The spring of Abu Ziâd. p.n.

عين العجيب   Ls   *'Ain el 'Ajeb.* The spring of the wonder.

عين العليق   Lt   *'Ain el 'Alik.* The spring of the bramble.

عين العنيزية   Mt   *'Ain el 'Aneizîyeh.* The spring of the 'Aneizîyeh (Arab clan name). *cf.* the powerful tribe of the 'Anazeh. The names come from عنز ' a goat.'

عين العرب   Kt   *'Ain el 'Arab.* The spring of the Arabs.

عين عريك   Ls   *'Ain 'Arîk.* The spring of the compactly-built one ; but *see* " Memoirs," under ' Archi.'

عين العصافير   Lt   *'Ain el 'Asâfîr.* The spring of the sparrows.

عين عطان   Lu   *'Ain 'Atân.* The spring of 'Atân. Heb. עיטם Etam. *See* " Memoirs."

عين البلد   Mt   *'Ain el Belled.* The spring of the town (100 yards north of the church at *Neby Samwîl*).

عين بيت عتاب   Ku   *'Ain Beit 'Atâb.* The spring of Beit 'Atâb. q.v. ; south-east of the village by main road.

عين بيت طلما   Lt   *'Ain Beit Tulma.* The spring of the house of the loaf.

عين بيت سوريك Lt   *'Ain Beit Sûrîk.* The spring of Beit Sûrik. q.v.

عين بنت نوح Ku   *'Ain Bint Nûh.* The spring of Noah's daughter; from a tradition. *See "Memoirs."*

عين البيرة Ms   *'Ain el Bîreh.* The spring of el Bireh (the fortress).

عين البركة Ku   *'Ain el Birkeh.* The spring of the pool at Beit 'Atâb.

عين البرك Lu   *'Ain el Burak.* The spring of the pools.

عين دير يسين Mt   *'Ain Deir Yesîn.* The spring of the convent of Yesin. *See Deir Yesîn.*

عين الدلبة (Lt) (Lu) (Ku)   *'Ain ed Dilbeh.* The spring of the plane-tree.

عين الامير Mt   *'Ain el Emîr.* The spring of the Emir; in a cave south-east of Neby Samwil.

عين فاغور Lu   *'Ain Fâghûr. See Khûrbet Faghûr.*

عين فارس Lu   *'Ain Fâris.* The spring of Fâris. p.n.; a horseman.

عين فروجة Lu   *'Ain Farûjeh.* The spring of the crevice.

عين فطير Ju   *'Ain Fatîr.* The spring of the fissure.

عين الفوقا Lt   *'Ain el Fôka.* The upper spring.

عين الفوار Lu   *'Ain el Fûwâr.* The spring of the fountain.

عين الحدبة Lu   *'Ain el Hadabeh.* The spring of the hammock.

عين حنية Lt   *'Ain Hanniyeh.* The spring of Anna.

عين حنطش Mu   *'Ain Hantash.* The spring of Hantash. p.n.

عين حوض Ku   *'Ain Haud.* The spring of the cistern. (Nt.) This latter is called the Apostles' fountain by Christians.

عين حوض كبريان Lu   *'Ain [el] Haud Kibriyân.* The spring of Cyprian's cistern.

عين حوبين Ku   *'Ain Hûbîn.* The spring of Hûbin; perhaps from *Habu* (Rododaphne), or from *Haban* (dropsy).

عين حميد Kt   *'Ain Humeid.* p.n. The spring of Humeid. p.n.

عين جادي Ms   *'Ain Jâdy.* The spring of saffron; also called عين النصبة *'Ain en Nasbeh,* the spring of the stone set up as an object of worship.

عين جاكوك   Mt   *'Ain Jakûk.* The spring of Jakûk. p.n. ; perhaps جقوق pl. of جق 'muting' (a bird). In the gardens south-east of *Neby Samwîl,* just west of *'Ain el Emîr.*

عين الجامع   Lu   *'Ain el Jâmiâ.* The spring of the mosque.

عين الجميل   Lt   *'Ain el Jemîl.* The beautiful spring.

عين جريوت   Ls   *'Ain Jeriût.* The spring of Jeriût ; perhaps from جرى *Jara,* 'to flow.'

عين الجنان   Ls   *'Ain el Jinnân.* The spring of the gardens.

عين الجويزة   Lu   *'Ain el Joweizeh.* The spring of the little walnut-tree.

عين الجوز   { Lt / Ms }   *'Ain el Jôz.* The spring of the walnut.

عين الجديدة   Lt   *'Ain el Judeideh.* The spring of the dykes or streaks in the rocks.

عين جفنة   Ls   *'Ain Jufna.* The spring of the large dish.

عين الجرون   Kt   *'Ain el Jurûn.* The spring of troughs.

عين كارم   Lt   *'Ain Kârim.* The spring of the dresser of the soil.

عين الكرود   Lt   *'Ain el Karûd.* The spring of the monkeys or goblins.

عين الكزبة   Ju   *'Ain el Kezbeh.* The spring of the Kezbeh, a tree with hard wood; but *see* " Memoirs," under ' Chezib.' Heb. כָּזָב 'lying.'

عين الخندق   Lt   *'Ain el Khanduk.* The spring of the fosse.

عين الخنزيرة   Ku   *'Ain el Khanzireh.* The spring of the sow ; it is below *'Ain Haud,* on the north-west near *Birket 'Atab.*

عين الخرجة   Lt   *'Ain el Kharjeh.* The outer spring.

عين الخاتولة   Kt   *'Ain el Khâtûleh.* The spring of the hare's form.

عين القدة   Ns   *'Ain el Kiddeh.* The spring of the fissure.

عين القبيبة   Ku   *'Ain el Kubeibeh.* The spring of the little dome, from a kubbeh near.

عين كليبة   Lu   *'Ain Kuleibeh.* The spring of the Beni Kuleib Arabs.

عين القصب   Kt   *'Ain el Kûsab.* The spring of reeds.

عين القسيس   Lu   *'Ain el Kûssîs.* The spring of the (Christian) priest.

عين المداد   Ku   *'Ain el Madâd.* The spring of manure or of extension.

عين المديق   Lu   *'Ain el Madîk.* The spring of the cattle which loathe forage; (perhaps ضيق 'narrow').

عين المدورة   Mt   *'Ain el Madowerah.* The round spring.

عين المحفور   Ku   *'Ain el Mahfûr.* The spring of the excavated place. It is also called عين ستي حسنا *'Ain Sitti Hasna,* 'the spring of our Lady Hasna.'

عين الأجور   Ku   *'Ain el Mâjûr.* The hired spring, or 'the spring of the flower-pot.'

عين المخلد   Ms   *'Ain el Makhled.* The perennial spring.

عين ملكة   Mt   *'Ain Malakah.* The spring of the angel.

عين المقصور   Lu   *'Ain el Maksûr.* The deficient spring.

عين مقتوش   Lt   *'Ain Maktûsh.* The spring of the exhilarated one.

عين المارودة   Lu   *'Ain el Mârûdeh.* Perhaps for مورودة *maurûdeh,* 'the frequented spring.'

عين المصون   Ms   *'Ain el Masûn.* The guarded spring.

عين مرجلان   Kt   *'Ain Merjelân.* The spring of the chaldron.

عين مرجلين   Lu   *'Ain Merjelîn.* The spring of the chaldron. *See 'Ain Merjelân above.*

عين مرجد   Ms   *'Ain Merjid* (according to others it is *Menjid*). The spring of the plateau, being situate at the head of a wâdy.

عين المزراب   Ms   *'Ain el Mezrâb.* The spring of the water-spout or gutter (south of Ram Allah, north of *'Ain Merjid*).

عين المحندس   Nt   *'Ain el Muhendis;* probably المهندس The spring of the architect.

عين المونة   Ku   *'Ain el Mûneh.* The spring of provisions.

عين منجد   *'Ain Munjed.* The spring of the helper.

عين مصباح   Ms   *'Ain Musbâh.* The spring of the lantern.

عين الناموس   Lu   *'Ain el Nâmûs.* The spring of the hunter's lodge; an old Arab word applied to stone huts and prehistoric stone dwellings throughout Arabia. In the modern language it means 'a mosquito.'

عين نيني   Js   *'Ain Nini.* The spring of Nini.

عين عُلّيق Lt 'Ain 'Olleik. The spring of brambles (just north of 'Ain Hanniyeh).

عين الرفيع Lt 'Ain er Rafiü. The high spring.

عين الرواس Lt 'Ain er Rüwüs. The spring of watercress.

عين الصيف Lu 'Ain es Saif. The spring of the summer.

عين صالح Lu 'Ain Sâleh. The spring of Sâleh. p.n.; 'righteous.'

عين سلمان Ls 'Ain Sehnân. The spring of Selmân. p.n.

عين سجما Kt 'Ain Sejma. The spring of flowing water.

عين الشامية Kt 'Ain esh Shâmiyeh. The Damascus or Syrian spring.

عين الشعطر Mt 'Ain esh Shâtir. The spring of Shâtir; perhaps شعطر 'cinnamon.' East of 'Ain el Emir.

عين شمس Jt 'Ain Shems. The spring of the sun. See "Memoirs," under 'Bethshemesh.'

عين الشرقية Lu 'Ain esh Sherkiyeh. The eastern spring.

عين الشيح Nu 'Ain esh Shîh. The spring of Shih, an aromatic herb; but see p. 90.

عين ستي حسنا Ku 'Ain Sitti Hasna. The spring of our Lady Hasna; 'Ain el Mahfûr. q.v.

عين ستي مريم Lt 'Ain Sitti Miriam. The spring of our Lady Mary, at 'Ain Kârim.

عين صوبا Lt 'Ain Sôba. The spring of the pile.

عين صوبية Ls 'Ain Sûbieh. The spring of the pile.

عين سور باهر Mu 'Ain Sûr Bâhir. The spring of Sûr Bâhir. q.v.

عين الصوان Mt 'Ain es Sûwân. The spring of flints.

عين السويدة Jt 'Ain es Sûweideh. The blackish spring; but see Jubb Sûweideh, p. 5.

عين الطاقة Lu 'Ain et Tâkah. The spring of the arch or shelf.

عين التنور Ku 'Ain et Tannûr. The springs of the oven or reservoir. See "Memoirs."

عين طرفيديا Ms 'Ain Tarfidia. The spring of Tarfidia. q.v.

عين التينة Lu 'Ain et Tineh. The spring of the fig-tree.

عين الطوالي   Mt   *'Ain el Tuwâly.* The spring of the long rock; *'Ain el Emîr.* q.v.

عين أم الشرايط   *'Ain Umm esh Sherâit.* The spring with the strips; at *Khûrbet Umm esh Sherît.* q.v.

عين أرطاس   Mu   *'Ain Urtâs.* The spring of Urtâs. q.v.

عين الوحش   Lu   *'Ain el Wahash.* The spring of the wild beast.

عين يالو   Mu   *'Ain Yâlô.* The spring of Yâlô. q.v.

عين زفيزيف   Mt   *'Ain Zefeizîf.* The spring of the red jujube-tree, east of *'Ain el Emîr.*

العيساوية   Mt   *El 'Aîsâwiyeh.* The place or sect of Jesus.

عتبة الغزلان   Mt   *'Akabet el Ghûzlân.* The mountain road or steep of the gazelles; west of the village *el 'Aîsawiyeh.*

عتبة الحريضية   Nu   *'Akabet el Hareidîyeh.* The mountain road or steep of the alkali makers.

عتبة الحطابة   Nu   *'Akabet el Hattâbeh.* The mountain road or steep of the wood-cutters.

عتبة النمر   Nu   *'Akabet en Nimr.* The mountain road or steep of the leopard; or perhaps 'of the abundant water.' *See Wâdy en Nimûr,* p. 37; near *Khûrbet el Haradân.*

عقور   Kt   *'Akûr.* 'Barren.'

علالي البنات   Kt   *'Alâli el Benât.* The upper chambers of the maidens.

العليليات   Ns   *El 'Aleilîât.* The upper chambers; caves high up in a precipice.

علار البصل   Ku   *'Allâr el Bûsl.* 'Allâr (p.n.) of the onions; also called

علار الفوقا   *'Allâr el Fôka.* The upper 'Allâr.

علار السفلة   Ku   *'Allâr es Siflch.* The lower 'Allâr.

العمور   Lt   *El 'Ammûr.* El 'Ammûr. *See Khûrbet 'Amr,* p. 268.

العمري   Lu   *El 'Amry.* El 'Amry. p.n. *See* last paragraph. The southern Kubbeh in Bittir; 'the ancient.'

عمواس   Js   *'Amwâs.* 'Amwâs. Greek Ἐμμαούς. The word is said by Josephus (B.J. iv, 1, 3) to mean a thermal spring; it has no meaning in Arabic.

عناتا   Nt   *'Anâta.* 'Anâta. p.n. Heb. עֲנָתוֹת *Anathoth.*

عنابة Js *'Anndbeh.* The 'jujube.' *See* "Memoirs."

العنزية Ls *El 'Anazîyeh.* El 'Anazîyeh.

المعنيزية Kt *El 'Ancizîyeh.* El 'Ancizîyeh. } *See 'Ain el 'Ancizîyeh,* p. 278.

عرب الوادية Nt *'Arab el Wâdîyeh.* The Arabs of the valley.

عراق العبّادة Ks *'Arâk el 'Abbâdeh.* The cliff, cavern, or buttress of the female devotee.

عراق عردا Ms *'Arâk 'Arda.* The cliff, bank or buttress of the mountain with water at the foot ; marked quarry ; west of *'Attâra.*

عراق الحمام Ks *'Arâk el Hamâm.* The cliff, cavern or buttress of the pigeons ; near B. Nûba there is a second place so called (Ku).

عراق اسمعين Kt *'Arâk Ismâîn.* The cliff, cavern, or buttress of Ishmael.

عراق الجمال Ku *'Arâk el Jemâl.* The cliff, cavern, or buttress of the camels ; a low cliff east of *Beit 'Atâb.*

عراق موسى Ns *'Arâk Mûsa.* The cliff, cavern, or buttress of Moses ; north side of *Wâdy Suweinît,* east of *Jebâ.*

عراق الرعيان Mt *'Arâk er Râîan.* The cliff, cavern, or buttress of the shepherds, south of Lifta.

عراق الشيخ اسمعين Lu *'Arâk esh Sheikh Ismâîn.* The cliff, cavern, or buttress of Sheikh Ishmael ; south of *'Ain el Madik.*

عراق الشمس Nt *'Arâk esh Shems.* The cliff, cavern, or buttress of the sun. *See* "Memoirs," under ' en Shemesh.'

عراق الغيرة Nt *'Arâk et Tîreh.* The cliff, bank, or buttress of the fortress.

عراق ام السلطان Nt *'Arâk Umm es Sultân.* The cliff, cavern, or buttress of the Sultan's mother ; west of *Sheikh 'Amber.*

الاربعين Kt *El Árbaîn.* The Forty.

العرقوب Ku *El 'Árkûb.* The winding track ; name of a district on a spur of the main chain.

عرتوف Jt *'Artûf.* 'Artûf. p.n.

عساكر Ms *'Asâkir.* Armies.

عسلين Jt *'Aslin.* Aslin. p.n. ; 'Honey.' *See Khûrbet el 'Asalîyeh,* p. 82.

عطارا Ms *'Attâra.* 'Attâra. p.n. Heb. עֲטָרוֹת

الطواق قبور الزمان الجديد   Mt   *'Atwâk Kabûr ez Zemân el Judîd.* Enclosures of the tombs of modern time. A modern graveyard near Burj es Tût.

عيون ابي محارب   Jt   *'Ayûn Abu Mehârib.* The springs of the father of the warriors ; also called, according to one native on the spot, *'Ayûn Kâra.* p.n.

عيون البجزة   Kt   *'Ayûn el Bejezeh.* The springs of Bejezeh. p.n.

عيون الخارجة   Jt   *'Ayûn el Khârjeh.* The outer springs.

عيون محمود   Jt   *'Ayûn Mahmûd.* Mahmûd. p.n.

عيون طريف   Jt   *'Ayûn Terîf.* The springs of the edge or extremity.

عيون التينة   Jt   *'Ayûn et Tîneh.* The springs of the fig-tree.

العزرية   Nt   *El 'Azerîyeh.* The place of Lazarus (Bethany).

باب الجبيا   Mu   *Bâb el Jebîa.* The gate of the cistern ; an opening in the aqueduct near Bâtn Fukkûs.

باب المضيق   Ns   *Bâb el Medîk.* The gate of the strait, close to the tombs south of Khŭrbet Haiyân.

باب المعلقة   Mt   *Bâb el Muâllakah.* The overhanging gate.

باب السكاك   Mu   *Bâb es Sikâk.* The gate to the paths. The cross roads west of Bethlehem on the road to Bâb Jâla. The side walls here begin.

باب الواد   Kt   *Bâb el Wâd.* The gate of the valley.

البلوع   Ms   *El Balûä.* The pond.

بلوط البدرية   Mt   *Ballût el Bedrîyeh.* The oak of the Bedriyeh ; (family name).

بلوط المدورة   Lu   *Ballût el Madowerah.* The round oak.

بلوط ام الشريط   Ku   *Ballût Umm esh Shŭrît.* The oak with the strips of rags. (A sacred tree covered with rags.)

باطن فقوس   Mu   *Bâtn Fakkûs.* The knoll of melons.

باطن الجرود   Ks   *Bâtn el Jeirûd.* The knoll of Jeirûd. *See Khŭrbet Jârûdîyeh.* p. 7.

باطن الخضر   Mu   *Bâtn el Khŭdr.* The knoll of el Khŭdr. q.v., p. 28

بابن المعيدة Lt *Bâtn es Saîdeh.* The knoll of the highlands.

البد Mu *El Bedd.* The idol, or idol temple.

البدية Mu *El Bediych.* El Bediych. p.n. *See Khûrbet el Bediych,* p. 83. The Sheikh's tomb in Sharafât is so called.

بيت عنان Ls *Beit 'Anân.* The house of 'Anân. p.n.

بيت عتاب Ku *Beit 'Atâb.* The house of 'Atâb; perhaps עיטם Etam.

بيت دقو Ls *Beit Dukku.* The house of Dukku. p.n.

بيت فجوس Kt *Beit Fajûs.* The house of Fajûs. p.n.; signifying 'haughty.'

بيت فصد Ju *Beit Fased.* The house of phlebotomy. *See* "Memoirs," under 'Ephes Dammim' אפס דמים.

بيت حنينا Mt *Beit Hannîna.* The house of Hannina. p.n.

بيت ايكا Ku *Beit Ika.* The house of Ika. p.n. *cf. Medinat el Aikah,* p. 131.

بيت اكسا Mt *Beit Iksa.* The house of Iksa. p.n.

بيت إزا Ls *Beit Izza.* The house of Izza. p.n.

بيت جالا Mu *Beit Jâla.* The house of Jâla. p.n.; perhaps from جال a parapet.

بيت الجمال Ju *Beit el Jemâl.* The house of camels.

بيت لحم Mu *Beit Lahm.* The house of meat. Heb. בית לחם 'house of bread.'

بيت لقيا Ks *Beit Likia.* The house of Likia. p.n.

بيت محصير Kt *Beit Mahsîr.* The house of Mahsir. Perhaps for محصور mahsûr, 'besieged.'

بيت ميس Kt *Beit Meis.* The house of the meis tree (*Cordia myxa*).

بيت نتوبا Lt *Beit Nakûba.* The house of the mountain pass.

بيت نتيف Ju *Beit Nettîf.* The house of Nettif. p.n.; (a camel whose hair comes off in patches).

بيت نوبا Ks *Beit Nûba.* The house of Nûba. p.n.

بيت ساحور Mu *Beit Sâhûr.* The house of magicians. Also called بيت ساحور النصارى *Beit Sâhûr en Nusâra;* (*Beit Sâhûr* of the Christians).

بيت ساحور المعتيقة Mt *Beit Sâhûr el 'Attkah.* The ancient Beit Sâhûr. Also called بيت ساحور الوادي *Beit Sâhûr el Wâdy* (Beit Sâhûr of the valley).

بيت ساقيا Lu *Beit Sâkia.* The house of irrigation.

بيت شنّا Js *Beit Shenna.* The house of Shenna. p.n.

بيت سيلا Ls *Beit Sîla.* The house of the stream.

بيت سيرا Ks *Beit Sîra.* The house of the fold.

بيت صفافا Mu *Beit Sûfâfa.* House of the summer-houses or narrow benches.

بيت سوريك اسوريق Lt *Beit Sûrîk.* The house of Sûrik. p.n.

بيت سوسين Jt *Beit Sûsîn.* The house of Sûsin. p.n. سوس liquorice, or a weevil ; سوسن a lily.

بيت تعامر Mu *Beit Tâmir.* The house of the T'âmirah Arabs.

بيت تول Kt *Beit Tûl.* The house of Tûl. p.n. ; meaning 'length.'

بيتونيا Ls *Beitûnia.* Beitûnia. p.n.

بيت عور الفوقآ Ls *Beit 'Ûr el Fôka.* The upper house of 'Ûr (Beth Horan).

بيت عور التحتآ Ks *Beit 'Ûr et Tahta.* The lower Beit 'Ûr.

بنو عامر Ks *Beni 'Âmir.* The Beni 'Âmir Arabs. *See* p. 268.

بنو حارث القبليّة {Ks Ls} *Beni Hârith el Kiblîyeh.* The southern Beni Hârith Arabs, sons of Hârith, an old noble Arab name (the Aretas of the New Testament).

بنو حسن Lt *Beni Hasan.* The Beni Hasan Arabs.

بنو مالك Ks *Beni Mâlik.* The Beni Mâlik Arabs.

برفليا Js *Berfilya.* Berfilya. p.n.

برنة Us *Berkah.* Berkah. p.n. ; meaning 'splitting,' 'sowing.' *See* " Memoirs."

بيار ابو ناذور Nu *Biâr Abu Nâdhûr.* The wells of Abu Nâdhûr. p.n.

بيارة خليل الله Ks *Biârct Khalîl Allah.* The wells of Abraham (the friend of God).

بدو Ls *Biddu.* Biddu. p n.

بئر العبد　Nu　*Bir el 'Abd*. The well of the slave.

بئر ابي سمرآ　Mu　*Bîr Abu Sumra*. The well with the brown colour.

بئر احمد　Nu　*Bîr Ahmed*. The well of Ahmed. p.n.

بئر العجمي　Nu　*Bir el 'Ajamî*. The well of the Persian.

بئر أقليدي　Ju　*Bîr Aklîdia*. The well of Aklídia. p.n.

بئر لعليليات　Nt　*Bîr el 'Aleilíât*. The well of the upper chamber.

بئر عليآ　Nu　*Bîr 'Alya*. The upper well.

بئر العمدان　Mu　*Bîr el 'Amdân*. The well of the columns.

بئر اميتة　Nu　*Bîr Ameîyeh*. Probably *Umeiyeh*. The well of the Ommaiyeh (the first dynasty of Caliphs of Damascus).

بئر العمود　Mu　*Bîr el 'Amûd*. The well of the column.

بئر عونآ　Mu　*Bîr 'Aûna*. The well of 'Aûna. p.n.

بئر عويطل　Nu　*Bîr 'Aweitil*. The well of the vacant places.

بئر عويستّة　Nu　*Bir 'Aweistyeh*. The well of the knife; but perhaps from عواس 'a night patrol.'

بئر عزير　Ms　*Bîr 'Azeir*. The well of 'Azeir. p.n. Ezra.

بئر البيضآ　Mu　*Bîr el Beida*. The white well.

بئر بيت بسا　Mu　*Bîr Beit Bassa*. The well of the house of the marsh.
or بصة

بئر بيت البيادر　Ks　*Bîr Beit el Beiâdir*. The well of the house of threshing floors.

بئر البيار　Ku　*Bîr el Bîâr*. The well of wells (a name obtained near *Khûrbet Deiry*).

بئر البرج　Nt　*Bîr el Burj*. The well of the tower.

بئر داود　Mu　*Bîr Dâûd*. David's well (at Bethlehem).

بئر الدير　Ms　*Bîr ed Deir*. The well of the monastery.

بئر العد　Nt　*Bîr el 'Edd*. The well of the perennial spring, or the old well (close to *'Ain Hand*, the so-called Apostle's fountain).

بئر ايوب　Mt　*Bîr Eyûb*. The well of Job. The Christians call it the well of Joab.

بئر فخذ الكول   Nu   *Bir Fakhedh el Kôl.* The well of the thigh of the hills (pronounced *Jôl* by the Arabs).

بئر الحبيطية   Mu   *Bir el Habeitiyeh.* The well of the Habaty Arabs (*Habatât beni Tamîm*).

بئر الحنجلية   Nu   *Bir el Hanjeliyeh.* The well of the Hanjaliyeh. p.n.; probably a dervish order.

بئر الحريضية   Nu   *Bir el Hareidiyeh.* The well of the alkali makers.

بئر حزمية   Nu   *Bir Hazmiyeh.* The girded well.

بئر الحلو   Jt   *Bir el Helu.* The sweet (water) well.

بئر الحطيمة   Js   *Bir el Heteimeh.* The well of rubbish; also called بئر اشتيوي *Bir Eshtawy.*

بئر الحمّام   Nu   *Bir el Hammâm.* The well of the bath, or the well of rain-water.

بئر الحمّص   Mu   *Bir el Hummus.* The well of the chick-pea.

بئر إخداش   Mu   *Bir Ikhdâsh.* The well of Ikhdâsh (scratching).

بئر اسطح   Mu   *Bir Istêh.* The well with the flat top.

بئر الجشورة   Mu   *Bir el Jashûrah.* The well of pasturing.

بئر الجبالي   Ns   *Bir el Jebâly.* The well of the Jebâly (mountaineer).

بئر الجوزة   Mt   *Bir el Jôzeh.* The well of the walnut-tree.

بئر قديسمو   Mu   *Bir Kadismu.* The well of Kadismu. p.n.; perhaps from קְדֹשִׁים 'holy men,' *i.e.*, the Magi (it is generally known as the 'well of the Magi').

بئر القنطارة   Jt   *Bir el Kantarah.* The well of the arch.

بئر التردية   Nu   *Bir el Keradiyeh.* The well of the Keradiyeh. p.n.; probably from *Karadah* (refuse of wool).

بئر القطا   Mu   *Bir el Katt.* The well of the crag.

بئر القياقبة   Mu   *Bir el Keiâkbeh.* The well of the Keikab trees.

بئر الترامة   Ju   *Bir el Kerrâmeh.* The well of the Kerrâmeh, a plant with long thin hollow branches.

بئر خلة المشمش   Mu   *Bir Khallet el Mishmish.* The well of the dell of apricots, near *Bir Wâr el Asad.*

بئر خلّة المقابر   Nt   *Bir Khallet el Mukâbir.* The well of the dell of the tombs.

بئر الخنزيرة   Nu   *Bîr el Khanezîreh.* The well of the swine.

بئر (ابى) خشيبة   Mu   *Bîr (Abu) Khashcibch.* The well of the muddy ground.

بئر الخربة   Mu   *Bîr el Khŭrbeh.* The well of the ruin.

بئر الكفّ   Mu   *Bir el Kuff.* The well of the palms of the hand. It is also called بئر مقطع الشختور *Bîr Muktà esh Shakhtûr* (the well of the cutting out of the one-masted boat).

بئر تمرة   Nu   *Bîr Kumrah.* The dusky white well.

بئر قرام غريب   Mu   *Bîr Kurâm Gharîb.* The well of the figured veil of the stranger; but *see Bîr el Kerrâmeh,* p. 289.

بئر القصب   Jt   *Bîr el Kûsab.* The well of reeds.

بئر اللتون   Nu   *Bîr el Lattûn.* The sweet well.

بئر الليمون   Ju   *Bîr el Leimûn.* The well of the lemon.

بئر لفتان   Nu   *Bîr Liftân.* The well of Liftân. Perhaps لفتان, from لفت 'turnips.'

بئر المحس   Mu   *Bîr el Mahas.* The well of currying leather. Perhaps محس 'the smooth-worn rope.'

بئر الملك   Ku   *Bîr el Malak.* The well of the angel. (Perhaps *el Melek;* 'of the king.')

بئر معين   Ks   *Bîr Mâin.* The well of the springs.

بئر المرمر   Mu   *Bîr el Marmar.* The well of marble.

بئر المصطبة   Mu   *Bîr el Mastabah.* The well of the bench. Near بئر اسطح *Bîr Istûh* ('the well with the flat top').

بئر الملو   Nu   *Bîr el Matû.* The well of the ears of millet, or of the water remaining at the bottom of a nearly empty well.

بئر المزراب   Mt   *Bîr el Mezrâb.* The well of the spout; south-east of *Neby Samwîl.*

بئر المسلم   Nt   *Bîr el Musillim.* The well of the saluter, between *Khŭrbet Kâwît* and *Khŭrbet ed Dikki.*

بئر المزّة   Ks   *Bîr el Mizzeh.* The well of hard limestone.

بئر المتيح   Mu   *Bîr el Mutîh.* The well of the unfortunate or fated one.

بئر النابت   Nu   *Bir en Nábit.* The well of vegetation.

بئر النغيق   Nu   *Bir en Nagheik.* The well of the raven's croaking.

بئر النهل   Ju   *Bir en Nahl.* The well of the drinking.

بئر نبالا   Ms   *Bir Nebála.* The well of apparatus.

بئر النفيس   Nu   *Bir en Nefís.* The precious well.

بئر النخيل   Mu   *Bir en Núkheil.* The well of the little palm, south of *Wády Umm el Kúláh.*

بئر الراس   Ks   *Bir er Rás.* The well of the hill-top.

بئر الرصاص   Nt   *Bir er Rasás.* The well of lead.

بئر الصلح   Nu   *Bir es Saláh.* The well of righteousness; near *Jófet el 'Akkás.* q.v.

بئر صالح   Nu   *Bir Sálch.* The excellent well, or Sáleh's well.

بئر الصليب   { Kt / Lu }   *Bir es Salíb.* The well of the Cross.

بئر صنع عطاآلله اللحام   Ku   *Bir Sanà 'Ataállah el Lehhám.* The well of the cistern of 'Atâ-allâh the Lehhâm (Flesher); family name.

بئر السبيل   Nu   *Bir es Sebíl.* The well of the wayside fountain.

بئر الشامي   Lt   *Bir esh Shámy.* The Damascene or Syrian well.

بئر الشيخة   Nu   *Bir esh Sheikhah.* The well of the female saint, near es Sineisil.

بئر الشنار   Nu   *Bir esh Shinnár.* The well of the Greek partridge.

بئر الصياح   Mt   *Bir es Siáh.* The well of streams of the shout; perhaps سياح 'surface water,' some fifty yards north of the church at Neby Samwil.

بئر السفلاني   Ju   *Bir es Sifláni.* The lower well.

بئر الست   Nt   *Bir es Sitt.* The lady's well.

بئر الصفصاف   Ju   *Bir es Súfsáf.* The well of the osier-willows.

بئر الصواني   Mu   *Bir es Siwáni.* The well of flints, near *Wády Samúrah.*

بئر الطاعون   Js   *Bir et Táánn.* The well of the plague.

بئر التقطق   Nu   *Bir et Taktah.* The well of the rattling stones.

بئر ام الفخط   Nu   *Bîr Umm el Fekhit.* The well with the split; near *Wâdy 'Alya.*

بئر ام الخنفس   Nu   *Bîr Umm el Khunfis.* The well with the beetle; near *Kabr Ghannam.*

بئر ام الزلق   Nu   *Bîr Umm ez Zelek.* The well of the slippery place; also called بئر سبسب *Bîr Sibsib,* the well of the open country; near *Bîr Hazmîyeh.*

بئر وعر الاسد   Mu   *Bîr Wâr el Asad.* The well of the rugged rocks of the lion.

بئر الوطاة   Ns   *Bîr el Wata.* The well of the level or trampled ground.

بئر ياسول   Mt   *Bîr Yâsûl.* The well of Yâsûl. p.n.; north of *Bîr Sâhûr el 'Atîkah.*

بئر الزاغ   Ju   *Bîr ez Zâgh.* The well of the hooded crow.

بئر الزرورة   Nu   *Bîr ez Zárûrah.* The well of the hawthorn; near *Bîr Hazmîyeh.*

بئر زين الدين   Mt   *Bîr Zein ed Dîn.* The well of Zein ed Din. p.n.; meaning the Adornment of the Faith.

بئر الزرع   Ju   *Bîr ez Zerâ.* The well of the sown corn.

بيرة   Ms   *Bîreh.* The well of the palace.

بركة الحمام   Mu   *Birket el Hümmâm.* Pool of the bath (below *Jebel Fureidîs,* on north side).

بركة الجوبا   Nu   *Birket el Jûba.* The pool of the reservoir (at el Jûba).

بركة مأملا   Mt   *Birket Mâmilla.* Pool of (St.) Mâmilla.

بتير   Lu   *Bittîr.* Bittir. p.n. Heb. בְּתָר and ביתתר Bether.

البقيع   Mt   *El Bükeiâ.* The plateau.

البرك   Lu   *El Burak.* The pools; commonly called Solomon's Pools.

البرج   {Ks / Ku}   *El Burj.* The tower.

برج الشيخ مرزوق   Ku   *Burj esh Sheikh Marzûk.* The tower of Sheikh Marzûk. p.n.; meaning 'provided for.'

برج التوت   Mt   *Burj et Tût.* The tower of the mulberries.

برقة   Ms   *Burkah.* The speckled ground.

البويرة   Ks   *El Buweirah.* The little pit.

الدوارة Ns *Ed Dawârah.* The houses.

الدحيشة Mu *Ed Deheisheh.* (Properly دحساً). The soft or reddish-black soil.

دير ابان Ku *Deir Abân.* The monastery of Abân. p.n.; perhaps אֶבֶן. It is the Christian site of Ebenezer.

دير ابي { قابوس
قبوس Kt *Deir Abu Kabûs.* The well of Abu Kabûs. A name of Noâmân Ibn Munzir (an ancient Arabian king).

دير ابي علي Ku *Deir Abu 'Aly.* The monastery of Abu 'Aly. p.n.

دير العمود Mu *Deir el 'Amûd.* The monastery of the column.

دير عصفور Ju *Deir Asfûr.* The monastery of sparrows.

دير العزر Lt *Deir el 'Azar.* The monastery of el 'Azar (*i.e.*, Lazarus).

دير البنات Lu *Deir el Benât.* The convent of the maidens.

دير ديوان Ns *Deir Dîwân.* The monastery of the Dîvân (sofa or council). Also called دير دبوان *Deir Dubwân.* p.n.

دير ايوب Kt *Deir Eyûb.* The monastery of Job.

دير الهوا Kt *Deir el Hawa.* The monastery of the wind. The place stands very high on a conical hill.

دير المحروق Mt *Deir el Mahrûk.* The burnt monastery.

دير نحلة Js *Deir Nahleh.* The monastery of the bee; but *see Wâdy en Nahl*, p. 37.

دير الصليب Mt *Deir es Salîb.* The monastery of the Cross.

دير شباب Jt *Deir Shebbab.* The monastery of the youth.

دير الشيخ Kt *Deir esh Sheikh.* The monastery of the Sheikh (elder or chief).

دير السد Nt *Deir es Sidd.* The monastery of the cliff.

دير الطاحونة Jt *Deir et Tâhûneh.* The monastery of the mill.

دير يسين Mt *Deir Yesîn.* The monastery of Yesîn. p.n.

دير الزيق Nu *Deir ez Zîk.* The monastery of the weft (of cloth). The word also means the sound of opening a door.

ظهر علين Kt *Dhahr 'Alein.* The ridge of 'Alein. p.n.; 'conspicuous.'

ظهر دير غنام   Mu   *Dhahr Deir Ghannâm.* The ridge of the monastery of shepherds, or of Ghannâm. p.n.; 'plunderer' (the hill between *Khârbet Umm Tôba* and *Bîr el Katt*).

ظهر مسعدة   Mu   *Dhahr Mesâdeh.* The ridge of Mesâdeh. p.n.; 'fortunate.' It should probably be مصعدة 'of the ascent.'

ظهر سريس   Kt   *Dhahr Sarîs.* The ridge of Sarîs. p.n. *See* p. 312. It is the hill of the trigonometrical point.

ظهر الشياح   Mt   *Dhahr esh Shîyâh.* The ridge of the cautious one; the hill just west of Bethany (*el 'Azirîyeh*).

ظهرة عابد   Lt   *Dhahret 'Âbid.* The ridge of the devotee.

ظهرة ابي رفا   Ns   *Dhahret Abu Rafâ.* The ridge with the elevation; west of *Mukhmâs.*

ظهرة دير العمود   Mu   *Dhahret Deir el 'Amûd.* The ridge of the monastery of the column. Hill of *Deir el 'Amûd.*

ظهرة الحدديّة   Nu   *Dhahret el Haddadîyeh.* The ridge of the Haddadîyeh. p.n.; probably حدادية *Haddâdîyeh,* 'blacksmiths.'

ظهرة الجنينة   Mu   *Dhahret el Jinneineh.* The ridge of the little garden, near *Beit Sâhûr,* on the east.

ظهرة جرون زعتر   Mu   *Dhahret Jurûn Zâter.* The ridge of the troughs, or threshing-floors of thyme.

ظهرة القضي   Nu   *Dhahret el Kady.* The ridge of death; or of the decisive man. *Kâât Keis.*

ظهرة القرع   Mu   *Dhahret el Karâ.* The ridge of gourds; near *Dhahret el Tabbâkh.*

ظهرة الغرياني   Mu   *Dhahret el Kreiâny.* The ridge of the aqueducts; near *Hindâzi.*

ظهرة خلة المغارة   Mu   *Dhahret Khallet el Mughârah.* The ridge of the dell of the cave; near *Dhahret et Tabbâkh.*

ظهرة الرتمية   *Dhahret er Retemîyeh;* near *Bîr Fakhedh el Kôl.* The ridge of the broom plant (*Genista rætam*), the juniper of the Bible, which grows in this part.

ظهرة الطباخ   Mu   *Dhahret et Tabbâkh.* The ridge of the cook; also called ظهرة ام الغزلان *Dhahret Umm el Ghûzlân,* the ridge of the gazelles.

دراءة العروض   Mu   *Drââh el 'Arûd.* A cubit in width. A name obtained near *el Keisarânîyeh*, applying to a piece of land. *See Hindâzi;* it is written also عروب 'sprouting plants,' or 'hard.'

اشوع   Kt   *Eshûâ.* Eshûâ. p.n.; (Joshua).

الغفر   Lu   *El Ghûfr.* The guard or toll; a watchtower.

الحبس   Lt   *El Habs.* The prison, or the church (mosque) endowment: a cell.

الحدب   Ls   *El Hadab.* The hummock; hillock east of *Khûrbet el Ibreij.* q.v.

الحاج عليان   Lu   *Hâj 'Aleiân.* The pilgrim 'Aleian. p.n.; a Kubbeh with fine trees, east of *Nehhâlîn. See* "Memoirs."

الحاج حسن الرأفاني   *Hâj Hasan er Rafâti.* The pilgrim Hasan of Râfât.

حوعان   Lu   *Hausân.* Hausân. p.n.; 'hovering round.'

حيطان الصنوبر   Ks   *Heitân es Sinôbar.* The walls of the cypress.

حيطان الوعر   Ks   *Heitân el Wâr.* The walls of the rugged rock.

هندازي   Mu   *Hindâzi.* The measurer. Close to this title, applying to a piece of ground, was obtained *Drââh el 'Arûd,* 'a cubit wide.' *See* above.

حزمة وقيل الحزمة   Ns   *Hizmeh* or *el Hizmeh.* The bundle. Perhaps from حزم 'rough ground.'

حفيرة { عرطوف عرطوف   Jt   *Hûfîret 'Artûf.* The excavation of 'Artûf (near that place).

حفير النبي بولس   *Hûfîyir en Neby Bûlus.* The excavations of the prophet Paul. Springs dug in the valley south of *Khûrbet en Neby Bûlus.*

حمام سليمان   Mu   *Hûmmâm Suleimân.* Solomon's bath; an old pool below Urtâs.

الحورية   Ks   *El Hûrîyeh.* El Hûrîyeh. p.n.; perhaps from خور 'the bottom of a well; or it may be from خَور 'a plane tree'; or خَور 'a houri' (maiden of paradise).

الحوطي   Mt   *El Hûty.* The walled. A well with a walled court round it (south of Lifta, near the main road).

الإمام عمر بن الخطاب   Mu   *Imâm 'Amr ('Omar) Ibn Khŭttâb.* The Imâm, *i.e.,* Caliph (خليفة). Omar Ibn el Khattâb (a great supporter, and afterwards successor of Mohammed). It is the *Kubbeh* in *Khŭrbet Umm Tôba.*

البحران   Ns   *El Jahrân.* The dens (rock-cut tanks).

جامع عمر بن الخطاب   Mu   *Jâmiâ 'Amr ('Omar) Ibn Khŭttab.* The mosque of Omar, son of Khŭttâb (the father-in-law of Mohammed, and afterwards Caliph), in *Beit T'âmîr.*

جبع $\left\{ \begin{matrix} Ns \\ Ku \end{matrix} \right\}$ *Jebâ.* Jebâ. p.n.; Heb. גִּבְעָה 'hill.'

جبل الالوافة   Mu   *Jebel el Atwâkeh.* The mountain of the enclosures; (hill above *'Ain Yâlo,* with dry stone walls).

جبل دير ابى نُور   Mt   *Jebel Deir Abu Thôr.* The mountain of Abu Thôr's convent.

جبل القدس   Ms   *Jebel el Kuds.* The mountain of Jerusalem.

جبل المكتّم   Ks   *Jebel el Mukŭttum.* The hidden mountain. Perhaps an error for جبل المتظم ' the quarried mountain.'

جبل السُنيتى   Mt   *Jebel es Sonneik.* The mountain of the white plastered house. The top is called باطن الهوا *Bâtn el Hawa,* 'knoll of the wind.'

جبل الطويل   Ms   *Jebel et Tawîl.* The long mountain.

جبيع   Ls   *Jebîâ.* The hill. *See* above, *Jebâ;* also "Memoirs," under ' Gibeah.'

جديرة   Ms   *Jedîreh.* The sheep-fold. Heb. גְּדֵרָה Gederah ; ' fold.'

جراش   Ku   *Jerâsh.* Jerâsh. p.n.

الجيب   Ms   *El Jîb.* El Jib. p.n.

جوفة العقّاش   Nu   *Jôfet el 'Akkâsh.* The hollow of the thick-bearded one. The last word is also the name of a herb and of the fruit of the Araka (الاراك), or *Paliurus aculeatus.*

جوفة الرتقة   Nu   *Jôfet er Ritkah.* The closed-up hollow.

الجوبآء   Nu   *El Jûba.* The water pits; a birkeh and some caves in a rock.

الجورة   Lt   *El Jûrah.* The water-hole.

جورة التمخ Nt *Jûrat el Kemkh.* The water-hole of pride (literally 'turning up the nose haughtily'); but it may be an error for قمح 'wheat.'

قاعة قيس Nu *Kâat Keis.* The court-yard or plain of Keis. p.n. *See Khûrbet Keis,* p. 234.

قبر غنام Nu *Kabr Ghannâm.* 'The grave of Ghannâm.' p.n.

قبر حلوة Mu *Kabr Helweh.* The grave of Helweh. A woman's name, 'sweet.'

القبو Lu *El Kabû.* The vault or cellar.

قبور البني اسرئيم Ns *Kabûr el Beni Isrâim.* 'The graves of the Children of Israel'; also called قبور العمالكة *Kabûr el 'Amâlikeh,* 'the graves of the Amalakites.'

قبور التفعل Mu *Kabûr et Tefâl.* The forged or artificial tombs.

القديرة Ns *El Kadeirah.* The little pot.

غدان المصري Nu *Kanân el Mâsri.* The peaks of the Egyptian.

تنات الكفار Mu *Kanât el Kuffâr.* The pagans' canal.

تناة وادي البئار Lu *Kanât Wâdy el Biâr.* The canal of the valley of the wells. The short western canal south of the Burak (Pools of Solomon).

قطمون وقيل دير القطمون Mt *Katamôn,* or *Deir el Katamôn.* Katamôn. p.n.; or the convent of Katamôn, from قطم 'to quarry.'

قطنة Lt *Katanneh.* Katanneh. p.n. *cf.* Heb. קָטֹן 'little.'

قطاة قنور Mt *Katât Kanwir.* Crags of the summit.

كبارا Lt *Kebâra.* The caper plant. In modern Arabic *Kebâra* means also a small kiln.

كفيرة Lt *Kefîreh.* The little village.

كفر {عتاب عقب} Ms *Kefr 'Akâb.* The village of the steep or mountain roads.

كفر ناتا وقيل خربة ابي صباح Ns *Kefr Nâta.* The village of Nâta. p.n.; also called *Khûrbet Abu Subbâh.* The village of Abu Subbâh, from a family which lived in the place at one time (Rob., late "Bib. Res.," p. 290).

كفر روطا Ks *Kefr Rât.* The village of the river.

شيال }
كفر { شيان
Ls *Kefr Shiyân.* The village of Shiyân. p.n.; also called *Kefr Shiyâl.* p.n. *Shiyân* means the plant dragon's blood.

كفر صوم
Lu *Kefr Sôm.* The village of fasting.

القيصرانيّة
Mu *El Keisarânîyeh.* The Cæsareum.

قلادة القطين
Ku *Kelâdet el Koteineh.* The necklace of cotton (or of the servants); also called قلادة القطينة *Kelâdet el Koteineh*, with the same meaning. The name is vulgarly applied to a string of dried figs.

الكنيسة
Js *El Kenîseh.* The Church.

كنيسة الرّعوات
Mu *Kenîset er Râwât.* The church of the shepherds. (The Mediæval Angelus ad Pastores.)

القرادية
{Mu}
{Nu} *El Kerâdiyeh.* The Kerâdiyeh قراد 'a tick'; a district name.

كسلة
كسلا
Kt *Kesla.* Kesla. p.n. Heb. כְּסָלוֹן Chesalon.

القطع
Ku *El Ketâ.* The separate part; a piece of ground near *Beit 'Atâb.*

خلّة عابد
Mt *Khallet 'Ábid.* The dell of the devotee; near *Bîr Sâhûr el 'Atîkah.*

خلّة الأبريكنة
Ns *Khallet el Abreikneh.* The dell of sorrel.

خلّة العصافير
Lu *Khallet el 'Asafîr.* The dell of sparrows; east of *'Ain Faghûr.*

خلّة البئر
Lt *Khallet el Bîr.* The dell of the well; near *Khûrbet Deir esh Sheikh.*

خلّة بئر الصّفصاف
Ju *Khallet Bîr es Sûfsâf.* The dell of the well of the osier-willow.

خلّة البطّة
Ju *Khallet el Buttah.* The dell of the duck.

خلّة الضبع
Ms *Khallet ed Dûbâ.* The dell of the hyena.

خلّة الفرس
Kt *Khallet el Faras.* The dell of the mare; (near *Khûrbet Marmîtah*).

خلّة حمد
Mu *Khallet Hamed.* The dell of Hamed. p.n.; (near *Khûrbet el Bedd Falûh*).

خلة الحمام Js *Khallet el Hammám.* The dell of the bath.

خلة القساءين Mu *Khallet el Kasámín.* The dell of the distributors (near *el Keisaráníyeh*).

خلة الخلف Ls *Khallet el Khŭlf.* The dell of opposition or succession.

خلة القطا Ls *Khallet el Kŭta.* The dell of the sand-grouse.

خلة المعلبة Ku *Khallet el Musellabeh.* The cross dell (at right angles to the valley).

خلة النتش Ms *Khallet en Netesh.* The dell of sprouting plants.

خلة السماقة Ju *Khallet es Semmákah.* The dell of the sumach-tree (*Rhus coriaria*, Linn.).

خلة الشامية Lt *Khallet esh Shámíyeh.* The Syrian dell (near *Khŭrbet Núh*).

خلة الشقيف Lt *Khallet esh Shukíf.* The dell of the cleft (north of *Khŭrbet Deir esh Sheikh*).

خلة الطيبة Ku *Khallet et Taiyibeh.* The goodly dell (near *Ballût Umm esh Shŭrít*).

خلة الطمع Kt *Khallet et Tamá.* The dell of covetousness, or hope.

خلة الزيتون Ju *Khallet ez Zeitûn.* The dell of olives.

الخان Ju *El Khán.* The Caravanserai.

خان مستى Ls *Khán Miska.* The inn of the waterworks.

خان الرام Ms *Khán (er Rám).* The inn of er Rám. p.n.

الخاروبة Js *El Khârûbeh* (properly *el Kharrûbeh*). The carob or locust tree (*Ceratonia siliqua*).

الخضر Lu *El Khŭdr.* El Khŭdr. *See* p. 28.

خراب الثريا Ku *Khŭráb eth Thureiya.* The ruins of the Pleiades, or soft soil.

خربة العبد Ku *Khŭrbet el 'Abd.* The ruin of the slave.

خربة عباد Ju *Khŭrbet 'Abbád.* The ruin of the devotee.

خربة ابرجان Ls *Khŭrbet Aberján.* The ruin of the towers.

خربة العبير Lu *Khŭrbet el 'Abhar.* The ruin of the mock orange (*Styrax officinalis*).

خربة ابي عدس Lu *Khŭrbet Abu 'Adas.* The ruin with the lentils.

خربة ابى عترة   Mu   *Khŭrbet Abu 'Atrah.* The ruin of Abu 'Atrah. p.n.; north of *Sûr Bâhir,* in the valley.

خربة ابى بريك   Mu   *Khŭrbet Abu Bureik.* The ruin with the little pool.

خربة ابى فريج   Ks   *Khŭrbet Abu Fureij.* The ruin of Abu Fureij. p.n.

خربة ابى حويلان   Nt   *Khŭrbet Abu Hûeilân.* The ruin of Abu Hûeilân. p.n.

خربة ابى قليبة   Lu   *Khŭrbet Abu Kuleibeh.* The ruin of Abu Kuleibeh. p.n.

خربة ابى ليمون   Mt   *Khŭrbet Abu Leimûn.* The ruin of Abu Leimûn ('him of the lemon'), from *Neby Leimûn.*

خربة ابى مكيري   Nu   *Khŭrbet Abu Makîrî.* The ruin of Abu Makireh. p.n. *cf. Kîreh,* p. 237.

خربة ابى عصيرة   Nt   *Khŭrbet Abu Maseirah.* The ruin with the fold. (*Sîr et Tabalâs.*)

خربة ابى محمد   Lt   *Khŭrbet Abu Muhammed.* The ruin of Abu Muhammed. p.n.

خربة ابى روس   Ns   *Khŭrbet Abu Rûs.* The ruin with the hill tops; on the top of a hill near edge of precipice.

خربة ابى صباح   Ns   *Khŭrbet Abu Sabbâh.* The ruin of Abu Sabbâh. p.n.; another name for *Kefr Nâta,* which see, p. 297.

خربة ابى سعد   Nt   *Khŭrbet Abu Sâd.* The ruin of Abu Sâd. p.n.

خربة ابى سود   Lu   *Khŭrbet Abu Sûd.* The ruin of Abu Sûd. p.n.

خربة ابى صوان   Nt   *Khŭrbet Abu Sûwân.* The ruin with the flints.

خربة عدسة   {Ms}{Mt}   *Khŭrbet 'Adaseh.* The ruin of lentils. Two ruins, either of which might be Adasa. (*See* "Memoirs.") The former is the most probable. It is said to be also called خربة لجة *Khŭrbet Lejjeh,* the ruins of the abyss or chasm.

خربة ابى زرور   Ls   *Khŭrbet Abu Zârûr.* The ruin with the hawthorn (marked R, west of *Beit Izza*).

خربة الاحمدية   Lu   *Khŭrbet el Ahmadiyeh.* The ruin of Ahmed's place or family.

خربة عيد   Ms   *Khŭrbet 'Aîd.* 'Aîd's ruin (a common Arabic proper name).

خربة عين الكنيسه Lu *Khŭrbet 'Ain el Kenîseh*. The ruin of the spring of the church.

خربة عين التوت Lt *Khŭrbet 'Ain et Tût*. The ruin of the spring of the mulberry.

خربة أقليديا Ju *Khŭrbet Aklîdia*. The ruin of Aklidia. p.n. *See* p. 288.

خربة العقد Ks *Khŭrbet el 'Akid*. The ruin of the vault or coping stone.

خربة الالون Mt *Khŭrbet el Âlaun*. The ruin of el Âlaun. p.n.

خربة عليآ Lu *Khŭrbet 'Alîa*. The upper ruin.

خربة علميطا Nt *Khŭrbet 'Almît*. The ruin of Almit. p.n. *cf.* Heb. עַלְמִית.

خربة العلي Ju *Khŭrbet el 'Aly*. 'Aly's ruin. p.n.

خربة عمران Ku *Khŭrbet 'Amrân*. The ruin of 'Amrân (Amram, the father of Moses). *See Neby Amrân*, p. 9.

خربة الأسد Ku *Khŭrbet el Asad*. The ruin of the bear. Also pronounced *Lesed* ('licking').

خربة عسقلان Ls *Khŭrbet 'Askalân*. The ruin of Ascalon.

خربة عطورة Ns *Khŭrbet 'Attûrah*. The ruin of 'Attûra. p.n. *cf. 'Attâra*, p. 284.

خربة عواد Ls *Khŭrbet 'Auwâd*. The ruin of 'Auwâd. p.n.; immediately west of *Beitûnia*, and east of *Khŭrbet Bîr esh Shâfa*.

خربة بلوطا الحالس Ms *Khŭrbet Ballût el Hâlis*. The ruin of the green oak.

خربة بردة Ks *Khŭrbet Baradah*. The cool ruin.

خربة باطن الصغير Lt *Khŭrbet Bâtn es Saghîr*. The ruin of the little knoll.

خربة البد Mu *Khŭrbet el Bedd*. The ruin of the idol or temple.

خربة بد فالوح Mt *Khŭrbet Bedd Fâlûh*. The ruin of the idol or temple of Fâlûh. p.n. ; ('prosperous').

خربة البدادين Kt *Khŭrbet el Beddâdein*. The ruin of the two idols.

خربة البدر Mt *Khŭrbet el Bedr* (from *Sheikh Bedr*). The ruin of el Bedr p.n. *See Sheikh Bedr*, p. 243.

خربة بيت جازا Lt *Khŭrbet Beit Jâza*. The ruin of the house of Jâza. p.n.

خربة بيت جيز Jt *Khŭrbet Beit Jîz*. The ruin of the house of the valley side. It is also called خربة صبرا *Khŭrbet Sŭbrah*, the ruin of the prickly pear.

خربة بيت مزمير   Mt   *Khŭrbet Beit Mizmîr.* The ruin of the house of the flute.

خربة بيت مزا   Lt   *Khŭrbet Beit Mizza.* The ruin of the house of good wine. It may however be a corruption of Heb. ה מֹצָה *Mozah*, 'spring head.' The place is surrounded with springs. *See* also *Bîr el Mizzeh*, p. 290, and "Memoirs," under 'Kŭlônieh.'

خربة بيت نوشف   Ks   *Khŭrbet Beit Nôshef.* The ruin of the house of Nôshef. p.n. ; 'one who dries.'

خربة بيت شباب   Ls   *Khŭrbet Beit Shebâb.* The ruin of the house of the youth.

خربة بيت سكاريا   Lu   *Khŭrbet Beit Skâria.* The ruin of Beit Skâria (the ancient Beth Zacharias).

خربة البيتوني   Ms   *Khŭrbet el Beitûni.* The ruin of the man of Beitûnia.

خربة بلد الفوقا   Ju   *Khŭrbet Belled el Fôka.* The ruin of the upper town.

خربة البيادر   Ns   *Khŭrbet el Blâdir.* The ruin of the threshing-floors.

خربة البيار   Ms   *Khŭrbet el Biâr.* The ruin of the wells.

خربة بيار لوقا   Mu   *Khŭrbet Biâr Lûka.* The ruin of the wells of Luke.

خربة بئر العد   Kt   *Khŭrbet Bîr el 'Edd.* The ruin of the perennial spring well.

خربة بئر الليمون   Ju   *Khŭrbet Bîr el Leimûn.* The ruin of the well of the lemon. *See Bîr el Leimûn*, p. 290.

خربة بئر الرصاص   Ms   *Khŭrbet Bîr er Rasâs.* The ruin of the lead well.

خربة بئر الشافع وقيل بئر خرابة الشافع   Ls   *Khŭrbet Bir esh Shâfâ,* or *Bîr Khŭrâbet esh Shâfâ.* The ruin of the well of esh Shâfâ, or the well of the ruin of esh Shâfâ. The last word may mean a narrow strip of land round a property left in order to invalidate the right of pre-emption by the owners of the adjoining estate. It may also be an error for *es Shafiî,* the name of the founder of one of the four great orthodox Modammedan sects.

خربة البقيع   Mt   *Khŭrbet el Bŭkeiâ.* The ruin of the open valley.

خربة بقيع ضأن   Nt   *Khŭrbet Bukeiâ Dhân.* The ruin of the vale of sheep.

خربة البريج   Ks   *Khŭrbet el Burcij.* The ruin of the little tower.

خربة البرج   Mt   *Khŭrbet el Burj.* The ruin of the tower.

خربة البَصَل   Kt   *Khŭrbet el Bŭsl.* The ruin of the onions.

خربة الدالي   Ks   *Khŭrbet ed Dály.* The ruin of the trailing vine.

خربة الدار   Ku   *Khŭrbet ed Dâr.* The ruin of the house (marked R, north of *Khŭrbet Duhy*).

خربة دار مصطفى   Ks   *Khŭrbet Dâr Mustafa.* The ruin of Mustafa's house.

خربة دارية   Ks   *Khŭrbet Dârieh.* The ruin of Dârieh. p.n. *See* "Memoirs," under 'Attaroth Adar.'

خربة الدوارة   Ns   *Khŭrbet ed Dawârah.* The ruin of the circuit.

خربة الدبيدبة   Ms   *Khŭrbet ed Debeidbeh.* The ruin of ed Debeidbeh. The word means a horse's hoofs on the stones. It may be دبابة 'little wells.'

خربة الدير   Lu   *Khŭrbet ed Deir.* The ruin of the monastery.

خربة دير عمر   Lt   *Khŭrbet Deir 'Amr.* The ruin of the monastery of Amr close to *Sheikh 'Amr.* q.v.

خربة دير داقر   Js   *Khŭrbet Deir Dâkir.* The ruin of the monastery of the gardener.

خربة دير حسّان   Ls   *Khŭrbet Deir Hassân.* The ruin of the monastery of Hassân. p.n.

خربة دير ابن عبيد   Nu   *Khŭrbet Deir Ibn 'Obeid.* The ruin of the monastery of the son of Obeid; also called *Mâr Theodosius.*

خربة دير { قلوس / قالوس }   Ks   *Khŭrbet Deir Kalûs.* The ruin of the convent of Kulûs. p.n. قلوس means 'ropes'; قلوس 'water increasing in a well.'

خربة دير الرهبان   Js   *Khŭrbet Deir er Rohbân.* The ruin of the monastery of the monks.

خربة دير سلّم   {Ms / Kt}   *Khŭrbet Deir Sellâm.* The ruin of the monastery of Sellâm.

خربة دير الشيخ   Lt   *Khŭrbet Deir esh Sheikh.* The ruin of the monastery of the Sheikh.

خربة ديري   Ku   *Khŭrbet Deiry.* The ruin of the property of the monastery (of *Deir esh Sheikh*).

خربة ذنب الكلب   Ks   *Khŭrbet Dheneb el Kelb.* The ruin of the dog's tail; just west of *Khŭrbet Beit Noshef.*

خربة الدكي Nt *Khŭrbet ed Dikki.* The ruin of the hillock; also called خربة السكة *Khŭrbet es Sikkeh,* 'ruin of the path.'

خربة الدرش Ks *Khŭrbet ed Dirish.* The ruin of ed Dirish. p.n.

خربة الدريمة Ls *Khŭrbet ed Dreihemeh.* The ruin of the enclosed gardens. *See* p. 139.

خربة ضمى Ku *Khŭrbet Duhy.* The ruin of barren ground.

خربة الدوير {Ls Ns} *Khŭrbet ed Duweir.* The ruin of the little monastery.

خربة ارها Ms *Khŭrbet Erha.* The ruin of aqueducts.

خربة عرما Kt *Khŭrbet 'Erma.* The ruin of the dam or dyke.

خربة ارزية Ms *Khŭrbet Erzîyah.* The ruin of cedars.

خربة العشي Ms *Khŭrbet el 'Eshshy.* The ruin of the nests.

خربة فعوش Ks *Khŭrbet Fáâûsh.* The ruin of Fáâûsh. p.n.; perhaps for فاعوس *fá'ûs,* 'a serpent,' 'mountain goat,' or 'a narrow necked bottle.'

خربة فاغور Lu *Khŭrbet Fâghûr.* The ruin of Fâghûr. p.n. Heb. פְּעוֹר *See* "Memoirs," under 'Peor' or 'Phagor.'

خربة فرج Mt *Khŭrbet Faraj.* The ruin of the cleft. *See 'Ain Farûjeh,* p. 279.

خربة فواقسة Mt *Khŭrbet Fâákseh.* The ruin of the melons.

خربة الفول Jt *Khŭrbet el Fûl.* The ruin of beans. *cf. Tell el Fûl.*

خربة الغشينة Ku *Khŭrbet el Ghasheinah.* The valley of the undressed stone.

خربة حبيق Ku *Khŭrbet Habeik.* The ruin of the herb pennyroyal.

خربة الحدبة Nt *Khŭrbet el Hadabeh.* The ruin of the hummock; just north-east of Bethany.

خربة الحداد Mt *Khŭrbet el Haddâd.* The ruin of the blacksmith.

خربة حديد Jt *Khŭrbet Hadîd.* The ruin of iron.

خربة الحدينة Ks *Khŭrbet Hadîtheh.* The new ruin.

خربة الحافي Ls *Khŭrbet et Hâfy.* The ruin of the barefoot ones.

خربة الحي Ns *Khŭrbet el Hat.* The ruin of the snake.

خربة حَيَّان Ns *Khŭrbet Haiyân.* The ruin of Haiyân. p.n.

خربة الحَيّة Ns *Khŭrbet el Haiyeh.* The ruin of the snake.

خربة الحُجَيْلَة Mu *Khŭrbet Hajeileh.* The ruin of the partridges (*Cacabis heyi*).

خربة الحاج حسن Jt *Khŭrbet el Hâj Hasan.* The ruin of the pilgrim Hasan. p.n.

خربة الحُمَام Js *Khŭrbet el Hamâm.* The ruin of the pigeon.

خربة حمدان Lu *Khŭrbet Hamdân.* The ruin of Hamdân. p.n.

خربة حظل Ku *Khŭrbet Hamdhal.* The ruin of the colocynth.

خربة حمّادة Kt *Khŭrbet Hammâdeh.* The ruin of Hammâdeh. p.n.; perhaps حمائة 'pasturage.'

خربة حنّا Ku *Khŭrbet Hanna.* The ruin of John.

خربة الحُرذان Nu *Khŭrbet el Haradân.* The ruin of the lizards.

خربة حاراش Lt *Khŭrbet Hârâsh.* The ruin of Hârâsh. *See* "Memoirs," under 'Arath.'

خربة حرفوش Ls *Khŭrbet Harfûsh.* The ruin of the viper.

خربة حاريسيس Kt *Khŭrbet Harsis.* The ruin of Harsis. p.n.; by the guard house at *Bâb el Wâd.*

خربة حَسَن Jt *Khŭrbet Hasan.* The ruin of Hasan. p.n.

خربة الهوا Ks *Khŭrbet el Hawa.* The ruin of the wind.

خربة حزور Mt *Khŭrbet Hazzûr.* The ruin of Hazzûr. p.n.; Heb. הָצוֹר *Hazor.*

خربة حبيلة Lu *Khŭrbet Hebeileh.* The ruin of the little terrace.

خربة حلابي Ks *Khŭrbet Hellâbi.* The ruin of the milkman; probably حلبي the man of Aleppo.

خربة حيبا Js *Khŭrbet Hîba.* The ruin of Hîba. p.n.

خربة حرشا Kt *Khŭrbet Hirsha.* The ruin of Hirsha. p.n.; perhaps Heb. הָרָשִׁים *Charashim. See* "Memoirs."

خربة الحوش Lt *Khŭrbet el Hôsh.* The ruin of the court-yard.

خربة حوبين Ku *Khŭrbet Hûbin.* The ruin of Hûbin. p.n.

2 R

خربة الحمام Ks *Khŭrbet el Hŭmmâm.* The ruin of the bath.

خربة ابى عواد Ks *Khŭrbet Ibn Auwâd.* The ruin of Auwâd.

خربة ابى بارك Ms *Khŭrbet Ibn Bârak.* The ruin of Ibn Bârak. p.n.; ('blessed').

خربة إقبالا Lt *Khŭrbet Ikbâla.* The ruin of prosperity; perhaps 'the southern ruin'; also called دير البنات *Deir el Benât,* the convent of the girls. *See* "Memoirs."

خربة العسا Ls *Khŭrbet Ilâsa.* The ruin of Ilâsa. p.n.; perhaps Ἐλεασά. *See* "Memoirs," under 'Eleasa.'

خربة اسم الله Jt *Khŭrbet Ism Allah.* The ruin of the name of God.

خربة الجامع Lu *Khŭrbet el Jâmiâ.* The ruin of the mosque.

خربة جبع Kt *Khŭrbet Jebâ.* The ruin of the hill.

خربة الجدير Ls *Khŭrbet el Jedeir.* The ruin of the walled enclosure, or sheepcote.

خربة جديرة Jt *Khŭrbet Jedîreh.* The ruin of the walled enclosure, or sheepcote. *See* "Memoirs." under 'Gederah.'

خربة الجمعة Mt *Khŭrbet el Jemâh.* The ruin of the gathering.

خربة جنابة الغربية Ju *Khŭrbet Jennâbet el Gharbîyeh.* The ruin of the west side.

خربة جنابة الشرقية Ju *Khŭrbet Jennâbet esh Sherkîyeh.* The ruin of the east side.

خربة جنعر Kt *Khŭrbet Jenâr.* The ruin of Jenâr; probably נוער 'a watchman.'

خربة الجرابة Kt *Khŭrbet el Jerâbeh.* The ruin of the plantation.

خربة جريوت Ls *Khŭrbet Jeriût.* The ruin of Jeriût. p.n.

خربة جنخدم Nu *Khŭrbet Jokhdhŭm.* The ruin of Jokhdhûm. p.n.

خربة جب الروم Nu *Khŭrbet Jubb er Rûm.* The ruin of the pit of the Greeks. Compare *Ferdûs er Rûm* at Jerusalem.

خربة الجبيعة Lt *Khŭrbet el Jubeiâh.* The ruin of the little hill.

خربة الجفير Ls *Khŭrbet el Jufeir.* The ruin of the half-filled well, or of the quiver.

خربة جفنة Ls *Khŭrbet Jufna.* The ruin of Jufna (the large dish).

خربة المجنجل Ks *Khûrbet el Junjul.* The ruin of Junjul. p.n. Perhaps an ancient Gilgal.

خربة جرفة Ju *Khûrbet Jurfa.* The ruin of the perpendicular bank (cut out by the torrent in the *débris* of a valley).

خربة جوريش Ku *Khûrbet Jûrish.* The ruin of Jûrish. p.n.

خربة الكبوش Ls *Khûrbet el Kabbûsh.* The ruin of Kabbûsh. p.n.

خربة كاكول Mt *Khûrbet Kâkûl.* The ruin of the herb pennyroyal. It may be an error for كعكولى 'soft stone.'

خربة قريتا Nt *Khûrbet Karrît.* The ruin of Karrit. p.n.; from قرط which means both 'an ear-ring,' and 'clover.'

خربة كبار Mu *Khûrbet Kebâr.* The ruin of the caper plant, or of the kiln.

خربة كفرتا Js *Khûrbet Kefrata.* The ruin of Kefrata. p.n.

خربة كفر راسى Ks *Khûrbet Kefr Râsy.* The ruin of the village of Râsy. p.n.

خربة كفر طاس Ms *Khûrbet Kefr Tâs.* The ruin of the village of Tâs (a cup).

خربة كفرورية Jt *Khûrbet Kefr Urich.* The ruin of Kefr Urich. (Kefrûrich.) p.n. *See* " Memoirs," under ' Cepararia.'

خربة القرينة Ls *Khûrbet el Kereina.* The ruin of the little peak.

خربة خلة العدس Ms *Khûrbet Khallet el 'Adas.* The ruin of the dell of lentils ; just west of *Khûrbet et Tireh.*

خربة خلة البيفآ Mu *Khûrbet Khallet el Beida.* The ruin of the white dell.

خربة خلة السدر Ns *Khûrbet Khallet es Sidr.* The ruin of the dell of the lotus tree.

خربة خلة الطرحة Mt *Khûrbet Khallet et Tarhah.* The ruin of the dell of the foundations.

خربة الخمسة Lu *Khûrbet el Khamaseh.* The ruin of the five.

خربة الخميس {Ls Mu} *Khûrbet el Khamîs.* The ruin of the fifth.

خربة الخان Ku *Khûrbet el Khân.* The ruin of the Caravanserai.

خربة خاتولة Kt *Khûrbet Khâtûleh.* The ruin of Khâtûleh. *See 'Ain el Khâtûleh,* p. 280.

خربة الخازوق Mt *Khûrbet el Khâzûk.* The ruin of the impaling stake.

خربة خير Kt *Khûrbet Kheir.* The ruin of good.

خربة الخيشوم   Ju   *Khŭrbet el Kheishûm.*   The ruin of the promontory.

خربة الخضرية   Ns   *Khŭrbet el Khădriyeh.*   The ruin of the chapel of el Khŭdr. See p. 28.

خربة قيافة   Ju   *Khŭrbet Kiâfa.*   The ruin of tracking footsteps.

خربة كيلا   Jt   *Khŭrbet Kîla.*   The ruin of Kila.   p.n.

خربة قديس   Lu   *Khŭrbet Kudeis.*   The ruin of the holy one.

خربة الكف   Mu   *Khŭrbet el Kuff.*   The ruin of the palm of the hand.

خربة الكرسنة   Ku   *Khŭrbet el Kursinneh.*   The ruin of the vetches.

خربة القصير   Lu   *Khŭrbet el Kŭseir.*   The ruin of the small palace.

خربة القصر   Kt   *Khŭrbet el Kŭsr.*   The ruin of the palace.

خربة القسيس   Mu   *Khŭrbet el Kŭssis.*   The priest's ruin.

خربة القصور   Lt   *Khŭrbet el Kŭsûr.*   The ruin of the palaces.

خربة اللحم   Ls   *Khŭrbet el Lahm.*   The ruin of meat.

خربة اللتاتين   Ls   *Khŭrbet el Lattâtîn.*   The ruin of those who triturate or pound.

خربة اللوز   Lt   *Khŭrbet el Lôz.*   The ruin of the almond.

خربة اللوزة   Lt   *Khŭrbet el Lôzeh.*   The ruin of the almond-tree.

خربة المحمى   Ls   *Khŭrbet el Mahmeh.*   The ruin of the fortified place.

خربة المخروم   Nu   *Khŭrbet el Makhrûm.*   The ruin of the mountain peak.

خربة ملكتحة   Ku   *Khŭrbet Malkat-hah.*   The ruin of Malkat-hah.   p.n.

خربة مرميتا   Kt   *Khŭrbet Marmîtah.*   The ruin of Marmitah.   p.n.

خربة المزار   Mu   *Khŭrbet el Mazâr.*   The ruin of the shrine ; just south of *Wâdy Abu Helweh* (marked R).

خربة ميتا   Ls   *Khŭrbet Meita.*   The ruin of low ground.

خربة مقيقا   Mt   *Khŭrbet Mekîka.*   The ruin of the bald patches.

خربة مناع   Ks   *Khŭrbet Menââ.*   The ruin of Menââ. *See Sitt Menâ.*

خربة المراغب   Mt   *Khŭrbet el Merâghib.*   The ruin of the open country.

خربة مرج الفكية | Kt | *Khŭrbet Merj el Fikich.* The ruin of the meadow of the joyous one, or of the fruit grower; perhaps it should be فقيه ' the doctor.'

خربة المشرفة | Jt | *Khŭrbet el Mesherfeh.* The ruin of the high place.

خربة مزموريا | Mu | *Khŭrbet Mezmûria.* The ruin of the pipes or flutes.

خربة المسمار | Kt | *Khŭrbet el Mismâr.* The ruin of the nail.

خربة المران | Lt | *Khŭrbet el Murân.* The ruin of the Murân; a tall slender tree, of which spears are made.

خربة المرضص | Nt | *Khŭrbet el Murŭssŭs.* The ruin of the place of the rubble; probably from the pavement of cobble stones in the ruins.

خربة المصري | Ms | *Khŭrbet el Mŭsry.* The ruin of the Egyptian.

خربة نبهان | Kt | *Khŭrbet Nabhân.* The ruin of vigilance.

خربة النحل | Mt | *Khŭrbet en Nahl.* The ruin of bees; it may be نهل 'drinking place,' a cistern being near it. But *see Wâdy en Nahleh*, p. 37.

خربة النبي بولس | Ju | *Khŭrbet en Neby Bûlus.* The ruin of Prophet Paul.

خربة ندي | Ls | *Khŭrbet Neda.* The ruin of dew.

خربة النجار | Mu | *Khŭrbet en Nejjâr.* The ruin of the carpenter.

خربة النياتة | Ku | *Khŭrbet en Niâtch.* The ruin of Niâtch. p.n.; perhaps نياتة ' desert ground.'

خربة نسية | Ms | *Khŭrbet Nisich.* The ruin of forgetfulness, or of credit (as opposed to ready money).

خربة نوح | Ku | *Khŭrbet Nûh.* The ruin of Noah. *See* "Memoirs"; also called خربة حريقة الكحالة *Khŭrbet Hariket el Kahhâleh,* ' the conflagration of the oculists.'

خربة ربع | Kt | *Khŭrbet Rabâ.* The ruin of the arbour or summer-house.

خربة الرغابنة | Nt | *Khŭrbet er Raghâbneh.* The ruin of the Raghâbneh, family name; (sing. *Raghbân*). This is a common form of Syrian proper name.

خربة ركوبس | Ks | *Khŭrbet Rakûbus.* The ruin of Rakûbus. p.n.

خربة الراس | {Ls}{Nt} | *Khŭrbet er Râs.* The ruin of the hill-top. A second, above *'Ain Haud.*

خربة راس ابي عيشة   Jt   *Khŭrbet Râs Abu 'Aisheh.* The ruin of the hill-top of the father of 'Aisheh ('Âyesha) ; *i.e.,* Abu Bekr.

خربة راس ابي مرة   Jt   *Khŭrbet Râs Abu Murrah.* The ruin of the hill-top of Abu Murreh. p.n. The *Beni Murreh* are a well known Arab tribe.

خربة راس العلوة   Mt   *Khŭrbet Râs el 'Alweh.* The ruin of the upper hill-top.

خربة راس الباد   Mt   *Khŭrbet Râs el Bâd.* The ruin of the conspicuous hill-top.

خربة راس المغار   Ls   *Khŭrbet Râs el Mŭghâr.* The ruin of the hill-top of the caves ; west of *el Kubeibeh.*

خربة راس الصنوبر   Ls   *Khŭrbet Râs es Sinôbar.* The ruin of the hill-top of the cypress ; just east of *Bîr 'Ûr et Tahta* (marked R).

خربة راس الطويل   Mt   *Khŭrbet Râs et Tawîl.* The ruin of the long top.

خربة الرمانة   Ls   *Khŭrbet er Rumâneh.* The ruin of the pomegranate.

خربة رويسون   Js   *Khŭrbet Ruweisûn.* The ruin of the hill-tops.

خربة السعيرة   Ku   *Khŭrbet es Sâireh.* The ruin of the flame; it is perhaps Heb. שַׁעֲרָיִם. *See* "Memoirs," under 'Shaaraim' and 'Sior.'

خربة الصلاح   Mt   *Khŭrbet es Salâh.* The ruin of es Salâh. *See Abu Salâh,* p. 278.

خربة سمونية   Kt   *Khŭrbet Sammûnieh.* The ruin of Semûnieh. p.n. *See* p. 115, Sheet V.

خربة سمويل   Mt   *Khŭrbet Samwîl.* The ruin of Samwil ; near *Neby Samwîl.*

خربة سناسين   Ku   *Khŭrbet Sanâsîn.* The ruin of Sanâsin. p.n. ; the word means the protuberant parts of the vertebral process.

خربة الصدر   Jt   *Khŭrbet es Sefâr.* The ruin of the dried plants.

خربة السملية   Lt   *Khŭrbet es Selamîyeh.* The ruin of the Selamîyeh family (marked R, west of *Khŭrbet el Lôz*) ; probably p.n.

خربة شعب ايلياس   Lt   *Khŭrbet Shâb Aîliâs.* The ruin of the mountain spur of Elias.

خربة الشغراب   Mu   *Khŭrbet esh Shaghrâb.* The ruin of tripping up (as wrestlers do).

خربة الشيخ ابرهيم   Kt   *Khŭrbet esh Sheikh Ibrahîm.* The ruin of the Sheikh Ibrahim. p.n.

خربة الشيخ سعد Ku *Khŭrbet esh Sheikh Sâd.* The ruin of the Sheikh Sâd : also called خربة ام اشتية *Khŭrbet Umm Eshteryah,* the ruin of the winter quarters.

خربة الشخخة Ku *Khŭrbet esh Shekhetah.* The ruin of esh Shekhetah. p.n. ; perhaps from the Heb. שַׁחַת 'a sepulchre,' as there are rock-cut sepulchres here.

خربة الشرقية Lu *Khŭrbet esh Sherkîyeh.* The eastern ruin.

خربة شوفا {Kt Lt} *Khŭrbet Shûfa.* The ruin of the harrow.

خربة شويكة Ju *Khŭrbet Shuweikeh.* The ruin of thistles. *See* "Memoirs," under 'Socoh.'

خربة الصياغ Kt *Khŭrbet es Siâgh.* The ruin of the goldsmiths.

خربة صير الغنم Mu *Khŭrbet Sîr el Ghanem.* The ruin of the sheepfold.

خربة الصومع Mt *Khŭrbet es Sômâ.* The ruin of the cloister or hermitage.

خربة صبيحة Mu *Khŭrbet Sŭbhah.* The reddish-white ruin.

خربة الصبر Lt *Khŭrbet es Sŭbr.* The ruin of the prickly pear.

خربة السكر Kt *Khŭrbet es Sukker.* The ruin of sugar.

خربة سوريك Jt *Khŭrbet Sûrik.* The ruin of Sûrik. Heb· שֹׂרֵק *Sorek.* *See* "Memoirs."

خربة الصوانة Ks *Khŭrbet es Sŭwâneh.* The ruin of flints.

خربة السويدية Ks *Khŭrbet es Sŭweidîyeh.* The ruin of the Sŭweidiyeh Arabs.

خربة سويكة Ms *Khŭrbet Suweikeh.* The ruin of Suweikeh. p.n.

خربة المنطورة Lu *Khŭrbet et Tantûrah.* The ruin of the peak (marked R) ; east of *Khŭrbet el Hebeileh.*

خربة طازة Lu *Khŭrbet Tâzah.* The ruin of Tâzah. p.n. ; ('fresh').

خربة التين Ks *Khŭrbet et Tin.* The ruin of the fig-tree.

خربة الطيرة Ms *Khŭrbet et Tîreh.* The ruin of the fort.

خربة ام العدس Ju *Khŭrbet Umm el 'Adas.* The ruin with the lentils.

خربة ام العمدان {Js Ms} *Khŭrbet Umm el Amdân.* The ruin with the columns. (A second close to *Khŭrbet ibn Bârak,* to the east.)

خربة ام العصافير   Mu   *Khŭrbet Umm el 'Asâfîr.* The ruin with the sparrows.

خربة ام الدجاج   Ku   *Khŭrbet Umm ed Dejâj.* The ruin with the fowls.

خربة ام الدرج   Jt   *Khŭrbet Umm ed Deraj.* The ruin with the steps.

خربة ام حارتين   Js   *Khŭrbet Umm Hâretein.* The ruin with the two hillocks, being a double ruin, half either side of the path.

خربة ام الجمال   Nt   *Khŭrbet Umm el Jemâl.* The ruin of the mother of camels (head of *Wâdy el Jemel*).

خربة ام جينا   Ju   *Khŭrbet Umm Jina.* The ruin of Umm Jina. p.n. *See* "Memoirs," under ' Engannim.' *See* Section A.

خربة ام القلعة   Lu   *Khŭrbet Umm el Kŭlâh.* The ruin of the mother of the castle.

خربة ام النتشة   Mu   *Khŭrbet Umm en Neteshah.* The ruin with the sprouting plants.

خربة ام الرجمان   Ks   *Khŭrbet Umm er Rujmân.* The ruin with the cairns.

خربة ام سريسة   Jt   *Khŭrbet Umm Sarîseh.* The ruin with the lentisk.

خربة ام الشريا   Ms   *Khŭrbet Umm esh Sherît.* The ruin with the strips. Also called خربة المصيون *Khŭrbet el Masîûne.* *See 'Ain Masiûn.* p. 281.

خربة ام الشنف   Lu   *Khŭrbet Umm esh Shukf.* The ruin of the mother of the cleft.

خربة ام طوبى   Mu   *Khŭrbet Umm Tôba.* The ruin with the good tree or water ; but *see* "Memoirs," under ' Metopa.'

خربة ام تونس   Ju   *Khŭrbet Umm Tûnis.* The ruin of Umm Tûnis. p.n.

خربة وادي عليين   Ju   *Khŭrbet Wâdy 'Alîn.* The ruin of Wâdy 'Alîn. p.n.

خربة وادى إدريس   Ns   *Khŭrbet Wâd Idrîs.* The ruin of the valley of Idris (Enoch).

خربة وادي صهيون   Mt   *Khŭrbet Wâdy Sahyûn.* The ruin of the valley of Zion.

خربة اليرموق   Ju   *Khŭrbet el Yarmûk.* The ruin of Yarmûk. p.n.

خربة اليهودي   Lu   *Khŭrbet el Yehûdi.* The ruin of the Jews.

خربة زبود   Kt   *Khŭrbet Zabbûd.* The ruin of Zabbûd. *See Khŭrbet Zebed,* p. 51.

خربة زعتوقة   Mu   *Khûrbet Zâkûkah.* The ruin of the young partridge.

خربة زانوع   Ju   *Khûrbet Zânûâ.* The ruin of Zânûâ, perhaps a corruption of the Heb. זָנוֹחַ Zanoah. *See* "Memoirs."

خربة الزعتر   Kt   *Khûrbet ez Zâter.* The ruin of thyme.

خربة الزيت   *Khûrbet ez Zeit.* The ruin of the oil; another name for *Khûrbet Harfûsh*, p. 305.

خربة زنوقلة   Kt   *Khûrbet Zunûkleh.* The ruin of Zunûkleh. p.n.

خربتا إبن السبع   Ks   *Khûrbetha Ibn es Sebâ.* The ruin (local form of the word) of the son of the wild beast, or 'of seven.'

القباب   Js   *El Kubâb.* The domes.

قبة راحيل   Mu   *Kubbet Râhîl.* Rachel's dome.

القبيبة   Ls   *El Kubeibeh.* The little dome.

القدس الشريف   Mt   *El Kuds esh Sherîf.* The noble or eminent holy place, JERUSALEM.

Under this head are collected alphabetically the notice of all names within the bounds of Major Wilson's $\frac{1}{10000}$ Survey of the Environs of Jerusalem.

احبال الكبريت   *Ahbâl el Kebrît.* The terraces of sulphur.

عين المدورة   *'Ain el Madowerah.* The round spring.

عين سلوان   *'Ain Silwân.* The spring of Siloam.

عين الصوان   *'Ain es Sûwân.* The spring of the flints.

عين ام الدرج   *'Ain Umm ed Deraj.* The spring with the steps.

العازرية   *El 'Aisawîyeh.* The place of Jesus.

عقبة الغزلان   *'Akabet el Ghûzlân.* The steep or mountain road of the gazelles.

عقبة الشيخ جراح   *'Akabet esh Sheikh Jerrâh.* The steep or mountain road of Sheikh Jerrâh. q.v.

القدس *El Kuds.*

عقبة الصوان   '*Akabet es Súwán.* The steep or mountain road of the flints.

ارض السمار   *Ard es Somár.* The brown ground.

عراق الطيرة   '*Arák et Tîreh.* The cliff or cavern of the fort.

باطن الهوا   *Bâtn el Hawa.* The knoll of the wind.

بئر ابى شلبك   *Bîr Abu Shalbek.* The well of Abu Shalbek. p.n.

بئر ايوب   *Bîr Eyûb.* Job's well. *See* "Memoirs."

بئر الحاج خليل   *Bîr el Hâj Khülil.* The well of the pilgrim Khülil.

بئر الحوارة   *Bîr el Hûwârah.* The well of the white marl.

بئر القاعة   *Bîr el Káah.* The well of the plain or court-yard.

بئر القوس   *Bîr el Kôs.* The well of the bow.

بئر المشارف   *Bîr el Meshârif.* The well of the high places.

بئر الرصاصية   *Bîr el Rasâsyeh.* The rubble well.

بئر الشيخ جراح   *Bîr esh Sheikh Jerráh.* The well of Sheikh Jerrâh, q.v.

بئر الثغرة   *Bîr eth Thogherah.* The well of the frontier or gap.

بئر اليهودية   *Bîr el Yehûdîyeh.* The well of the Jews; also called بئر شمعون الصديق *Bîr Shem'ôn es Saddîk,* the well of Simon the Just, being near his tomb.

بئر زيتونات الحويلة   *Bîr Zeitûnat el Haweileh.* The well of the olives of the mansion, or of the neighbour-hood.

بركة ماملا   *Birket Mâmilla.* The pool of (St.) Mamilla; from a church which once stood near.

بركة ستي مريم   *Birket es Sitti Miriam.* The pool of Our Lady Mary.

القدس *El Kuds.*

بركة السلطان *Birket es Sultân.* The Sultan's pool.

دير التقديس بابيلا *Deir el Kaddîs Bâbíla.* The monastery of St. Mamilla.

دير مار نقولا *Deir Mâr Nikôla.* The monastery of St. Nicholas.

دير الصليب } *Deir es Salîb.* The monastery of the Cross ;
دير المصلبة } or *Deir el Musellabeh.*

فردوس الروم *Ferdûs er Rûm.* The paradise of the Greeks.

حق الدم *Hakk ed Dûmm,* for حقل الدم *Hakal ed dûm.* The field of blood ; חֲקַל דְּמָא 'Aceldama.'

الهاوية *El Hâwîeh.* The gulf or abyss.

البيذمية *El Heidhemîych.* The Heidhimiyeh (Der-vishes). p.n. Lieut. Conder says "the name is said by the Sheikh of the *Heidhe-miyeh* Derwishes, and by Sheikh Yûsef, son of Asâd Fishfish, of Jerusalem, to be called in Arabic 16th century writings, البيرمية *el Heiremîych,* of which the present name is a corruption. The older name is derived from Jeremiah, the place having been called Jeremiah's grotto since the 15th century. This information is obtained from Mr. S. Bergheim, of Jerusalem. The Jews point out this place as the בית הסקיל, 'house of stoning,' the place of public execution. This information is derived from Dr. Chaplin, of Jerusalem."

حمّام طبرية *Hŭmmâm Tubarîych.* The bath of Tiberius ; also called حمّام داود *Hummâm Dâûd,* the bath of David, probably because of its proximity to *Neby Dâûd.*

جبل دير ابي ثور *Jebel Deir Abu Tôr.* The mountain of the monastery of Abu Thor. *See Sheikh Ahmed et Toreh,* p. 318.

القدس *El Kuds.*

جبل السنينة *Jebel es Sonneik.* The mountain of the white-plastered house.

جبل الطور *Jebel et Tôr.* The mountain of the mount.

قبور السلاطين *Kabûr es Salatûn (Salâtîn).* The tombs of the Sultans, or قبور الملوك *Kabûr el Molûk.* The tombs of the kings.

القعدي *El Kâdi.* The place of sitting—the tradition being that Christ sat here.

خلة العجوز *Khallet el 'Ajûz.* The dell of the old woman. *See Wâdy Sheikh el 'Ajûz,* p. 37.

خلة القصب *Khallet el Kŭsab.* The dell of reeds.

خلة الطرحة *Khallet el Tarhah.* The dell of the foundations.

الخلوة *El Khelweh.* The secluded spot.

خربة ابى وعير *Khŭrbet Abu Wâir.* The ruin of the rugged rocks.

خرير البدر *Khŭrbet el Bedr.* The ruin of el Bedr; from *Sheikh el Bedr.* q.v.

خربة {الخميس / خميش *Khŭrbet Khamîs.* The ruin of the fifth; or *Khŭrbet Khamîsh,* the ruin of the scratcher.

خربة الخازوق *Khŭrbet el Khâzûk.* The ruin of the impaling stake.

خربة الصلاح *Khŭrbet es Salâh.* The ruin of rectitude; perhaps connected with the Aramaic צלח as the name seems frequently applied to quarried rock.

قبة العبد *Kubbet el 'Abd.* The dome of the slave.

قبة الربعين *Kubbet el Arbaîn.* The dome of the Forty; *scil.,* martyrs.

كرم احمد *Kurm Ahmed.* The vineyard of Ahmed.

كرم الشيخ *Kurm esh Sheikh.* The vineyard of the Sheikh.

التدس   *El Kuds.*

قصر اشنار   *Kûsr Ishenâr.* Schneller's house by the asylum; so called from the owner, of whose name it is a corruption. *Kûsr,* 'a palace,' is used in Palestine for any house.

قصر التاعة   *Kûsr el Kââh.* The house of the plain.

قصر الخطيب   *Kûsr el Khatîb.* The notary's house.

قصر الكرمة   *Kûsr el Kurmeh.* The house of the vineyard.

قصر القطب   *Kûsr el Kûtb.* The house of el Kŭtb. p.n.; meaning 'the pole'; it belongs to a Jerusalem Moslem family of the name.

قصر المفتي   *Kûsr el Mufti.* The house of the Mufti.

قصر الشهابي   *Kûsr esh Shehâbi.* The house of esh Shehâbi, an old Jerusalem family, to which the Mufti above-named belonged.

مغارة العدوية   *Mûghâret el 'Adawîyeh.* The cave of the 'Adawiyeh (family name).

مغارة العنب   *Mughâret el 'Anab.* The cave of grapes.

المسقوبية   *El Muskôbîyeh.* The Muscovite establishment; 'the Russian buildings.'

النبي داود   *Neby Dâûd.* The prophet David.

رأس ابو حلاوي   *Râs Abu Hâlawi.* The hill-top with the Halâwi (a thorny plant).

رأس المدبسة   *Râs el Madbaseh.* Hill-top place of *Dibs* (grape syrup). It may be from the Heb. דביסה 'a hill.'

رأس المشارف   *Râs el Meshârif.* The hill-top of the high places.

رأس النادر   *Râs en Nâdr.* The hill-top of the projecting part of a mountain.

رأس السلام   *Râs es Sillâm.* The hill-top of es Sillâm. p.n.: probably Sellâm Ibn Kaisar, one of the companions of Mohammed, who was Governor of Jerusalem under the Caliph Mo'âwiyeh.

القدس　*El Kuds.*

رأس ام الطلع　*Râs Umm et Tala.* The hill-top of the ascent. A road leads up to it.

رجوم البهيمة　*Rŭjûm el Behîmeh.* The cairns of the beast; perhaps from بهيمة 'a rock.' These are heaps of flints gathered to clear the land.

رجم الكراكير　*Rŭjm el Kahakîr.* The cairn of the stone heaps.

رؤس مصاعدة سيدنا عيسى　*Râs Mesââdet Sidna 'Aisa.* The hill-tops of the ascension of our Lord Jesus.

الشيخ احمد الثورة　*Sheikh Ahmed et Toreh.* Sheikh Ahmed of the bull; also called ابو تور *Abu Tôr*, the man with the bull. His real name was *Sheikh Shehâb ed Dîn el Mukŭddasi.* *See* Besant and Palmer's "Jerusalem," p. 432. He lived in the time of Saladin.

الشيخ بدر　*Sheikh Bedr.* Sheikh Bedr. p.n.

الشيخ الدجاني　*Sheikh ed Dejâni.* The Sheikh of Beit Dejan. *See* "Memoirs," under *Sheikh Abu Zeitûn.*

الشيخ جراح　*Sheikh Jerrâh.* Sheikh Jerrâh. p.n.; meaning surgeon.

الشيخ كامر وقيل النبي كامر　*Sheikh Kâmir* or *Neby Kâmir.* The Sheikh or Prophet Kâmir. p.n.; perhaps the Heb. בכיר 'a priest.'

الشيخ القراشي (القرشي)　*Sheikh el Korâshi.* The Sheikh of the Koreish, the tribe to which Mohammed belonged, and the guardians of the Kaábeh or temple at Mecca.

الشيخ المنسي　*Sheikh el Mensi.* The forgotten Sheikh.

الشيخ محمد الحنبلي　*Sheikh Muhammed el Hanbeli.* Sheikh Mohammed of the Hanbeli sect.

الشيخ محمد العلمى　*Sheikh Muhammed el 'Alami.* Sheikh Mohammed el 'Alami. p.n.

القدس   *El Kuds.*

الشيخ محمّد   *Sheikh Muhammed el Mujâhid.* Sheikh
المجاهد   Mohammed the Mujâhid, *i.e.,* one who
             takes part in a 'holy war.' Close by in
             the enclosure of the Heidheniyeh are
             also—

الشيخ بروح الدين   *Sheikh Barûh ed Dîn.* Sheikh Barûh ed
                  Dîn. p.n.

الشيخ السلطان   *Sheikh es Sultân Ibrahîm.* The Sheikh
ابرهيم   Sultân Ibrahîm.

الشيخ سلمان   *Sheikh Selmân el Fârsi.* The Sheikh
الفارسي   Selmân the Persian (one of the com-
             panions of the Prophet).

سلوان   *Silwân.* Siloam.

رهبة العدوية   *Rahibat el 'Adawiyeh.* The 'Adawiyeh.
              p.n., 'hermitage.'

رهبة بنت حسن   *Rahibat Bint Hasan.* The hermitage of
              the daughter of Hasan. Also called—

مغارة العدوية   *Mughâret el 'Adawiyeh.* The cave of the
              'Adawiyeh. p.n.; vaults, called the retreat
              of Saint Pelagia.

طنطور فرعون   *Tantûr Fer'ôn.* Pharaoh's peak (Absolom's
              pillar).

الطور وقيل كفر الطور   *Et Tôr,* or *Kefr et Tôr.* The mount, or the
                     village of the mount. *See Jebel et Tor,*
                     p. 316.

وادي دير السنيق   *Wâdy Deir es Sonneik.* The valley of the
                convent of the white plastered house.
                See Jebel es Sonneik, p. 316.

وادي الحمارة   *Wâdy el Hamârah.* The valley of the
              asses.

وادي الجوز   *Wâdy el Jôz.* The valley of the walnut.

وادي قدوم   *Wâdy Kadûm.* The valley of Kadum;
            eastern or foremost.

القدس *El Kuds.*

وادي النار — *Wâdy en Nâr.* The valley of fire; it is also called وادي ستي مريم *Wâdy Sitti Miriam*, the valley of our Lady Mary (from a Moslem tradition); وادي ايوب *Wâdy Eyûb*, from *Bîr Aiyûb*, q.v.; and وادي فرعون *Wâdy Fer'ôn*, valley of Pharaoh, from the *Tantûr Fer'ôn*, near Silwân.

وادي الربابة — *Wâdy er Rabâbeh.* The valley of the lute.

وادي السهل — *Wâdy es Sahel.* The valley of the plain.

وادي السمار — *Wâdy es Samâr.* The brown valley, or the valley of nightly converse.

وادي طبل — *Wâdy Tabal.* The valley of the drum.

وادي ام احمد — *Wâdy Umm Ahmed.* The valley of the mother of Ahmed. p.n. *See Kurm Ahmed.*

وادي ام العنب — *Wâdy Umm el 'Anab.* The valley with the grapes.

وادي ام واديين — *Wâdy Umm Wâdein.* The valley of the mother of two valleys.

وادي الولي — *Wâdy el Wely.* The valley of the saint or saint's tomb. The Kubbeh or 'dome' of Sheikh Bedr stands at the head of the valley.

زحويلة — *Zahweileh.* 'Steep.' *See* "Memoirs," under 'Zoheleh,' Section A.

الزويقة — *Ez Zurreika.* The little town.

كنّ ساسين   Ku   *Kuf Sanâsîn.* The palm (of the hand), *i.e.*, flat place of Sanâsin. *See Khŭrbet Sanâsin*, p. 310 ; but it may be a corruption of سلاسين, literally ' chains,' *i.e.*, streamlets or canals.

قلعة ابي داموس   Ns   *Kŭlât Abu Dâmûs.* The castle of the hunter's lurking place.

قلعة الغولة   Lu   *Kŭlât el Ghûleh.* The ghoul's castle. It is also called قلعة صباح الخير *Kŭlât Subâh el Kheir,* the castle of ' good morning ' ; a salutation to the ghoul. *See* " Memoirs."

قلعة الصوان   Mt   *Kŭlât es Sûwân.* The castle of flint ; so-called from a tradition. *See* " Memoirs."

قلعة الطنورة   Ks   *Kŭlât et Tantûrah.* The castle of the peak.

القليعية   Mu   *El Kŭlâiyeh.* The little castle.

قلونية (قلونيا)   Lt   *Kŭlônieh.* The Latin Colonia.

تلنديا   Ms   *Kulŭndia.* Kulŭndia. p.n.

قرن ابي زيتون   Nu   *Kŭrn Abu Zeitûn.* The peak with olives ; near *Kabr Ghannâm.*

قرن رشيدة   Nu   *Kŭrn Rŭsheideh.* The peak of Rŭsheideh ; *i.e.*, members of the *Rŭshâidîyeh* tribe.

قرون ابي سرحان   Ku   *Kŭrûn Abu Sŭrhân.* The peaks of Abu Sŭrhân. p.n. ; meaning the father of the wolf.

قرون الحجر   Mu   *Kŭrûn el Hajr.* The peaks of the rocks.

قرية العنب   Lt   *Kŭryet el 'Enab.* The town of grapes. It is almost always called simply *el Kuryeh* by the natives.

قرية سعيدة   Lt   *Kŭryet Sâideh.* The town of Sáideh. p.n.

كشتين   Ns   *Kushtîn.* Kushtin. p.n.

قصر علي   Nt   *Kŭsr 'Aly.* The house of 'Aly. Also called *Khŭrbet 'Arkûb es Suffa,* 'the ruin of the winding mountain pass of the row or ridge.'

قصر عويص   Mu   *Kŭsr 'Aweis.* The ruin of difficult ground.

قصر البدوية   Mt   *Kŭsr el Bedawîyeh.* The house of the Bedawin.

2 T

قصر بيطرو Lu *Ḳŭsr Beitru.* The house of Peter (Meshullam); situated between *'Ain el Ḳŭssis* and the *Burak* (a modern house so called from its owner).

قصر البراءميا Mt *Ḳŭsr el Brâmia.* The house of Abramius; architect of the Greek convent.

قصر ابن يمين Mu *Ḳŭsr Ibn Yemîn.* Benjamin's house; a country villa.

قصر جدع Mu *Ḳŭsr Jedâ.* The house of the young man.

قصر الخضر Mt *Ḳŭsr el Khŭdr.* The house (or tower) of el Khŭdr. *See* p. 28.

قصر الشيخ Mt *Ḳŭsr esh Sheikh.* The house of the Sheikh.

القسطل Lt *Ḳŭstŭl. Castellum* or *Castale.*

لالرون Jt *Lâtrôn.* Apparently *Latronis;* being called by the 16th century writers *Castellum Boni Latronis.* In the official list it is called الاترون *el Atrôn;* and by the peasants sometimes رطلون, *Ratlôn. See* "Memoirs."

لفتا Mt *Lifta.* Lifta. p.n.

معصر الخميس Jt *Máaser el Khamîs.* The fifth wine-press.

ماحة Mt *Mâlhah.* The salt-pan. *See* "Memoirs," under 'Manahath' and 'Manocho.'

مار الياس Mu *Mâr Eliâs.* Saint Elias.

مراح الدينار Mu *Marâh ed Dînâr.* The nightly resting-place (sheepfold) of the dinâr (an ancient coin).

الماوية Mu *El Máwieh.* The place of shelter (caves by a road).

مرج التمس Ls *Merj el Kummus.* The immersed meadow; north of *Jebâ.*

مرج الزرور Ks *Merj ez Zârûr.* The meadow of hawthorns; near *Wâdy Mûsa.*

مغاير الرتغبر Nu *Mŭghâîr er Retaghbir.* The caves of er Retaghbir. p.n.

مغارة ابي عيسى Mu *Mŭghâret Abu 'Aîsa.* The cave of Abu 'Aîsa; it is west of *Bir Sufâfa.*

مغارة البطيخ Ns *Mŭghâret el Bâtikh.* The cave of the melon; east of *Jebâ.*

مغارة بئر الحسوتا Ku *Mŭghâret Bîr el Hasûta.* The cave of the well of Hasûta. p.n. *cf.* Heb. הסותה a place of refuge. *See* "Memoirs," under 'Beit Atab.'

مغارة داود Nu *Mughâret Dâûd.* David's cave.

مغارة الحمّام Ms *Mughâret el Hûmmâm.* The cave of the bath; below *Kefr 'Akûb.*

مغارة الجحش Ns *Mughâret el Jahash.* The cave of the ass; south side of *Wâdy en Nimr.*

مغارة الجعي Ns *Mughâret el Jâi.* The cave of el Jâi. p.n.; ('muddy').

مغارة الجرون Mu *Mughâret el Jurûn.* The cave of the troughs or 'threshing floors,' which are near it.

مغارة القريس Ku *Mughâret el Kareis* (properly *Kurrais*). The cave of nettles.

مغارة القتلى Mu *Mughâret el Kutla.* The cave of the slain; near *Khûrbet esh Shaghrâb.*

مغارة المّارية Mt *Mughâret el Mâsâriyeh.* The cave of the money (literally 'small coins').

مغارة شعب الشرقي Mt *Mughâret Shâb esh Sherky.* The cave of the eastern spur; west of *'Anâta.*

مغارة الصوان Nt *Mughâret es Sûwân.* The cave of flints; at *Khûrbet Abu Sûwân.* q.v.

مغارة ام فنون Ms *Mughâret Umm Ifnûn.* The cave with the by-ways.

مغارة ام التوائمين Ku *Mughâret Umm et Tûeimîn.* The cave of the mother of the twins.

المحدّد Mu *El Muhaddad.* The boundary; near Bethlehem, on the south.

مقام الامام علي (ابن ابي طالب) Kt *Mukâm el Imâm 'Aly (Ibn Abu Tâleb).* The shrine of the Imâm-Ali Ibn Abu Tâleb. See p. 216.

مخماس Ns *Mûkhmâs.* The Heb. מִכְמָשׁ Michmash.

مخرش Ns *Mûkhrûsh.* Scratching.

المناطر Ls *El Munâtir.* The watch-towers.

المنطار Ls *El Muntâr.* The watch-tower.

المرقاقة Ku *El Murkâka.* The rolling-pin (north of *Beit 'Atâb*).

المسكوبية Mt *El Muskôbiyeh.* The Muscovite establishment (the Russian building). See p. 317.

النبي دابيال Lu *Neby Daniâl.* The prophet Daniel.

النبي داود Mt *Neby Dâûd.* The prophet David.

النبي ليمون Lt *Neby Leimûn.* The prophet Lemon. It is a sacred tree, but not a lemon ; the name is perhaps a corruption of Lemuel or Eleemon.

النبي سمويل Ms *Neby Samwîl.* The prophet Samuel.

النبي شوع *Neby Shûa.* The prophet Joshua ; in Eshûa. q.v.

النبي طرفيني Ks *Neby Turfîni.* The prophet Turfini. *See Râs Turfîni,* p. 241.

النبي عور *Neby 'Ûr.* The prophet 'Ûr. p.n.; in *Beit 'Ûr.* q.v.

النبي يمين *Neby Yemîn.* The prophet Yemin (Benjamin), at *Beit 'Anân.*

نحلين Lu *Nehhalîn.* Nehhalin. p.n. ; from נַחַל a water-course.

نجم الورور Js *Nejm el Awerwer.* The star of the bee-eaters (*Merops apiaster,* Lin.).

رافات {Jt/Ms} *Râ-fât.* Râfât. p.n.

الرام Ms *Er Râm.* Er Râm ; from the Heb. הָרָמָה Ramah, 'the hill.' In Arabic the word means 'stagnant water.'

رام الله Ms *Râm Allah.* The hill of God.

الرأس Lu *Er Râs.* The hill-top ; also أس ابي عمار, *Râs Abu 'Ammâr,* the hill-top of Abu 'Ammâr. p.n. *See* p. 268.

رأس العبد Ks *Râs el 'Abd.* The hill-top of the slave.

رأس ابي زياد Ns *Râs Abu Ziâd.* The hill-top of Abu Ziâd. p.n. ; south of *Mukhmâs.*

رأس العقوب Nt *Râs el 'Akûb.* The hill-top of the 'Akûb. The word means 'supplanting' or 'succeeding' (*cf. Yakûb,* Jacob) ; it may however be connected with عقبة 'a steep mountain road,' or with عقاب 'an eagle.'

رأس عرقوب الصفا Nt *Râs 'Arkûb es Suffa.* The hill-top of the winding path of the ridge.

رأس العيازيرة Nt *Râs el 'Ayâzireh.* The hill-top of the people of *el 'Azirîyeh.* q.v.

رأس البلوع Mu *Râs el Balûa.* The hill-top of the gulley-hole.

رأس الحوار Mt *Râs el Huwâr.* The hill-top of white marl ; near *Wâdy el Khûlf.*

رأس الكرامة Mt *Rás el Kerámi.* The hill-top of the tree stumps.

رأس خميس Mt *Rás Khamís.* The hill-top of the fifth ; near *Katát Kanzúir.*

رأس الخروبة Nt *Rás el Kharrúbeh.* The hill-top of the locust tree.

رأس خاطر Jt *Rás Khátir.* The hill-top of Khátir. The word amongst other meanings signifies 'one who moves from one place to another.'

رأس المدور Mt *Rás el Madáwar.* The round hill-top ; north of *el 'Aisáwíyeh.*

رأس المعيقلة Nt *Rás el Máeikleh.* The hill-top of the fortresses.

رأس المشارف Mt *Rás el Meshárif.* The hill-top of the high places.

رأس المكبر Mt *Rás el Mukabbir.* The hill-top of the proud man.

رأس النادر Ks *Rás en Nádir.* The hill-top of the projecting part of a mountain.

رأس الشبابون Mu *Rás esh Shababún.* The hill-top of esh Shababún. p n

رأس الشهادة Mt *Rás esh Shehádeh.* The sharp-pointed hill-top.

رأس الشرفة Lu *Rás esh Sherifeh.* The high or imminent hill-top.

رأس شوفا Lt *Rás Shúfa.* The hill-top of the harrow.

رأس السلام Mt *Rás es Sillám.* The hill-top of es Sellám. p.n.; near *Wády Suleim.* See p. 317.

رأس تبالية Mu *Rás Tabálieh.* The hill-top of herbs used for seasoning west of *Kúsr Ibn Yemín.*

رأس الطبيب Mt *Rás et Tabíb.* The hill-top of the physician; east of *Wády 'Aisa.*

رأس الطاحونة Ms *Ras et Táhúneh.* The hill-top of the mill.

رأس الزمبي Nt *Rás ez Zamby.* The hill-top of ez Zamby. p.n.

ركبة حراسة Mu *Rikbet Harâseh.* The summit of the guard ; near *Kabr Heleceh.*

رجم عفانة Lt *Rujm 'Afâneh.* The cairn of rottenness ; it is of a crumbling character.

رجم ابى شويكة Ms *Rujm Abu Shuweikeh.* The cairn with the thistles.

رجم عطية Mt *Rujm 'Atíyeh.* The cairn of 'Atíyeh. p.n.

رجم البَرِيش Lt　*Rujm el Barîsh.* The cairn of the ground covered with variegated herbage.

رجم الدير Lt　*Rujm ed Dîr.* The cairn of the circle.

رجم الدريبة Lu　*Rujm ed Dûrîbeh.* The cairn of the little road.

رجم الحليك يعني حيك بنت سلطان الفنش Js　*Rujm el Heik* (*yâni Heik Bint Sultan el Fenîsh*). The cairn of the spindle ; *i.e.*, the spindle of the daughter of the Sultan of the Fenish. *See* "Memoirs."

رجو التهتير Mt　*Rujm el Kahakîr.* The cairn of the stone heaps.

رجم لوكار Js　*Rujm Lûkâr.* The cairn of Lûkâr. p.n.

رحم الشيخ سليمان Js　*Rujm esh Sheikh Suleimân.* The cairn of Sheikh Suleimân. p.n.

رجم الطارود Lt　*Rujm et Târûd.* The cairn of the projection or prominent peak. *See et Târûd,* p. 245.

سهل جبع Ns　*Sahel Jebâ.* The plain of Geba.

الصلاح Ku　*Es Salâh.* Rectitude. *See Abu Salâh,* p. 278.

سريس وقيل سارس Kt　*Sarîs* or *Sârîs.* p.n. ; perhaps *Sores,* Σάρις. *See* "Memoirs.'

السبع رجوم Mu　*Es Sebâ Rujûm.* The seven cairns.

سلبيط Js　*Selbît.* Selbît. p.n.

صطاف وقيل صاطاف Lt　*Setâf* or *Sâtâf.* Setâf or Sâtâf. p.n.

شعفاط Mt　*Shâfât.* Shâfât. p.n. The village is said by the peasantry to have been named after a king Shâfât (perhaps Jehoshaphat).

شعب صلح Ms　*Shûb Salâh.* The spur of Salâh. *See Abu Salâh,* p. 278.

شعيب الرباح Nu　*Shâîb er Rabâh.* The spur of the kid.

شهاب الدين Ks　*Shehâb ed Dîn.* Shehâb ed Din. p.n.; shooting star of the faith. According to Moslem legend, the shooting stars are firebrands, which the angels hurl at the devils when the latter are found eaves-dropping at the gates of heaven.

الشيخ عبد العزيز Lt　*Sheikh 'Abd el 'Azîz.* Sheikh 'Abd el 'Aziz. The servant of the Glorious One.

الشيخ ابو غوش   Lt   *Sheikh Abu Ghôsh.* Sheikh Abu Ghôsh ; head of a well-known family of robbers, who ravaged the country some thirty years ago.

الشيخ ابو حسن   Kt   *Sheikh Abu Hasan.* Sheikh Abu Hasan. p.n. ; at *Khŭrbet Jendr.*

الشيخ ابو هلال   Ju   *Sheikh Abu Helâl.* Sheikh Abu Helâl, of the *Beni Helâl* tribe.

الشيخ ابو فاطمة   Ju   *Sheikh Abu Fâtmeh.* Sheikh Abu Fâtmeh (father of Fâtimeh).

الشيخ ابو زيد البستاني   Lu   *Sheikh Abu Zeid el Bustâny.* Sheikh Abu Zeid, the gardener. p.n. ; (above the gardens of Bittir).

الشيخ ابو الزيتون   Ls   *Sheikh Abu 'z Zeitûn.* The Sheikh (Sheikh's tomb) with the olive. *See* "Memoirs."

الشيخ احمد ابو ركبة   Ns   *Sheikh Ahmed Abu Rikbeh.* Sheikh Ahmed of the hill-summit.

الشيخ احمد العمرى   Ku   *Sheikh Ahmed el 'Amri.* Sheikh Ahmed el 'Amri (family name).

الشيخ احمد البخارى   Lt   *Sheikh Ahmed el Bokhâry.* Sheikh Ahmed of Bokhâra ; shown half-mile east of *Khŭrbet el Lôz.*

الشيخ احمد الحوباني   Ku   *Sheikh Ahmed el Hûbâni.* Sheikh Ahmed of Khŭrbet Hûbin. q.v.

الشيخ احمد الكركي   Lt   *Sheikh (Ahmed) el Kereky.* Sheikh Ahmed of Kerek.

الشيخ عيسى   Js   *Sheikh 'Aisa.* Sheikh Jesus ; in 'Annâbeh.

الشيخ علي   Jt   *Sheikh 'Aly.* Sheikh 'Aly. p.n.

الشيخ العجمي   Kt   *Sheikh el 'Ajamî.* The Persian Sheikh.

الشيخ عنبر   Nt   *Sheikh 'Amber.* Sheikh 'Amber. p.n. ; (ambergris).

الشيخ عمار   Ns   *Sheikh 'Ammâr.* Sheikh 'Ammâr. p.n. *See* p. 268.

الشيخ عمرو   { Ls / Lt }   *Sheikh 'Amr.* Sheikh 'Amr. p.n.

الشيخ العمري   Ju   *Sheikh el 'Amri.* Sheikh el 'Amri ; family tribal name.

الشيخ البريج   Ks   *Sheikh el Bureij.* The Sheikh of the little tower.

الشيخ غازي   Ku   *Sheikh Ghâzy.* The warrior Sheikh.

الشيخ حسن   { Ks / Ls }   *Sheikh Hasan.* Sheikh Hasan. p.n. ; in *Khŭrbetha Ibn es Sebâ.*

| الشيخ حسين | Ms | *Sheikh Husein.* | Sheikh Husein. p.n. ; at *er Râm.* |

الشيخ حسين Ms *Sheikh Husein.* Sheikh Husein. p.n. ; at *er Râm.*

الشيخ ابرهيم Kt *Sheikh Ibrahîm.* Sheikh Abraham ; in *Beit Fajûs.*

الشيخ اسمعين Ku *Sheikh Ismaîn.* Sheikh Ishmael ; at Jebâ.

الشيخ جراح Mt *Sheikh Jerrâh.* Sheikh Jerrâh. p.n. ; (surgeon).

الشيخ كامر Mt *Sheikh Kâmir.* Sheikh Kâmir. *See* p. 318.

الشيخ خليفة Nu *Sheikh Khalîfeh.* Sheikh Khalîfeh. p.n. ; (successor or Caliph).

الشيخ محمود Lu *Sheikh Mahmûd.* Sheikh Mahmûd. p.n. ; in *Hausân.*

الشيخ منصور Ns *Sheikh Mansûr.* Sheikh Mansûr. p.n. ; (' victor ').

الشيخ معروف Ku *Sheikh Mârûf.* Sheikh Mârûf. p.n. ; (' well-known,' or ' kindness ').

الشيخ معلى Js *Sheikh Muâllâ.* Sheikh Muâllâ. p.n. ; (' lofty ').

الشيخ معنس Ku *Sheikh Muânnis.* Sheikh Muânnis. p.n. *See* p. 218.

الشيخ مفرح Kt *Sheikh Muferrih.* Sheikh Muferrih. p.n. (' exhilarated.')

الشيخ مصطفى Lt *Sheikh Mustafa.* Sheikh Mustafa. p.n. ; the northern *Kubbeh* at Sóba.

الشيخ نذير Jt *Sheikh Nedhîr.* Sheikh Nedhîr. One who makes a vow (Nazarite).

الشيخ نجم Ms *Sheikh Nejm.* Sheikh Nejem. p.n. ; (' star').

الشيخ عبيد {Js Lt} *Sheikh 'Obeid.* Sheikh 'Obeid. p.n. There is a third in *'Ain Kârim.*

الشيخ سعد Nu *Sheikh Sâd.* Sheikh Sâd. p.n. ; (' fortunate ').

الشيخ الصالحي Ju *Sheikh es Sâlihî.* Sheikh es Sâlihî ; *i.e.*, a client of the Salâh ed Din (Saladin's) family.

الشيخ سمط Jt *Sheikh Samat.* Sheikh Samat. The word means a thread on which beads are strung.

الشيخ سلامة Ku *Sheikh Selâmeh.* Sheikh Selâmeh. p.n. ; in *Jebâ.*

الشيخ سلمان {Ks Kt} *Sheikh Selmân.* Sheikh Selmân. p.n. *See* p. 319.

الشيخ شيبان Ms *Sheikh Sheibân.* Sheikh Sheibân. p.n. ; (' hoary')

الشيخ الصحاب   Kt   *Sheikh es Schâb.* The Sheikh of the companions of the Prophet.

الشيخ صباح   Ku   *Sheikh Subâh.* Sheikh Subâh. p.n.; perhaps Hasan es Sûbbâh, founder of the sect of Assassins; better known as *Sheikh el Jebel*, 'the old man of the mountains.' (In *'Ellâr el Bûsl.*)

الشيخ سليمان   Js   *Sheikh Suleimân.* Sheikh Suleimân. p.n.

الشيخ السلطان بدر   Kt   *Sheikh es Sultân Bedr.* Sheikh es Sultân Bedr. p.n.

الشيخ زيد   Jt   *Sheikh Zeid.* Sheikh Zeid. p.n.; (at *Khûrbet Beit Jîz*).

شرافات (شرفات)   Mu   *Sherâfât.* Battlements.

شقفان   Mu   *Shûkfân.* The clefts; (a cliff facing south).

السفلة   Ku   *Es Sifleh.* The low ground.

سلوان   Mt   *Silwân.* Siloam. *See el Kuds.*

السناسل السلاسل   Nu   *Es Sineisil* (for *Silsisil*). The streams. Literally, 'chains.'

صير التبلس   Nt   *Sîr et Tabalâs.* The fold of Tabalâs. p.n.; (a cave and tree).

الست حنية   Ks   *Sitt Hannîyeh.* The Lady Anne (near *'Ain 'Arîk*, which is a Christian village on the south-west).

الست منع   Ks   *Sitt Menâ.* The Lady Menâ. p.n.

الست زهرا   Ks   *Sitt Zahra.* The Lady Zahra. p.n.; (Venus).

صوبا   Lt   *Sôba.* The heap. No doubt from the rocky mound on which it stands.

صفا   Ks   *Sûffa.* In rows. *See* "Memoirs."

السلطان ابرهيم   {Mt Lt}   *Sultân Ibrahîm.* Sultân Ibrahîm; a *kubbeh* in *Shâfât*; and another, the southern one, at Sôba.

سور باهر   Mu   *Sûr Bâhir.* The wall of Bâhir. p.n.; (prominent).

صرعة   Jt   *Sûrâh.* p.n. Heb. צרעה 'Zoreah.'

الصوانة   Mu   *Es Sûwâneh.* The flints (near *Bîr Istîh*).

تل ابى ناذور   Nu   *Tell Abu Nâdhûr.* The mound of Abu Nâdhûr. p.n.; ('father of vows'). It is perhaps an error for ناطور *Natûr*, 'watchman.'

تل العسكر   Ns   *Tell el 'Askar.* The mound of the army.

تل الغول　Mt　*Tell el Fûl.* The mound of the bean.

تليليا　Mt　*Tellîla.* The little mound.

تل القبوس　Nt　*Tell el Kabûs.* The mound of el Kabûs. *cf. Deir Abu Kabûs.* Abu Kabûs was the name of the son of el Mundhir, King of Hira.

تل خربة الرمل　Lt　*Tell Khûrbet er Rûml.* The mound of the ruin of sand.

تل القوقة　Ks　*Tell el Kôkah.* The mound of cock-crowing.

تل مريم　Ns　*Tell Miriam.* Mary's mound.

تل النصبة　Ms　*Tell en Nasbeh.* The mound of the idol. *Nûsb* is the name given to stones which were set up in pagan times as objects of worship.

تل الصوان　Ns　*Tell es Sûwân.* The mound of flints.

تبنا　Ku　*Tibna.* Straw; probably a corruption of Heb. הִּמְנָה *Timnah. See* "Memoirs."

الطيرة　Ls　*Et Tîreh.* The fort.

الطور　Mt　*Et Tôr.* The mount. *See* under *el Kuds.*

ام ليسون وقيل أم الّيسون　Mu　*Umm el Âisûn,* or *Umm Leisûn.* Umm Leisûn. p.n.

ام العمدان　Ku　*Umm el 'Amdân.* The place with the columns.

ام عراق　Nu　*Umm 'Arâk.* The place with the cliff, near *Khûrbet el Haradân.*

ام الذياب　Ju　*Umm edh Dhiâb.* The place of the wolves.

ام الرؤوس　Ku　*Umm er Rûs.* The place with the hill-top (two ruins so-called, each on a hill-top facing one another).

ام السميقات　Js　*Umm es Semmeikât.* The place with the sumach trees.

ام الشيخ　Ls　*Umm esh Sheikh.* The mother of the Sheikh.

ام الصور　Js　*Umm es Sûr.* The place with the rock.

ام الطلع　Nu　*Umm et Talâ.* The place with the ascent.

أرطاس　Mu　*Urtâs.* Urtâs. p.n.

وادي الأبيدة　Mt　*Wâdy el Abbeideh.* The valley of el Abbeideh; perhaps بيدا 'desert.'

وادي ابي علي Mu *Wâdy Abu 'Aly*. The valley of Abu 'Aly. p.n.

وادي ابي حلوة Mu *Wâdy Abu Helweh*. The valley of Abu Helweh. p.n.; (father of sweetness).

وادي ابي هندي Nt *Wâdy Abu Hindi*. The valley of Abu Hindi. p.n.; (father of the Indian).

وادي ابي سفيرة Mt *Wâdy Abu Sefeirah*. The valley of Abu Sufeirah. p.n.; (father of the little journey), near *Katât Kanûir*.

وادي احمد Mu *Wâdy Ahmed*. Ahmed's valley. p.n.

وادي العين Nu *Wâdy el 'Ain*. The valley of the spring; by *'Ain Sûr Bâhir*.

وادي عَيّاد Ms *Wâdy 'Aiyâd*. Valley of feasters.

وادي عين العرب Kt *Wâdy 'Ain el 'Arab*. The valley of the Arab's spring; by *'Ain el 'Arab*.

وادي عين عريك Ls *Wâdy 'Ain 'Arîk*. The valley of 'Ain 'Arik. q.v., p. 278.

وادي عين الرفيع Lt *Wâdy 'Ain er Rafiâ*. The valley of 'Ain er Rafiâ. q.v., p. 282.

وادي عين التيّة Kt *Wâdy 'Ain et Taiyeh*. The valley of the spring of wandering.

وادي عيسى Mt *Wâdy 'Aîsa*. Jesus' valley.

وادي العقل Ks *Wâdy el 'Akil*. The valley of the fortress.

وادي علكة Kt *Wâdy 'Alakah*. The valley of the lentisk.

وادي علي Kt *Wâdy 'Aly*. The valley of 'Aly. cf. *Imâm '.Ily*, p. 323.

وادي عليا Nu *Wâdy 'Alya*. The valley of 'Alya. See *Bîr '.Alyâ*, p. 288.

وادي عامر Ls *Wâdy 'Âmir*. The valley of 'Âmir. p.n. See p. 268. There is another (Ku) near *Beit Ika* and *el Burj*.

وادي عمّار Mt *Wâdy 'Ammâr*. The valley of 'Ammâr. See p. 268.

وادي عناتا Nt *Wâdy 'Anâta*. The valley of 'Anâta. q.v.

وادي العراق Ls *Wâdy el '.Arâk*. The valley of the cliff, near *el Munâtir*.

وادي عراق مازل Nt *Wâdy '.Arâk Mâzil*. The valley of the cliff of Mâzil. p.n.

وادي عراق الوعر Ns *Wâdy '.Arâk el Wâr*. The valley of the cliff of the rugged rocks; south-east of *Mûkhmâs*.

وادي العريس Nu *Wâdy el 'Areis.* The valley of the bridegroom.

وادي العسكر Ms *Wâdy el 'Askar.* The valley of the soldier.

وادي عسيس Ns *Wâdy 'Asîs.* The valley of the night patrol.

وادي العوج Mu *Wâdy el 'Aûwaj.* The crooked valley (close to *Khûrbet Mezmûria*).

وادي العويني Ls *Wâdy el 'Aweiny.* The valley of el 'Aweiny. p.n.; from عون 'help,' or 'flocks' (near *Khûrbet el Lahm*).

وادي العيون Ku *Wâdy el 'Ayûn.* The valley of springs.

وادي باب الشعب Ns *Wâdy Bâb esh Shâb.* The valley of the entrance of the spur (east of *Jebâ*).

وادي البان Nu *Wâdy el Bân.* The valley of the ban tree (*Salix Ægyptiacus*); near *Bîr 'Alya*.

وادي بردية Ns *Wâdy Baradîyeh.* The valley of cool places (an open valley with olives, east of *Mûkhmâs*).

وادي بيت حنينا Mt *Wâdy Beit Hannina.* The valley of Beit Hannina. q.v.

وادي بجاجلي Lu *Wâdy Bejâjli.* The valley of Bejâjli. p.n.

وادي بتير Lu *Wâdy Bittîr.* The valley of Bittîr. q.v.

وادي بزاي Lt *Wâdy Bûzâi.* The valley of the nutmegs.

وادي البريج Ks *Wâdy el Burej.* The valley of the little tower (near *Khûrbet el Bureij*).

وادي الدعبوب Nu *Wâdy ed Dâbûb.* The valley of the beaten track, or of the black ants.

وادي الدالي Lt *Wâdy ed Dâly.* The valley of the vine, or of the bucket drawer; (east of *Kefîreh*).

وادي الدشيش Mu *Wâdy ed Dashîsh.* The valley of wheaten broth.

وادي دويت Ms *Wâdy Daweit.* The valley of Daweit. Perhaps for دوية *Dawîyet* (the name by which the Knights Templars were known).

وادي { دحيشة } { دحيشى } Mu *Wâdy Deheisheh.* The valley of ed Deheisheh. q.v., p. 293. It is also called *Wâdy ed Deheishimy.*

وادي الدير Ms *Wâdy ed Deir.* The valley of the monastery.

وادي ديَر سلام   Ms   *Wâdy Deir Sellâm.* The valley of the convent of Sellâm. See *Râs es Sillâm,* p. 317.

وادي الذياب   Mt   *Wâdy edh Dhîâb.* The valley of the Wolves (a clan).

وادي الضياع   Mu   *Wâdy ed Diâa.* The valley of farms.

وادي الدلبة   Ku   *Wâdy ed Dilbeh.* The valley of 'Ain Dilbeh. q.v., p. 279.

وادي دبار   Ls   *Wâdy Dubbâr.* The valley of Dŭbbâr; perhaps *Dŭbâr,* trenches for irrigation.

وادي الدم   Ms   *Wâdy ed Dum[m].* The valley of blood.

وادي فخد   Mu   *Wâdy Fakhed.* The valley of the thigh; (perhaps فخت *Fakhit,* fissure).

وادي فوكين   Lu   *Wâdy Fûkîn.* The valley of Fûkin. p.n.

وادي الغدير   Lt   *Wâdy el Ghadîr.* The valley of the swamp.

وادي غنيم   Mu   *Wâdy Ghaneim.* The valley of Ghaneim. p.n.; near *Khŭrbet Bedd Fâlûh,* p. 301.

وادي الغراب   Kt   *Wâdy el Ghŭrâb.* The valley of the raven.

وادي الحداد   Ks   *Wâdy el Haddâd.* The valley of the blacksmith.

وادي الحافي   Ms   *Wâdy el Hâfi.* The valley of the barefooted man.

وادي الحمار   Kt   *Wâdy el Hamâr.* The valley of the ass which would apply.

وادي الحمود   Ms   *Wâdy [el] Hammûd.* The valley of [el] Hammûd. p.n.; south-west of *Râfât.*

وادي الحريق   Nt   *Wâdy el Harîk.* The valley of the conflagration.

وادي حلاس   Ku   *Wâdy Helâs.* The valley of verdure.

وادي الحوت   Kt   *Wâdy el Hôt.* The valley of the fish.

وادي الحمام   Ks   *Wâdy el Hŭmmâm.* The valley of the bath.

وادي الحمص   Mu   *Wâdy el Hummŭs.* The valley of the chick-pea.

وادي ابن عيد   Mt   *Wâdy Ibn 'Aîd.* The valley of Ibn 'Aid. p.n.; the upper part of *Wâdy 'Anâta.*

وادي اقبالا   Lt   *Wâdy Ikbâla.* The valley of Ikbâla. See *Khŭrbet Ikbâla,* p. 306.

وادي اميش   Ls   *Wâdy Imeish.* The valley of the maish trees (*Zizyphus Palinrus*).

وادي إسمعين Kt *Wâdy Ismâîn.* The valley of Ishmael.

وادي المجاعر Js *Wâdy el Jââr.* The valley of the hyena, or of dung.

وادي جبع Ns *Wâdy Jebâ.* The valley of Jebâ. q.v., p. 296.

وادي الجمل Nt *Wâdy el Jemel.* The valley of the camel.

وادي جريوت Ls *Wâdy Jeriût.* The valley of Jeriût. *See Khǔrbet Jeriût.* p. 306.

وادي جيج Lt *Wâdy Jîj.* The valley of Jij, *i.e.*, the noise made in calling camels. It is close to *Khǔrbet el Hôsh.* *i.e.*, 'the ruin of the court-yard,' or 'enclosure.'

وادي جليان Ms *Wâdy Jiliân.* The valley of Jiliân. p.n.

وادي الجندي Ku *Wâdy el Jindi.* The valley of the soldier.

وادي جرجوس Mu *Wâdy Jiriûs.* The valley of George. p.n. It leads up to *el Khudr. See* p. 28.

وادي القاعة Mu *Wâdy el Kââh.* The valley of the court or plain.

وادي قاقوسة Mu *Wâdy Kâkûsah.* The valley of Kâkûsa ; perhaps قاقوزة 'a goblet.'

وادي قنية Ju *Wâdy Kanieh.* The valley of reeds or canals.

وادي الكلب Ms *Wâdy el Kelb.* The valley of the dog.

وادي القيقبة Ls *Wâdy el Keikabeh.* The valley of the Keikabeh ; a tree so called.

وادي الترقعة Ku *Wâdy el Kerkâh.* The valley of the echo of falling trees.

وادي الخليل Jt *Wâdy el Khalîl.* The valley of Hebron (el Khâlil = Abraham).

وادي الخناديق Ks *Wâdy el Khaneidik.* The valley of the fosses ; valley immediately south of *Birket Likia.*

وادي الخارجة Jt *Wâdy el Khârjeh.* The outer valley.

وادي الخوخ Mu *Wâdy el Khôkh.* The valley of plums ; near *Khǔrbet Khallet el Beida.*

وادي الخلف Mt *Wâdy el Khǔlf.* The valley of el Khûlf. p.n. *See Khallet Khǔlf,* p. 299.

وادي خشخاش Kt *Wâdy Khushkhush.* The valley of the poppy.

وادي القبلة Ks *Wâdy el Kibleh.* The valley of the south.

وادي القبلية	Ms	*Wâdy el Kiblîyeh.* The southern valley (near *el Jîb*).
وادي القطن	Ns	*Wâdy el Kotn.* The valley of cotton (south-cast of *el Khŭrbeh*).
وادي القطنة	Kt	*Wâdy el Kotneh.* The valley of the cotton tree.
وادي القمح	Ks	*Wâdy el Kumh.* The valley of wheat.
وادي قريقع	Js	*Wâdy Kureika.* The valley of the noise of falling stones.
وادي الكروم	Mt	*Wâdy el Kurûm.* The valley of vineyards (south of *Birket Iksa*).
وادي اللجام	Nt	*Wâdy el Lehhâm.* The valley of the butcher.
وادي لوزة	Mu	*Wâdy Lôzeh.* The valley of the almond tree.
وادي محبوس	Ls	*Wâdy el Mahbûs.* The valley set apart as a religious endowment. It may also mean 'the valley of the prisoner.'
وادي المراح	Ns	*Wâdy el Marâh.* The valley of the night-resting place; *i.e.*, 'fold for cattle' (close to *el Jâhran* on the west).
وادي مارود	Ls	*Wâdy el Mârûd.* The valley of Mârûd. p.n.; ('rebellious').
وادي المعروج	Ks	*Wâdy el Mârûj.* The valley of the ascent (a path goes up it).
وادي المسرجة	Ks	*Wâdy el Masrajeh.* The valley of the lamp.
وادي المزار	Mt	*Wâdy el Mazâr.* The valley of the shrine; from the *Mazâr* at *Khŭrbet el Burj*, at the head of the valley.
وادي المدينة	Ns	*Wâdy el Medîneh.* The valley of the city.
وادي الملاعب	Ks	*Wâdy el Melââb.* The spring of the play-house.
وادي المقتلي	Ks	*Wâdy el Miktely.* The valley of the place of slaughter.
وادي المريج	Ls	*Wâdy el Mureij.* The valley of the little meadow (south-west of *Sheikh Abu Zeitûn*).
وادي موسى	Ks	*Wâdy Mûsa.* The valley of Moses (north of *Heitân es Sinôbar*).
وادي المحر	Lu	*Wâdy el Musŭrr.* The valley of the place of pebbles.
وادي المطلق	Jt	*Wâdy el Mŭtluk.* The valley of the free range.

وادي النحلة	Mu	*Wâdy en Nahleh.* The valley of the bee; but *cf. Wâdy en Nahl,* p. 119.
وادي النهير	Ju	*Wâdy en Nahîr.* The valley of the river.
وادي النار	Nu	*Wâdy en Nâr.* The valley of fire.
وادي النجيل	Ku	*Wâdy en Najîl.* The valley of the Najil (a plant like sorrel).
وادي الندا	Js	*Wâdy en Neda.* The valley of dew.
وادي النمل	Ls	*Wâdy en Neml.* The valley of the ants.
وادي النتيف	Ns	*Wâdy en Netîf.* The valley of dripping.
وادي النمر	Ns	*Wâdy en Nimr.* The valley of abundant water. It is near the great spring *'Ain Fârâh. See Wâdy en Nimûr,* p. 37.
وادي النخيلة	Nt	*Wâdy en Nŭkheileh.* The valley of the little palm-tree.
وادي النسر	Ku	*Wâdy en Nŭsr.* The valley of the eagle.
وادي النسورة	Kt	*Wâdy en Nŭsûrah.* The valley of eagles.
وادي الرعيان	Mu	*Wâdy er Râîân.* The valley of the shepherds; near *Beit Sâhûr,* the traditional site of the shepherds' plain, near *Keníset er Râwât,* p. 298.
وادي الرديدة	Ns	*Wâdy er Redeidey.* The valley of the turnips.
وادي روابي	Nt	*Wâdy Rûâbeh.* The valley of hillocks.
وادي الرمانة	Ku	*Wâdy er Rummâneh.* The valley of the pomegranate.
وادي صهيون	Mt	*Wâdy Sahyûn.* The valley of Sion.
وادي سعيد	Ku	*Wâdy Sâîd.* The valley of Sâîd. p.n.; '(Felix').
وادي صالح	Mu	*Wâdy Sâleh.* The valley of Sâleh. p.n.; near *Ķŭsr Jedâ,* p. 322.
وادي السمورة	Mu	*Wâdy es Samûrah.* The valley of the acacias, or the brown valley.
وادي سلمان	Ks	*Wâdy Selmân.* The valley of Selmân. p.n.; probably Selmân el Farsi, the companion of the Prophet.
وادي السنام	Nt	*Wâdy es Senâm.* The valley of the camel's hump.
وادي الشبابون	Mu	*Wâdy esh Shababûn.* The valley of esh Shababûn. *See Râs esh Shababûn,* p. 325.

وادي شعيب   Mu   *Wády Shaíb.* The valley of the little spur; south of *Khŭrbet el Bedd.*

وادي الشامي   Lu   *Wády esh Shámy.* The Syrian valley; near *Khŭrbet Fŭghŭr.*

وادي الشباب   Ls   *Wády esh Shebáb.* The valley of youth.

وادي الشمارين   Lt   *Wády esh Shemmárín.* The valley of esh Shemmárin. p.n.; either from *Shammár,* one who girds up his loins; or *Shamár,* 'fennel.'

وادي الشقاقي   {Ms}{Ns}   *Wády esh Shukáki.* The valley of the clefts.

وادي السكة   {Lu}{Nt}   *Wády es Sikkeh.* The valley of the path; near *Sheikh 'Amber.*

وادي السمار   Mt   *Wády es Somár.* The brown valley.

وادي سليم   {Nt}{Ju}   *Wády Suleim.* The valley of Seleim. p.n.; also the name of a tree (the *Mimosa flava*).

وادي السنط   Ls   *Wády es Sŭnt.* The valley of the acacia.

وادي صورة   Nu   *Wády Sûrah.* The valley of the picture, or of the rock.

وادي صرار   {Jt}{Lt}   *Wády Sŭrár.* The valley of pebbles.

وادي سويكة   Js   *Wády Suweikeh.* The valley of Suweikeh. p.n.; perhaps سويقة 'a little market.'

وادي السوينيطا   Ns   *Wády Suweinít.* The valley of the acacias. *See* "Memoirs," under 'Valley of Thorns.'

وادي تبلاس   Nt   *Wády Tabalás.* The valley of Tabalás. *See Sŭr el Tabalás,* p. 329.

وادي الطاحونة   Mu   *Wády et Tŭhŭneh.* The valley of the mill.

وادي الطاقة   Ls   *Wády et Tákah.* The valley of the arch or shelf.

وادي التين   {Mu}{Ku}   *Wády et Tin.* The valley of figs; there is a second near *Sheikh Muánnis.*

وادي ام العصافير   Mu   *Wády Umm el 'Asáfir.* The valley with the sparrows.

وادي ام الخارزة   Nu   *Wády Umm el Khárzeh.* The valley with the Kharazeh (a plant like sorrel). The word also means 'a bead.'

وادي ام القلعة   Mu   *Wády Umm el Kŭláh.* The valley with the castle.

وادي ام الثليثات   Nu   *Wády Umm eth Theleithát.* The valley of tamarisks.

وادي أرطاس   Mu   *Wâdy Ûrtâs.* The valley of Ûrtâs. q.v.

وادي الوعر   Nu   *Wâdy el Wâr.* The valley of the rugged rocks.

وادي الورد   {Ku} {Mu}   *Wâdy el Werd.* The valley of roses, or of going down to water. *See 'Ain Yerdeh*, p. 265. A second of the name occurs further west. The valley near Jerusalem is culti vated with roses.

وادي ياسول   Mt   *Wâdy Yâsûl.* The valley of Yâsûl. p.n.

وادي زكريآ   Ju   *Wâdy Zakariya.* The valley of Zakariya; named from the village.

وادي زعيط   Ls   *Wâdy Zâît.* The valley of balsam; vulgar Arabic for سعيط The word also means 'snuff.'

وادي زاوية   Nt   *Wâdy Zâwich.* The valley of the corner or hermitage; north of *Abu Dîs.*

وادي زريق   Ns   *Wâdy Zereik.* The grey valley; south of *Jeba*, north of *Kabûr Beni Isrâîm.*

وادي زعرب   Mt   *Wâdy Zimry.* The valley of the piper.

وعر ابي غازي   Mu   *Wâr Abu Ghâzy.* The rugged rocks of the father of the warrior; south of *Bîr Istêh.*

وعرة الحمير   Jt   *Wâret el Hamîr.* The rugged rocks of the asses; west of *Khûrbet el Fûl.*

وعرة الخروبة   Mu   *Wâret el Kharrûbeh.* The rugged rocks of the locust tree; close to *Râs el Balûa* on the east.

الوردانة   Ns   *El Werdâneh.* The watering place; east of *Mukhmâs.*

الولجة   Lu   *El Welejeh.* The bosom of the hill; so called from its position.

يالو   Ks   *Yâlô.* Yâlô. p n.; Heb. אילון (place of the fallow deer), Ajalon.

زكريآ   Ju   *Zakariya.* Zacharias. *See* "Memoirs."

زيتونة النبى   Mt   *Zeitûnet en Neby.* The Prophet's olive; on the *Râs el Mukabbir.* *See* the tradition given in the "Memoirs."

SHEET XVIII.

اجار العز   Ou   *Ahjâr el 'Azz.* The hard rocks.

عين عروس   Qs   *'Ain 'Arûs.* The spring of the bride.

عين الدوك   Os   *'Ain ed Dûk.* The spring of ed Dûk (Δωκ). See "Memoirs," under ' Docus.'

عين الفشخة   Pu   *'Ain el Feshkhah.* The spring of el Feshkah. p.n.

عين حجة   Qt   *'Ain Hajlah.* The spring of the partridge.

عين لجحكير   Pt   *'Ain el Jcheiyir.* The spring of the little den.

عين القلت   Os   *'Ain el Kelt.* The spring of the mountain hollow or tarn.

عين النويعمة   Os   *'Ain en Nûeiâmeh.* The spring of the soft soil. It is also called نوايع (plural of نبع), with the same meaning).

عين التنور   Pu   *'Ain et Tannûr.* Spring of the oven or reservoir.

عين وادي فارة   Nt   *'Ain Wâdy Fârah.* The spring of the valley of the mouse.

عقبة الحطابة   Nu   *'Akabet el Hattâbeh.* The steep or mountain road of firewood carriers.

عقبة المعارجة   Nu   *'Akabet el M'ârajeh.* The steep or mountain road of steps.

عقبة الشريف   Ps   *'Akabet esh Sherif.* The steep or mountain road of the noble or eminent one.

عقوبة الاكراد   Ot   *'Akûbet el Akrâd.* The ascent of the Kurds.

العليليات   Os   *'El 'Aleiliât.* The upper chambers (a cave with three entrances).

عرب العبيدية   {Nu / Ou}   *'Arab el 'Abbeidiyeh.* The Abbeidiyeh (serf) Arabs (the servants of the Mar Saba monastery).

عرب العبيد   {Os / Ps}   *'Arab el 'Abîd.* The 'Abîd (slave) Arabs.

عرب ابى نصير   Os   *'Arab (el) Abu Nûseir.* The Abu Nûseir Arabs (*nuseir* means a 'little eagle').

2 X 2

عرب احتيمات {Nt / Ot} 'Arab el Heteimât. The Heteimât Arabs. p.n.

عرب صواحرة الوادي {Ot / Nu} 'Arab Sâwaharet el Wâdy. The Arabs of the deserts of the valley.

عراق ابى القرع   Nt   'Arâk Abu 'l Karâ. The cliff with the bald patch.

عراق الدير   Nt   'Arâk ed Deir. The cliff, cavern, or buttress of the monastery.

عراق ابرهيم   Nt   'Arâk Ibrahîm. The cliff, cavern, or buttress of Abraham.

عراق سعيد   Ns   'Arâk Sâîd. The cliff, cavern, or buttress of Sâîd. p.n.

ارض المقام   Ps   Ard el Mûkâm. The land of the station (shrine); applied to the neighbourhood of Mûkâm el Imâm 'Aly. See p. 323.

بحر لوط   Bahr Lût. The sea of Lot (usually called بحيرة لوط Baheirat Lût, Lot's lake). The Dead Sea.

بيت جبر الفوقاني   Os   Beit Jûbr el Fôkâni. The upper Beit Jûbr.

بيت جبر التحتاني   Js   Beit Jûbr et Tahtâni. The lower Beit Jûbr. The house of Jûbr. p.n.; meaning to 'set a broken bone'; but it may be a corruption of Κυπρος. See "Memoirs," under 'Cyprus.'

بلم الضيف   Ot   Belâm ed Deif. The belâm of the guest. Belâm is the name of a plant. The name applies to a hill-side covered with bellâm (Poterium spinosum).

بلاوة الذهيبان   Pt   Belâwet edh Dheheibân. The misfortune or trial of the passers by. The point commonly visited by travellers.

بئار المعزية   Nu   Biâr el Mâziyeh. The wells of the Mâziyeh (goat herd) Arabs.

بئر ابى هندي   Nt   Bîr Abu Hindi. The well of the father of the Indian.

بئر ابى كلاب   Nu   Bîr Abu Kelâb. The well of Abu Kelâb (the father of dogs). An Arab tribal name.

بئر ابى شعلة   Ou   Bîr Abu Shuâleh. The well with the cavern; Heb. שׁעל. The word is used in this sense throughout the country, and this well is close to a great cavern. See "Memoirs."

بئر العضوا   Nu   Bîr el 'Adwa. The well of the thorny tree.

بئر العمارة   Ou   Bîr el 'Ammâra. The well of the edifice.

بئر الاطرش   Os   Bîr el Âtrash. The well of the deaf man.

بئر البقوق Nu *Bir el Bakkûk.* The well of bugs, or sand-flies.

بئر الضيقة Nu *Bir ed Dîkah.* The narrow well.

بئر الفخت Ot *Bir el Fukhteh.* The well of the crack (in the rock). It is called also بئر ابى العون *Bir Abu el 'Aûn*, the 'well of the father of the flock.'

بئر الغزالة Nu *Bir el Ghůzâleh.* The well of the gazelle.

بئر حيدر Ps *Bir Heider.* The well of Heider (a name of Imâm Aly).

بئر ابرهيم Nu *Bir Ibrahîm.* Ibrahim's well.

بئر ابتانى Nu *Bir Ibtâni.* The well of Ibtâni. p.n.

بئر الجميع Nt *Bir el Jemâi.* The well of the gathering.

بئر جوفة العلندة Nu *Bir Jôfet el 'Alandah.* The well of the hollow of the thorn tree.

بئر القطار Ou *Bir el Katâr.* The well of drops, or of the train or string (of camels) ; it is supplied by an aqueduct.

بئر القطاية Nu *Bir el Kattiyeh.* The hewn well.

بئر الخليل الجديد Nt *Bir el Khaleil (el Judîd).* The new well of the gap.

بئر الخاربة Ou *Bir el Khârbeh.* The ruined well.

بئر القريمزان Nu *Bir el Kureimzân.* The well of the crimson places ; probably from the colour of the marl.

بئر القرن Ot *Bir el Kůrn.* The well of the peak.

بئر القطفى Nu *Bir el Kůtâfi.* The well of the St. John's wort.

بئر المخروم Nt *Bir el Makhrûm.* The well of the mountain peak.

بئر الملاق Ot *Bir el Malâki.* The well of the *Meleki* stone (املس a smooth stone).

بئر معلى Nu *Bir Muálla.* The well of the watering place.

بئر مخيبة Nu *Bir Mukheibeh.* The hidden well; in the corner of a valley.

بئر الراوضة Ot *Bir er Râwâwă.* The well of the Râwâwâ (name of an ancient Arab tribe).

بئر السنافى Nu *Bir es Safâfi.* The well of the camel cropping dry herbage.

بئر الصلاح Nu *Bîr es Saláh.* The well of es Saláh. *See Abu Saláh,* p. 278.

بئر السوق Nu *Bîr es Sûk.* The well of the market. *See* " Memoirs," under ' Zuk.'

بئر السكيرية Nu *Bîr es Sukeiríyeh.* The well of the dam.

بئر السوسة Ot *Bîr es Sûseh.* The well of the liquorice plant.

بئر الثمائل Nu *Bîr eth Themâil.* The well of the shallow water-pits.

بئر أمهات العقارب Nu *Bîr Umât el 'Akârib.* The well of the mothers of scorpions.

بئر ام الفوس Ou *Bîr Umm el Fûs.* The well with the pickaxes. (Hewn in soft rock; a square reservoir.)

بئر ام الثليثات Nu *Bîr Umm eth Theleithât.* The valley of the place with the tamarisks.

بئر ام الزلط Ou *Bîr Umm ez Zallat.* The well with the pebbles.

بئر الويبدة Ot *Bîr el Weibedeh.* The well of the mountain pass.

بئر الزرع Ot *Bîr ez Zera.* The well of sown corn.

بركة حرب Os *Birket Harb.* The pool of Harb. p.n.; meaning ' battle,' or ' a spear.'

بركة جلجولية Ps *Birket Jiljûlieh.* The pool of Gilgal. *See* " Memoirs."

بركة موسى Ps *Birket Mûsa.* The pool of Moses. *See* " Memoirs."

بيارة ام طير Ot *Biyâret Umm Teir.* The wells with the birds.

البقيع Ou *El Bukeiâ.* The little plateau.

برج الحمار Nu *Burj el Hammâr.* The tower of the asses.

بويبات الهوا Ot *Buweibât el Hawa.* The gates of the wind.

دبوس العبد Ot *Dabbûs el 'Abd.* The club of the slave.

دار مقتل الغوير Nu *Dâr Maktal el Ghuweir.* The house (region) of the place of slaughter in the hollow (from the hermits' caves in the rocks).

دار ام قلعة Ou *Dâr Umm Kŭlâh.* The house (region) with the castle (a hill-top without buildings).

دير القلت Os *Deir el Kelt.* The monastery of el Kelt. *See 'Ain el Kelt,* p. 339.

دير مكلك Ot *Deir Mukelik.* The monastery of Mukelik. p.n.

درب الرواجيب Pt *Derb er Rilájib.* The road of the Rejeb pilgrims (used by the pilgrims visiting *Neby Músa*).

ظهرة ذنيب عير Pt *Dhahret Dheneib 'Aiyir.* The ridge of the tail of the wild ass; a precipice down which a torrent flows in winter.

ظهرة الهوا Ot *Dhahret el Hawa.* The ridge of the wind.

ظهرة ابن رشيد Ot *Dhahret Ibn Rushíd.* The ridge of the son of Rushid. p.n.; probably from the Rusheidiyeh Arabs.

ظهرة الثنية Ps *Dhahret eth Theniyeh.* The ridge of the mountain road. *See* also " Memoirs."

الذويات Ps *Edh Dhaweiát.* The withered herbage.

دبيبة خوث Nu *Dubeibet Khóth.* The little soft patch; applied to ground near a well.

اريحا Ps *Ertha.* Eríhâ; the Heb. יְרִיחוֹ Jericho.

الغور *El Ghôr.* The hollow (lowland or depression). It applies also to the higher plain or upper level of the Jordan valley, in which is *ez Zôr.*

الغورانية Qs *El Ghôrániyeh.* The place of the Arabs of the *Ghor.*

الحديدون Ot *El Hadeidûn.* El Hadeidûn. p.n.; from حدّ 'border.' ' limit.'

حجر الأصبح Pt *Hajr el Asbáh* or *Subáh.* The reddish-white stone; but *see* " Memoirs."

حجر المنبتة Pt *Hajr el Minbeteh.* The stone of the vegetation.

حتو الضراحي Pu *Hakku ed Dráhy.* The rugged valley side of the kicking horse, or of the assailant.

حتو مطامير Pu *Hakku Metámír.* The rugged valley side of underground granaries (singular مطمور ' Metamore ').

حسن الراعي Pt *Hasan er Ráái.* Hasan the shepherd.

الحنو Qt *El Henû.* The bend.

حنو ابي احميد Ps *Henû Abu 'l Hameid.* The bend of Abu 'l Hamid. p.n.

الحرمة Nu *El Hirmeh.* The sanctuary, or forbidden ground.

الحصن Ns *El Hosn.* The fortress; a rocky precipice.

حديب الزوار   Ot   *Hudeib ez Zûwâr.* The travellers' hillock.

حربه الحطابة   Nu   *Hurubbet el Hattâbeh.* The cistern of the firewood carriers.

جبل الكتيف   Ot   *Jebel Ekteif.* The mountain of the shoulder.

جبل جوفة شرف   Ot   *Jebel Jôfet Sherif.* The mountain of the hollow of the noble one, or of the eminent place.

جبل الكهمو   Ot   *Jebel ed Kahmûm* or *Kahmûn.* The mountain of the
جبل الكهمون   unripe grape, or of 'the eye' كهم. Its other name is
                       الحثرورة *el Hathrûrah,* which is connected with حثر 'a grape-stone just opening,' or with حثرة a round piece of earth. The title is perhaps connected with the tradition of vineyards in the *Bukeia.*

جبل قرنطل   Ps   *Jebel Kŭrŭntŭl.* For *quarantana,* 'forty days'; from the tradition of the Temptation. *See* "Memoirs."

الجحيير   Pt   *El Jeheiyir.* The little den. *cf.* 'Ain el Jeheiyir, p. 339.

جلجل الغوراني   Nu   *Jiljil el Ghôrâni.* The circle of the Arabs of the Ghor or lowland. The word *Jiljil* جلجل is interesting, but has no connection with *Gilgal.*

جسر ابى غبوس   Ps   *Jisr Abu Ghabbûsh.* The bridge of Abu Ghabbûsh. p.n.

جسر الدير   Os   *Jisr ed Deir.* The bridge of the monastery.

جوفة الاصلع   Pt   *Jôfet el Aslà.* The hollow of the bald (man).

جوفة حميمة   Ou   *Jôfet Humeimeh.* The hollow of the little bath.

جوفة المغيرة   Ou   *Jôfet el Mugheiyirah.* The changed hollow; or perhaps 'of the caves.'

جوفة ام سيبة   Ou   *Jôfet Umm Seibeh.* The hollow with the flowing water.

جوفة زبن   Pt   *Jôfet Zeben.* The hollow of Zeben. p.n.; probably from رابية a hill in a valley.

قبر فرس خليل   Nu   *Kabr Faras Khŭlil.* Grave of Khŭlil's mare. He was the ancestor of the Tââmireh Arabs.

قبور الدواري   Nu   *Kabûr ed Dawâri.* The graves of the people of Dawâr. *See* "Memoirs."

قبور المدادي   Ot   *Kabûr el Madâdi.* The tombs of the Medâdi. p.n.; meaning 'ink,' 'extension,' 'a road,' or 'sect.' *See* "Memoirs."

فيقور العورانيّة — Os — *Kakakūr el Ghôrânîyeh.* The stone heaps of the Arabs of the Ghor. q.v.

قناة موسى — Os — *Kanât Mûsa.* The aqueduct of Moses. *See* "Memoirs."

قطاع موسى — Ns — *Katât Mûsa.* The crags of Moses. *See* "Memoirs."

خليل ابو رضف — Nt — *Khaleil Abu Radf.* The dells with the hot stones. Heb. רֶצֶף, applied to a long stony valley.

خلّة العميرة — Ns — *Khallet el 'Abharah.* The dell of the mock orange (*Styrax officinalis*).

خان الاحمار وقيل خان السهل — Nt — *Khân el Ahmâr.* The Caravanserai, or the red rocks; also called *Khân es Sahel,* the Caravanserai of the plain, from the plain in which it stands, and the red rocks.

خان حثرورة — Ot — *Khân Hathrûrah.* The Caravanserai of Hathrûrah. *See Jebel el Kahmûm,* p. 344.

خور ابو ذاحي — Ps — *Khaur Abu Dhâhy.* The trench of the barren land.

خور البلاوي — Ot — *Khaur el Belâwi.* The trench of the Belâwi; named after an Arab tribe.

خور التنف — Qs — *Khaur el Kutûf.* The trench of the St. John's wort.

خور التمرار — Ps — *Khaur et Tumrâr.* The hollow of the passage; being the place where *Wâdy Kelt* is usually crossed.

خور الطرفة — Qt — *Khaur et Turfah.* The hollow of the tamarisk-tree.

خربة ابي لحم — Ps — *Khŭrbet Abu Lahm.* The ruin of the father of meat; perhaps of the boxing-match, لحمة, from the tradition of a battle fought here.

خربة فارة — Ns — *Khŭrbet Fârah.* Ruin of the mouse. *See* "Memoirs," under 'Parah' (but perhaps Heb. פָּרָה 'the heifer'). It is also called *Khŭrbet et Tell,* 'the ruin of the mound.'

خربة جنجس — Nu — *Khŭrbet Jinjîs.* The ruin of Jinjis. p.n.

خربة قاقون — Ps — *Khŭrbet Kâkûn.* The ruin of Kâkûn; perhaps قاقم 'ermine.'

خربة قمران — Pu — *Khŭrbet Kumrân.* The ruin of the greyish spot (being situate on white marl).

خربة كرم عتراد — Ou — *Khŭrbet Kurm 'Atrâd.* The ruin of the vineyard of 'Atrâd. p.n.

2 Y

خربة المجبر Ps *Khŭrbet el Mejjir.* The ruin of the place to which water is brought (from the aqueduct leading to it).

خربة المشرب Ot *Khŭrbet el Meshrâb.* The ruin of the long strip.

خربة مرد Ou *Khŭrbet Mird.* The ruin of Mird. p.n.; meaning 'insolence,' 'contumacy.' Heb. מרד 'lofty.' The name Nimrod comes from this root, and is traditionally connected with the ruin. *See* "Memoirs," under ' Mons Mardes.'

خربة السمرة Ps *Khŭrbet es Sŭmrah.* The brown ruin. *See* "Memoirs," under ' Zemaraim.'

خربة السمرة Ou *Khŭrbet es Sŭmrah.* The brown ruin.

خربة الزرانيق Ou *Khŭrbet ez Zerânîk.* The ruin of the rivulets.

قبة رمانة Ns *Kubbet Rummâmaneh.* The dome of the Rŭmmôns; it is the plural of رمون .

التويسرة Pt *El Kŭeiserah.* The hampers or baskets. The name is applied to a pool in the rocky bed of the torrent.

كرم ابي طبق Ou *Kurm Abu Tŭbk.* The vineyard with the terrace.

كرم العجز Ou *Kurm el 'Ajaz.* The vineyard of helplessness.

القرن Nu *El Kŭrn.* The horn or peak.

قصر حجلة Pt *Kŭsr Hajlah.* The tower of the partridge (Hoglah). *See* '*Ain Hajlah,* p. 339.

قصر اليهود Qs *Kŭsr el Yehûd.* The tower of the Jews.

معقر الجاموس Os *Mââkkir el Jâmûs.* The place of houghing the buffalo. A buffalo is said there to have broken its leg.

مباع عبيد التتن Ou *Mabââ 'Obeid et Titn.* The market of the slave of tobacco. The Arabs cultivate a little tobacco here.

مخاضة حجلة Qt *Makhâdet Hajlah.* The food of the partridge (Hoglah). *See* '*Ain Hajlah,* p. 339.

المخروق Os *El Makhrûk.* The fissure in the rocks of *Wâdy Kelt.*

مكسر الحصان Ot *Maksar el Hisân.* The place of killing of the horse.

مكسر اسمعين Nu *Maksar Ismâin.* The place of killing Ismail; an Arab executed by the Government under Ibrahim Pacha.

المندسي Os *El Mandesi.* p.n. *See Basset el Mandesî,* p. 160.

مار سابا وقيل دير   Nu   *Mâr Sâba.* Saint Sabas; also called *Deir ez Sîk*, the
السيق   monastery of the ravine.

مَرَدّ حاني (ابي   Pt   *Maradd Hâni (Abu Seleb).* The curved winding (the
سلَب)   father of spoil or robbery). A winding ascent of the road.

مسبع العيرنة   Pu   *Masebâ el 'Airneh.* The remote covert.

مسبع جورة   Pu   *Masebâ Jûrat el Beid.* The covert of the hollow of
البيض   eggs.

مسطح البرودة   Pt   *Mastah el Barûdeh.* The plain of the cool place (a favourite
   camping place of the Arabs).

المدش   Qt   *El Medesh.* The travelling ground; applied to the delta at
   Jordan mouth.

ميدان العبد   Ps   *Meidân el 'Abd.* The plain or exercise ground of the
   slave. *See* " Memoirs."

مثور السدري   Pt   *Mekûr es Sidri.* The holes of the lotus; a hollow between
   rocks with water and a lotus tree (*Zizyphus lotus*).

مَلَّاحة رشيدِيَّة   Qt   *Mellâhet Rusheidîyeh.* The salt pond of the Rusheideh
   (Arabs).

مَلَّاحة امّ احداب   Qt   *Mellâhet Umm Ehdeib.* The salt pond with the slender
   branches. There is much drift wood near, of which a
   little tent has been made.

المجثع   Os   *El Mijthâ.* El Mijthâ. p.n.

المستى   Nt   *El Miska.* The waterworks. It is also called مسكى
   Muskâ (the tank).

مَعَرس خطمة   Pt   *Muârras Khatmeh.* The evening resting place of the
   summit; high ground used for camping on by the Arabs.

مغاير ابي زلَف   Nu   *Mûghâir Abu Zelef.* The caves with the cisterns.

مغاير المَّاري   Ou   *Mûghâir el Makâri.* The caves of the hillocks, or of the
   pots, or of places to which water flows.

مغاير المَّانسية   Nu   *Mûghâir el Mûânsîyeh.* The caves of the familiar ones.

مغاير الزبد   Nu   *Mûghâir er Rûbûd.* The caves of ashes; perhaps ربض
   'to lie down' (cattle).

مغاير الشمس   Nt   *Mûghâir esh Shemls.* The caves of sunlight.

مغارة ابي طبق   Ou   *Mûghâret Abu Tûbk.* The cave of the terraced rock.

مغارة اكتاف   Pt   *Mŭghâret Ektcif.* The caves of the shoulders.

مغارة الدياسة   Pu   *Mŭghâret ed Diâsch.* The cave of the people of Abu Dis, or of those who tread out corn.

مغارة الجثة   Os   *Mŭghâret el Jeththch.* The cave of the corpse ; from a corpse found there.

مغارة المشرعب   Os   *Mŭghâret el Meshrâb.* The cave of the long strip. *See* p. 346.

مغارة الشمس   Os   *Mŭghâret esh Shems.* The cave of the sun ; so called because facing to sunrise.

مغارة الصقيع   Os   *Mŭghâret es Sikyâ.* The cave of the hoar-frost.

مغارة ام الحمام   Ot   *Mŭghâret Umm el Hŭmmâm.* The cave with the bath.

مغارة ام الحسن   Ns   *Mŭghâret Umm el Hasn.* The cave of the mother of Hasan. p.n.

مغر السليطة   Os   *Mŭghr es Salîtah.* The cave of sesame oil. It is also the name of an Arab tribe.

مغر الزيت   Os   *Mŭghr ez Zeit.* The cave of olive oil.

مجن ابى راس   Ou   *Mujn Abu Râs.* The hard place of Abu Râs. p.n. (father of a head or hill-top) ; a cave on the hill-side.

مقام الامام على بن ابى طالب   Ps   *Mukâm el Imâm 'Aly (Ibn Abu Tâleb).* The shrine of the Imâm 'Aly ibn Abu Tâlib. *See* " Memoirs." *See* also under ' Khŭrbet Abu Lahm,' in " Memoirs," a note on the chapel of the apparition of Michael.

المخرص   Ot   *El Mukharras.* Probably مخرص the reservoir or canal.

المنطار   Ou   *El Mŭntâr.* The watch-tower.

المسيلقة   Ot   *El Museilkah.* The scaling ladders.

مؤذن البلال   Ps   *Mŭwedhdhen el Belâl.* The place where *Belâl* called to prayer. Belâl was Mohammed's favourite muezzin or crier. *See* " Memoirs."

النبي موسى   Pt   *Neby Mûsa.* The prophet Moses. *See* " Memoirs."

نقب ابو صراج   Ps   *Nŭkb Abu Serâj.* The passage with the cement (the name of the aqueduct along *Kŭrŭntŭl*).

نقب الاسفار   Os   *Nŭkb el Asfâr.* The pass of journeys, or of books.

نقب فشخة   Pu   *Nŭkb Feshkhah.* The pass of Feshkhah. p.n.; *see Râs Feshkhah,* p. 349.

نقب تمران	Pu	*Nŭkb Kumrân.* The pass of Kumrân. p.n. *See Khŭrbet Kumrân,* p. 345.
نقب الربعي	Pu	*Nŭkb er Rubâi.* The pass of spring herbage.
نقيب الخيل	Pu	*Nŭkeib el Kheil.* The little pass of horses.
نصب الميدان	Ps	*Nŭsb el Meidân.* The erected stone, or ' high place ' of the open place or exercise ground.
نصب المقاصد	Ou	*Nusb el Mukeisdeh.* The high place of objects or aims (a hill with caves).
نصب ام سيبة	Ou	*Nusb Umm Seibeh.* The high place of Umm Seibeh. *See Jôfet Umm Seibeh,* p. 344.
نصيب عويشيرة	Ps	*Nuseib 'Aweishireh.* The little high place or erected stone of the clans. *See* " Memoirs," under ' Ebal.'
نصيب النوارة	Ot	*Nuseib en Nuwârah.* The little high place of the flower.
النصرانية	Ot	*En Nusrânîyeh.* The Christians (a pool).
عش الغراب	Ps	*'Osh el Ghurâb.* The raven's nest.
راس فشخة	Pu	*Râs Feshkhah.* The headland of Feshkhah. p.n. ; *cf.* the Aramaic פשח to break to pieces. The foot of the cliff is covered with a mass of broken fragments of rock fallen from above.
راس مواقيف	Pu	*Râs Mŭâkif.* The headland of those who are stopped ; or of the religious endowments (mosque property).
راس الطويل	Ns	*Râs et Tawîl.* The long hill-top.
راس ام ديسيس	Nt	*Râs Umm Deisîs.* The hill-top with the holes to creep into (numerous caves in the side).
رجم البحر	Qt	*Rujm el Bahr.* The cairn of the sea.
رجم حليسة	Qt	*Rujm Haleiseh.* The cairn of the verdure.
رجم الخياري	Nu	*Rujm el Khiâri.* The cairn of the cucumbers (close to Mar Sâba).
رجم المحوافت القبلية	Pt	*Rujm (el Mehawâfet) el Kibliyeh.* The southern cairn (of the boundary).
رجم المحوافت الشمالية	Pt	*Rujm (el Mehawâfet) esh Shemâliyeh.* The northern cairn (of the boundary).

رجم المغيثر
(probably المغافر)* — I's — *Rujm el Moghcifir.* The cairn of the pardoned, *i.e.,* 'deceased persons'; or it may mean an esculent vegetable substance like honey. The place is also called *Tell el Kursi,* 'mound of the throne.'

رجم رهيف — Ot — *Rujm Rehif.* The sharp-pointed cairn.

رجم الوعير — Ot — *Rujm el Wàir.* The cairn of rugged rocks.

سادة الحرمل — I's — *Sàdet el Hirmil.* The cliffs of rue.

سهلة الميدان — {Os} {I's} — *Sahlet el Meidàn.* The plain of the open place or exercise ground.

سحصول حميد — Pt — *Sahsùl Hameid.* The slip of little Hamed (an Arab boy who broke his neck here). Applies to a pointed crag jutting out of the cliff.

السقطرى — Pu — *Es Sakatri.* Either *es Sukutri,* an inhabitant of the island of Sukutra, in the Indian Ocean, from which bitter aloes and dragon's blood were imported, or *es Sikitrî,* meaning a *savant.* It is locally said to mean the bitter (water).

سرابيط — Pu — *Seràbit.* The heights, or summits.

شعب الحزيم — Ot — *Shàb el Hazìm.* The spur of the backbone.

شعب القريني — Ns — *Shàb el Kureini.* The spur of Tell Kureini. q.v.

شعب النقع — Ot — *Shàb el Makà.* The spur of the stagnant water.

شعب السويد — Pt — *Shàb es Sùweid.* The blackish spur; but *see Jubb Suweid* p. 5.

شعبة الاخراف — Ps — *Shàbet el Akhràf.* The spur of the sheep.

شعيب المثوار — Ou — *Shàib el Methwàr.* The spur abounding in bulls.

الشيخ مسيف — Nu — *Sheikh Maseiyif.* The sworded Sheikh.

شجرة الاثلة — Ps — *Shejeret el Ithleh.* The tamarisk tree (*T. palasii*); a fine tamarisk.

الشريعة — {Qs} {Qt} — *Esh Sherîàh.* The watering-place. The Jordan. In literary Arabic it is called اردن *Urdunna.*

شنخ الضبع — {I's} {Os} — *Shukh ed Dubà.* The staling of the hyena.

---

* The *alif* ا is in many words pronounced with the *imàleh* in Syria; that is, it takes the sound of *ei.*

سد عطية	Pu	*Sidd 'Atiyeh.* The cliff of 'Atiyeh. p.n.; 'gift.' The name applies to the marl banks below the cliffs.
سدرات العويسي	Qt	*Sidrāt el 'Aweisi.* The lotus of the difficult ground.
سواد البيض	Ou	*Sūwād el Beid.* The cave of the eggs. *Sūwād* literally means the dark interior of anything, and is locally used for a cave.
طاحونة الهوا	Ps	*Tâhûnet el Hawa.* The windmill. It is, however, a fortress. See "Memoirs," under 'Kŭrŭntŭl.'
طاحونة المجير	Ps	*Tâhûnet el Mejjir.* The mill of the place to which water is brought. Supplied by an aqueduct.
طلعة الدم	Ot	*Talât ed Dum[m].* The ascent of blood. See "Memoirs."
طلعة السلامة	Ou	*Talât es Salâmeh.* The ascent of safety.
طلعة السمرآ	Ot	*Talât es Sûmra.* The brown ascent.
طواحين السكر	Ps	*Tawâhîn es Sukker.* The sugar mills. See "Memoirs."
تل ابي هندي	Ps	*Tell Abu Hindi.* The mound of Abu Hindi. p.n.
تل ابي زلف	Ps	*Tell Abu Zelef.* The mound with the cisterns.
تل العرايس	Ps	*Tell el 'Arâis.* The mound of the bride; but it may be connected with عرايس 'a thicket,' being amid the thorn groves.
تل البريكة	Ps	*Tell el Bureikeh.* The mound of the little pool.
تل دير غنام	Ps	*Tell Deir Ghannâm.* The mound of the convent of Ghannâm. p.n.; 'shepherd or spoiler.'
تل درب الحبش	Ps	*Tell Derb el Habash.* The mound of the road of the Abyssinian.
تل الجرن	Ps	*Tell el Jurn.* The mound of the trough.
تل الكوس	Ps	*Tell el Kôs.* The mound of accumulated sand-hills.
تل القريني	Ns	*Tell el Kureini.* The mound of the little peak.
تل المحفورية	Ps	*Tell el Mahfûriyeh.* The mound of the excavated place.
تل المصنع	Ps	*Tell el Masnâ.* The mound of the trough, or reservoir; also called *Tell el Asmar,* the brown mound.
تل المطلب	Ps	*Tell el Matlab.* The mound of the object of search.

تل مَحْلَل　Pt　*Tell Muhalhal.* The mound of the chief (a grave).

تل الرشيدية　Qt　*Tell er Rusheidiyeh.* The mound of the Rusheidiyeh Arabs.

تل السمرات　Ps　*Tell es Samarát.* The mound of the thorny trees (*Spina Ægyptiaca*).

تل السلطان　Ps　*Tell es Sultán.* The mound of the Sultan; also called *Tell el 'Ain* (i.e., 'Ain es Sultán). The mound of the spring (i.e., of the Sultan's spring). q.v.

التليلات　Ps　*El Teleilát.* The little mounds.

تلول ابي العليق　Ps　*Tellûh Abu el 'Aleik.* The mounds where the barley grows; barley grows round them.

طور ام صيرة　Os　*Tôr Umm Sírah.* The hill with the sheepcote.

ثغرة الدبر　Ot　*Thoghret ed Debr.* The pass of the rear. The word perhaps retains a trace of Debir. Josh. xv, 7.

الطبقة　Os　*Et Tûbakah.* The terrace (or table-land).

طبق عمرية　Ou　*Tûbk 'Amriyeh.* The terrace of el 'Amriyeh. p.n. ('Amr's family). A barren, deserted track, at a low level. Lieutenant Conder notices that the word closely represents the Heb. עֲמֹרָה *Ghomorrah* [Gen. x, 19]. (*See* Note, "Quarterly Statement," October, 1876, p. 168.)

طبق القنيطرة　Pt　*Tûbk el Kaneiterah.* The terrace of the arch.

طبق السمارة　Ou　*Tûbk es Sammárah.* The terrace of those who converse by night.

طبق ام تينيس　Ou　*Tûbk Umm Keinîs.* The table-land with the game (or of the hunter).

طويل العقبة　Ps　*Tuweil el 'Akabeh.* The peak or long ridge of the steep or mountain road.

طويل الذياب　Os　*Tuweil edh Dhiâb.* The peak of edh Dhiâb (tribal name; 'wolves').

ام الاوتاد　Ou　*Umm el Aûtâd.* The mother of tent-pegs (an old camp).

ام انخولة　Qs　*Umm Enkhôla.* The mother of Enkhôla. p.n.; perhaps from نخل 'a palm tree.'

ام الحا دم
OR
ام الها دم
Os — *Umm Ilha Dumm.* This name is evidently incorrect, and has no meaning whatever in either of the forms given. The first word means 'mother'; the last has been rendered by the officers 'blood,' in which case it should be written دم, without the doubled final radical. Lieut. Conder calls attention to the similarity of sound to the Latin *Maledommim* used by Jerome ("Onom.," s. v, *Adommim*), for Heb. מַעֲלֵה אֲדֻמִּים ; but it is very unlikely that the latter Hebrew name should assume such a form as that in question.

ام الرجوم
Ot — *Umm er Rujûm.* The mother of cairns.

ام سيرة
Os — *Umm Sirah.* The mother of the sheepfold.

ام الطلعة
Os — *Umm et Talâh.* The mother of the ascent.

وادي العبد
Nu — *Wâdy el 'Abd.* The valley of the slave.

وادي ابي عبيدة
Ps — *Wâdy Abu 'Abîdeh.* The valley where the Abid Arabs dwell.

وادي ابى ضبع
Ot — *Wâdy Abu Dubâ.* The valley with the hyenas.

وادي ابي جرنين
Os — *Wâdy Abu Jurnein.* The valley with the troughs.

وادي ابي رتمة
Os — *Wâdy Abu Retmeh.* The valley with the broom plant (the juniper of the Bible).

وادي العضوا
Nu — *Wâdy el 'Adwa.* The valley of the thorny tree. *See* p. 340.

وادي اخشيبة
Ou — *Wâdy Akhsheibeh.* The muddy valley.

وادي العليآ
Ou — *Wâdy el 'Aliâ.* The upper valley.

وادي عمرية
Ou — *Wâdy 'Amriyeh.* The valley of 'Amriyeh. *See Tûbk Amriyeh,* p. 352.

وادي العصى
Ns — *Wâdy el 'Asa.* The valley of rebellion.

وادي الاصلع
Pt — *Wâdy el Aslâ.* The valley of the bald man. *See Jôfet Aslâ,* p. 344.

وادي العوج
Nt — *Wâdy el 'Awaj.* The crooked valley.

وادي العويصية
Qt — *Wâdy el 'Aweisîyeh.* The difficult valley.

وادي الدكاكين
Ot — *Wâdy ed Dekâkîn.* The valley of shops (caves and tombs are so called by the natives). *See* p. 255.

وادي الدبور
Ot — *Wâdy ed Dubbâr.* The valley of wasps (*Wâdy es Sidr*).

2 Z

وادي اكتيف   Ot   *Wâdy Ektcif.* The valley of the shoulder.

وادي فارة   Nt   *Wâdy Fârah.* The valley of the mouse. *See* p. 345.

وادي حجر   Nu   *Wâdy Hajr.* The valley of the rock.

وادي حليسة   Ot   *Wâdy Haleisch.* The valley of verdure.

وادي الحريق   Ns   *Wâdy el Harîk.* The valley of the conflagration.

وادي الخزيم   Pt   *Wâdy el Hazîm.* The valley of the backbone. *See Shûb el Hazîm,* p. 350.

وادي الحرمة   Nu   *Wâdy el Hirmeh.* The valley of the sanctuary ; containing sacred tombs (*Kabûr ed Dawâri*).

وادي جرفان   Ou   *Wâdy Jerfân.* The valley of banks of *débris* cut through by the winter torrents.

وادي الجيئة   Ps   *Wâdy el Jiththeh.* The valley of the corpse.

وادي جوفة زبن   Pt   *Wâdy Jôfet Zeben.* The valley of Jôfet Zeben. q.v., p. 344.

وادي جريف غزال   Pt   *Wâdy Joreif Ghüzâl.* The valley of the banks of the gazelle.

وادي قاقون   Ns   *Wâdy Kâkûn.* The valley of Kâkûn. *See* p. 345.

وادي القنطيطرة   Pt   *Wâdy el Kaneiterah.* The valley of the arch.

وادي القلت   Os   *Wâdy el Kelt.* The valley of the tarn. *See* p. 339.

وادي خيران اجعي   Ou   *Wâdy Kheirân Ajâi.* The valley of the trenches or hollows of dung.

وادي قمران   Pu   *Wâdy Kumrân.* The greyish valley. *See Khürbet Kumrân,* p. 345.

وادي لويطا   Os   *Wâdy Lûcit.* The valley of wounding with an arrow ; from the killing of a man in it, according to the natives. The word written لواط but pronounced with the *imâleh* (*see* note, p. 350), as here given, also means sodomy.

وادي المدورة   Ot   *Wâdy el Madôwerah.* The valley of the round rock.

وادي مكرفة نطم   Pt   *Wâdy Makarfet Kattûm.* The valley of the sniffing of Kattûm. p.n. ; the Arabs give as the origin of this name that a certain Kattûm, 'vulture' or 'glutton,' fell on his head here from his horse.

وادي المقاري   Ou   *Wâdy el Makâri.* The valley of hillocks; or of nodules of earth. The ground is covered with hard black flint nodules.

وادي مكمن النايم   Ou   *Wâdy Makman en Nâim.* The valley of the ambuscade of sleeper.

وادي المخروم   Nu   *Wâdy el Makhrûm.* Valley of the mountain peak. *See* p. 341.

وادي الماقوق   Os   *Wâdy el Mâkûk.* The valley of the bald patch.

وادي المرازة   Ot   *Wâdy el Marâzeh.* The valley of the place of the shepherd (a local word).

وادي معزن   Ou   *Wâdy Mâzn.* The valley of apportioning lots.

وادي مطامير   Ou   *Wâdy Metâmir.* Valley of subterranean granaries.

وادي مذبح عيّاد   {Os} {Ps}   *Wâdy Medhbâh 'Aiyâd.* The valley of the slaughter of 'Aiyâd; a common Arabic proper name. The pilgrim road to Jericho comes down this valley.

وادي المفجر   Os   *Wâdy el Mefjir.* The valley of el Mefjir. *See Khûrbet Mefjir,* p. 346.

وادي مصاعدة عيسى   Ps   *Wâdy Mesâadet 'Aisa.* The valley of the ascent of Jesus. *See* "Memoirs."

وادي النفاخ   Nt   *Wâdy el Mufâkh.* The valley along which the wind whistles.

وادي مكلك   Ot   *Wâdy Mukelek.* The valley of Mukelek. p.n. *See Deir Mukelik,* p. 343.

وادي النار وقيل وادي الراهب   Ou   *Wâdy en Nâr.* The valley of fire; also called *Wâdy er Râhib,* the Monk's valley.

وادي نويعمة   Ps   *Wâdy Nûeiâmeh.* The valley of soft soil. *See 'Ain Nûeiâmeh,* p. 339.

وادي النخيلة   Nt   *Wâdy en Nukheileh.* The valley of the small palm-tree.

وادي عش الغراب   Ps   *Wâdy 'Osh el Ghûrâb.* The valley of the raven's nest.

وادي رجعان   {Os} {Ot}   *Wâdy Rijân.* The valley of 'stagnant water,' or 'of returning.' There are two on the sheet. In both cases a road crosses and re-crosses them several times.

وادي رماني   Ot   *Wâdy Rumâni.* The Roman valley. It has a Roman milestone in it.

وادي رمّامنة Os   *Wâdy Rummâmaneh*. The valley of the pomegranate.

وادي السمّارة Pu   *Wâdy es Sammârah*. The valley of those who hold nightly converse.

وادي شاموط Os   *Wâdy Shâmût*. The speckled valley; or 'the valley of the large pot.'

وادي السدر Ot   *Wâdy es Sidr*. The valley of the lotus (*Zizyphus lotus*).

وادي سقيع Ns   *Wâdy Sikyâ*. The valley of hoar-frost.

وادي السنيسلة Ns   *Wâdy es Sineisileh*. The valley of the streams.

وادي السوينيت Ns   *Wâdy es Suweinît*. The valley of the acacias.

وادي السكوت Os   *Wâdy es Sukût*. The valley of silence.

وادي طلعة الدم Ot   *Wâdy Talât ed Dumm*. The valley of the ascent or ravine of blood.

وادي الطارة Os   *Wâdy et Târah*. The valley of the circle (*vulgo* for الطا ' a gorge').

وادي التيسون Ps   *Wâdy et Teisûn*. The valley of he-goats.

وادي ثغرة رحوة Nu   *Wâdy Thoghret Rahwah*. The valley of the pass or frontier of the high ground.

وادي طويل الذياب Ps   *Wâdy Tuweil edh Dhiâb*. The valley of Tuweil edh Dhiâb. q.v., p. 352.

وادي ام البويمات Ot   *Wâdy Umm el Bûeimât*. The valley of the mother of little owls.

وادي ام هوا Nu   *Wâdy Umm Hawa*. The valley of the mother of the wind.

وادي ام القينيس Ou   *Wâdy Umm el Keinîs*. The valley of Umm el Keinis. See *Tûbk Umm Keinîs*, p. 352.

وادي ام صرج Nu   *Wâdy Umm Serj*. The valley of Umm Serj. See *Nûkb Umm Serâj*, p. 348.

وادي ام سيرة Ou   *Wâdy Umm Sîrah*. The valley of Umm Sirah. q.v., p. 353.

وادي ام الصيارة Ou   *Wâdy Umm es Siyârah*. The valley with the sheepcote.

وادي ام الطلعة Ot   *Wâdy Umm et Talâh*. The valley of Umm et Talâh. q.v., p. 353.

وادي ام الثليثات	Nu	*Wâdy Umm eth Theleithât.* The valley of Umm eth Theleithât. *See Bîr Umm eth Theleithât*, p. 342.
وادي زرب الخربيان	Pu	*Wâdy Zerb el Khreiân.* The valley of the channels of the trenches.
وعر الاخشيبة	Ou	*Wâr el Akhsheibeh.* The rugged rocks of Wâdy el Akhsheibeh. q.v.
وعر الزرانيق	Ou	*Wâr ez Zerânîk.* The rocks of the runlets; from the numerous little water-worn channels down the face of the soft cliff.
الوشاشة	Ot	*El Washâsheh.* El Washâsheh. p.n.; perhaps from وش *wish, vulgo* for وجه *wajh*, 'face.' It is a hill-side with ruins of dry stone walls.
زقومت سرحان	Ps	*Zakkûmat Sirhân.* The balsam trees of Sirhân (wolf), an Arab hero killed by the *Adwân.*
الزور	It	*Ez Zôr.* The depression.

SHEET XIX.

ابو الكأس Dx *Abu el Kâs.* The father of the cup.

علي المنطار Dx *'Aly el Muntâr.* 'Aly of the watch-tower.

عرب الحناجرة Gx *'Arab el Hanâjerch.* The Arabs of the Hanâjerch (gullet) tribe.

عسقلان Ev *'Askalûn.* Ascalon. Heb. אשקלון

باب الدارون Dx *Bâb ed Dârûn.* The gate of Darum (leading to the town so-called, now *Deir el Belah*). *See below.*

بيت دردس Ew *Beit Dŭrdîs.* The house of Dŭrdis. p.n.

بيت حنون Ew *Beit Hanûn.* The house of Hanûn. p.n.

بيت لاهي Dw *Beit Lâhi.* The house of Lâhi. p.n.

دير البلح وقيل دير مار جريوس Gx *Deir el Belah.* The monastery of date palms; also called *Deir Mâr Jiriûs,* Monastery of St. George. It is the Crusading Darum, which Marino Sanuto (1322 A.D.) translates Domus Græcorum, House of the Greeks, *i.e.,* دير الروم *Deir er Rûm.* Foucher of Chartres (*circa* 1100 A.D.) gives the same derivation.

دير سنيد Ew *Deir Sineid* (or *Esneid*). Monastery of the declivity.

غزة Dw *Ghŭzzeh.* Gaza or Azzah. Heb. עזה (Deut. ii, 23). The following names occur in the immediate neighbourhood of the town, as shown on the special plan :—

باب الدارون *Bâb ed Dârûn.* The gate of Darum. *See above.*

بئر خليل *Bîr Khuleiyil.* Well of the little dell.

بيارة افراس *Biyâret Afrâs.* The well of mares.

بيارة البقارة *Biyâret el Bukkârah.* The well of the cowherds.

غزّة *Ghūzzeh.* Gaza—*continued.*

بئارة البرهان   *Biyâret el Berhân.* The wells of the evidence or demonstration.

بئارة الدرج   *Biyâret ed Deraj.* The well of steps.

بئارة الغباري   *Biyâret el Ghabâri.* The well of the dusty one.

بئارة الجرن   *Biyâret el Jurn.* The well of the trough.

بئارة قرتش   *Biyâret Kûrkûsh.* The well of the pulse.

بئارة القصر   *Biyâret el Kûsr.* The well of the house.

بئارة معتوق   *Biyâret Mâtûk.* The emancipated house, or family palace.

بئارة المراجعة   *Biyâret el Merâjah.* The well of returning.

بئارة عبيّة   *Biyâret 'Obbeiyeh.* The well of boasting.

بئارة صبرة   *Biyâret Sûbrah.* The well of the prickly pears (*Ficus Indicus*).

بئارة ام الليمون   *Biyâret Umm el Leimûn.* The well with the lemons.

بئارة وحشة   *Biyâret Wahasheh.* The wild or lonely well.

بئارة يمين   *Biyâret Yemîn.* The well of Yemin. p.n. ; (Ibn Yemin, 'Benjamin').

حارة الطافي   *Hâret et Tûfen.* The prison quarter (north).

حارة السجيّة   *Hâret es Sejjiyeh.* The quarter of flat-roofed clay-plastered houses (سبح). The houses here are of mud.

حارة الزيتون   *Hâret ez Zeitûn.* The quarter of olive trees. This is the southern quarter.

حارة الدرج   *Hâret ed Deraj.* The quarter of the steps ; to the west.

جامع الكبير   *Jâmiâ el Kebîr.* The great mosque.

غزَّة *Ghŭzzeh.* Gaza—*continued.*

مأذنة علي مروان *Mâdhnet 'Aly Mirwân.* The minaret of 'Aly Mirwân. p.n.; the traditional tomb of Samson. (Merwân was the name of the eighth Ommaiyid Caliph.)

مأذنة ابن عثمان *Mâdhnet Ibn 'Othmân.* The minaret of the son of 'Othmân.

مأذنة كاتب ولايات *Mâdhnet Kâteb Wilâyât.* The minaret of the secretary of state for the provinces.

مأذنة المحكمة *Mâdhnet el Mehkemeh.* The minaret of the Government house (or court).

مأذنة سيدنا هاشم *Mâdhnet Sidna Hâshem.* The minaret of our Lord Hâshem; the Prophet's grand-father, who is buried here.

السراي *Es Serâi.* The governor's house.

الشيخ نبك *Sheikh Nabak.* Sheikh Nabak. p.n.

الشيخ شعبان *Sheikh Shâbân.* Sheikh Shâbân. p.n.

—   --   —

حربية Ev *Herbieh.* Herbieh. p.n.; from حرب 'to take flight.'

جامع ابو برجاس *Jâmiâ Abu Berjâs.* The mosque of Abu Berjâs. p.n. This is in Jebâlieh.

جبالية Dw *Jebâlieh.* The mountaineers.

الجسر Ew *El Jisr.* The bridge.

الجورة Ev *El Jûrah.* The hollow; probably Chaldaic יגור 'Yagur.' *See* "Memoirs."

قضا غزة *Kada Ghŭzzeh.* The district of Gaza.

القشاني Dw *El Kîshâni.* El Kishâni. Persian ornamental tiles are so called, from the town of Kîshân, where they are made.

الخضرة Ev *El Khŭdrah.* El Khŭdrah. *See* p. 28. (In Ascalon.)

خربة العدار Dx *Khŭrbet el 'Adâr.* The ruin of 'Adâr. p.n.; (عدار 'sailor.')

خربة البئر Ex *Khŭrbet el Bîr.* The ruin of the well.

خربة البرجليّة   Dx   *Khŭrbet el Burjalîyeh.* The ruin of el Burjaliyeh. p.n.

خربة حوادي   Dx   *Khŭrbet Hawādi.* The ruin of Hawâdi (the word means hind-legs).

خربة انصيرات   Cx   *Khŭrbet Inseirât.* The ruin of *Inseirât* (apparently from انصار 'the ansârs,' or helpers of Mohammed; but it is more probably an error for نسيرات *Nuseirât*, the Arabs of the tribe of Abu Nuseir.

خربة الخصاص   Ev   *Khŭrbet el Khesâs.* Ruin of booths or reed huts.

خربة كوفية   Ex   *Khŭrbet Kûfeh.* Cufic ruin, or the ruin of the silk hand-kerchief.

خربة منصورة   Dx   *Khŭrbet Mansŭrah.* The ruin of Mansûrah. *See* p. 9.

خربة المشرفة   Dx   *Khŭrbet el Meshrefeh.* The ruin of the high place.

خربة الناموس   Dx   *Khŭrbet en Nâmûs.* The ruin of the hunter's lodge. The word is a very old one, and is applied throughout Arabia to stone buildings of a primitive type. In modern Arabic *nâmûs* means 'a mosquito,' and the confusion has given rise to many stories amongst the Bedawin, especially in Sinai.

خربة الرسيم   Ex   *Khŭrbet er Reseim.* The ruin of the vestiges of buildings.

خربة الشلوف (الشالوف)   Dx   *Khŭrbet esh Shelûf.* The ruin of the waterfall or declivity.

خربة المشراف   Ev   *Khŭrbet esh Sherâf.* The ruin of the pinnacle.

خربة الصيحان   Ex   *Khŭrbet es Sihân.* The ruin of the courtyard, صحن.

خربة الصيرة   Dx   *Khŭrbet es Sîreh.* The ruin of the fold.

ميدان ابو زيد   Dx   *Meidân Abu Zeid.* The open place or exercise ground of Abu Zeid. p.n.; a racecourse.

المشاهرة   Dw   *El Meshâherah.* The monthly hiring. The root شهر means also to be renowned.

مشهد سندنا الحسين   Ev   *Mesh-hed Sidna el Husein.* The shrine of our Lord Husein, son of Ali, a chief saint of the Shiah sect.

المينة   Dw   *El Mîneh.* The harbour, from לִמֵינָה (Chaldee) from the Greek λιμήν. The word is of universal use in Arabic.

النزلة   Dw   *En Nŭzleh.* The settlement, or hamlet.

3 A

راس ابى اميرة   Ev   *Râs Abu Ameireh*. The hill-top with the road-mark, or sign-post أمارة.

رسم العطاونة   Ex   *Resm el 'Atâwineh*. The vestige of the leather dresses, or of those who halt by a watering place.

رسم الغاربي   Ex   *Resm el Gharby*. The western ruin.

رسم الشرقي   Ex   *Resm esh Sherky*. The eastern ruin.

رجال الاربعين   Dw   *Rijâl el Arbâîn*. The forty men.

الشيخة نقية   Dw   *Sheikhah Nakîyeh*. The female Sheikh, or Saint Nakiyeh. p.n.; (pure).

الشيخ عباس   Dx   *Sheikh 'Abbâs*. Sheikh 'Abbâs. p.n.

الشيخ احمد   Cw   *Sheikh Ahmed*. Sheikh Ahmed. p.n.

الشيخ عجلين   Cw   *Sheikh 'Ajlîn*. Sheikh 'Ajlîn. p.n.; (hasty).

الشيخ حسن   Dw   *Sheikh Hasan*. Sheikh Hasan. p.n.

الشيخ محمد الموصلي   Ev   *Sheikh Muhammed el Mûsli*. Sheikh Muhammed of Mosul.

الشيخ منعم   Ew   *Sheikh Munâm*. Sheikh Munâm; ('favouring,' or 'wealthy').

الشيخ نبهان   Dx   *Sheikh Nebhân*. Sheikh Nebhân. p.n.; ('vigilant').

الشيخ راشد   Cx   *Sheikh Râshid*. Sheikh Râshid. p.n.; ('orthodox').

الشيخ رضوان   Dw   *Sheikh Redwân*. Sheikh Redwân. p.n.; (pleasing to God). It is the name of the angel who guards Paradise, according to the Kor'ân.

الشيخ شوباني   Cx   *Sheikh Shûbâni*. The Sheikh of Shûba. q.v.

شجرة غنايم   Cx   *Shejeret Ghanâîm*. The tree of plunder.

شجرة ام فظين   Cx   *Shejeret Umm Kadhein*. The tree of Umm Kadhein. p.n.

تل احمر   Dx   *Tell el Ahmar*. The red mound.

تل العجول   Cx   *Tell el 'Ajjûl*. The mound of the calf, or of the hastener. *See* "Memoirs."

تل نجيد   Cx   *Tell Nujeid*. The mound of the rising ground.

التينة   Cu   *Et Tîneh*. The fig-tree.

تلول الحمرة   Ex   *Tellûl el Humrah*. The red mounds.

تَمْرَة Ew *Tŭmrah.* Tŭmrah. p.n.; also called بيت دمرة *Beit Dimreh.* p.n.; by the peasantry.

أُمّ عَامِر Cx *Umm 'Âmir.* The mother of 'Âmir; a name given to the hyena.

واد بني خميس {Dx}{Ew} *Wâd Beni Khamîs.* The valley of the Beni Khamis Arabs.

وادي الباحة Ex *Wâdy el Bâha.* The valley of the gulf, pool, plantation, or wide area.

وادي فعيلس Ex *Wâdy Fuâilis.* The valley of Fuâilis. p.n.

وادي غزة Cx *Wâdy Ghâzzeh.* The valley of Gaza.

وادي الحليب Ew *Wâdy el Halîb.* The valley of milk.

وادي الحسي Ev *Wâdy el Hesy.* The valley of water accumulated in a sandy place.

وادي الحمرة Ex *Wâdy el Humrah.* The valley of the red rocks.

وادي المقدمة Dx *Wâdy el Mukaddemeh.* The fronting (or eastern) valley.

وادي النخابير Dx *Wâdy en Nŭkhâbîr.* The valley of Nukhâbir; perhaps an inversion for نخاريب *nŭkhârîb,* 'fissures.'

وادي الصافية Ew *Wâdy es Sâfieh.* The valley of smooth stones or pure water.

وادي الصيحان Ex *Wâdy es Sîhân.* The valley of the court-yard.

SHEET XX.

عين بكره Iv   *'Ain Bakrah.* The spring of the virgin, or first-born.

عين القنيطرة Gw   *'Ain el Kaneiterah.* The spring of the little arch.

عين مسيجد Iv   *'Ain Museijid.* The spring of the mosques.

عين السد Gw   *'Ain es Sidd.* The spring of the dam, or barrier.

عيطون التحتا Ix   *'Aitûn et Tahta.* The lower 'Aitûn. p.n.; cf. *Rasm el Atâwineh,* p. 362; the root being the same.

اخفات الزهارات Hw   *Akhjât ez Zahârât.* The hidden places of flowers.

العمدان Iw   *El 'Amdân.* The columns.

عرب العمرين IIv   *'Arab el 'Amarîn.* The Ammarin Arabs. *See* p. 268.

عرب الجبارات Gw   *'Arab el Jubârât.* The Jubârât Arabs. p.n.

عرب السواركي Gx   *'Arab es Sûârki.* The Sûârki Arabs. p.n.

عراق ابى العمد Jw   *'Arâk Abu el 'Amed.* The cavern or cliff with the pillars.

عراق ابى مزبلة Iv   *'Arâk Abu Mizbeleh.* The cavern with the dung-heap: used as a stable, at *Beit Jibrîn* by *Bâb el Medineh.*

عراق الاسلمة Iv   *'Arâk el Asalmeh.* The cavern of the Sillam trees; near *Beit Jibrîn,* close to *Khûrbet ez Zemmâr.*

عراق الفنش Iv   *'Arâk el Fenish.* Cavern of the Fenish. p.n.; meaning 'slack.'

عراق فرهود Iv   *'Arâk Ferhûd.* The cavern of the kid.

عراق حالة {Iv / Hx}   *'Arâk Hâla.* The cavern of the fuller's beetle, or of soft earth.

عراق حليل Jv   *'Arâk Heleil.* The cavern of the crescent.

عراق الخارب IIv   *'Arâk el Khârab.* The ruined cavern.

عراق الخيل Iv *'Arâk el K'heil.* The cavern of horses or of cavalry.

عراق الماء Iv *'Arâk el Mâ.* The cavern of water.

عراق المنشية Hv *'Arâk el Menshiyeh.* The cliff of the place of growth.

عراق المقطع Iv *'Arâk el Mûktâ.* The quarried cavern.

عراق الشارة Iv *'Arâk esh Shârah.* The cavern of the figures (from the figures on the wall).

عراق الشيخ Iv *'Arâk esh Sheikh.* The cavern of the Sheikh (near *Sheikh Barrâk*).

عراق الشرف Iv *'Arâk esh Sherif.* The cavern of the nobleman (at *Beit Jibrîn*, near *Khûrbet Hehsheh*).

عراق شوبك Iv *'Arâk Shôbak.* The cavern of fennel.

عراق شويدان Gv *'Arâk Suweidân.* The cavern of Suweidân. p.n.

عراق الزاغ Iv *'Arâk ez Zâgh.* The cavern of the crow.

ارض المقحز {Hx / Hw} *Ard el Mak-haz.* The land of the leap.

عيون قصابة Hx *'Ayûn Kŭssâbah.* The springs of reeds.

عيون الحسي Gw *'Ayûn el Hesy.* The springs of el Hesy. *See Wâdy el Hesy*, p. 363.

باب المدينة Iv *Bâb el Medîneh.* The gate of the city. *See* "Memoirs," under ' Beit Jibrin.'

بيت عفة Gv *Beit 'Affeh.* The house of 'Affeh. p.n.; ('chaste ').

بيت أمير Iw *Beit Emîr.* The house of the prince.

بيت جرجة Ev *Beit Jerjah.* The house of the highway (near the main road).

بيت جبرين Iv *Beit Jibrîn.* The house of Jibrin. (Aramaic בית גוברין *Beth Geborim*, ' the house of great men.')

بيت لي Jw *Beit Leyi.* The house of Leyi. p.n.; ('twisting ').

بيت مدسوس Gw *Beit Madsûs.* The house into which one creeps.

بيت الروش Jx *Beit er Rûsh.* The house of er Rûsh. p.n.

بيت طيما Fv *Beit Tîma.* The house of Tima. p.n.

بيت تتن Gw *Beit Titn.* House of tobacco.

البيارة Ix   *El Beiyârah.* The well.

بستان الفنش Iv   *Bestân el Fenish.* The garden of the Fenish. p.n.

بيار حالا Iv   *Biâr Hâla.* The wells of the fuller's beetle, or of soft earth ; by *'Arâk Hâla.*

بئر ابي عريض Gw   *Bîr Abu 'Arîdeh.* The well of the petition.

بئر ابي بابين Iw   *Bîr Abu Bâbein.* The well with the two mouths (or doors).

بئر العجمي Iv   *Bîr el 'Ajami.* The well of the Persian ; by *Beit Jibrîn,* on the north by the camp.

بئر الحربية IIv   *Bîr el Harbîyeh.* The well of the warriors ; at *'Arâk el Menshîyeh.*

بئر الحمام Iv   *Bîr el Hûmmâm.* The well of the bath ; south-west of the Kûlâh at *Beit Jibrîn.*

بئر الجعفري Hv   *Bîr el Jâfari.* The well of Jaafer. p.n.

بئر جلمة Fw   *Bîr Jelameh.* The well of Jelameh. *See Khûrbet Jelameh,* p. 371.

بئر الجبارات Hx   *Bîr el Jubârât.* Well of the Jubârât (Arabs).

بئر قطام Iw   *Bîr Kattâm.* The well of the quarryman.

بئر الكنيسة Iv   *Bîr el Kenîseh.* The well of the Church ; near *Sandahanna.*

بئر القلعة Iv   *Bîr el Kûlâh.* The well of the castle ; close to the castle of *Beit Jibrîn.*

بئر مليتا Gv   *Bîr Melita.* The well of Melita. *See Khûrbet Melita,* p. 373.

بئر مغاير صرار Hw   *Bîr Mâghâir Sûrrâr.* The well of the caves of pebbles.

بئر صميل IIv   *Bîr Summeil.* The well of Summeil. *See Sûmmeil,* p. 274.

بئر الطيطح Hv   *Bîr et Teitch.* The well of et Teitch. p.n.

بئر ام جديع Iv   *Bîr Umm Judeiâ.* Well mother of cutting.

بئر الوسطانية Hv   *Bîr el Wustîyeh.* The middle well.

بئر ياسين Iw   *Bîr Yasîn.* The well of Yâsin. p.n.; also called *Bîr esh Sheikh,* the well of the Sheikh.

بئر الزمار Iv   *Bîr ez Zemmâr.* The well of the pipers ; near *Khûrbet ez Zemmâr.*

البيرة Jx   *El Bîreh.* The fortress.

بركة الحمرآ Iv   *Birket el Hámra.* The red pool.

بربرة Ev   *Bŭrberah.* 'Chattering'; 'barbarians.'

بزير Fw   *Bureir.* The little wilderness.

برج البيارة Ix   *Burj el Beiyárah.* The tower of the wells.

البرجلية Hv   *El Bŭrjaliyeh.* El Bŭrjaliyeh. p.n.

الدوايمة Iw   *Ed Dawáimeh.* The little Dóm trees.

دير خاروف Iw   *Deir Khárúf.* The monastery of sheep.

دير محسن Ix   *Deir Muhcisin.* The monastery of good deeds.

دير الموس Ix   *Deir el Mûs.* The monastery of the penknife.

دير نخاس Iv   *Deir Nakhkhás.* The monastery of the cattle drover.

دير الشاطر Jv   *Deir esh Shátir.* The monastery of the sharper.

ذكرين البردان Iv   *Dhikerîn el Baradân.* Dhikerin the cool. *See Khŭrbet Dhikerîn*, p. 269.

اجة Fv   *Ejjeh.* Confusion.

الفالوجة Gv   *El Fálûjeh.* Fálûjeh. p.n. *cf.* Heb. גֶּלֶג a stream.

غياضة Iv   *Gheiyádah.* The thicket.

الحبس Iw   *El Habs.* The prison, or the religious endowment (mosque property). *Habs* is equivalent to وقف *wakf.*

حبور سيسمح Hw   *Habûr Seisamakh.* The level grounds of Seisamakh. p.n. حبور is the plural of حبر.

الحاج سالم Iv   *El Háj Sálim.* The pilgrim Sálim.

حمرة الدالية Hx   *Hamret ed Dálieh.* The red ground of the trailing vine.

حتة Gv   *Hatteh.* Hatteh. p.n.

الحنو Hx   *El Henû.* The bend.

حنو تمش Hv   *Henû Kumsh.* The bend where things collect.

حوج Fw   *Hûj.* Hûj. *See Neby Hûj*, p. 377.

حليقات Fv   *Huleikát.* The circles. *cf.* Heb. חֶלְקָה 'a field.'

الحمرة Hw   *El Humurah.* The red.

خربة الخروف   Iv   *Hurubbet el Kharûf.* The tank of the sheep.

جسير   IIv   *Ijseir.* The little bridge.

جبل أبى حَضيرش   Ix   *Jebel Abu Huteirish.* The mountain of Abu Huteirish. p.n. ; ('the father of romping ').

الجميزه   Gv   *El Jimmeizeh.* The sycamore.

جرف الخليل   IIv   *Jorf el Khaleil.* The bank of the little dell.

جورة الخوات   Iv   *Jûrat el Khawât.* The hollow of the murmuring torrent ; the valley east of *el Gheiyâdah.*

قنان مغامس   Jw   *Kanân Meghâmis.* The peaks of the thickets.

قنان السرو   Gx   *Kanân es Seru.* The peaks of the cypress.

كوكبة   Fv   *Kaukabah.* The word means 'star,' 'mountain,' and 'donjon.'

الكنيسة   Iv   *El Kenîseh.* The church. Also called صَنْدَحنّ *Sanda-hanna,* 'St. Anne.' *See* " Memoirs."

كرتيا   Gv   *Keratîya.* Keratiya. p.n. It may be كُرَثة 'thick entangled grass'; but *cf.* ברתי the Cherethites (1 Sam. xxx, 14), who appear to have lived in Philistia.

خلّة الأعرج   Jv   *Khallet el 'Aâraj.* The dell of the lame one ; south-east of *Khûrbet el Kabbârah.*

خلّة ايلياس   Iv   *Khallet Aîlias.* The dell of Elias ; west of *Khûrbet Sâfieh.*

خلّة الحملة   Ix   *Khallet el Hamleh.* The dell of the load.

خلّة القيسية   Iv   *Khallet el Keisîyeh.* The dell of the Keis Arabs ; south of *Deir esh Shâtir.*

خلّة مرط السيل   IIv   *Khallet Mert es Seil.* The dell of the *débris* of the stream.

خلّة المغيطية   Jv   *Khallet el Mugheitîyeh.* Dell of the soft and fertile place. Upper part of *Wâdy Beit 'Alâm.*

خلّة ام ناب   Iw   *Khallet Umm Nâb.* The dell of 'the mother of a fang'; but it should perhaps be *Menâb,* the path to a watering place ; south of *Râs Abu Hattam.*

خلّة الزاغ   Jv   *Khallet ez Zâgh.* The dell of the crow ; west of *Sheikh 'Abdallah.*

خلال الماء Iv — *Khelâl el Má.* The dells of water.

جربة ابرقا Iw — *Khûrbet Abraka.* The ruin of speckled ground.

خربة ابو عزام IIv — *Khûrbet Abu 'Azâm.* The ruin of the maker of dams or dykes.

جربة ابى غيث IIx — *Khûrbet Abu Gheith.* The ruin of Abu Gheis. p.n.

خربة ابى الملسمة Ix — *Khûrbet Abu Mulassamah.* The ruin of Abu Mulassamah. p.n. ; meaning 'keeping on the road.'

خربة ابى الرخيم Jv — *Khûrbet Abu Rekheim.* The ruin of the father of the vultures ; (dim.).

خربة ابى سهويلة Ix — *Khûrbet Abu Sihweileh.* The ruin of Abu Sihweileh. p.n.

خربة عيطون Jx — *Khûrbet 'Aitûn.* The ruin of 'Aitûn. *See* p. 364.

خربة عجلان Gw — *Khûrbet 'Ajlân.* The ruin of Eglon. Heb. עֶגְלוֹן

خربة الأقرع Fw — *Khûrbet el Akrâ.* The ruin of the bald man or rock.

خربة عمودة Fv — *Khûrbet 'Amûdeh.* The ruin of the pillar.

خربة العرب Iw — *Khûrbet el 'Arab.* The ruin of the Arab.

خربة عراق ابى الحسين IIw — *Khûrbet 'Arâk Abu el Husein.* The ruin of the cavern of Abu Husein. p.n.

خربة العتر Iv — *Khûrbet el 'Atr.* The ruin of el 'Atr. p.n.; meaning sacrificing ; also the name of an idol, and of a dwarf thorny shrub.

خربة العترية Iv — *Khûrbet el 'Atarîyeh.* The ruin of el 'Atariyeh. *See* last entry.

خربة البها Ex — *Khûrbet el Baha.* The ruin of splendour.

خربة البهلوان Gw — *Khûrbet el Bahlawân.* The ruin of the wrestler.

خربة بكرة Iv — *Khûrbet Bakrah.* The ruin of the virgin or first-born.

خربة الباشا Iw — *Khûrbet el Bâsha.* The ruin of the Pacha.

خربة البحيرة Hw — *Khûrbet el Beheirah.* The ruin of the lake.

خربة البيضاء Iv — *Khûrbet el Beida.* The white ruin.

3 B

خربة بيرام   Iw   *Khŭrbet Beirâm.* The ruin of Beirâm (the Turkish festival so called).

خربة بيت لجوس   Ev   *Khŭrbet Beit Lejûs.* The ruin of the house of Lejûs.   p.n.

خربة بيت مامين   Gv   *Khŭrbet Beit Mâmîn.* The ruin of the house of Mâmîn; so called from *Neby Mâmîn.*

خربة بيت مرسم   Ix   *Khŭrbet Beit Mirsim.* The ruin of the house of Mirsim. p.n.; meaning 'ambling.'

خربة البناوي   IIw   *Khŭrbet el Benâwy.* The ruin of el Benâwi. p.n.; from بنآ a builder.

خربة بناية   Ix   *Khŭrbet Benâyeh.* The ruin of the building.

خربة بدغش   Ix   *Khŭrbet Bidghush.* The ruin of Bidghush.   p.n.

خربة البئر   Iw   *Khŭrbet el Bîr.* The ruin of the well.

خربة بشر   Jw   *Khŭrbet Bishir.* The ruin of Bishir.   p.n.; meaning 'joyous.'

خربة برناطة   Iv   *Khŭrbet Bornâtch.* The ruin of the hat. Connected perhaps with the Aramaic בִּירְנְתָא 'a palace.'

خربة البصل   Iv   *Khŭrbet el Bŭsl.* The ruin of the onion.

خربة بطيحة   Gx   *Khŭrbet Butcihah.* The ruin of the marsh.

خربة دهنا   Jw   *Khŭrbet Dahneh.* The ruin of the desert.

خربة دير سعد   Iw   *Khŭrbet Deir Sâd.* The ruin of monastery of Sâd. p.n.; ('fortunate').

خربة الظبية   Iv   *Khŭrbet edh Dhubeiyeh.* The ruin of the antelope or fawn.

خربة الدروسة   Jv   *Khŭrbet ed Drûseh.* The ruin of the obliterated paths.

خربة دلدوب   Fv   *Khŭrbet Dŭldŭb.* The ruin of Dŭldûb.   p.n.

خربة عجس الراس   Fv   *Khŭrbet 'Ejjis er Râs.* The ruin of the hinder part of the hill-top.

خربة ارزة   Fv   *Khŭrbet Erzeh.* The ruin of the cedar.

خربة فصاصة   Iv   *Khŭrbet Fassâsah.* The ruin of the setters of precious stones in a ring.

خربة فطاطة   IIv   *Khŭrbet Fattâtah.* The ruin of the pug-nosed ones.

خربة فهيدي   Iw   *Khŭrbet Fuheidy.* The ruin of the little lynx.

خربة فوليتة Iv   *Khŭrbet Fûliyeh.*   The ruin of beans.

خربة قرط IIv   *Khŭrbet Furut.*   The ruin of the hill or sign-post.

خربة الفوارة IIx   *Khŭrbet el Fůwârah.*   The ruin of the fountain.

خربة الغبية Iv   *Khŭrbet el Ghobeiyeh.*   The ruin of the thicket.

خربة الببور Hw   *Khŭrbet el Habŭr.*   The ruin of the level lands.

خربة الحاج عيسى Iv   *Khŭrbet el Hâj 'Aîsa.*   The ruin of the pilgrim 'Aisâ (Jesus or Emmanuel).

خربة حمدة Iv   *Khŭrbet Hamdeh.*   The ruin of Hamdeh.   p.n.

خربة حزانة Iw   *Khŭrbet Hazzâneh.*   The ruin of the unfortunate ones, or of the ' rugged ground.'

خربة الهزارة Gw   *Khŭrbet el Hazzârah.*   The ruin of the nightingales.

خربة حبرا Iw   *Khŭrbet Hebra.*   The ruin of el Hebra.   p.n. حبر means 'a Jewish doctor.'

خربة البشة Iv   *Khŭrbet el Heshsheh.*   The ruin of alacrity.

خربة حوران Ix   *Khŭrbet Hôrân.*   The ruin of marshes.

خربة هوج Fw   *Khŭrbet Hûj.*   The ruin of Hûj.   *See Neby Hûj.*

خربة الحمام Fw   *Khŭrbet el Hůmmâm.*   The ruin of the bath.

خربة حراب ذياب Fx   *Khŭrbet Hurâb Diâb.*   The ruin of the tanks of Diâb. The last word is probably ذياب *Dhiâb,* 'wolves,' the well-known Arab family name.

خربة الحسيمية   *Khŭrbet el Huseimiyeh.*   The ruin of the Huseimiyeh; name of a family or sect.

خربة الحسينات Iv   *Khŭrbet el Huseinât.*   The ruin of the Huseinât (family name).

خربة الجبو Iw   *Khŭrbet el Jebŭ.*   The ruin of the watering trough.

خربة جيمر Ix   *Khŭrbet Jeimar.*   The ruin of Jeimar.   p.n.; from جمرة a fire-brand.

خربة جلمة Fw   *Khŭrbet Jelameh.*   The ruin of the hill.

خربة جماعة Gx   *Khŭrbet Jemmâmeh.*   The ruin of abundance, or of reservoirs.

خربة جَنّتا   Iw   *Khŭrbet Jenneta.* The ruin of the garden ; Heb. גִנַּת.

خربة المجلس   Gv   *Khŭrbet el Jils.* The ruin of the high rugged ground.

خربة المجندي   Fx   *Khŭrbet el Jindy.* The ruin of the soldier.

خربة الجوزة   Jw   *Khŭrbet el Jôzeh.* The ruin of the walnut.

خربة جعيثني   Fx   *Khŭrbet Juáithiny.* The ruin of Juáitheneh. *See Khŭrbet Játhûn,* p. 48.

خربة الجبارات   Hx   *Khŭrbet el Jubârât.* The ruin of the Jubârât Arabs.

خربة الجديدة   Gv   *Khŭrbet el Judeiyideh.* The ruin of the dykes, *i.e.,* stripes of a different colour in a rock. There is a second of the

Iv   name on the Sheet (Iv).

خربة جويجة   Iw   *Khŭrbet Juweijah.* The ruin of the breasts or prows.

خربة جوتي   Jx   *Khŭrbet Juwet.* The inner ruin.

خربة القبّارة   Jv   *Khŭrbet el Kabbârah.* The ruin of the grave-diggers, or of the hunter's lantern.

خربة القاضي   Gx   *Khŭrbet el Kâdy.* The ruin of the judge.

خربة القنيطرة   Hw   *Khŭrbet el Kaneiterah.* The ruin of the little arch.

خربة كشكاية   Iw   *Khŭrbet el Kashkaliyeh.* The ruin of the Kashkaliyeh ; perhaps 'beggars,' from كشكول the tray or cup in which a beggar gathers scraps.

خربة قماص   Fv   *Khŭrbet Kemás.* The ruin of galloping.

خربة قرقرة   Iw   *Khŭrbet Kerkerah.* The ruin of smooth soft soil.

خربة الخروع   Iv   *Khŭrbet el Kherwâ.* The ruin of the castor-oil plant.

خربة الخريسة   Jw   *Khŭrbet el Khoreisah.* Probably the Heb. חֹרֶשׁ a 'thicket.'

خربة الخلف   Iv   *Khŭrbet el Khŭlf.* The ruin of succession or opposition.

خربة الخصاص   Gw   *Khŭrbet el Khŭsâs.* The ruin of reed huts.

خربة الكفخة   Fx   *Khŭrbet el Kefkhah.* The ruin of beating with a stick, or 'of white butter.'

خربة القوقا   Iv   *Khŭrbet el Kûka.* The ruin of the owl.

خربة قمحة   Iv   *Khŭrbet Kŭmhah.* The ruin of wheat.

خربة القصر   Jw   *Khŭrbet el Kŭsr.* The ruin of the house or palace.

خربة قصابة   Hx   *Khŭrbet Kŭssábah.* The ruin of the butchers.

خربة اللحم   Iw   *Khŭrbet el Lahm.* The ruin of meat; perhaps the Heb. לַחְמָם.

خربة لسن   Fw   *Khŭrbet Lasan.* The ruin of stinging.

خربة المدورة   Iw   *Khŭrbet el Madôwerah.* The round ruin.

خربة الخمز   Hw   *Khŭrbet el Mak-haz.* The ruin of the leap.

خربة المنصورة   Hy   *Khŭrbet el Mansûrah.* The ruin of Mansûrah. *See* p. 9.

خربة مرشان   Fw   *Khŭrbet el Marashân.* The ruin of the ground excoriated by rain.

خربة مجادل   Ix   *Khŭrbet Mejâdil.* Ruin of watch-towers.

خربة مجدلة   Iw   *Khŭrbet el Mejdeleh.* The ruin of the watch-tower.

خربة مليطا   Gv   *Khŭrbet Melita.* The plastered ruin.

خربة مرعش   Iv   *Khŭrbet Merâsh.* The ruin of Merâsh. *cf.* Heb. מְרֹאשָׁה.

خربة مرتينيا   Ix   *Khŭrbet Mertinia.* The ruin of Mertinia. p.n.

خربة مرط السيل   Hv   *Khŭrbet Mert es Seil.* The ruin of the *débris* of the stream.

خربة المصادي   Iw   *Khŭrbet el Mesâdi.* The ruin of the covert.

خربة مغيسل   Hx   *Khŭrbet Mugheisil.* The ruin of the place of washing.

خربة المكمين   Hx   *Khŭrbet el Mukeimin.* The ruin of the ambuscades.

خربة منظرة البغل   Fx   *Khŭrbet Muntaret el Baghl.* The ruin of the watch-tower of the mule.

خربة المرماخ   Hx   *Khŭrbet el Murmâkh.* The ruin of the clump (of bushes).

خربة مران   Ix   *Khŭrbet Murrân.* The ruin of Murrân. p.n.; probably connected with the Beni Murrah Arabs.

خربة المساجد   Iv   *Khŭrbet el Meseijid.* The ruin of the mosques.

خربة المصرة   Iv   *Khŭrbet el Musirreh.* The ruin of the persevering ones.

خربة نجد   Fw   *Khŭrbet Nejed.* The ruin of Nejed (high ground); from the village of Nejed.

خربة النصراني   Ix   *Khŭrbet en Nusrâny.* The ruin of the Christian.

خربة رافا   Jv   *Khŭrbet Râfa.* The ruin of Râfa. p.n.

خربة الرايع   Jw   *Khŭrbet er Râiâ.* The ruin of the handsome man.

خربة الرسم   {Iw Fx Hx}   *Khŭrbet er Resm.* The ruin of the vestige of buildings. There are several of these.

خربة الرسوم   Iw   *Khŭrbet er Resûm.* The ruin of the vestiges (of buildings).

خربة الرز   Iw   *Khŭrbet er Roz.* The ruin of rice.

خربة الرجلتن   Hx   *Khŭrbet er Rnjliyeh.* The pedestrians; but perhaps from رجلى 'stony ground.'

خربة رومية   Iw   *Khŭrbet Rûmiyeh.* The ruin of the Greeks.

خربة رمانة   Iw   *Khŭrbet Rummâneh.* The ruin of pomegranates.

خربة صافية   Iv   *Khŭrbet Sâfieh.* The ruin of bright smooth stones; or pure clear water.

خربة سامي   Fv   *Khŭrbet Sâmy.* The ruin of Neby Sâmy. q.v.

خربة صندحنة   Iv   *Khŭrbet Sandahannah.* The ruin of St. Anne or St. John; near *Tell Sandahannah.*

خربة الشأة   Iv   *Khŭrbet esh Shâh.* The ruin of the sheep.

خربة شالخة   Hv   *Khŭrbet Shalkhah.* The ruin of Shalkhah. p.n.; meaning either 'cleaving with a sword,' or 'vulva.'

خربة شعرتا   Fw   *Khŭrbet Shârata.* The ruin of the thick foliage.

خربة الشمسانيات   Iv   *Khŭrbet esh Shemsâniyât.* The ruin of the Sampson family.

خربة شعلية   Iw   *Khŭrbet Shuâliyeh.* The ruin of the cavern. Compare *Bîr Abu Shuâleh,* p. 340.

خربة شقاتية   Iw   *Khŭrbet Shŭkâkieh.* The ruin of clefts.

خربة الشقاق   Iw   *Khŭrbet esh Shŭkkâk.* The ruin of the cleaver.

خربة شتيوي   Gx   *Khŭrbet Shntciwy el 'Oseiby.* The ruin of the winter quarters of el 'Oseiby. p.n.

خربة سمبس   Fv   *Khŭrbet Simbis.* The ruin of Simbis. p.n.; meaning speed.

خربة سومرة   Jx   *Khŭrbet Sômerah.* The brown ruin, or the ruin of those who hold nightly converse.

خربة سعيد   Iv   *Khŭrbet Suáid.* The ruin of Suáid. p.n.; diminutive of Sád (Felix).

خربة السقيفة   Iw   *Khŭrbet es Sukeiyifeh.* The ruin of the roofed-in chamber.

خربة السكرية   Hw   *Khŭrbet es Sukríyeh.* The ruin of the sugar factories.

خربة الصميلي   Gw   *Khŭrbet es Summeily.* The ruin of the man of Summeil.

خربة الصورة   Iv   *Khŭrbet es Súrah.* The ruin of the picture.

خربة صرار   Hw   *Khŭrbet Sŭrrâr.* The ruin of pebbles.

خربة تنر   Gw   *Khŭrbet Tannar.* The ruin of ovens or reservoirs.

خربة طوطا   Jw   *Khŭrbet Tút.* The ruin of the chimney or funnel.

خربة ام عميدات   Gx   *Khŭrbet Umm 'Ameidât.* The ruin with the small pillars

خربة ام عمود   Hv   *Khŭrbet Umm 'Amúd.* The ruin with the pillar.

خربة ام بغلة   Ix   *Khŭrbet Umm Baghleh.* The ruin with the mules; perhaps a corruption of בִּעְלָה

خربة ام باطية   Gx   *Khŭrbet Umm Bâtieh.* The ruin with the large bowl.

خربة ام البكار   Gw   *Khŭrbet Umm el Bikâr.* The ruin of the mother of maidens.

خربة ام دبكل   Hx   *Khŭrbet Umm Dabkal.* The ruin of the mother of the bear.

خربة ام حارتين   Ix   *Khŭrbet Umm Hâretein.* The ruin with the two quarters or divisions.

خربة ام كلخة   Gx   *Khŭrbet Umm Kelkhah.* Ruin with the Kelkha (a plant like fennel or hemlock).

خربة ام خشرم   Ix   *Khŭrbet Umm Khŭshram.* The ruin of the soft stone.

خربة ام مالك   Iv   *Khŭrbet Umm Mâlak.* The ruin of Umm Mâlak. p.n.; (meaning 'mother of the owner').

خربة ام الميس   Jw   *Khŭrbet Umm el Meis.* The ruin with the Meis tree (Cordia Myxa).

خربة ام معرف   Hw   *Khŭrbet Umm muârrif.* The ruin of Umm Muârrif. p.n.: 'mother of the informer.'

خربة ام عشيش   Jv   *Khŭrbet Umm 'Osheish.* The ruin with the little nest.

خربة ام رجوم   Gw   *Khŭrbet Umm Rujŭm.* The ruin with the cairns.

خربة ام الشقف   Ix   *Khŭrbet Umm esh Shŭkf.* The ruin with the cleft. Another name for *Deir Muheisin.*

خربة ام السويدة   Iw   *Khŭrbet Umm es Suweideh.* The ruin of the blue thrush. Lieutenant Conder says: "According to the Bedawin of Jericho this is the name of *Pterocincla cyaneus* (Linn.)." The word means 'blackish.'

خربة ام طانون   Ew   *Khŭrbet Umm Tâbûn.* The ruin with the pit where they cover up fire to prevent its dying out.

خربة ام الطلعة   Iw   *Khŭrbet Umm et Talâh.* The ruin with the ascent.

خربة وادي صابر   Iw   *Khŭrbet Wâdy Sâbir.* The ruin of Wâdy Sâbir, q.v.

خربة الوحشية   Gw   *Khŭrbet el Wahashîyeh.* The ruin of the wild place.

خربة الويبدة   Ix   *Khŭrbet el Weibedeh.* The ruin of the pass.

خربة زارع   Gx   *Khŭrbet Zârâ.* The ruin of the sower.

خربة زيدان   Gw   *Khŭrbet Zeidân.* The ruin of Zeidân. p.n.

خربة الزمار   Iv   *Khŭrbet ez Zemmâr.* The ruin of the pipers.

خربة زحيلية   Fx   *Khŭrbet Zuheilikah.* The ruin of rolling or slipping.

القبيبة الشرقية   Iw   *El Kubeibeh (esh Sherkîyeh).* The little dome (the eastern).

كدنا   Iv   *Kudna.* Kudna. p.n.

القلعة   Iv   *El Kŭlâh.* The castle at *Bîr Jibrîn.*

قلعة الفنش   Gv   *Kŭlât el Fenish.* The castle of the Fenish. p.n.

مغر صرار   Hw   *Mŭghr Sŭrrâr.* The caves of pebbles.

منطار سومرة   Jx   *Muntâr Sômerah.* The watch-tower of Sômerah. See *Khŭrbet Sômerah*, p. 374.

منطرة القنيطرة   Hw   *Muntaret el Kuneiterah.* The watch-tower of the little arch.

باحية المجدل   {Gv / Hv}   *Nâhiet el Mejdel.* The commune of Mejdel.

بعلية   Ev   *Nâlia.* Nâlia. p.n.; either meaning 'in the form of a horse-shoe,' or from نعل sterile hard ground.

النبي دانيال　Fw　*Neby Dániál.* The Prophet Daniel.

النبي حام　Gv　*Neby Hám.* The Prophet Ham.

النبي هوج　Fw　*Neby Húj.* Perhaps the Prophet Og (though this is generally written in Arabic عوج).

النبي جبرين　Iv　*Neby Jibrín.* The Prophet Gabriel, near *Bir Jibrín.*

النبي كامل　Gv　*Neby Kámil.* The perfect prophet.

النبي مامن　Gv　*Neby Mámin.* The Prophet Mámin; ('trusty').

النبي صالح　Gv　*Neby Sáleh.* The Prophet Sáleh; ('righteous').

النبي سامي　Fv　*Neby Sámy.* Perhaps Prophet Shem; (though this is usually written سام). The form given in the list means 'lofty.'

نجد　Fw　*Nejed.* Highland.

رعناء　Iv　*Rána.* Rána. p.n.; ('foolish').

راس ابي حلتم　Iw　*Rás Abu Haltam.* The hill-top of Abu Haltam. p.n.

رسم العذرا　Iw　*Resm el 'Adhrá.* The vestiges of the virgin.

رسم اقتيش　Iw　*Resm Akteish.* The vestiges of Akteish. p.n.

رسم عامر　Jw　*Resm 'Âmir.* The vestiges of 'Âmir. p.n.

رسم حزانة　Iw　*Resm Hazzáneh.* The vestiges of Hazzáneh. *See Khúrbet Hazzáneh.* p. 371.

رسم المجارحي　Iw　*Resm el Mejárhi.* The vestiges of the combatants; (مجارحة 'wounding each other.')

رسم الرشيدات　Iw　*Resm er Rusheidát.* The vestiges of the Rusheidát Arabs.

رسم الشتاق　Ix　*Resm esh Shúkkák.* The vestige of the cleaver.

رسم ام بغلة　Ix　*Resm Umm Baghleh.* The vestiges of Umm Baghleh. *See Khúrbet Umm Baghleh,* p. 375.

الرجلية　Ix　*Er Rujlíyeh.* The pedestrians, or the stony land رجلي.

رجم العزازمة　Jv　*Rujm el 'Azázimeh.* The cairn of the Azázimeh Arabs.

رجم الدربي　Hv　*Rujm ed Derbî.* The cairn of the roadster.

3 C

رجم "تغندول	Jv	*Rujm el Kandôl.* The cairn of the thorn-tree (*Spartium aspalathoides*).
رجم السعاة	Hw	*Rujm es S'â.* The cairn of the tax-gathers.
سهل برجلتة	Hv	*Sahel Bûrjaliyeh.* The plain of Bûrjaliyeh.
سهل حمرة برير	Fw	*Sahel Hamret Bureir.* The plain of the red ground of Bureir. q.v., p. 367.
سهل السبطي	Ix	*Sahel es Sabtî.* The plain of the tribesman.
سطح برير	Fx	*Satch Bûrber.* The flat ground of Bûrber ; probably named from the village Bûrberah. q.v.
الشيخ عبدالله	Jv	*Sheikh 'Abdallah.* Sheikh Abdallah. p.n.
الشيخ ابو جاعد	Jw	*Sheikh Abu Jâad.* Sheikh Abu Jâad. p.n. ; meaning the man with the curly or crisp locks.
الشيخ ابو مسلم	Fv	*Sheikh Abu Musillim.* Sheikh Abu Musellim. p.n.; 'father of the saluter or of the preserver.'
الشيخ ابو رذان	Hv	*Sheikh Abu Raddân.* Sheikh Abu Raddân. p.n.; ('spinning').
الشيخ ابو ربيع	Iv	*Sheikh Abu Robîa.* Sheikh Abu Robîa. p.n.
الشيخ احمد ابوطوق	Ix	*Sheikh Ahmed Abu Tôk.* Sheikh Ahmed with the collar.
الشيخ احمد العريني	Hv	*Sheikh Ahmed el 'Areinî.* Sheikh Ahmed of the high place.
الشيخ علي الدوايمي	Iw	*Sheikh 'Aly (ed Duwâimî).* Sheikh Ali of Duweimeh. See p. 367.
الشيخ عمر	Iv	*Sheikh 'Amr.* Sheikh 'Amr.
الشيخ برّاك	Jv	*Sheikh Barrâk.* Sheikh Barrâk. p.n.; ('blessed').
الشيخ ابرهيم	Iw	*Sheikh Ibrahîm.* Sheikh Ibrahîm. p.n.; at *Khûrbet Sandahannah.*
الشيخ محمود	Iv	*Sheikh Mahmûd.* Sheikh Mahmûd. p.n.
الشيخ محمد	Fv	*Sheikh Muhammed.* Sheikh Mohammed. p.n.
الشيخ محمد القباقبة	Fv	*Sheikh (Muhammed) el Kebâkbeh.* Sheikh Mohammed the chatterbox.

الشيخ محمد المجاهد Gw *Sheikh Muhammed el Mujâhid.* Sheikh Mohammed the champion (taking part in a holy war).

الشيخ الشعيب Iv *Sheikh esh Shâib.* The Sheikh of the little spur of the hill ; from its position. *cf. Shuaib* = Jethro, p. 132. It is near *Tell Sandahannah.*

الشيخ شحادة ابو سل Hv *Sheikh (Shehâdeh) Abu Sell.* Sheikh Shehâdeh Abu Sell. p.n. ; meaning the beggar with the drawn sword.

الشيخ التيم Iv *Sheikh et Teim.* The Sheikh of et Teim. *See* p. 245.

سمسم Fw *Simsim.* Sesame.

السوق Iv *Es Sûk.* The market.

سميل Hv *Summeil.* Summeil. p.n. ; hard and withered.

الطبلة Iw *El Tableh.* The drum.

تل ابی دلاخ Gx *Tell Abu Dilâkh.* The mound of the father of fatness.

تل ابی الشقف Gw *Tell Abu esh Shâkf.* The mound with the cleft.

تل الاقرع Iw *Tell el Akra.* The bald mound.

تل بيت مرسم Ix *Tell Beit Mirsim.* The mound of Beit Mirsim. q.v.

تل برنایا Iv *Tell Bornât.* The mound of Bornât. *See Khûrbet Bornâteh,* p. 370.

تل الدوير Iw *Tell ed Daweir.* The mound of the small monastery.

تل الهوا Fv *Tell el Hawa.* The mound of the wind.

تل الحسي Gw *Tell el Hesy.* The mound of the water collecting in sandy soil.

تل هديون Gx *Tell Hudeiweh.* The mound of gifts (هداون).

تل إدبيس Gw *Tell Idbis.* The mound of Idbis. p.n. ; either from دَبَس 'black,' or دِبْس 'grape syrup.'

تل الجديدة Iv *Tell el Judeiyideh.* The mound of the 'dykes,' or streaks in the rock.

تل خراقة Hw *Tell Kharâkah.* Mound of rags or tatters.

تل مجادل Ix *Tell Mejâdil.* The mound of watch-towers.

تل الملیحة Hx *Tell el Muleihah.* The mound of the salt-pan.

تل النجيلة   Hw   *Tell en Nejileh.* The mound of the nejileh (a plant).

تل الصحرآ   Ev   *Tell es Sahra.* The mound of the desert.

تن صندحنة   Iv   *Tell Sandahannah.* The mound of St. Anne.

طور دمرة   Ew   *Tôr Dimreh.* The mound of Dimreh. *See Beit Tûmrah,* p. 363.

طور فرطا   Hv   *Tôr Furut.* The mount of Furut. *See Khûrbet Furut,* p. 371.

طور الهري   Hv   *Tôr el Hiry.* The mount of the granary ; a cliff formed apparently by a landslip.

طور خنزيرة   Iw   *Tôr Khandzîreh.* The mount of the swine ; a hill east of *Khûrbet Beiram,* with a cave.

طويل الشعير   Ew   *Tuweil esh Shâîr.* The hill-peak or long ridge of barley.

ام بابين   Iv   *Umm Bâbein.* The place with two entrances. *See Bîr Abu Bâbein,* p. 366.

ام لاقس   Gw   *Umm Lâkis.* The place of the itch.

وادي العبد   {Gx / Ix}   *Wâdy el 'Abd.* The valley of the slave.

وادي ابى الاعرج   Jv   *Wâdy Abu el 'Âdraj.* The valley of the father of the lame one.

وادي ابى العيون   Gv   *Wâdy Abu el 'Ayûn.* The valley with the springs.

وادي ابى دلّخ   Gx   *Wâdy Abu Dilâkh.* The valley of the father of fatness.

وادي ابى حنا   Ix   *Wâdy Abu Henna.* The valley where the henna grows. (The plant used by ladies in the East to tinge their fingers red.)

وادي ابى ربيع   Iv   *Wâdy Abu Robiâ.* The valley of Abu Robiâ. p.n. It may be rendered 'the place where spring pasturage is to be found.'

وادي العرب   Iw   *Wâdy el 'Arab.* The valley of the Arabs.

وادي بيت علام   Iv   *Wâdy Beit 'Alâm.* The valley of the house of the land-mark.

وادي البئارة   Ix   *Wâdy el Beiârah.* The valley of wells.

وادي البياض   Iw   *Wâdy el Bâad.* The white valley.

وادي البيرة   Fv   *Wâdy el Bîreh.* The valley of el Bireh (the palace) ; from the ruin of that name.

وادي بيسيا Hᵥ *Wâdy Bîsia.* The valley of Bisia. p.n.; perhaps from بيس 'difficulty,' 'confusion.'

وادي بربر Gx *Wâdy Bŭrber.* The valley of Bŭrber. *See Bŭrberah.*

وادي البطم ${ \text{Iw} \atop \text{Ix} }$ *Wâdy el Butm.* The valley of the terebinth.

وادي دباغة Hᵥ *Wâdy Dabbâghah.* The valley of the tanners.

وادي الدروش Jᵥ *Wâdy ed Drŭsch.* The valley of ed Drŭsch. *See Khŭrbet ed Drŭsch,* p. 370.

وادي الفرس Ix *Wâdy el Faras.* The valley of the mare.

وادي فضيل Iᵥ *Wâdy Fedeil.* The valley of Fedeil. p.n. ('accomplished,' 'excellent ').

وادي فضيلس Ex *Wâdy Fuâilis.* The valley of Fuâilis. p.n.

وادي الغار Fx *Wâdy el Ghâr.* The valley of the depression.

وادي الغويط Gᵥ *Wâdy el Ghŭeit.* The valley of the low lying, or naturally irrigated ground.

وادي الغفر Iw *Wâdy el Ghŭfr.* The valley of the escort.

وادي غزة Gᵥ *Wâdy Ghŭzzeh.* The valley of Gaza.

وادي الهبور Hw *Wâdy el Habûr.* The valley of the level lands.

وادي حجر المخزوق Hᵥ *Wâdy Hajr el Makhzûk.* The valley of the perforated stone; an old millstone lies in it.

وادي حزانة Iw *Wâdy Hazzâneh.* The valley of Hazzâneh. *See Khŭrbet Hazzâneh,* p. 371.

وادي الحسي ${ \text{Fw} \atop \text{Gw} }$ *Wâdy el Hesy.* The valley of the water accumulated in sand.

وادي الجبار Fᵥ *Wâdy el Jabbâr.* The valley of the tyrant.

وادي جمامة Gw *Wâdy Jemmâmeh.* The valley of collections, or of reservoirs.

وادي الجلس Gᵥ *Wâdy el Jils.* The valley of el Jils. *See Khŭrbet el Jils,* p. 372.

وادي الجزاير Iw *Wâdy el Jizâir.* The valley of islands.

وادي جرف الرزيقي Ix *Wâdy Jorf el Ruzeiky.* The valley of the bank of Ruzeiky. p.n.; (رزق 'provision,' or رزيغي *Ruzeighy,* 'miry.')

وادي الجديدة Ix *Wâdy el Judeiyideh.* The valley of the dyke, or streaks in the rock.

وادي القنيطرة IIw *Wâdy el Kaneiterah.* The valley of the small arch.

وادي التشب Iw *Wâdy el Kashb.* The valley of poison (also the name of a plant).

وادي قلا Gv *Wâdy Kella.* The valley of Kella. p.n. Perhaps كلّا pasturage.

وادي قماس Fv *Wâdy Kemâs.* The valley of Kemâs. *See Khûrbet Kemâs,* p. 372.

وادي قرقرة Hv *Wâdy Kerkerah.* The valley of soft smooth soil.

وادي ترتيطي Gx *Wâdy Kerkîty.* The valley of the fragment or section, the usual form is قرقوطة ; it is a modern Arabic word.

وادي خضرآ IIx *Wâdy Khâdra.* The green valley.

وادي الخليل IIv *Wâdy el Khaleil.* The valley of the little dell.

وادي الكوفخة Fx *Wâdy el Kôfkhah.* The valley of Kôfkhah. *See Khûrbet Kôfkhah,* p. 372.

وادي قرام Jw *Wâdy Korrâm.* The valley of stumps.

وادي القمح Ix *Wâdy el Kûmh.* The valley of wheat.

وادي قمحة Iv *Wâdy Kûmhah.* The valley of wheat.

وادي قصيب IIx *Wâdy Kûseib.* The valley of the little reed.

وادي قصابة Hx *Wâdy Kûssâbah.* The valley of the butchers.

وادي مجادل Hx *Wâdy Mejâdil.* The valley of watch-towers.

وادي مليتا Gv *Wâdy Melîta.* The valley of Melita. *See Khûrbet Melita,* p. 373.

وادي مرتينيا Ix *Wâdy Mertînia.* The valley of Mertinia. p.n.; near the ruin of the same name.

وادي المفرد Gv *Wâdy el Mufârred.* The separated valley. The word is applied to hills standing apart. *See Muferredât es Sebâ.*

وادي المليحة Hx *Wâdy el Muleihah.* The valley of the salt-pan.

وادي مران Ix *Wâdy Murrân.* The valley of Murrân. *See Khûrbet Murrân,* p. 373.

وادي المسيجد Iv *Wâdy el Museijid.* The valley of the mosques.

وادي النعاس	IIx	*Wády en Nás.* The valley of drowsiness.
وادي الندا	IIw	*Wády en Neda.* The valley of dew.
وادي النفاخ	Ix	*Wády en Neffakh.* The valley of the blower.
وادي نخبار المغل	Fx	*Wády Nukhbár el Baghl.* The valley of the fissure of the mule. خبار probably by inversion from نجراب a hole or fissure in a rock.
وادي الرجليّة	IIx	*Wády er Rujliyeh.* The valley of *er Rujliyeh.* q.v.
وادي الرخريخ	Iv	*Wády er Rukhreikh.* The valley of soft mud.
وادي الرمل	Fw	*Wády er Rúml.* The valley of sand.
وادي صابر	Iv	*Wády Sábir.* The valley of prickly pear.
وادي السعدة	Gx	*Wády es Sádeh.* The valley of es Sádeh. p.n. ; ('Felicia').
وادي السّدبي	Jv	*Wády es Setjí.* The valley of the distaff.
وادي الشهوان	Jx	*Wády esh Shahwán.* The valley of falcons.
وادي الشاور	Iv	*Wády esh Sháúr.* The valley of Sháúr. p.n. ('counsel') : another name for upper part of *Wády Umm el Haiyát.*
وادي الشيخ براك	Jv	*Wády esh Sheikh Barrák.* The valley of Sheikh Barrák. q.v.
وادي الشلاليّين	IIx	*Wády esh Shelláleiún.* The valley of the two cascades.
وادي الشقاق	Iw	*Wády esh Shŭkkák.* The valley of the clefts (or chasm).
وادي السّدّ	Gx	*Wády es Sidd.* The valley of the cliff, dam, or barrier.
وادي سمسم	Iw	*Wády Simsim.* The valley of Sesame.
وادي السكريّة	IIw	*Wády esh Sukríyeh.* The valley of the sugar manufactories.
وادي الصورة	Iv	*Wády es Súrah.* The valley of the pictures.
وادي الطبقة	Gw	*Wády et Tŭbakah.* The valley of the terrace.
وادي طقيش	Hv	*Wády Tŭkkish.* The valley of Tŭkkish. p.n.
وادي ام عود	IIy	*Wády Umm 'Amúd.* The valley of the place with the columns.
وادي ام فطيرة	Jw	*Wády Umm Fatíreh.* The valley of the place where the sheep is sacrificed on the feast called *'Íd el Fitr* or Beirám.

وادي ام الحيّات　Iv　*Wâdy Umm el Haiyât.* The valley with the snakes.

وادي ام حجر　Gx　*Wâdy Umm Hajr.* The valley with the rocks.

وادي ام خشرم　Ix　*Wâdy Umm Khûshram.* The valley of Umm Khûshram. *See* p. 375.

وادي ام بتس　Fw　*Wâdy Umm Lâkis.* The valley of Umm Lâkis. *See* p. 380.

وادي ام معرف　Hw　*Wâdy Umm Muârrîf.* The valley of Umm Muârrîf. *See Khûrbet Umm Muârrîf,* p. 375.

وادي ام رجوم　Gw　*Wâdy Umm Rujûm.* The valley with the cairns.

وادي الوصية　Iv　*Wâdy el Wasîyeh.* The valley of the commandment, or testament.

وادي الواوية　Iw　*Wâdy el Wâwîyeh.* The valley of the jackals.

وادي زردم المصري　Iw　*Wâdy Zerdûm el Musrî.* The valley of strangling of the Egyptian.

وادي الزيادة　Gx　*Wâdy ez Ziâdeh.* The valley of overflowing.

زيتا　Iw　*Zeita.* Oil.

SHEET XXI.

ابو نجيم   Mv   *Abu Nejeim.* The father of the little star.

ابو الروازن   Jv   *Abu er Rûâzin.* The father of crevices.

عين ابي جبارة   Kw   *Ain Abu Jebârah.* The spring of Abu Jebârah; probably from the Jebârah Arabs.

عين ابي خيط   Kx   *'Ain Abu Kheit.* The spring of Abu Kheit. p.n.; meaning the father of the thread, or small stream.

عين ايتيقة   Kw   *'Ain Aîtikah.* The spring of Aîtikah; perhaps عتيقة 'ancient.'

عين العاصي   Lw   *'Ain el 'Âsi.* The spring of the 'Âsi; ('rebellious'). *See* p. 157.

عين العوينات   Kv   *'Ain el 'Aweinat.* The spring of the grounds flooded by rain: عرانات pronounced with the *imâleh.* *See* Note, p. 350.

عين بحّة   Kw   *'Ain Bahhah.* The spring of the pool, or open space.

عين بقّار   Kv   *'Ain Bakkâr.* The spring of cowherd.

عين البيضآ   Lx   *'Ain el Beida.* The white spring.

عين البيض   Kv   *'Ain el Beiyid.* The whitish spring.

عين البستان   Kw   *'Ain el Bestân.* The spring of the garden.

عين البويرب   Lx   *'Ain el Bûeiry.* The spring of the little hole.

عين الدالية   Kw   *'Ain el Dâlieh.* The spring of the trailed vine (vineyards exist near it).

عين الدارات   Mw   *'Ain ed Dârât.* The spring of houses, or circles in the open desert.

عين الذروة النوتآ   Lv   *'Ain edh Dhirweh (el Foka).* The upper spring of the summit.

3 D

عين الذروة التحتآ   'Ain edh Dhirweh (et Tahta). The lower spring of the summit.

عين الدلبة   Kx   'Ain ed Dilbeh. The spring of the plane-tree.

عين اصحا   Lw   'Ain Es-ha. The spring of Es-há (the word means 'clearing up,' [used of the weather] or 'getting sober.'

عين ايوب   Lv   'Ain Eyûb. Job's (or Joab's) spring.

عين فرتة   Kw   'Ain Firâh. The spring of the mountain top; marked 'A at Khârbet Firâh.

عين الفريديس   Kx   'Ain el Fureidîs. The spring of the little paradise.

عين الفوار   Kx   'Ain el Fuirâr. The spring of the fountain.

عين حمدة   Mv   'Ain Hamdeh. The spring of Hamdeh. p.n.

عين الحتة   Lv   'Ain el Hassch. The spring of gravel.

عين الهاوية   Lv   'Ain el Hawiyeh. The spring of the chasm.

عين الهجري   Kx   'Ain el Hejeri. The spring of the fugitive.

عين ابن سليم   Lw   'Ain Ibn Islîm. The spring of Selim's son. p.n.; at Khârbet Ibn Islîm.

عين جدور   Lv   'Ain Jedâr. The spring of the walls, or of the herbs growing in the sand.

عين الجديدة   Lw   'Ain el Judeideh. The new spring, or the spring of the dyke or streak in the rock. See "Memoirs."

عين تعيدة   Lw   'Ain Kâideh. The spring of the sitter (female).

عين القناة   Lw   'Ain el Kana. The spring of the aqueduct.

عين كنار   Kw   'Ain Kanâr. The spring of the fruit of the lotus-tree, or of 'the margin.'

عين كارب   Kw   'Ain Kârib. The spring of the sower, or of the man who ties up the bucket in a well.

عين كشكلة   Lw   'Ain Kashkaleh. The spring of Kashkaleh. See p. 372.

عين خير الدين   Lw   'Ain Kheir ed Dîn. The spring of Kheir ed Din. p.n.: ('the elect of the Faith').

عين كويزيبا   Lv   'Ain Kûeiziba. The spring of Kûeiziba. See Khârbet Kûeizîba, p. 400.

عين المتوف Kw '*Ain el Kúf.* The spring of the margin.

عين كوفين Lv '*Ain Kúfin.* The spring of Kúfin. q.v.

عين المجنون Kw '*Ain el Mejnún.* The spring of the madman.

عين الماجور Kx '*Ain el Májúr.* The spring of the flower-pot.

عين مرزوقة Kw '*Ain Marzúkah.* The spring of the (woman) provided for by God.

عين المزروق Lw '*Ain el Mezrúk.* The spring of the (man) provided for by God.

عين الناطوف Mv '*Ain en Nátúf.* The dripping spring ; a small spring dripping from a precipice close to *Mughárel Músa.*

عين الرشراش Lv '*Ain er Rishrásh.* The spring of the willow (*Agnus castus*).

عين سعبية Kw '*Ain Sábiych.* The spring of the Sábiych. p.n. ; ('difficult').

عين السادة Lx '*Ain es Sáddeh.* The spring of the cliffs or dams.

عين السخري Mv '*Ain es Sakhri.* The spring of the forced labourer ; perhaps سخرى 'rocky.'

عين صارة Lw '*Ain Sárah.* The spring of the copse.

عين شخاخ ابي ثور Kx '*Ain Shekhákh Abu Thôr.* The spring of the staling of the bull, *i.e.*, defiled by oxen.

عين السيهانية Lx '*Ain es Sihánīych.* The spring of running water.

عين التينة Kv '*Ain et Tíneh.* The spring of the fig-tree.

عين ام ركبة Lx '*Ain Umm Rukbeh.* The spring with the acclivity.

عين النقر Kw '*Ain el Unkúr.* The spring of perforations.

عين وادي الشنار Lv '*Ain Wády esh Shinnár.* The spring of the valley of the Greek partridge.

عين الزرقآ Kw '*Ain ez Zerka.* The azure spring.

عين زبود Lv '*Ain Zúbbúd.* The spring of foam. *cf. Khúrbet Zebed,* p. 51.

عقبة زيدان Nw '*Akabet Zeidán.* The steep or mountain road of Zeidán. p.n.

عرب الجِّهبالين {Mx} *Arab el Jâhalîn* (properly *Jehhâlîn*). The Jâhalin Arabs.
{Nx} p.n.; meaning ignorant.

عرب التعامِرة {Mw} *'Arab et T'âmireh.* The T'âmireh Arabs. p.n.; cultivating.
{Nv} They are not true Bedawin, and are also called عرب الصبيح *Arab es Subeih.* p.n.

العوجآء Lx *El 'Auja.* The crook; near *Rujm el Fahjeh.*

عراق الزان Kw *'Arâk ez Zân.* The cavern or cliff of ez Zân (perhaps *Zânî* 'adulterer'). A cave in a cliff.

عرقان المدنيّة Jv *'Arkân el Medenîyeh.* The cavern of the people of Medina; caverns near *Umm Burj.*

عرقان الطراد Lv *'Arkân et Trâd.* The caverns of et Trâd (perhaps Tarrâd, a spacious place).

بلّوطة الجدور Lv *Ballûtet el Jedûr.* The oak of Jedûr; north of Jedûr. *See 'Ain Jedûr,* p. 386.

بلّوطة القسّيس Kv *Ballûtet el Kůssîs.* The oak of the (Christian) priest.

بلّوطة سبتا Kw *Ballûtet Sebta.* The oak of rest. The traditional oak of Mamre.

بلّوطة الشيخ Kv *Ballûtet esh Sheikh.* The oak of the Sheikh.

بلّوطة اليرزة Lv *Ballûtet el Yerzeh.* The oak of el Yerzeh. p.n.

بيت علام Jw *Beit 'Alâm.* The house of the land-mark.

بيت اولا Kv *Beit 'Aûlâ.* The house of Aulâ. p.n.

بيت عوّا Jw *Beit 'Aûwâ.* The house of 'Aûwâ (barking like a dog).

بيت البان Jw *Beit el Bân.* The house of the ben-tree.

بيت فِجّار Lv *Beit Fejjâr.* The house of the debauchees. But it may be from فجرة an open valley with water in it, such as exists here.

بيت كانون Lv *Beit Kânûn.* The house of the hearth.

بيت كاحيل Kw *Beit Kâhel.* The house of the occulist.

بيت خيران Lv *Beit Kheirân.* The house of Kheirân. p.n.

بيت خليل الرّحمن Lv *Beit el Khůlîl (er Rahman).* The house of the friend of the Merciful One, *i.e.,* Abraham. This is the early Christian site of Mamre.

بيت لام Jv *Beit Lâm.* The house of Lâm. p.n.

بيت مكدوم Jw *Beit Makdûm.* The house of Makdûm. p.n. It should probably be either مخدوم *Makhdûm*, 'served,' *i.e.*, 'respected,' or مقدوم *Makdûm*, 'put forward.'

بيت نصيب Jv *Beit Nûsib.* The house of Nusib. p.n.; (meaning portion, fortunes).

بيت سلّة Kv *Beit Sûllûh.* The house of Sûllûh. p.n.

بيت صور Lv *Beit Sûr.* The house of rock. Heb. בית צור *Beth zur.*

بيت أمر Lv *Beit Ummar.* The house of Ummar. p.n. *See* "Memoirs."

بيت زعتا Lv *Beit Zâta.* The house of Zâta. p.n.

بني نعيم Lv *Beni Nâim.* Sons of the affluent one.

بنار الديرات Lx *Biâr ed Deirât.* The wells of the convents; at *Khûrbet Deirât.*

بنارة وادي الصور Jv *Biâret Wâdy es Sûr.* The wells of Wâdy es Sûr. q.v.

بئر ابو الحمام Lx *Bir Abu el Hamâm.* The well with the pigeons; at *Khûrbet Abu Hamâm.*

بئر ابي قبّة Kv *Bir Abu Kabsch.* The well of Abu Kabsch. p.n. قبس 'to obtain light for a fire from another.'

بئر ابي المثور Jw *Bir Abu el Mekûr.* The well with the water-holes.

بئر ابي شبّان Lx *Bir Abu Shebbân.* The well of the father of the youths.

بئر العينازية Mv *Bir el 'Ainâzîyeh.* The well of the 'Ainâziyeh. p.n.; *cf.* '*Ain el 'Anâziyeh,* p. 278. عنز she goats.

بئر العلقة Kx *Bir el 'Alakah.* The well of leeches.

بئر علّا Mv *Bir 'Alla.* The well of repeated drinking.

بئر المريّة Lx *Bir el 'Amriyeh.* The well of 'Amriyeh. p.n. *See* p. 268.

بئر العروس Kv *Bir el 'Arûs.* The well of the bridegroom.

بئر عزيز Kx *Bir 'Azîz,* at *Khûrbet 'Aziz.* The well of 'Aziz. p.n.; ('precious').

بئر بنت دغيم Jw *Bir Bint Dugheim.* The well of the daughter of Dugheim. p.n.; meaning blackish colour.

بئر البطّة Lx   *Bîr el Búttah.* The well of the duck.

بئر دار حسين عيد Lv   *Bîr Dâr Husein 'Aîd.* The well of the house of Husein 'Aîd.

بئر داود Jw   *Bîr Dâûd.* David's well.

بئر الدحية Nw   *Bîr ed Dehîyeh.* The well of the intelligent one, or of the misfortune.

بئر دير صامت Jw   *Bîr Deir Sâmat.* The well of the convent of the silent man ; at *Khŭrbet Deir Sâmat.*

بئر الدلبة {Lv / Lw}   *Bîr ed Dilbeh.* The well of the plane tree. *See 'Aîn ed Dilbeh*, p. 279.

بئر الفوار {Lv / Jw}   *Bîr el Fûwâr.* The well of the fountain.

بئر الحاجّ رمضان Lv   *Bîr el Hâj Ramadân.* The well of the pilgrim Ramadhân. p.n.

بئر هرون Jv   *Bîr Harûn.* The well of Aaron. Also called بئر امّ برج *Bîr Umm Burj* (the well of Umm Burj) ; q.v. ; and بئر الحنو *Bîr el Henû* (the well of the bend) ; north of *Umm Burj.*

بئر الحسوة Mv   *Bîr el Heswch.* The well of sipping.

بئر حباحب Mv   *Bîr Hŭbâhib.* The well of trickling, or of melons.

بئر هديهد Jw   *Bîr Hudeihid.* The well of hoopoes (*Upupa epops*).

بئر الحمّام Jv   *Bîr el Hŭmmâm.* The well of the bath ; at *Khŭrbet el Hŭmmâm.*

بئر جابر Lv   *Bîr Jâbr.* The well of Jâber. p.n. ; meaning ' repairer,' ' bone-setter.'

بئر الجرادات Lw   *Bîr el Jerâdât.* The well of locusts.

بئر جرن الكنف Jw   *Bîr Jurn el Kuf.* The well of the trough of the margin ; at *Beit 'Anwa.*

بئر الكنوب Mv   *Bîr el Kanûb.* The well of flower-buds.

بئر الخنزير Lv   *Bîr el Khanzîr.* The well of the pig.

بئر الكسي Lv   *Bîr el Kisi.* The well of the rear.

بئر القوس Kv   *Bîr el Kôs.* The well of the bow.

بئر القصير Nw   *Bîr el Kŭseir.* The well of the little castle.

بئر المالح   Kw   *Bîr el Mâleh.* The salt well.

بئر مرينا   Lv   *Bîr Marrîna.* The well of Marrina. p.n.

بئر المشرفة   Mw   *Bîr el Meshrefeh.* The well of the high place.

بئر المنية   Nw   *Bîr el Minyeh.* The well of el Minyeh. p.n.

بئر المؤنسة   Mw   *Bîr el Mûneseh.* The familiar well; also called *Bîr el Jerâdât.* q.v., p. 390.

بئر النجد   Kv   *Bîr en Nejed.* The well of the highland.

بئر النصار   Mx   *Bîr en Nûssâr.* The well of Nûssâr. p.n.; meaning 'helper.'

بئر النصارى   Lw   *Bîr en Nûsâra.* The Christian's well.

بئر رخمة   Mv   *Bîr Rakhamah.* The well of the Egyptian vulture.

بئر الرامة   Lw   *Bîr er Râmeh.* The well of er Râmeh. q.v.

بئر الرشيدية   Mw   *Bîr es Rusheidîyeh.* The well of the Rusheidiyeh Arabs.

بئر السبيل   Kw   *Bîr es Sebîl.* The well of the wayside fountain.

بئر السفلي   Jw   *Bîr es Sifli.* The lower well.

بئر السير   Mw   *Bîr es Sîr.* Well of the fold.

بئر السكيرية   Nw   *Bîr es Sukeirîyeh.* The well of the dam; it is partly formed by a masonry wall.

بئر السوقية   Mw   *Bîr es Sûkîyeh.* The well of the market folk.

بئر تكوش   Lv   *Bîr Takkûsh.* The well of Takkûsh. p.n.

بئر الطويل   Lw   *Bîr el Tawîl.* The long well (in deep).

بئر الثريا   Nw   *Bîr Thureiya.* The well of the Pleiades.

بئر مالحة الطرفآ   Nv   *Bîr (el Mâlhah) et Turfah.* The well of the salt marsh of the tamarisk.

بئر ام العوسج   Mx   *Bîr Umm el 'Ausej.* The well with the thorn-tree (*Lycium Europeum*); in the wâdy of the same name.

بئر ام الدرج   Lv   *Bîr Umm ed Deraj.* The well with the steps.

بئر ام فطيرة   Jx   *Bîr Umm Fatîreh.* The well where the sheep is slaughtered at the Îd. el Fitr. *See Wâdy Umm Fatîreh,* p. 383.

بئر ام البجعد   Nv   *Bîr Umm el Jâd.* The well with the moist ground; it may also mean the mother of the curly-headed man.

بئر ام جدي   Nw   *Bîr Umm Jidy.* The well with the kid.

بئر ام سلمان   Mv   *Bîr Umm Selmân.* The well of the mother of Selmân. p.n.

بئر وادي البيار   Kv   *Bîr Wâdy el Biâr.* The well of the valley of wells (*Wâdy el Biâr*). q.v.

بئر وادي الزلقة   Jv   *Bîr Wâdy ez Zelefeh.* The well of the valley of the cistern.

بئر الوعر   Mv   *Bîr el Wâr.* The well of the rugged rocks.

بئر الزعفران   Mv   *Bîr ez Zâferân.* The well of Saffron.

بركة العروب   Lv   *Birket el 'Arrûb.* The well of el 'Arrûb. p.n.; عروب a woman who loves her husband.

بركة السلطان   Lw   *Birket es Sultân.* The Sultan's pool; in Hebron.

البقعة   {Jx Lw}   *El Bŭkâh.* The open valley.

بقعة التقوع   Mv   *Bŭkât et Tekûa.* The vale of Tekoa. *See Khŭrbet Tekûa,* p. 402.

برج بيت ناصيف   Kv   *Burj Beit Nâsîf.* The tower of Beit Nâsîf. *See Khŭrbet Beit Nâsîf.*

برج حسكة   Kw   *Burj Haskeh.* The tower of the Caltrop (a prickly plant).

برج الصور   Lv   *Burj es Sûr.* The tower of the rock; at Beit Sûr.

بطمة وادي الصور   Kv   *Butmet Wâdy es Sûr.* The terebinth of valley of the rock.

دويربان   Kw   *Daweirbân.* Daweirbân. p.n.; perhaps 'the little house of the ben-tree.' A hill-top near Hebron.

الديدورا   Kv   *Ed Deidûra.* Ed Deidûra. p.n.

الدير   Mw   *Ed Deir.* The monastery.

دير الاربعين   Lw   *Deir el Arbâin.* The monastery of the Forty (Martyrs of Cappadocia).

دير العسل   Jx   *Deir el 'Asl.* The monastery of honey.

دير الدومة   Jx   *Deir ed Dômeh.* The monastery of Dômeh. q.v.

دير الموس   Jv   *Deir el Mûs.* The monastery of the razor.

ظهر ابي رمان   Kw   *Dhahr Abu Rummân.* The ridge with the pomegranates.

ظهر قلقس Lw *Dhahr Kilkis.* The ridge of the potato.

ظهر الطويل Lw *Dhahr et Tawîl.* The long ridge; not written on plan; south of *Khûrbet el 'Addeisîyeh.*

ظهرة العتبة Mw *Dhahret el 'Atcibah.* The ridge of the threshold or steps.

ظهرة فاتح صدره Mv *Dhahret Fâtch Sidru.* The ridge of the man who opens his breast.

ظهرة الحدّادة Mv *Dhahret el Haddâdeh.* The boundary ridge.

ظهرة الكوّا Mx *Dhahret el Kôlah.* The ridge of undulating ground; possibly the Heb. הר ח חבילה *Hill Hachilah.*

ظهرة المشرفة Mw *Dhahret el Meshrefeh.* The ridge of the high place.

ظهرة السوقية Mx *Dhahret es Sûkîyeh.* The ridge of the market-place.

ظهرة أم مغور Mv *Dhahret Umm Mughûr.* The ridge with the caves; or perhaps مغُر *Mukûr*, water-holes.

الدومة Jx *Ed Dômeh.* The dome palm; Heb. דוּמָה *Dumah.*

دورا Kw *Dûra.* Dûra. p.n.; Heb. אֲדוֹרַיִם *Adoraim.*

فاثورة الزير Mv *Fathûret ez Zir.* The large dish or salver of Zir (one of the Tâmireh Arabs); a hill.

فقيقيس Jx *Fukeikis.* Melons. A ruin.

الحبس Lw *El Habs.* The prison, or the mosque property.

الحدب Kn *El Hadab.* The hummock.

حجر السمرا Nw *Hajr es Sumra.* The brown stone; a land-mark.

الحرم Lw *El Haram.* The sanctuary; in Hebron.

حرملي Mv *Harmely.* Rue; a plot of ground. The name has not been put on the map. It applies west of *Jebel Fureidîs.*

حنو الارنب Nw *Henû el Erneb.* The bend of the hare.

حرابي البيض Mv *Herâbi el Beid.* The white cisterns.

حلحول Lw *Hûlhûl.* Hûlhûl. p.n. Heb. חַלְחוּל *Halhul.*

خربة ابي سلهوب Nx *Hurubbet Abu Selhûb.* The cistern of Abu Selhûb. p.n. (father of the tall one).

خربة أم القلب Nx *Hurubbet Umm el Kuleib.* The cistern of the old well.

3 E

خربة أم الثراقة   Nv   *Hurubbet Umm et Terâfeh.* The cistern of Umm et Terâfeh. p.n.; (signifying the woman or tribe with a long line of ancestry).

إذنا   Jw   *Idhna.* Lower.

جامع ابن عثمان   Lw   *Jâmiâ Ibn 'Othmân.* The mosque of the son of 'Othmân.

جبل ابي بجيم   Mv   *Jebel Abu Nejeim.* The mountain of Abu Nejeim; by the mosque so named. *See* p. 385.

جبل فريديس   Mv   *Jebel Fureidîs.* The mountain of the little paradise.

جبل خليل الرحمن   *Jebel Khŭlîl (er Rahmân).* The mountain of Hebron.

البلاجل   Lw   *El Jilâjil.* The circuits; a winding road.

جوفة الدويرة   Nw   *Jôfet ed Duweiyirah.* The hollow of the little circle.

جوفة الغزال   Nw   *Jôfet Ghŭzâl.* The hollow of the gazelle.

جوفة المفزرة   Nw   *Jôfet el Mufezzerah.* The fissured hollow.

البليدة   Kv   *El Juleideh.* The hard rock.

جورة الجمل   Lx   *Jûrat el Jemel.* The hollow of the camel.

قاعة الزيتونة   Lv   *Kâât er Zeitûneh.* The plain or courtyard of the olive tree.

قبر عبيان   Nw   *Kabr 'Abeiyân.* Grave of 'Abeiyân. p.n.; ('modern').

قبر غنامة   Nv   *Kabr Ghannâmeh.* The grave of Ghannâmeh. p.n.; a woman.

قبر حبرون   Lw   *Kabr Hebrûn.* The grave of Hebron.

قبر شيخ   Nw   *Kabr Sheikh.* The Sheikh's tomb.

قبر يسّى   Lw   *Kabr Yessa.* The tomb of Jesse; at *Deir el Arbâîn.*

قنان الدير   Lx   *Kanân ed Deir.* The peaks of the monastery.

قبان الحدادية   Lx   *Kanân el Haddâdîyeh.* The peaks of the boundaries.

قنان إسقير   Mv   *Kanân Iskîr.* The peaks or Iskîr. p.n.; but *see Wâdy Abu Sekeir,* p. 409.

قنان الغميب   Nv   *Kanân el Keheib.* The peaks of the dusky mountain.

قنان الملاقي   Lx   *Kanân el Malâki.* The peaks of smooth stone.

ننان رجم قذاح Nw — *Kanán Rujm Kůddáh.* The peaks of the cairn of the potter, or of the flint-stone for striking fire.

قنان الصبيخة Mv — *Kanán es Subhah.* The peaks of reddish-white.

قنان الزعفران Mv — *Kanán ez Záferán.* The peaks of saffron.

قطائا عزون Lw — *Katát 'Azzún.* The crags of the wild olive.

الكنيسة Jw — *El Kenîseh.* The church. At *'Khŭrbet Beit 'Ainûn.*

خلة الاعرج Jv — *Khallet el 'Aáraj.* The dell of the lame man.

خلة ابى ناذور Nv — *Khallet Abu Nádhûr.* The dell of Abu Nâdhùr. *See Tell Abu Nádhûr,* p. 329.

خلة ابى شلبى Jv — *Khallet Abu Shelebeh.* The dell of the dandy.

خلة العملة Jv — *Khallet el 'Amleh.* The dell of labour. The word may also mean ' deceit,' and ' wages.'

خلة الدفي Lx — *Khallet ed Dafi.* The dell of warmth and comfort.

خلة الحرم Lw — *Khallet el Haram.* The dell of the sanctuary (near the mosque of Neby Yûnis).

خلة الحمص Kx — *Khallet el Hummŭs.* The dell of the chick-pea.

خلة الجوخ Jv — *Khallet el Jûkh.* The dell of the stream that carries away the banks.

خلة القرما Jv — *Khallet el Kerma.* The dell of tree-stumps.

خلة مزهر Lx — *Khallet Mizhir.* The dell of hospitality (literally, ' kindling a fire to attract a guest '; near *Mukám el Khŭdr*).

خلة المصلى Kx — *Khallet el Musálla.* The dell of the praying place.

خلة النعجة Lw — *Khallet en Nájeh.* The dell of the ewe.

خلة سراج Lw — *Khallet Seráj.* The dell of the lamp.

خلة السرارة Jv — *Khallet es Surárah.* The dell of good soil.

خلة ام الشقحان Lx — *Khallet Umm esh Shuk-hán.* The dell of Umm esh Shuk-hán. p.n.; ' mother of a reddish-white colour '; near *Khŭrbet el Miyeh.*

خراس Kv — *Kharás.* Kharás. p.n.

خروفا Lx — *Kharûfa.* The sheep.

خشم الكرم　Mx　*Khashm el Kurm.* The promontory of the vineyard.

خشم صفرا لوندى　Nx　*Khashm Sŭfra Lawundi.* The promontory of Sŭfra Lawundi. p.n.

خشم ام درج　Mx　*Khashm Umm Deraj.* The promontory with the steps.

خلال خروايا　Lx　*Khelâl Kherwât.* Perhaps خروات the dells of the castor-oil plants.

خلايل ابى بيص　Lw　*Khelâyil Abu Beid.* The dells of the white place.

الخليل الرحمن　Lw　*El Khŭlil (Khŭlil er Rahmân).* The friend (of the Merciful One), *i.e.,* Abraham ; the modern name of Hebron.

[*The following are names of places in and close to Hebron, not all written on the Map.*]

باب الزاوية　*Bâb ez Zâwieh.* The gate of the corner.

بركة السلطان　*Birket es Sultân.* The Sultan's pool.

حبال الرياح　*Habâl er Riâh.* The terraces of the winds.

الحرم　*El Haram.* The sanctuary.

حارة باب الزاوية　*Hâret Bâb ez Zâwieh.* The quarter of the gate of the corner or hermitage.

حارة الحرم　*Hâret el Haram.* The quarter of the sanctuary ; also called حارة القلعة *Hâret el Kŭlâh,* the quarter of the castle.

حارة المشرقى　*Hâret el Mesherky.* The eastern quarter.

حارة الشيخ　*Hâret esh Sheikh.* The quarter of the Sheikh ('Aly Bukka).

الجوالدية　*El Jawâlidiyeh.* The Jawâlidiyeh. p.n.; probably name of a dervish order; it is the mosque of *Emîr Abu Saïd Sanjar.*

جامع ابن عثمان　*Jâmia Ibn 'Othmân.* The mosque of the son of 'Othmân.

جامع شيخ على بكا　*Jâmia Sheikh 'Aly Bukka.* The mosque of Sheikh 'Aly Bukka.

قب الجانب　*Kubb el Jânib.* The dome on the hillside ; a hill on which the quarantine house stands.

خرب ابی زید   Jv   *Khŭrâb Abu Zeid.* The ruins of Abu Zeid. p.n.

خربة عبدة   Kx   *Khŭrbet 'Abdeh.* The ruins of 'Abdeh (Eboda).

خربة ابی الضبع   Lw   *Khŭrbet Abu ed Dubâ.* The ruin of the father of the hyena.

خربة ابی الحمام   Lx   *Khŭrbet Abu el Hamâm.* The ruin of the father of the dove.

خربة ابی ریشة   Lw   *Khŭrbet Abu Rîsheh.* The ruin of the father of the feather.

خربة ابی الشوك   Kv   *Khŭrbet Abu esh Shôk.* The ruin of the father of thorns.

خربة ابی السلاسل   Jv   *Khŭrbet Abu es Silâsil.* The ruin of the father of chains, or of streams.

خربة العدیسیة   Lw   *Khŭrbet el 'Addeisîyeh.* The ruin of the lentil-fields.

خربة عید الما   Jv   *Khŭrbet 'Aid el Mâ.* The ruin of the feast of water; also called خربة عید المیة *Khŭrbet 'Aid el Mîyeh,* 'feast of one hundred.' *See* "Memoirs."

خربة علین   Kv   *Khŭrbet 'Alin.* The ruin of 'Alin. p.n.

خربة عطوس   Jv   *Khŭrbet 'Atôs.* The ruin of sneezing.

خربة اوساط   Jw   *Khŭrbet Aûsât.* The ruin of the middle place.

خربة اوساطین   Jw   *Khŭrbet Aûsâtein.* The ruin of the two middle places.

خربة عزیز   Kx   *Khŭrbet 'Azîz.* The ruin of Aziz. p.n.; ('precious'). Heb. כפר עזיז Caphar Aziz.

خربة بقار   Kv   *Khŭrbet Bakkâr.* The ruin of the cowherd.

خربة بعارة   Kw   *Khŭrbet Bâ'rneh.* The ruin of the he-camels.

خربة بیت عینون   Lw   *Khŭrbet Beit 'Ainûn.* The ruin of the house of 'Ainûn Chaldee pl. עֵינִין 'springs.' Heb. בֵית עֲנוֹת Beth Anoth.

خربة بیت عمرا   Kx   *Khŭrbet Beit 'Amra.* The ruin of the house of 'Amra. p.n. *See* p. 268.

خربة بیت باعر   Jw   *Khŭrbet Beit Bâar.* The ruin of the house of camel dung.

خربة بیت ناصیف   Kv   *Khŭrbet Beit Nâsif.* The ruin of the house of Nâsif. p.n.

خربة بیت صاویر   Lv   *Khŭrbet Beit Sâwir.* The ruin of the house of Sâwir. p.n.

خربة بیت شعار   Lv   *Khŭrbet Beit Sh'âr.* The ruin of the tent (literally 'the house of hair'; Bedouin tents being so-called).

خربة بتيوس Kv *Khŭrbet Beiyûs.* The ruin of Beiyûs. p.n.

خربة بني دار Lx *Khŭrbet Beni Dâr.* The ruin of the sons of the house. Another name for Khŭrbet Yûkin.

خربة بئرين Lx *Khŭrbet Bîrein.* The ruin of two wells.

خربة بسم Kx *Khŭrbet Bism.* The ruin of smiling.

خربة البس Jv *Khŭrbet el Biss.* The ruin of the cat.

خربة بريكوت Lv *Khŭrbet Breikût.* The ruin of Breikût. Heb. בְּרָכוֹת 'blessings.' *See* "Memoirs," under 'Valley of Berachah.'

خربة البويب Lx *Khŭrbet el Bueib.* The ruin of the small door.

خربة البصل Jv *Khŭrbet el Bŭsl.* The ruin of onions.

خربة الدير Kv *Khŭrbet ed Deir.* The ruin of the monastery.

خربة الديرات Lx *Khŭrbet ed Deirât.* The ruin of houses or monasteries.

خربة دير رازي Kx *Khŭrbet Deir Râzi.* The ruin of the monastery of the man of Rai (the ancient Rhages in Persia).

خربة دير صامت Jw *Khŭrbet Deir Sâmat.* The ruin of the monastery of the silent man.

خربة الدلب Lv *Khŭrbet ed Dilb.* The ruin of plane-trees.

خربة الدلبة Kx *Khŭrbet ed Dilbeh.* The ruin of the plane-tree.

خربة إمرا Jx *Khŭrbet Emra.* The ruin of Imra. p.n.

خربة أرنبا Kv *Khŭrbet Ernebah.* The ruin of the hare.

خربة اصحا وقيس
خربة حطمان Lw *Khŭrbet Es-hâ.* The ruin of Eshâ. p.n.; also called *Khŭrbet Hotmân,* the ruin of rubbish.

خربة فطوم Jv *Khŭrbet Fattŭm.* The ruin of Fattûm. p.n.; ('weaned').

خربة فرعة Kw *Khŭrbet Firah* The ruin of the mountain-top or declivity.

خربة فرجاس Jw *Khŭrbet Firjâs.* The ruin of Firjâs. p.n.

خربة الفريديس Lv *Khŭrbet el Fureidîs.* The ruin of the little paradise.

خربة غنائم Lx *Khŭrbet Ghanâim.* The ruin of spoils; also called خربة ام العمد *Khŭrbet Umm el 'Amed,* the ruin with the pillars.

خربة غرابة   Kv   *Khŭrbet Ghŭrâbeh.* The ruin of the raven.

خربة حاكورة   Kw   *Khŭrbet Hâkûrah.* The ruin of the small plot of ground.
الحاكورة تنطعة صغيرة من الارض وهو من اصطلاح العامة

خربة حوران   Jv   *Khŭrbet Hôrân.* The ruin of low-lying ground where water stagnates.

خربة الحمّام   Jv   *Khŭrbet el Hŭmmâm.* The ruin of the bath.

خربة هرائب البيض   Mv   *Khŭrbet Hurâib el Beid.* The ruin of the white cisterns.

خربة إنسل   Lw   *Khŭrbet Insil.* The ruin of Insil. p.n.

خربة إسطبول   Lx   *Khŭrbet Istabûl.* The ruin of the stable. *See* "Memoirs," under ' Aristobulias.'

خربة جالا   Kv   *Khŭrbet Jâla.* The ruin of Jâla. Probably Heb. גִּלֹה ' Giloh.'

خربة جمرورة   Jv   *Khŭrbet Jemrûrah.* The ruin of Jemrûrah. p.n. *See* "Memoirs," under ' Gemaruris.'

خربة الجرادات   Mw   *Khŭrbet el Jerâdât.* The ruin of the locusts.

خربة جمرين   Kv   *Khŭrbet Jimrîn.* The ruin of Jimrin. p.n. (جمرة ' a fire-brand ').

خربة الجوف   Kx   *Khŭrbet el Jôf.* The ruin of the hollow.

خربة جبر   Jv   *Khŭrbet Jŭbr.* The ruin of repairing, force, or bone-setting.

خربة كفر   Kv   *Khŭrbet Kafir.* The ruin of the village.

خربة كنعان   Kw   *Khŭrbet Kan'ân.* The ruin of Canaan. Heb. כְּנַעַן

خربة تنيا   Jv   *Khŭrbet Kanya.* The ruin of the store.

خربة القط   Lv   *Khŭrbet el Katt.* The ruin of the crag.

خربة قيزين   Lw   *Khŭrbet Keizîn.* The ruin of round sand-hills (قيزان).

خربة كفر جور   Jx   *Khŭrbet Kefr Jôr.* The ruin of the village of oppression.

خربة قرما   Jv   *Khŭrbet Kerma.* The ruin of tree-stumps.

خربة خلة الدار   Lw   *Khŭrbet Khallet ed Dâr.* The ruin of the dell of the house.

خربة خلة الحمرآ   Mv   *Khŭrbet Khallet el Hŭmra.* The ruin of the red dell.

خربة الخنازير Lx — *Khŭrbet el Khanâzîr.* The ruin of the swine.

خربة خروف Jv — *Khŭrbet Kharûf.* The ruin of the sheep. Possibly the Heb. חרוף (1 Chron. xii, 5).

خربة خريسا Lx — *Khŭrbet Khoreisa.* The ruin of Khoreisa. perhaps Heb. חֹרֶשׁ זִיף the 'wood of Ziph.' *See* "Memoirs," under 'Choresh Ziph.'

خربة خرسا Kx — *Khŭrbet Khorsa.* The ruin of Khorsa. p.n. *See* last paragraph.

خربة الخريطون Mv — *Khŭrbet el Khureitûn.* The ruin of (Saint) Chariton.

خربة قبرة Kw — *Khŭrbet Kibreh.* The ruin of Kibreh. p.n. *cf.* قبر 'a tomb.'

خربة قيلا Kv — *Khŭrbet Kîla.* The ruin of Kila. p.n. *cf.* Heb. קְעִילָה Keilah.

خربة قلقس Lx — *Khŭrbet Kilkis.* The ruin of potatoes.

خربة الكوم Jw — *Khŭrbet el Kôm.* The ruin of heaps.

خربة القطن Jv — *Khŭrbet el Kotn.* The ruin of cotton.

خربة كويزيبا Lv — *Khŭrbet Kueiziba.* The ruin of Kueiziba. p.n.; possibly the Heb. כֹּזֵבָא Chozeba.

خربة الكفير Lx — *Khŭrbet el Kufeir.* The ruin of the little village.

خربة كوفين Lv — *Khŭrbet Kûfîn.* The ruin of Kûfîn. p.n.

خربة الكرمة Kx — *Khŭrbet el Kurmah.* The ruin of the vineyard.

خربة كورزا Jx — *Khŭrbet Kûrza.* The ruin of Kûrza (a pine cone).

خربة القصعة Jw — *Khŭrbet el Kŭsah.* The ruin of the bowl.

خربة كسبر Kv — *Khŭrbet Kŭsbŭr.* The ruin of Coriander.

خربة القصر Mx — *Khŭrbet el Kŭsr.* The ruin of the palace or house.

خربة قصة Jv — *Khŭrbet Kŭssah.* The ruin of plaster.

خربة الخبية Jv — *Khŭrbet el Makhbîyeh.* The hidden ruin.

خربة ماماس Lv — *Khŭrbet Mâmâs.* The ruin of Mâmâs. p.n.

خربة مرينا Lv — *Khŭrbet Marrina.* The ruin of Marrina. p.n.

خربة مرسيع Lv — *Khŭrbet Marsîâ.* The ruin of Marsîâ. p.n.

خربة المحامي Jw *Khûrbet el Mehâmi.* The ruin of fortified places.

خربة مجدل باغ Kx *Khûrbet Mejdel Bâa.* The ruin of the watch-tower with the gutter.

خربة المنية Mv *Khûrbet el Minyeh.* The ruin of el Minyeh. p.n.

خربة المية Lx *Khûrbet el Miyeh.* The ruin of the hundred.

خربة الورق Jw *Khûrbet el Mûrak.* The ruin of the leafy place. This is also the Arabic form of the Frank name Maurice.

خربة الناقية Jw *Khûrbet en Nâkieh.* The ruin of the cleanser or sifter.

خربة النمرة Lw *Khûrbet en Nimreh.* The ruin of the abundant water ; a spring exists here.

خربة النصارى Lw *Khûrbet en Nûsâra.* The ruin of the Christians ; from a tradition.

خربة الربية Kx *Khûrbet er Rabîyeh.* The ascending or conspicuous ruin.

خربة ربوض Lx *Khûrbet Râbûd.* The ruin of the animal's lair.

خربة رقة Lx *Khûrbet Rakâh.* The ruin of Rakâh. p.n. ; the word means a large tree like the plane-tree, and also a certain medicinal herb.

خربة الرسم Jv *Khûrbet er Resm.* The ruin of the trace or relic.

خربة الربيعة Mv *Khûrbet er Robiâh.* The ruin of spring herbage.

خربة ربا Jv *Khûrbet Rubba.* The ruin of Rubba. p.n. ; possibly the Heb. רַבָּה Rabbath.

خربة سلمى Lx *Khûrbet Salma.* The ruin of Salma (female p.n.).

خربة صمعآ Jx *Khûrbet Samâh.* The lofty ruin.

خربة سانوطا Lw *Khûrbet Sânût.* The ruin of acacias.

خربة سبع Jv *Khûrbet Sebâ.* The ruin of the lion.

خربة سنابرة Jv *Khûrbet Senâbreh.* The ruin of Sennâbrah (Sennabris).

خربة سراسير Kw *Khûrbet Serâsîr.* The ruin of cold winds. صراصرة *Serâsirah* was a name given to the Nabathæan inhabitants of Syria.

خربة شبرقة Jw *Khûrbet Shebrakah.* The ruin of cutting in pieces.

خربة الشيخ مذكور Jv *Khûrbet esh Sheikh Madhkûr.* The ruin of Sheikh Madhkûr. q.v., p. 407. *See Sheikh Madhkûr.*

3 F

خربة شنّة   Lv   *Khŭrbet Shennch.* The ruin of the old leathern bottle.

خربة الشروي   Jw   *Khŭrbet esh Sherwî.* The ruin of esh Sherwi. p.n. (شرو 'honey').

خربة سكّة   *Khŭrbet Sikkeh.* The ruin of the path.

خربة السمآ   Jw   *Khŭrbet es Smieh.* The ruin of the mark or sign.

خربة صرة   Jx   *Khŭrbet Sirreh.* The ruin of cold.

خربة سوبا   Jw   *Khŭrbet Sûba.* The ruin of Sûba. p.n.

خربة صبيح   Jv   *Khŭrbet Subih.* The reddish-white ruin.

خربة السويطي   Jw   *Khŭrbet es Sŭeity.* The ruin of es Sûweiti; from سوط. which means 'a whip,' or 'remains of water in a well.'

خربة الصفا   Kv   *Khŭrbet es Sŭfa.* The ruin of the bright or smooth rock.

خربة طلعة منحوتة   Lx   *Khŭrbet Talât Manhûteh.* The ruin of the ascent hewn out of the rock.

خربة طاوس   {Jx / Jv}   *Khŭrbet Tâûwâs.* The ruin of the peacock.

خربة تقوع   Mv   *Khŭrbet Tekûa.* The ruin of Tekûa. Heb. תְּקוֹעַ *Tekoa.*

خربة تل البيضآ   Jv   *Khŭrbet Tell el Beida.* The ruin of the white mound.

خربة طن ابرن   Lv   *Khŭrbet Tenn Ibrinn.* The ruin of the ringing sound of Ibrinn. p.n.

خربة الطبقة   Lv   *Khŭrbet et Tŭbîkah.* The ruin of the terrace.

خربة تفوح   Lx   *Khŭrbet Tŭffûh.* The ruin of Tŭffûh. q.v., p. 408. *See Tŭffûh.*

خربة ام العمدان   Jv   *Khŭrbet Umm el 'Amdân.* The ruin with the pillars.

خربة ام العمد   Kx   *Khŭrbet Umm el 'Amed.* The ruin with the pillars.

خربة ام الصفة   Kx   *Khŭrbet Umm el Asfeh.* The ruin with the caper plant.

خربة ام حلسة   Mx   *Khŭrbet Umm Halasch.* The ruin with the verdure.

خربة ام الخنازير   Jv   *Khŭrbet Umm el Khanâzîr.* The ruin with the swine.

خربة ام اللوز   Jv   *Khŭrbet Umm el Lôz.* The ruin with the almonds.

خربة أم سلموني   Lv   *Khŭrbet Umm Salamôni.* The ruin of the mother of Salamôni. p.n.

خربة الوابدة   Lx   *Khŭrbet el Wâbedeh.* The ruin of the gorge.

خربة وادي القطع   Lw   *Khŭrbet Wâdy el Kutâ.* The ruin of the valley of the cutting.

خربة الوزيا   Mv   *Khŭrbet el Wezla.* The ruin of the assembly.

خربة يقين   Lx   *Khŭrbet Yûkîn.* The ruin of Yûkin. p.n.; meaning 'certain.'

خربة زعتوقة   Jv   *Khŭrbet Zâkûkah.* The ruin of the young partridge.

خربة زعطوطا   Mx   *Khŭrbet Zâtût.* The ruin of Zâtût (either from زعط 'to sound,' or 'strangle,' or from زعوط snuff).

قود غنائم   Mx   *Kôd Ghanâim.* Leading the sheep.

الكرمل   Lx   *El Kurmul.* Carmel. Heb. כַּרְמֶל Carmel.

قرن الجاموس   Mv   *Kŭrn el Jâmûs.* The Buffaloe's Horn (a peak).

قرون العينازية   Mv   *Kŭrûn el 'Ainâziyeh.* The peak of Ainâziyeh Arabs. *See* '*Ain el 'Anêtziyeh,* p. 278.

قرون أم السلالم   Mv   *Kŭrûn Umm es Silâlem.* The peaks of the rock with the staircase.

القصر   Jw   *El Kŭsr.* The palace.

قصر عنتر   Mv   *Kŭsr 'Antar.* The palace of 'Antar : a famous ancient Arab hero and poet.

قصر اصلايين   Lv   *Kŭsr Islâiyîn.* The palace of Islâiyin ; probably صلبان name of a certain leguminous plant.

قصر ثغرة رشيدية   Mw   *Kŭsr Thoghret Rusheidiyeh.* The palace of the gap or pass of the Rushâid Arabs.

قصر أم ليمون   Mv   *Kŭsr Umm Leimûn.* The palace with lemons.

الملحة   Nv   *El Malhah.* The salt.

مانعين   Kv   *Mânâin.* Mânâin. p.n.; from منع Manâ, 'to prevent,' or 'to be inaccessible.'

مراح معلة   Mv   *Marâh Maâllah.* The fold of the drinking place.

مراح سمور   Lx   *Marâh Sammûr.* The fold of the nightly converse.

مَعرَّش بعرة   Jv   *Márrash Bárah.* The hut of camel's dung.

مذبح الشيخ صالح   Kv   *Medhbâh Sheikh Sâleh.* The altar of Sheikh Sâleh ; also called ناقة النبي صالح *Náket en Neby Sâleh.* "The she-camel of the prophet Sâleh." Sâleh, according to the Kor'ân, ch. vii, v. 70, was a prophet who was sent to the people of Thamûd. His sign was a she-camel miraculously produced from the rock, which the people were commanded to respect and not to injure. Disobeying the command, they were visited with condign punishment. *See* my translation of the Kor'ân, Vol. I, p. 147, note.

المغيضة   Nw   *El Megheidhah.* The place with little water.

المجد   Jx   *El Mejed.* Glory.

المنية   Mv   *El Meníyeh.* Death.

المرجمة   *El Merjameh.* The cairn, or place of stoning.

مرج غزيز الغربي   Kx   *Merj 'Azîz el Gharby.* The western meadow of 'Aziz ; west of Khŭrbet 'Azîz. q.v.

مرج غزيز الشرقي   Kx   *Merj 'Azîz esh Sherky.* The eastern meadow of 'Aziz.

مرج دومة   Jx   *Merj Dômeh.* The meadow of Domeh. q.v., p. 393.

مغاير الذناني   Nw   *Mŭghâir edh Dhenâti.* The caves of edh Dhenâti. p.n.

مغاير حلمنة   Mx   *Mŭghâir Halâmŭnneh.* The caves of Hâlamŭnneh. p.n. ; north of Dhahret el Kôla.

مغاير خليل   Mw   *Mŭghâir Khŭlîl.* The caves of Hebron ; overlooking the desert, one and a-half miles north of Beni Nâûn. This name is not on the plan.

مغارة ابى الصوان   Mx   *Mŭghâret Abu es Sûwân.* The cave of the place of flints.

مغارة بئر طقطق   Mv   *Mŭghâret Bîr Taktak.* The cave of the plashing well.

مغارة المعصى   Mv   *Mŭghâret el Mâsa.* The cave of the rebellion ; also called مغارة خريطون *Mŭghâret Khureitûn,* the cave of Khureitûn ; from the ruin of that name. q.v., p. 400.

مغارة الشرق   Lv   *Mŭghâret esh Sherk.* The eastern cave.

مغارة صفا   Kw   *Mŭghâret Sûfa.* The cave of Sŭffa. *See Sŭffa,* p. 329.

مغارة سواحل   Nw   *Mŭghâret Suweihil.* The caves of the coasts.

مغارة الثعلا   Mx   *Mŭghâret eth Thâla.* The cave of the vixen.

مقام [ال]عيسى   Lv   *Mukâm [el] 'Aîsa.* The station of Jesus ; in the village of Siäir.

مقام الخضر   Kx   *Mukâm el Khŭdr.* The station of el Khŭdr. *See* p. 28.

المنظار   Kx   *El Mântâr.* The watch-tower.

متّن حصائاً   Nw   *Mutukh Hŭsâsah.* The *débris* of the gravel. *See Wâdy Hŭsâsah.*

المطويف   Kx   *El Mutuwif.* The edge ; from its position.

النبي لوط   Lw   *Neby Lût.* The prophet Lot.

النبي متّى   Lv   *Neby Metta.* The prophet Matthew. *See* "Memoirs."

النبي نعمان   Jw   *Neby Nâmân.* The prophet Nâmân. *See* "Memoirs," under 'Nephsah Neemana.'

النبي نوح   Kw   *Neby Nûh.* The prophet Noah.

النبي يقين   Lx   *Neby Yŭkîn.* The prophet Yŭkin. p.n. ; ('certain ').

النبي يونس   Lw   *Neby Yûnis.* The prophet Jonah.

نوبا   Kv   *Nûba.* Nûba. p.n. ; probably נב Nob, 'a top.' *See* "Memoirs."

نزلة الذيبة   Lv   *Nuzlet edh Dhîbeh.* The lair of the she-wolf.

الرحيّة   Kx   *Er Rahîyeh.* The coiled-up snake ; but it may be from رحى 'a millstone'; also 'a plot of ground about a mile square, above the water level'; it is a large ruin.

رخمة   Mv   *Rakhmeh.* A vulture ; but perhaps connected with رخام *Rukhâm*, white marble.

الراعة   Lw   *Er Râmeh.* Er Râmeh ; the Heb. רמה 'hill.' The full name is رامة الخليل, *Râmet el Khŭlil*, Rama of Hebron.

راموز   Lv   *Râmuz.* The mystery ; a plot of ground.

رأس البياث   Kx   *Râs el Biâth.* The hill-top of dispersion.

رأس ظهر كفورة   Nw   *Râs Dhahr Kafûrah.* The top of the desert ridge.

رأس اسحق   Kv   *Râs Es-hak.* Probably the hill-top of Isaac.

رأس الجبعرة   Lw   *Râs el Jâbreh.* The hill-top of the bowl. It is near '*Ain Kashkaleh*, q.v., p. 386, and *Râs Jŭher.*

رأس الجمجمة   Lv   *Râs el Jemjemeh.* The hill-top of the skull.

رأس جوهر Lw *Rás Júher.* The hill-top of Júher. p.n. جَوْهر 'substance,' 'jewel'; 'the wavy appearance on a damascened sword-blade.'

رأس القاضي Lv *Rás el Kâdy.* The hill-top of the judge.

رأس الكوفة Mx *Rás el Kúfeh.* The hill-top of the margin.

رأس صرّة Jx *Rás Sirreh.* The hill-top of intense cold; or of the purse.

رأس طورة Lv *Rás Tôra.* The top of the mountain.

رأس الينبوع Kw *Rás el Yanbúâ.* The hill-top of the perennial spring.

رسم اسمعين Jv *Resm Ismâîn.* The vestige of Ishmael. p.n.

رسم ام البماجم Jw *Resm Umm el Jemâjim.* The vestige with the skulls.

رسم الواوي Jv *Resm el Wâwy.* The vestige of the jackal.

رجم ابى هلال Kx *Rujm Abu Helâl.* The cairn of Abu Helâl. p.n. ('father of the crescent'); probably from the Beni Helâl Arabs.

رجم ابى زميترة Nw *Rujm Abu Zumeitir.* The cairn of Abu Zumeitir. p.n.

رجم علي Mx *Rujm 'Alei.* The cairn of the high place.

رجم باروق Lx *Rujm Bârûk.* The cairn of Bârúk. p.n.

رجم الدبابة Nw *Rujm ed Debbâbeh.* The cairn of the moveable hut.

رجم الدير Kx *Rujm ed Deir.* The cairn of the monastery.

رجم الفحجة Lx *Rujm el Fahjeh.* The cairn of el Fahjeh. p.n.; ('pigeon-toed').

رجم حنظل Jw *Rujm Handhal.* The cairn of colocynth.

رجم جمعة Mw *Rujm Jemâh.* The cairn of the gathering.

رجم المتخ Nw *Rujm el Mutukh.* The cairn of the *débris.*

رجم النقب Jx *Rujm en Nŭkb.* The cairn of the pass.

رجم ريا Lx *Rujm Reiya.* The cairn of quenching thirst; or of sweet fragrance.

رجم ام خير Mx *Rujm Umm Kheir.* The cairn of Umm Kheir. p.n.

رجم ام السّطا Lx *Rujm Umm es Sata.* The cairn of the mother of the assault.

صافا Lv *Sáfa.* Sáfa. p.n.; 'clear,' 'bright.'

سهل ابي الغزيلات Mx *Sahel Abu el Ghazeiyilát.* The plain with the gazelles.

سهة المتيردة Nw *Sahlet el Muteiredeh.* The plain of the highway. (مطرود); perhaps from عتيرد 'mutual opposition.'

سيف ابي رماح Mw *Seif Abu Rummáh.* The sword of the lancer. *Sif* means 'a coast line.'

سيل العروب Lv *Seil el 'Arrûb.* The stream of 'Arrûb. *See Birket el 'Arrûb.* p. 392.

سيل الدلية Kx *Seil ed Dilbeh.* The stream of the plane-tree.

السكيرية Nw *Es Sekeiriyeh.* The dam (a pool).

شعب الزرور Mv *Shâb ez Zárûr.* The hill-spur of the hawthorn. It is near *Kûsr 'Antar.*

الشيخ ابو عرقوب Kw *Sheikh Abu 'Arkûb.* The Sheikh of the winding track over a mountain.

الشيخ ابو الجياش Kw *Sheikh Abu el Jiâsh.* The Sheikh Abu Jiâsh; perhaps from جيش 'an army,' or 'a strong stream.'

الشيخ عبد القادر Kw *Sheikh 'Abd el Kâdir.* Sheikh 'Abd el Kâdir. p.n.; meaning 'servant of the Almighty.'

الشيخ احمد العبد Jw *Sheikh Ahmed el 'Abd.* Sheikh Ahmed the slave.

الشيخ علان Jx *Sheikh 'Allân.* Sheikh 'Allân. p.n.

الشيخ علي بكا *Sheikh 'Aly Bukka.* Sheikh 'Aly the weeper. In the suburb of Hebron; called *Hâret esh Sheikh,* 'the Sheikh's quarter.'

الشيخ الاربعين Lv *Sheikh el Arbâîn.* The Sheikh of the Forty (Martyrs).

الشيخ حميد الفليت *Sheikh Hameid (el Felik).* Sheikh Hameid the wonderful.

الشيخ ابرهيم الهدمي *Sheikh Ibrahîm (el Hidmy).* Sheikh Ibrahim el Hidmy; *i.e.,* member of the Heidemiyeh derwish order.

الشيخ مذكور Jv *Sheikh Madhkûr.* Sheikh Madhkûr. p.n.; ('mentioned').

الشيخ مغازي Kw *Sheikh Meghâzi.* The warrior Sheikh; in *Beit Kâhel.*

الشيخ مساعد Jw *Sheikh Mesâád.* Sheikh Mesâád; p.n.; ('assister').

الشيخ ربيع Jx *Sheikh Rabiâ.* Sheikh Rabiâ. p.n.; ('spring').

الشيخ سلمان Jv *Sheikh Selmân.* Sheikh Selmân, p.n.; probably Selmân el Fârsi, one of the companions of Mohammed.

الشيوخ Lw *Esh Shiûkh.* The Sheikhs (a village).

شونة الرصيفية Nv *Shûnet er Raseifiyeh.* The solid granary.

سعير Lv *Siâir.* Siâir. p.n.; the Heb. צִיעֹר Zior.

السيميآء Kx *Es Sîmia.* The mark or sign.

سير معين Lw *Sîr Maîn.* The fold of the spring (a sheepfold by a spring).

سيرة البلاع Lw *Sîret el Bellâa.* The fold of the pond.

صوريف Kv *Sûrif.* Sûrif. p.n. *See* "Memoirs."

ستجة Jx *Sûtjeh.* The weaver's spool.

الطيبة Kw *Et Taiyibeh.* The goodly water or land.

طلعة المجاز Nw *Talât el Mejâz.* The ascent of the crossing (of a road, &c.).

تل الزيف Lx *Tell ez Zif.* The mound of Ziph. Heb. זִיף.

ترقومية Kw *Terkûmieh.* Tricomias.

ثغرة الضبع Mw *Thoghret ed Dubâ.* The gap or fissure of the hyena.

طبلة حمرآء Nw *Toblet Hŭmra.* The red drum.

طور ابى على Lw *Tôr Abu 'Aly.* The mountain of Abu 'Aly. p.n.

طور ابى ظهير Mw *Tôr Abu Dhaheir.* The mountain of the little ridge.

طور الدبابة Nw *Tôr ed Debbâbeh.* The mountain with the moveable hut or sand-hill (near *Wâdy el Merukkuk*).

طبق الميدآء Nv *Tŭbk el Meida.* The extreme or opposite terrace.

تفوح Kw *Tŭffûh.* Tŭffûh. The Heb. בֵּית תַּפּוּחַ 'Beth Tappuah.'

ام العمد Nv *Umm el 'Amed.* The mother of columns.

ام برج Jv *Umm Burj.* The mother of the tower.

ام القبور Mx *Umm el Kabûr.* The mother of tombs.

ام سكتين Jx *Umm Sakketein.* The mother of two paths.

ام سويد Jv *Umm Suweid.* The mother of black things; or of a little water.

وادي الاعور Lw · *Wády el Áàwir*. The valley of the one-eyed man.

وادي العبهر Kv · *Wády el 'Abhar*. The valley of the mock orange (*Syrxa officinalis*).

وادي عبيان Mw · *Wády 'Abeiyán*. The valley of 'Abeiyán. p.n. *See Kab 'Abeiyán*, p. 394.

وادي ابي اعرج Jv · *Wády Abu Aá'raj*. The valley of Abu Aá'raj. p.n. , 'the father of the lame man.'

وادي ابي عياش Mw · *Wády Abu 'Aiyásh*. The valley of Abu 'Aiyásh. p.n. , father of the debauchee.

وادي ابي العروق · *Wády Abu el 'Aruk*. The valley with the roots, 'spurs' or 'beaten tracks' (or 'veins').

وادي ابي ذياب Lx · *Wády Abu Dhiáb*. The valley of Abu Dhiáb. p.n. , ('father of wolves').

وادي ابي حداد Ka · *Wády Abu Hidád*. The valley of Abu Hidád ; ('father of mourning').

وادي ابي حرش Lx · *Wády Abu Hirsh*. The valley of Abu Hirsh. p.n. , ('stupid').

وادي ابي خشيبة Lx · *Wády Abu Khasheibeh*. The valley with the mud.

وادي ابي الخيل Jv · *Wády Abu el Kheil*. The valley with the horses.

وادي ابي الثمرا Kx · *Wády Abu el Kumra*. The valley with the pale colour.

وادي ابي نجيم Mv · *Wády Abu Nejeim*. The valley of Abu Nejeim. q.v.

وادي ابي رجب Lv · *Wády Abu Rajab*. The valley of Abu Rajab. p.n.

وادي ابي سكير Kv · *Wády Abu es Sekeir*. The valley of Abu Sekeir. p.n. ('the man with the hawk') ; perhaps وادي ابو سكير the valley with the dam.

وادي ابي زناخ Kx · *Wády Abu Zennákh*. The valley of Abu Zennákh. In Arabic the root means 'to be rancid' (oil, &c.), but it is probably connected with the Heb. צנח broken ground. *See Josh.* xv, 34, 56.

وادي ابي زعيتر Nw · *Wády Abu Zumeitir*. The valley Abu Zumeitir. p.n.

وادي العدود Lx · *Wády el 'Adúd*. The valley of perennial springs.

وادي الافرنج Jw · *Wády el Afránj*. The valley of the Franks.

3 G

وادي عين حمدة Mo · *Wâdy 'Ain Hamdeh.* The valley of Hamdeh's spring ; p.n.

وادي عين زأر Jw · *Wâdy 'Ain Zâr.* The valley of the spring where canes grow.

وادي العنزية Jx · *Wâdy el 'Anaziyeh.* The valley of 'Anaiziyeh Arabs. *See 'Ain el 'Aneiziyeh, p. 278.*

وادي العرب Kv · *Wâdy el 'Arab.* The valley of the Arabs.

وادي الرقبة Mw · *Wâdy el Arkebeh.* Perhaps for رقبة, the valley of watches.

وادي العروب Mv · *Wâdy el 'Arrûb.* The valley of el 'Arrûb. *See Seil el 'Arrûb, p. 407.*

وادي عويدة Nw · *Wâdy 'Aweideh.* The valley of 'Aweideh. p.n.

وادي العوينات Kv · *Wâdy el 'Aweinât.* The valley of flooded lands.

وادي عزيز Jw · *Wâdy 'Aziz.* The valley of 'Aziz. *See Khûrbet 'Aziz, p. 397.*

وادي البصة Kw · *Wâdy el Basseh.* The valley of the marsh.

وادي بصاص Mv · *Wâdy Bassâs.* The valley of marshes.

وادي البطيخة Mv · *Wâdy el Batîkhah.* The valley of the water-melon.

وادي بيت فجار Lv · *Wâdy Beit Fejjâr.* The valley of Beit Fejjâr. q.v.; p. 388.

وادي بيت سكاريا Lv · *Wâdy Beit Skâria.* The valley of Beit Skâria. *See Khûrbet Beit Skâria, p. 302.*

وادي البئار {Kv Lv} · *Wâdy el Biâr.* The valley of wells.

وادي بئر الخنزير Lv · *Wâdy Bîr el Khanzir.* The valley of the well of the swine.

وادي بئر السفلي Jw · *Wâdy Bîr es Sifli.* The valley of the lower well.

وادي بئر السويدة Jv · *Wâdy Bîr es Sûweideh.* The valley of the well of Sûweideh. *See Jubb Sûweid, p. 5.*

وادي البويب Mx · *Wâdy el Bûeib.* The valley of the little door.

وادي البطنية Jx · *Wâdy el Butniyeh.* The valley of the interior or lining.

وادي دنون Mv · *Wâdy Dannûn.* The valley of Dannûn. p.n.

وادي الدبابة Nw · *Wâdy ed Debbâbeh.* The valley of the moveable hut, or sand hill.

وادي الدغيمة Nv · *Wâdy ed Degheimeh.* The valley of Degheimeh. p.n. (blackish).

وادي الظلّ Kv — *Wády edh Dhŭl.* The valley of shade.

وادي الدلبة Kx — *Wády ed Dilbeh.* The valley of the plane tree.

وادي الدور Lv — *Wády ed Dör.* The valley of the circle.

وادي الدروسة Jv — *Wády ed Drüseh.* The valley of obliterated traces of buildings.

وادي الامير Lv — *Wády el Emir.* The valley of the prince.

وادي الفحص Qx — *Wády el Fahas.* The valley of the search.

وادي فرة Kw — *Wády Firah.* The valley of the mountain-top.

وادي فخاري Lx — *Wády Fokhári.* The valley of the potter.

وادي الفقرا Lv — *Wády el Fokra.* The valley of the poor.

وادي فريديس Mv — *Wády Fureidis.* The valley of the little paradise.

وادي غنيم Mv — *Wády Ghaneim.* The valley of plunder.

وادي الغار Mu — *Wády el Ghár.* The valley of the hollow in the desert.

وادي الغول Mv — *Wády el Ghŭl.* The valley of the ghoul (vampire).

وادي احمام Jw — *Wády el Hamâm.* The valley of doves.

وادي حمامة Lx — *Wády Hamámeh.* The valley of the dove.

وادي الحنية Lx — *Wády el Haniyeh.* The crooked valley.

وادي حواريث Kw — *Wády el Hawáirith.* The valley of plough lands.

وادي حواس Kv — *Wády Hewâs.* The valley of verdure.

وادي الحنو Lx — *Wády el Henú.* The valley of the bend.

وادي حيش Mw — *Wády Hiâsh.* The valley of the enclosures or copses.

وادي الهندي Lv — *Wády el Hindi.* The valley of the Indian.

وادي البرم Jx — *Wády el Hirim.* The valley of purslain.

وادي الحمام Jv — *Wády el Hŭmmâm.* The valley of the bath.

وادي الحمص Kv — *Wády el Hummŭs.* The valley of the chick-pea.

وادي حصاصآ Nw — *Wády Hŭsásah.* The valley of gravel.

وادي ابن سليم Lw — *Wády Ibn Islim.* The valley of the son of Selim.

وادي جابر Jv — *Wády Jábr.* The valley of the repairer or bone-setter.

وادي جالا   Kᵛ   *Wâdy Jâla.* The valley of Jâla. *See Khûrbet Jâla,* p. 399.

وادي جدور   Kᵛ   *Wâdy Jedûr.* The valley of walls.

وادي الجمرورة   Jᵛ   *Wâdy el Jemrûrah.* The valley of Jemrûrah. *See Khûrbet el Jamrûrah,* p. 399.

وادي الجرادات   Lʷ   *Wâdy el Jerâdât.* The valley of locusts.

وادي الجرفان   {Nx / Lx}   *Wâdy el Jerfân.* The valley of perpendicular banks cut out by the stream.

وادي الجيحار   Mᵛ   *Wâdy el Jḥâr.* The valley of dens.

وادي الجسر   Mᵛ   *Wâdy el Jisr.* The valley of the bridge.

وادي جسر السوق   Jʷ   *Wâdy Jisr es Sûk.* The valley of the bridge of the market.

وادي الجوز   Lʷ   *Wâdy el Jôz.* The valley of the walnut.

وادي جبّ إبلان   Mᵛ   *Wâdy Jubb Iblân.* The valley of the pit of the bath بلّان 'Balneum.'

وادي القاضى   Lʷ   *Wâdy el Kâdy.* The valley of the judge.

وادي قفورة   Nʷ   *Wâdy Kufûrah.* The valley of the desert.

وادي قعيدة   Kʷ   *Wâdy Kâïdeh.* The valley of Kâïdeh. *See 'Aïn Kâïdeh,* p. 386.

وادي قدير   Kʷ   *Wâdy Kedîr.* The valley of Kâdir (below *Sheikh Abd el Kâder*). q.v., p. 407.

وادي قيس   Jx   *Wâdy Keis.* The valley of the Keis Arabs.

وادي قلافا   Kᵛ   *Wâdy Kelâfa.* The valley of Kelâfa. p.n. قلافة means 'husk,' or 'bark'; also 'caulking a boat.'

وادي قلمة   Jᵛ   *Wâdy Kelmah.* The valley of the pen or writing-reed.

وادي الخالديات   Lʷ   *Wâdy el Khâldiât.* The valley of el Khâldiât (property of the Khalid family).

وادي الخروف   Jᵛ   *Wâdy el Kharûf.* The valley of the sheep.

وادي خنتيس   Lʷ   *Wâdy Khenciyis.* The receding valley.

وادي {خراش / خرسا}   Jx   *Wâdy Kherâsh,* or *Khorsa.* The valley of scratching. For *Khorsa,* see *Khûrbet Khorcisa,* p. 400.

وادي الخليل   {Jx / Lʷ}   *Wâdy el Khûlîl.* The valley of Hebron.

وادي خريطون   Mᵛ   *Wâdy Khureitûn.* The valley of Khureitûn. *See Khûrbet Khureitun,* p. 400.

وادي الكف Nv *Wâdy el Kuf.* The valley of the palm of the hand.

وادي تف الذهب Nx *Wâdy Kuff edh Dhaheb.* The valley of the hill of gold.

وادي الكلاب Jx *Wâdy el Kilâb.* The valley of dogs.

وادي القروش Lw *Wâdy el Kûrûsh.* The valley of piastres.

وادي القرية Mw *Wâdy el Kuryeh.* The valley of the town ; the road to Bein Nâim leads up it.

وادي القطع Lw *Wâdy el Kuta.* The valley of the cutting.

وادي الملك وبس Kv *Wâdy el Malak.* The valley of the angel. It is also called
وادي جمرين *Wâdy Jimrin,* from *Khûrbet Jimrin.* q.v., p. 399.

وادي الملقي Mx *Wâdy el Malâki.* The valley or smooth stone.

وادي الملحة Nv *Wâdy el Malhah.* The salt valley.

وادي المنشر Kw *Wâdy el Manshar.* The spreading (open) valley.

وادي مراح العجل Mv *Wâdy Marâh el 'Ajel.* The valley of the fold for calves.

وادي مرينا Lv *Wâdy Marrina.* The valley of Marrina. p.n.

وادي مرسيع Mv *Wâdy Marsiâ.* The valley of Marsia. *See* p. 400.

وادي معصرة الدم Mv *Wâdy Mâsret ed Dûm[m].* The valley of the squeezing of blood ; (probably from the red streaks in the rock).

وادي المنازل Jv *Wâdy el Menâzil.* The valley of mansions, camps, or stations.

وادي المنقع Mv *Wâdy el Menkâ.* The valley of the swamp.

وادي المرج Kw *Wâdy el Merj.* The valley of the meadow.

وادي المرتق Nw *Wâdy el Merukkuk.* The valley of the mischief maker ; (the word also means ‘thinned’).

وادي المشتى Mw *Wâdy el Mesheiti.* The valley of the winter quarters.

وادي المزيرعة Kv *Wâdy el Mezeirah.* The valley of sown land.

وادي المعلق Nv *Wâdy el Muâllak.* The overhanging valley.

وادي المغاير Lw *Wâdy el Mûghâir.* The valley of caves.

وادي مقطع الجص Nv *Wâdy Mukta' el Juss.* The valley of the gypsum quarry.

وادي مطر Kw *Wâdy Mûtr.* The valley of rain.

وادي النمر Mx *Wâdy en Nimr.* The valley of the leopard.

وادي النصاري {Jv Lw} *Wâdy en Nâsâra.* The valley of the Christians.

وادي الرعي Lv *Wâdy er Râi.* The valley of pasturing cattle.

وادي الركبان Mv *Wâdy er Rekebân.* The valley of the riders.

وادي الريشة Jw *Wâdy er Rîsheh.* The valley of the feather.

وادي الرشراش Kv *Wâdy er Rishrâsh.* The valley of the willow. *See* p. 387.

وادي رجم الخليل Mx *Wâdy Rujm el Khŭlîl.* The valley of Abraham's cairn.

وادي الرواسة Lw *Wâdy er Rûwâseh.* The valley of the heads or hill-tops.

وادي السادة Lx *Wâdy es Sâddeh.* The valley of the cliffs or dams.

وادي صافا Kv *Wâdy Sâfa.* The valley of Sâfa ; from the village so-called. *See* p. 407.

وادي السبع Kw *Wâdy es Sebà.* The valley of the lion, or of the seven. *cf.* Beersheba.

وادي سنابرة Jv *Wâdy Senâbreh.* The valley of Senâbrah (Sennabris).

وادي شهادة Lx *Wâdy Shehâdy.* The valley of witness.

وادي الشيخ {Kv Lv} *Wâdy esh Sheikh.* The valley of the Sheikh.

وادي الشرق Lv *Wâdy esh Sherk.* The eastern valley.

وادي الشرقية Jv *Wâdy esh Sheikîyeh.* The eastern valley.

وادي الشنار Lv *Wâdy esh Shinnâr.* The valley of the Greek partridge.

وادي سعير Lv *Wâdy Siâîr.* The valley of Siâir. q.v., p. 408.

وادي السيحانية Mx *Wâdy es Sîhânîyeh.* The valley of es Sihâniyeh. p.n. *See* '*Ain es Sîhânîyeh,* p. 387.

وادي السيميآ Jw *Wâdy es Sîmieh.* The valley of the mark or sign.

وادي الصفر Jw *Wâdy es Sŭffar.* The valley of empty houses.

وادي السوقية Nx *Wâdy es Sûkîyeh.* The valley of the market-folk.

وادي سقر Lv *Wâdy Sŭkr.* The valley of the hawk.

وادي سقرة Nw *Wâdy Sŭkrah.* The valley of the hawk ; (probably connected with the Beni Sakr, or the Sakâirât Arabs).

وادي الصور Kv *Wâdy es Sûr.* The valley of the rock.

وادي السويد	Kw	*Wâdy es Suweid.* The black valley; but *see Jubb Suweid,* p. 5.
وادي السويدية	Mw	*Wâdy es Suweidiyeh.* The valley of the Suweidiyeh Arabs; but *see* last paragraph.
وادي التعامرة	Nv	*Wâdy et Tâmireh.* The valley of the Tâmireh Arabs.
وادي الثعلا	Mx	*Wâdy eth Thâla.* The valley of the vixen.
وادي التفاح	Lw	*Wâdy et Tûffâh.* The valley of apples; near *Tûffûh.* q.v., p. 408.
وادي ام العوسج	Mw	*Wâdy Umm el 'Ausej.* The valley with the thorn-tree.
وادي ام حذوة	Jx	*Wâdy Umm Hadhweh.* The valley of Umm Hadhweh. p.n.; (meaning 'opposite').
وادي ام الجعد	Nv	*Wâdy Umm el Jâd.* The valley of Umm el Jâd. p.n.; meaning the woman with crisp or curly hair.
وادي ام خيّرة	Mx	*Wâdy Umm Kheiyirah.* The valley of Umm Kheiyirah. p.n.
وادي الوعر	Mx	*Wâdy el Wâr.* The valley of the rugged rocks.
وادي اليهودي	Kv	*Wâdy el Yehûdi.* The Jew's valley.
وادي الزعفران	Mv	*Wâdy ez Zâferân.* The valley of saffron.
وادي الزيتون	Kv	*Wâdy ez Zeitûn.* The valley of olives.
وادي الزلفة	Jv	*Wâdy ez Zelefeh.* The valley of the cistern.
يتّا	Lx	*Yutta.* Yutta; the Heb. יֻטָּה *Juttah.*

SHEET XXII.

ابو القباع Ow *Abu el Kubâa.* The father of the skull-cap.

عين العريجة Ox *'Ain el 'Areijeh.* The spring of el 'Areijeh. p.n.; from خرج 'to ascend,' 'to be lame,' or 'to turn aside and halt at a station.'

عين الغوير Ov *'Ain el Ghuweir.* The spring of the little hollow.

عين جدي Ox *'Ain Jidy.* The spring of the kid. Engedi.

عين سدير Ox *'Ain Sideir.* The spring of the small lotus-tree.

عين التراب Oy *'Ain et Trâbeh.* The spring of soil or earth.

عرب الكعابنة {Nx} {Ox} *'Arab el K'âbneh.* The Arabs of the Kaabah (the shrine at Mecca).

عرب الرشايدة {Nx} {Ow} *'Arab er Rushâideh.* The Arabs of the Rushâideh tribe. p.n.

عرب التعامرة {Nw} {Ov} *'Arab et T'âmireh.* The Ta'âmireh (cultivating) Arabs.

بئر البجالية Xw *Bîr el Bejjâliyeh.* The well of the Bejjaliyeh. p.n.

بئر الجديلة Nv *Bîr el Jedeilah.* The well of the little rivulet.

بئر قرن الحجر Nv *Bîr Kûrn el Hajr.* The well of the peak of rock.

بئر المنوا Nv *Bîr el Menwa.* The well of Menwa. p.n.

بئر المنقوشية Ow *Bîr el Mûnkûshiyeh.* The well of the engraved things.

بئر الصوانة Nv *Bîr es Suwâneh.* The well of flints.

الدرجة Ow *Ed Derajeh.* The steps.

حجر ابى الظهور Ox *Hajr Abu edh Dhahûr.* The rock with the ridges.

حجر دبكن دبكل دبقل دبلك } Nv *Hajr Dabkan.* This is written in four different ways; but as the Arabs never confound the letters ق and ك it is difficult to say which is right. It is probably دبكل 'deformed' or 'stinking.' *See* "Memoirs."

الخثرورة Ox *El Hathrûrah.* The nodules of earth. *See Khân Hathrûrah,* p. 345.

جسّارة Ow *Jessârah.* Jessârah. The words means 'daring,' and 'building a bridge.'

جوفة الدواعرة Ow *Jôfet ed Dwaâârah.* The hollow of the Dawâârah Arabs, who are here buried. *See* "Memoirs."

جوفة المقاد Nv *Jôfet el Makâd.* The hollow of leading.

جوفة النمعي Ov *Jôfet en Nehâi.* The hollow of en Nehâi ; perhaps نعى 'flaying a sheep'; the form given is a combination of letters impossible in Arabic.

فايدة حمامة Nx *Kâidet Hamâmeh.* The ridge of the pigeon.

خشم الخثرورة Nv *Khashm el Hathrûrah.* The promontory of the nodules of earth. *See el Hathrûrah,* above.

خشم المقدّم Ow *Khashm el Mukaddem.* The fronting promontory.

خشم صغر الصانع Nx *Khashm Sûfra es Sâni.* The promontory of the empty house of the artizan.

خربة القصير Mv *Khûrbet el Kuseir.* The ruin of the little house or palace.

قلعة البوارديّة *Kûlât el Bûârdiyeh.* The castle of the gunners. (A large stone on the ascent from '*Ain Jidy* to the cliff above ; so called because it is a favourite mark.)

قرن الحجر Ov *Kûrn el Hajr.* The peak of the rock.

قصر العريجة Ox *Kûsr el 'Areijeh.* The house of 'Areijeh. *See 'Ain el 'Areijah,* p. 416.

مذبح سعد عبيدة Ow *Medhbah Saîd 'Obeideh.* The place of the slaughter of Saîd 'Obeideh. p.n.

مغاير الزرانيق Ov *Mughâir ez Zerânîk.* The caves of the water channels.

مغارة مغسل العدة Nu *Mûghâret Mughussil el 'Addah.* The cave of the washing-place of utensils.

مغارة النصرانيّة Nw *Mûghâret en Nasrânîyeh.* The cave of the Christian woman.

مغارة الشقف Ow *Mûghâret esh Shukf.* The cave of the cleft.

مخرس دلال Nx *Mûkhrûs Delâl.* The place where guides are dumb.

نقب عين جدي Ox *Nûkb 'Ain Jidy.* The mountain pass of Engedi. *See* p. 416.

نقب المربعة Ow *Nŭkb el Merabbáh.* The square pass.

نقب سدير Ox *Nŭkb Sideir.* The pass of the little lotus-tree.

نصب حلحول Nv *Nusb Halhûl.* The memorial stone of the prince.

رأس مرصد Ox *Râs Mersid.* The hill-top or headland of the observatory.

رأس المنقوشية Ow *Râs el Munkûshîyeh.* The headland of the engravings.

رأس نقب حمار Ow *Râs Nukb Hamâr.* The headland of the pass of asses.

رأس الشقف Ow *Râs esh Shukf.* The headland of the cleft.

الرويكبة Nv *Er Ruwikbeh.* The summits; it applies to a prominent top.

رجم الكرات Nw *Rujm el Kurrât.* The stone of the attacks.

رجم الناقة Nw *Rujm en Nâkeh.* The cairn of the she-camel.

رجم النويتا Ow *Rujm en Nûeita.* The sailors' cairn.

شعيب جرو Nv *Shaib Jerro.* The spur (or ridge) of Jerro. p.n.

شكارة النجار *Shekâret en Nejjâr.* The carpenter's plot of ground. The name applies to a flat plateau with deep gorges on either side, in which are torrents.

سن حافظ *Sinhâfedh.* Literally 'the tooth of one who remembers.' The word *Hâfidh* is technically used for 'one who knows the Korân by heart'; but the name in question, if correctly given, is probably a corruption from an older one.

تل الجرن Ox *Tell el Jurn.* The mound of the trough.

وادي العريجة *Wâdy el 'Areijeh.* The valley of 'Areijah. *See 'Ain el Areijah,* p. 416.

وادي البسة Nv *Wâdy el Bassah.* The valley of the marsh.

وادي الدرجة Ow *Wâdy ed Derajeh.* The valley of steps.

وادي الغوير Ov *Wâdy el Ghuweir.* The valley of the little hollow.

وادي حصاصة {Nw / Ow} *Wâdy Husâsah.* The valley of gravel.

وادي جرفان Ov *Wâdy Jerfân.* The valley of banks cut out by the torrent (plural of جرف *Jorf*).

وادي الكلب Nx *Wâdy el Kelb.* The valley of the dog.

وادي الخبيرة Nx *Wâdy el Khubera.* The valley of soft soil.

وادي الكرّات   Nw   *Wády el Kurrát.* The valley of attacks.

وادي بيّداة مسلّم   Ov   *Wády Mámát Musillim.* The valley of the desert of Musellem. p.n.

وادي المخّوّمة   Ox   *Wády el Mekhôremeh.* Perhaps المكوّمة 'the pile of earth, stones, &c.'

وادي المشاش   Ov   *Wády el Meshish.* The valley of the waterpits.

وادي المعلّق   Nv   *Wády el Muállak.* The overhanging valley.

وادي المقبرة   Nw   *Wády el Mukeiberah.* The valley of the cemetery. An Arab cemetery exists in it.

وادي مقطع الجبس   Nw   *Wády Makta el Juss.* The valley of the quarry of gypsum.

وادي المطيردة   Nw   *Wády el Muteirdeh.* The valley of the high road. *See Sahlet el Muteirdeh,* p. 407.

وادي نويتا   Nw   *Wády Nueita.* The valley of the sailors.

وادي سن ابريك   Nv   *Wády Senn Abreik.* The valley of the tooth or ridge of the little pool. There are wells in this valley.

وادي الشقف   Ow   *Wády esh Shákf.* The valley of the cleft.

وادي سدير   Ox   *Wády Sideir.* The valley of the little lotus-tree.

أم القلعة   Ov   *Umm el Kúlah.* The mother of the castle.

SHEET XXIII.

عرب العزازمة $\left\{\begin{array}{l}\text{Dz}\\\text{Ez}\end{array}\right\}$ *'Arab el 'Azâzimeh.* The Azâzimeh Arabs. p.n.

عرب الطرابين $\left\{\begin{array}{l}\text{Bz}\\\text{Cz}\\\text{Ba}\end{array}\right\}$ *'Arab et Terâbîn.* The Terâbîn Arabs. p.n.

بايكة ابي المعيلك Dy *Bâiket Abu Mâîlik.* The cattle-shed of Abu Muailik. p.n.

بايكة الصانع Ey *Bâiket es Sâna.* The cattle-shed of the artizan.

خربة العصيفرية Dy *Khûrbet el 'Aseiferiyeh.* The ruin of the bird's haunt.

خربة القطشان Dy *Khûrbet el Kutshân.* The ruin of Kutshân. p.n.

خربة المندور Dy *Khûrbet el Mendûr.* The ruin of the threshed corn.

خربة شعرتا Dy *Khûrbet Shârta.* The ruin of thick foliage.

خربة ام جرار Dy *Khûrbet Umm Jerrâr.* The ruin of Umm Jerrâr. p.n.; meaning ' pots.' Probably the Heb. גרר Gerar.

خربة ام رجال Dy *Khûrbet Umm Rijl.* The ruin with the rivulets.

خربة زمارة Ey *Khûrbet Zummârah.* The ruin of the flute ; or of the harlot.

المنخيلة Dy *El Munkheileh.* The palm groves ; or of the sifted grain.

الشلالة Ae *Esh Shellâleh.* The waterfall or cascade.

تل جمة $\left\{\begin{array}{l}\text{Dy}\end{array}\right.$ جما *Tell Jemmeh* (or *Jemma*). The mound of the reservoir.

وادي الباحا Ey *Wâdy el Bâha.* The valley of the plantation, pool, or open space.

وادي غزة $\left\{\begin{array}{l}\text{Dy}\\\text{Dz}\end{array}\right\}$ *Wâdy Ghûzzeh.* The valley of Gaza.

وادي شعرتا Dy *Wâdy Shârta.* The valley of Shârta. *See Khûrbet Shârta, above.*

وادي الشريعة $\left\{\begin{array}{l}\text{Dy}\\\text{Ey}\end{array}\right\}$ *Wâdy esh Sheriâh.* The valley of the watering place.

وادي السبع Ae *Wâdy es Seba.* The valley of the seven ; from *Bîr es Seba* (Beersheba), p. 421.

SHEET XXIV.

عين كحمة Hq '*Ain Koḥleh.* The spring of manganese.

عناب Jy '*Anâb.* The grapes or jujube. Heb. עֲנָב *Anâb.*

عرب العزيزة {Ga / Ha} '*Arab el 'Azizimeh.* The 'Az'izimeh Arabs. p.n.

عرب الكديرات Iz '*Arab el Kedeirât.* The Arabs of the Kedeirât tribe.

عرب التائهت {Gy / Gz} '*Arab et Teeiha.* The Arabs of the desert of the Tih, or 'wandering.'

عرق الأبرق Iz '*Irk el Abrek.* The speckled vein, or foot of rock.

عسيلة Jy '*Aseileh.* Aseileh. p.n.; from عسل 'honey'; perhaps from عسل 'oleander.'

عيون السعدة Gy '*Ayûn es Sâdeh.* The springs of Sâdeh. p.n.

عيون الشريعة Fy '*Ayûn esh Sheriah.* The springs of the watering place.

بئار السبع Ha *Biâr es Sebâ.* The seven wells, or the wells of Sheba : Beersheba.

بئر أبو خف Iq *Bir Abu Khuff.* The well of Abu Khuff. p.n.; meaning 'the man with the boot.'

بئر البستان Iq *Bir el Bestân.* The well of the garden.

بئر فطس Fq *Bir Futeis.* The well of the flat-nosed one.

بئر خويلفة Iq *Bir Khuweilfeh.* The well of the water-drawers, or of the successors.

بئر المقرونة Iz *Bir el Makrûneh.* The adjacent well.

بئر المشاش Ha *Bir el Meshâsh.* The well of the water-pits.

بئر المذبح Hz *Bir el Mazbeh.* The well of el Mezbih 'the pit' at-pit

بئر الساقطي Iz *Bir es Sakity.* The well of falling down.

بئر سليمان ابى شارب   Gy   *Bir Suleimân Abu Shârib.* The well of Solomon, the man with a moustache. He was Sheikh of the Teiâhah Arabs.

بئر زبالة   Hy   *Bir Zubâlah.* The well of the dung-heap.

دير الغاوى   Iy   *Deir el Ghâwy.* The convent of the wanderer.

دير الهوى   Jy   *Deir el Hawa.* The monastery of the air.

دير سعيدة   Jy   *Deir Sâideh.* The monastery of Sâideh. p.n.; ('Felicia').

الجابرى   Iy   *El Jâbry.* El Jâbry. p.n.; from جبر to set bones.

جورة المقرى   {Iy Iz}   *Jûrat el Mikrôh.* The hollow of the place where water collects.

خلّة النقب   Iy   *Khallet en Nûkb.* The dell of the pass.

جشم البتير   Hz   *Khashm el Buteiyir.* The promontory of el Buteiyir. p.n.; from بتر to dock or cut off.

خربة ابى جرة   Fy   *Khûrbet Abu Jerrah.* The ruin of Abu Jerrah. p.n.; meaning the man with the jar.

خربة ابى خف   Iy   *Khûrbet Abu Khuff.* The ruin of Abu Khuff. p.n. *See Bir Abu Khuff,* p. 421.

خربة ابى رزق   Hy   *Khûrbet Abu Rizik.* The ruin of Abu Rizik. p.n.; *i.e.,* the man who is provided for.

خربة ابى رقيق   Fy   *Khûrbet Abu Rukeiyik.* The ruin of the land liable to floods.

خربة ابى رشيد   Hy   *Khûrbet Abu Rusheid.* The ruin of Abu Rusheid. p.n. *cf. Arab er Rushâideh,* p. 416.

خربة ابى سمارة   Gz   *Khûrbet Abu Samârah.* The ruin of the man who holds nightly converse.

خربة ابى تلول المذبح   Ia   *Khûrbet Abu Tellûl el Medhbah.* The ruin of the place where are the mounds of the altar, or of the place of slaughter.

خربة براتا   Gy   *Khûrbet Barrâta.* The ruin of Barrâta. p.n. It may be from برات *bur-at,* which, like ناموس *nâmûs,* means a 'hunter's lurking place.'

خربة بئر السبع   Ha   *Khûrbet Bir es Sebâ.* The ruin of the well of the seven (the Heb. בארשבע 'Beersheba'). *See* "Memoirs."

خربة بريدة   Iy   *Khûrbet Bureideh.* The ruin of Bureideh. p.n.; meaning either 'cool,' or 'a stage of the post.'

خربة عرق   Ey   *Khûrbet 'Erk.* The ruin of the foot of the rock.

خربة فنس   Fy   *Khûrbet Futeis.* The ruin of the flat-nosed man.

خربة الحاج عواد   Gy   *Khûrbet el Hâj 'Awwâd.* The ruin of the pilgrim 'Awwâd. p.n.

خربة حورا   Jz   *Khûrbet Hôra.* The ruin of the plane-tree.

خربة الجبين   Hz   *Khûrbet el Jubbein.* The ruin of the two pits.

خربة قناص   Iy   *Khûrbet Kannâs.* The ruin of the hunter

خربة قاووقة   Fy   *Khûrbet Kâuwûkah.* The ruin of Kâuwûkah. p.n.; perhaps قوق 'a barren rock.'

خربة الكسيح   Gy   *Khûrbet el Kesih.* The ruin of the well-cleaner.

خربة خويلفة   Iy   *Khûrbet Khuweilifeh.* The ruin of Khuweilifeh. *See Bîr Khuweilifeh,* p. 421.

خربة اللقية   Iz   *Khûrbet el Lekiyeh.* The ruin of the encounter.

خربة المجيدلات   Iy   *Khûrbet el Mujeidilât.* The ruin of the watch-towers.

خربة المويلح   Hz   *Khûrbet el Muweilih.* The ruin of Muweilih. *See Bîr el Muweilih,* p. 421.

خربة العمري   Hz   *Khûrbet el 'Omry.* The ruin of el 'Omry. p.n.

خربة الرأس   Iz   *Khûrbet er Râs.* The ruin of the hill-top or headland.

خربة الشلندي   Iy   *Khûrbet esh Shelendy.* The ruin of the Shelendy (a sort of boat).

خربة تات ريط   Jy   *Khûrbet Tât-Reit.* The ruin of Tât-Reit. p.n. The name is not Arabic, and is probably a mistake.

خربة ام عاذرة   Ey   *Khûrbet Umm 'Âdrah.* The ruin with the manure.

خربة ام البقر   Gy   *Khûrbet Umm el Bakr.* The ruin with the cows.

خربة ام بطينة   Iz   *Khûrbet Umm Buteineh.* The ruin with the little knolls.

خربة ام الرمامين   Iy   *Khûrbet Umm er Râmâmîn.* The ruin with the pomegranates.

خربة ام صوانة   Iy   *Khûrbet Umm Sûwâneh.* The ruin with the flints.

خربة الوطن   Ha   *Khûrbet el Wâtn.* The ruin of home.

خربة زعق   Iy   *Khûrbet Zâk.* The ruin of driving cattle, or of terrifying.

خربة زبالة. | Hy | *Khŭrbet Zubâlah.* The ruin of the dung-heap.

قرنة غزالة. | Iy | *Khŭrbet Ghŭzâleh.* The horn (or peak) of the gazelle.

قصور المحافظة. {Iz}{Jz} | *Kŭsûr el Mehâfedheh.* The palaces of protection.

محطا العفش | Iz | *Mehatt el Hafsh;* properly *'Afsh.* The place of putting down the luggage (the Arabs bring things thus far and then hand them over to the Fellahin).

مفردات السبع | Ha | *Muferredât es Sebâ.* The isolated hills of (Beer) Sheba ; low broken hills.

مغارة النصراني | Iy | *Mŭghâret en Nŭsrâny.* The cave of the Christian.

راس النقب | Iy | *Râs en Nŭkb.* The top of the pass.

رسم ابى حنّا | Iy | *Resm Abu Henna.* The vestiges of buildings with the henna (a red dye used by women for colouring the finger tips).

رسم ابى جروان | Iz | *Resm Abu Jerwân.* The traces of buildings of Abu Jerwân. p.n. ; (the man with the wine).

رسم المقصر | Iy | *Resm el Miksar.* The traces of the buildings of the fuller's beetle.

رسوم البتير | Hz | *Resûm el Buteiyir.* Ruins of Buteiyir. *See Khashm el Buteiyir.* p. 422.

رجم الذيب | Hy | *Rujm edh Dhib.* The cairn of the wolf.

رجم الحمص | Jy | *Rujm el Hummâs.* The cairn of the chick-pea.

رجم جريدة | Jy | *Rujm Jureideh.* The cairn of the troop.

رجم قطيطا | Jy | *Rujm Kuteit.* The cairn of the cat ; or of the crag.

سهل ام بطين | Iz | *Sahel Umm Butein.* The plain with the little knoll.

الشيخ ابو خروبة | Jy | *Sheikh Abu Kharrubeh.* The Sheikh of the Carob-tree (or locust-tree, *Ceratonia siliqua*).

تل ابى هريرة | Fy | *Tell Abu Hureireh.* The mound of Abu Hureireh. p.n. (The man with the kitten.) He was one of the companions of the Prophet.

تل خويلفة | Iy | *Tell Khuweilfeh.* The mound of Khuweilfeh. p.n. *See Bîr Khuweilfeh,* p. 421.

تل السقاطي	Iz	*Tell es Sakáty.* The mound of falling down.
تل السبع وقيل تل المشاش	Ia	*Tell es Seba*, or *Tell el Meshásh.* The mound of Sheba ; or mound of the water-pit.
تل الشريعة	Gy	*Tell esh Sherâh.* The mound of the drinking place.
تل أم بطين	Iz	*Tell Umm Butein.* The mound of Umm Butein. *See Sahel Umm Butein*, p. 424.
طويل ابي جرول	Hz	*Tuweil Abu Jerual.* The long peak or ridge of the stony ground.
أم ديمنة	Jy	*Umm Deimneh.* Umm Deimneh. p.n.; possibly Heb. מַדְמַנָּה Madmannah.
وادي ابي خف	Hy	*Wády Abu Khuff.* The valley of Abu Khuff. *See Bír Abu Khuff*, p. 421.
وادي عربيد	Jy	*Wády 'Arbíd.* The valley of the noisy drunkard.
وادي عوجان	Iz	*Wády 'Aujân.* The valley of curves.
وادي براتا	Gy	*Wády Barráta.* The valley of Barráta. *See Khûrbet Barráta*, p. 422.
وادي البطم	Iz	*Wády el Butm.* The valley of the terebinth.
وادي اثميثة	Iy	*Wády edh Dhikah.* The narrow valley.
وادي فطّاس	Iy	*Wády Fattás.* The valley of the flat-nosed one.
وادي فطيس	{Ey} {Fy}	*Wády Futeis.* The valley of the flat-nosed one.
وادي غزالة	Hy	*Wády Ghûzâleh.* The valley of the gazelle.
وادي إتمي	Iz	*Wády Itmy.* The valley of the wild olive.
وادي الجابري	Iy	*Wády el Jâbry.* The valley of el Jâbry. q.v., p. 422.
وادي الكرتيطي	Fy	*Wády el Kerkity.* The valley of small fragments (a vulgar Arabic word).
وادي الخليل	Iz	*Wády el Khûlîl.* The valley of Hebron. *See el Khûlîl*, p. 396.
وادي خويلفة	Hy	*Wády Khuweilfeh.* The valley of Khuweilfeh. *See Bír Khuweilfeh*, p. 421.

3 I

وادي كحلة   Hy   *Wâdy Kohleh.* The valley of manganese.

وادي المقصر   Iy   *Wâdy el Miksar.* The valley of el Miksar. *See Resm el Miksar*, p. 424.

وادي نخبار البغل   Fy   *Wâdy Nukhbâr el Baghl.* The valley of the fissure of the mule. *See* p. 383.

وادي السعدة   Gy   *Wâdy es Sâdeh.* The valley of es Sâdeh. p.n.; ('Felicia').

وادي سعوة   Iz   *Wâdy Sâweh.* The valley of amplitude.

وادي السبع   {Fa} {Ga}   *Wâdy es Sebâ.* The valley of es Sebâ. *See Bîar es Sebâ*, p. 421, and *Khŭrbet Bîr es Sebâ*, p. 422.

وادي الشريعة   Fy   *Wâdy esh Sherîâh.* The valley of the watering places; also called وادي بشكة *Wâdy Bashkah;* but the last only means in modern Arabic 'a different valley.'

وادي الشومار   Jy   *Wâdy esh Shômâr.* The valley of fennel.

وادي الشلطان   Iy   *Wâdy es Sultan.* The valley of the Sultan.

وادي ام برغوث   Hz   *Wâdy Umm Baraghûth.* The valley with the fleas.

وادي ام الرمامين   Iy   *Wâdy Umm er Rŭmâmîn.* The valley of Umm er Rŭmâmin. *See Khŭrbet Umm er Rŭmâmîn*, p. 423.

وادي ام صيرة   Iy   *Wâdy Umm Sîrah.* Valley mother of the fold.

وادي زبالة   Gy   *Wâdy Zubâlah.* The valley of the dung-heap.

SHEET XXV.

ابو الروازن Jy *Abu er Rûâzin.* Father of crevices. *See* p. 385.

عناب الصغيرة Jy *'Anâb es Sûghîreh.* The lesser Anab. *See 'Anâb,* p. 421.

عرب الظلام {Ka/Lz} *'Arab edh Dhullâm.* The Dhullâm Arabs.

عرب الجبالين {Ly/My} *'Arab el Jâhalin* (properly *Jehhâlîn*). The Jehâlin Arabs.

عرقوب دعيس Ly *'Arkûb Dâîs.* The winding mountain path of the sand-heap.

عرقوب الدريبات Kz *'Arkûb ed Dereijât.* The winding mountain path of the steps ; named from the ruin of *ed Dereijât.*

باحة غوين Ky *Bâhet Ghuwein.* The area of the quarrelsome man.

باط افرع Jy *Bât el Akra.* The bald Bât. p.n.; perhaps باطية 'a large basin'; near *Bîr 'Abd el Hâdy.*

البناية Ky *El Benâyeh.* The building.

بئار الغوران Jy *Biâr el Ghôrân.* The wells of the hollows.

بئار المشاش Ja *Biâr el Meshâsh.* The wells of the water-pits.

بئر عبد الهادي Jy *Bîr 'Abd el Hâdy.* The well of 'Abd el Hâdy. p.n.; 'servant of the Guider.'

بئر ابي نجيم Jy *Bîr Abu Nujjeim.* The well of Abu Nujeim. p.n.; 'father of the little star.'

بئر الباحة Jy *Bîr el Bâhah.* The well of the open space or pool ; near *'Anâb es Sûghîreh.*

بئر البص Ly *Bîr el Bûs.* The well of the marsh

بئر الدوائسة Jz *Bîr ed Dawâîseh.* The well of tramplings ; or of the crowd.

بئر الذباذب Ly *Bîr ed Debâdib.* The well of the little wells. From ذباب.

312

بئر دير اللوز   Ky   *Bir Deir el Lôz.* The well of the monastery of the almond-tree.

بئر الدمي   Jy   *Bir ed Dâmy.* The bloody well.

بئر الدويرية   My   *Bir ed Duweiriyeh.* The well of the people or estates of the little monastery. This well is in *Wâdy esh Sherky.*

بئر الغفرآة   Jy   *Bîr el Ghûfrah.* The well of the escort.

بئر الجسراوي   Ky   *Bîr el Jisrâwy.* The well of el Jisrâwy. p.n.; either from the village of *Jiseir*, or from جسر a bridge.

بئر قصعة سعيد   Ky   *Bîr Kasât Saîd.* The well of the hole or dish of Sâid; name of a man who is said to have fallen into it. قصعة means 'a dish,' or 'a Jerboa's hole.'

بئر الخالدية   Ly   *Bîr el Khâldiyeh.* The well of the Khâlidiyeh. p.n.; *see Wâdy el Khâldiyât, p. 412.*

بئر خلة صالح   Ly   *Bîr Khallet Sâleh.* The well of the dell of Saleh. *See Medhbah Sheikh Sâleh, p. 404.*

بئر الخان   Ky   *Bîr el Khân.* Well of the Caravanserai.

بئر معين   Ky   *Bîr Maîn.* The well of the spring.

بئر الملاقي   Ky   *Bîr el Malâki.* The well of the smooth stone.

بئر المهجب   Ky   *Bîr el Mihjab.* The well of el Mihjab. p.n.; from محجب 'to beat with a stick,' or 'to hasten.' It may be an error for محجب 'secluded.'

بئر المقيبرات   Mz   *Bîr el Mukeiberât.* The well of the cemeteries.

بئر النحل   Jy   *Bîr en Nahl.* The well of bees; but *see Wâdy en Nahl, p. 37.*

بئر النطاف   Jy   *Bîr en Nettâf.* The well of droppings.

بئر سالم السلامين   Ky   *Bîr Sâlim es Silâmein.* The well of Sâlim of the two salutations. p.n.

بئر الشرقي   My   *Bîr esh Sherky.* The eastern well.

بئر الشقفان   Ly   *Bîr esh Shûkfân.* The well of the clefts; caves exist near it.

بئر الصوفي   Ky   *Bîr es Sûfy.* The well of the Sûfi. The Sûfis are Mohammedan mystics of the Shiah sect; they are supposed to derive their names from صوف *sûf*, 'wool,' from their dress. They are principally found in Persia.

بئر السرار Ky — *Bîr es Sûrâr.* The well of pebbles.

بئر السُّطْحا Jy — *Bîr es Sut-ha.* The well of the flat surface.

بئر السويد Jy — *Bîr es Suweid.* The brackish well; but *see Jubb Suweid,* p. 5.

بئر امّ صوانة My — *Bîr Umm Siwânch.* The well with the flints.

بئر امّ الاخوص Ly — *Bîr Umm Ikhwas.* The well of Umm el Ikhwas, p.n.; الخوص means 'to come into leaf' (a plant). The word may be الخوص which means to take camels one by one to the water, and not allow them to crowd together.

بئر الزيز Ly — *Bîr ez Zîz.* The well of squills.

الدير Jy — *Ed Deir.* The monastery.

دير رافات Ky — *Deir Râ-fât.* The monastery of Râ-fât.

دير الشمس Ky — *Deir esh Shems.* The monastery of the sun.

دير زانوتا Jy — *Deir Zânûta.* The monastery of Zânûta. *See Khŭrbet Zânûta,* p. 431.

دريجات Kz — *Dereijât.* The little steps.

الظاهرية Jy — *Edh Dhâheriyeh.* The village on the ridge, or the 'apparent village.' *See* "Memoirs."

ظهرة العرايمة Nz — *Dhahret el 'Arâimeh.* The ridge of the dyke.

ظهرة حميدة My — *Dhahret Hameideh.* The ridge of Hameideh, p.n.

ظهرة جمعت My — *Dhahret Jemât.* The ridge of gatherings.

الغرا Ja — *El Ghŭrra.* The 'blaze' (white mark on a horse's forehead): a white chalk hill.

غوين الفوقآ Ky — *Ghuwein el Fôka.* The upper Ghuwein. *See Bâhet Ghuwein,* p. 427.

غوين التحتآ Ky — *Ghuwein et Tahta.* The lower Ghuwein. *See* last paragraph.

حجر السخائن Ky — *Hajr es Sakhâin.* The stone of Sakhâin. *Sakhâin* سخائن is the plural of سخينة *Sakhîna,* 'gruel,' a nick-name applied to the Kureish, the tribe to which Mohammed belonged. The root means to be warm, especially applied to water.

الخُنَيَّرة　My　*El Hudeirah.* The little enclosure; حنـ means any settled abode, as opposed to nomad life. Heb. חָצוֹר *Hazor.*

تنان السيف　Ly　*Kanán el Aseif.* The peaks of el Aseif; perhaps اسف 'the caper plant,' 'hyssop.'

تنان محمّد　Jy　*Kanán Muhammed.* The peaks of Mohammed. p.n.

تنان ودادي　Jy　*Kanán Wadády.* The peaks of Wadády. p.n.

خلّة الدخان　Jy　*Khallet ed Dokhân.* The dell of millet (*Holcus dochna,* Linn.).

خلّة إبن صالح　Jy　*Khallet Ibn Sâleh.* The dell of Ibn Sâleh. p.n.

خلّة جربا　Jy　*Khallet Jarba.* The dell of the plantation; near *Tôr es Sádân.*

خشم العير　My　*Khashm el 'Aîr.* The promontory of the wild ass.

طويل الحمارة　　*Tuweil el Hamârah.* The peak or ridge of the ass.

خشم بيوض　My　*Khashm Beiyûd.* The promontory of Beiyûd. *See below.*

خشم بطيح　My　*Khashm Buteih.* The promontory of the low-lying ground with gravelly soil, where water collects.

خشم أم السويد　My　*Khashm Umm es Sûweid.* The promontory of Umm Sûweid. *See Jubb Sûweid,* p. 5.

خرابة　Ly　*Khŭrâbeh.* Ruins.

خربة الاصفر　Ly　*Khŭrbet el Asfir.* The ruin of Isfir. p.n.; perhaps from صفر 'empty dwellings.'

خربة عتير　Ky　*Khŭrbet 'Attîr.* The ruin of 'Attîr. p.n. *See* "Memoirs." (عتيرة means a sheep sacrificed to idols in the month Rejeb.)

خربة بيوض　Lz　*Khŭrbet Beiyûd.* The ruin of the white place.

خربة بئر العد　Ly　*Khŭrbet Bîr el 'Edd.* The ruin of the perennial well.

خربة الدواسة　Jz　*Khŭrbet ed Dawâseh.* The ruin of the trampling, or of the crowd.

خربة دير اللوز　Ky　*Khŭrbet Deir el Lôz.* The ruin of the monastery of the almond.

خربة الاميرة　Ly　*Khŭrbet el Emîreh.* The ruin of the princess.

خربة النخيت Ly *Khûrbet el Fekhît.* The ruin of the fissure.

خربة جنبة Ly *Khûrbet Janbah.* The ruin on the side.

خربة جديبة Lz *Khûrbet Jedeibeh.* The ruin of barrenness.

خربة قويريس Ly *Khûrbet Kûeiwîs.* The ruin of archers. From قوس .

خربة القريتين Ly *Khûrbet el Kureitein.* The ruin of the two towns.

خربة قبر اعبية Jz *Khûrbet Kabr 'Aâbeiyeh.* The ruin of the tomb of 'Aâbeiyeh. p.n.

خربة معين Ly *Khûrbet Maîn.* The ruin of the spring.

خربة منازل Ly *Khûrbet Menâzil.* The ruin of the settlements, or stations.

خربة المركز Ly *Khûrbet el Merkez.* The ruin of the centre.

خربة المشاش Ja *Khûrbet el Meshâsh.* The ruin of the water-pits.

خربة الملح Ka *Khûrbet el Milh.* The ruin of salt. *See* "Memoirs."

خربة معيد Ly *Khûrbet Muâîyîd.* The ruin of the feaster.

خربة سلنطح Jz *Khûrbet Salantâh.* The ruin of Salantâh. p.n.

خربة سعوة Jz *Khûrbet Sâweh.* The ruin of Sâweh. p.n. *See Tell es Sâweh,* p. 433.

خربة شويكة Ky *Khûrbet Shuweikeh.* The ruin of Shuweikah (thorns). Heb. שׂוכה Socoh.

خربة الطيبة Lz *Khûrbet et Teibeh.* The goodly ruin.

خربة التواني Ly *Khûrbet et Tûâny.* The ruin of delay.

خربة زانوتا Jy *Khûrbet Zânûta.* The ruin of Zânûta. p.n.; Heb. זנוח Zanoah.

خريبة نبة Ly *Khûreibet Inbeh.* The ruins of vigilance.

كسيفة Kz *Kuseifeh.* The ruin of dross or scoria.

قصر خلة المردوم Jy *Kûsr Khallet el Mardûm.* The house of the dell of the stopped-up place.

مكحول Kz *Mak-hûl.* Anointed with collyrium.

مغاير المتعور Jy *Mûghâir el Makâwer.* The excavated caves.

مغاير ريزة Jy   *Mŭghâir Rubzeh.* The caves of the fat-tailed rams.

مغاير أم صيرة Jy   *Mŭghâir Umm Sîrah.* The caves of Umm Sirah. q.v., p. 434.

مغارة غفرآ Jy   *Mŭghâret Ghŭfrah.* The cave of the escort.

مغارة الجراب Ky   *Mŭghâret el Hurâb.* The cave of the tanks.

مغارة إبن صالح Jy   *Mŭghâret Ibn Sâleh.* The cave of the son of Sâleh.

مغارة الخروعة Jy   *Mŭghâret el Khirwâh.* Perhaps خروا the cave of the castor-oil plant; near *Sahlet Umm et Talla.*

مغارة المحوّط Jy   *Mŭghâret el Mehaûwat.* The cave of the walled place.

مغارة النحل Jy   *Mŭghâret en Nahl.* The cave of the bees. *See Bîr en Nahl,* p. 428.

مغارة الطلّ Jy   *Mŭghâret et Till.* The cave of dew.

مغارة أمّ الحيران Jy   *Mŭghâret Umm el Hîrân.* The cave of the maze.

المغرقة Ly   *El Mughrakah.* The immersed. The name here applies to a large boulder in a valley.

مقطعة احيا Jy   *Mukatât Ahya.* The place of the cutting off of John (the Baptist). *See* "Memoirs."

المطلّة Jz   *El Mutallah.* The outlook (a cairn.)

رافات Ky   *Râ-fât.* Râ-fât. p.n.

راس الجليمة Mz   *Râs el Jelamch.* The top of the hill.

راس أم رقبة Mz   *Râs Umm Rukbeh.* The hill-top with the watch station.

رجم البقرة My   *Rujm el Bakarah.* The cairn of the cow.

رجم إبن بسما Ly   *Rujm Ibn Basma.* The cairn of Ibn Basma. p.n.; ('son of a smile').

رجم الحمرآ Ly   *Rujm el Humra.* The red cairn.

رجم خرازمية Mz   *Rujm Kherâzmîyeh.* The cairn of the Kharezmians.

رجم النياس Ky   *Rujm en Niâs.* The cairn of en Niyâs. p.n.

رجم السويف Ly   *Rujm es Sûeif.* The cairn of the little sword.

رجم أم العرايس Ly   *Rujm Umm el 'Arâis.* The cairn of the mother of brides.

رجم حماتة Ly   *Rujm Humeitah.* The cairn of the mountain fig.

رجوم أم خروبة   Lz   *Rujûm Umm Kharrûbeh.* The cairns by the locust-tree (*Ceratonia siliqua*).

رقبة الزفرية   Ny   *Rukbet ez Zafariyeh.* The watch-station of the beacon.

الصفي   My   *Es Safiyi.* The bright or clear.

سهل فرعة   Ka   *Sahel Farâh.* The plain of Farâh. *See Wâdy Farâh*, p. 210.

سهلة أم التلا   Jy   *Sahlet Umm et Talla.* The plain of Umm et Talla. q.v.

السموع   Ky   *Es Semûâ,* Es Semûâ. p.n. Heb.•אשתמיה (Josh. xv. 50) ; or אשתמיע *Eshtemoh* or *Eshtemoa.*

شعب البطم   Ly   *Shâb el Butm.* The spur of the terebinth.

شيخ احمد الغماري   Jy   *Sheikh Ahmed el Ghamâry.* Sheikh Ahmed, 'the stupid.'

شجرة ابي حداد   Ky   *Shejeret Abu Hidâd.* The tree of the boundary place ; a village boundary.

سوسية   Ly   *Susieh.* Abounding in the liquorice plant.

تل عراد   Lz   *Tell 'Arâd.* The mound of 'Arâd. Heb. ערד ' 'Arad.'

تل الغر   Ja   *Tell el Ghŭr.* The mound of el Ghŭrra. q.v.

تل السعوة   Jz   *Tell es Sâweh.* The mound of es Sâweh. *See Khŭrbet Sâweh*, p. 431.

تل اليواني   Ly   *Tell et Tûâny.* The mound of delay.

طور السعدان   Jy   *Tôr es Sâdân.* The mount of the Sâdân (a plant on which camels love to pasture).

طويل البطحية   My   *Tuweil el Butâhiyeh.* The peak or ridge of the open gravelly soil where water collects.

طويل الحامدة   Ky   *Tuweil el Hâmedeh.* The peak or ridge of Hâmideh. p.n.

طويل الحمارة   My   *Tuweil el Hŭmârah.* The ridge of the she-ass.

طويل المحذي   Jz   *Tuweil el Mahdhy.* The opposite peak or ridge.

طويل الشيح   My   *Tuweil esh Shîh.* The peak or ridge of Shîh (an aromatic plant ; *Artemisia Judaica*).

أم الدرج   Jy   *Umm ed Derej.* The mother of steps (a quarry where the rock remains in steps).

أم العبيرة   Lz   *Umm el 'Abharah.* The place of the mock orange (a plant ; *Styrax officinalis*).

3 K

ام القصب	Jy	*Umm el Kûsab.* The place of reeds.
ام السير	Ky	*Umm es Seir.* The place of the folds.
ام سيرة	Jy	*Umm Sîrah.* The place of the fold.
ام الطلى	Jy	*Umm et Talla.* Probably مطلة 'the look-out.'
وادي ابي حداد	Ky	*Wâdy Abu Hidâd.* The valley of the boundary. *See Shejeret Abu Hidâd*, p. 433.
وادي ابي نجيم	Jy	*Wâdy Abu Nujjeim.* The valley of Abu Nujeim. p.n. *See* p. 427.
وادي الاكسيس	Jy	*Wâdy el Aksîs.* The valley of pemmican; east of *edh Dhâheriyeh.*
وادي العلالى	Ky	*Wâdy el 'Alâly.* The valley of the upper chamber (*i.e.*, a cave high up in a cliff). *cf. 'Alâly el Benât*, p. 283.
وادي العمائر	Ky	*Wâdy el 'Amâir.* The valley of bee-hives.
وادي عناب	Jy	*Wâdy 'Anâb.* The valley of 'Anâb. q.v., p. 427.
وادي عربيد	Jy	*Wâdy 'Arbîd.* The valley of the noisy drunkard.
وادي عرموشة	Ly	*Wâdy 'Armûsheh.* Perhaps for *'Armûseh.* The valley of rocks.
وادي الاصفر	Ly	*Wâdy el Asfir.* The valley of Asfir. q.v.
وادي العسيرة	Ly	*Wâdy el 'Asîreh.* The difficult valley.
وادي عتير	Jy	*Wâdy 'Attîr.* The valley of 'Attir. *See Khûrbet 'Attîr*, p. 430
وادي الباحة	Jy	*Wâdy el Bâhah.* The valley of the pool or open space.
وادي البطم	Jz	*Wâdy el Butm.* The valley of the terebinth.
وادي البطمة	Ky	*Wâdy el Butmeh.* The valley of the terebinth tree.
وادي البطنة	Jy	*Wâdy el Butneh.* The valley of the interior.
وادي ذباذب	Ky	*Wâdy Debâdib.* The valley of little wells.
وادي دير اللوز	Ky	*Wâdy Deir el Lôz.* The valley of the monastery of the almond.
وادي الدنانة	My	*Wâdy ed Denâneh.* The valley of the wine jars.

وادي الدريجات Kz *Wády ed Dereiját.* The valley of the little steps. *See ed Dereiját,* p. 429.

وادي الاميرة Ly *Wády el Emíreh.* The valley of the princess.

وادي فضول Ly *Wády Fedúl.* The exuberant valley.

وادي الغماري Jy *Wády el Ghamáry.* The valley of Sheikh Ahmed el Ghamáry. q.v., p. 433.

وادي الغوران Jy *Wády el Ghórán.* The valley of the hollows.

وادي غوين Ky *Wády Ghuwein.* The valley of Ghuwein. *See Báhet Ghuwein,* p. 427.

وادي الجبور Jz *Wády el Habúr.* The valley of the level lands.

وادي حكوة Ry *Wády Hakkúweh.* The valley of the side of a rugged narrow pass, or of a rugged place overhanging a stream.

وادي الحسيني Ry *Wády el Hoseiny.* The valley of el Huseini. p.n.; from *Husein,* the Shiah martyr.

وادي الحضائر My *Wády el Hudáir.* The valley of enclosures. *See el Hudeirah,* p. 430, of which word it is the plural.

وادي الهراب Ky *Wády el Hurúb.* Valley of tanks.

وادي الحويكرات My *Wády el Huweikarát.* The valley of fields (cultivated enclosures).

وادي ابن صالح Jy *Wády Ibn Sáleh.* Valley of the son of Sáleh. p.n.

وادي جحيش Ky *Wády Juheish.* The valley of the asses.

وادي الخان Ky *Wády el Khán.* The valley of the Caravanserai.

وادي خربة الطيبة Lz *Wády Khúrbet et Teibeh.* The valley of Khúrbet et Teibeh. q.v.

وادي الخصيبية My *Wády el Khuseibiyeh.* The valley of abundant herbage.

وادي قويسيس Ly *Wády Kúeiwis.* The valley of Kúeiwis. *See Khúrbet Kúweiwis,* p. 431.

وادي قريتين Jz *Wády Kureitein.* The valley of Kureitein. *See Khúrbet Kureitein,* p. 431.

وادي القسيس Ky *Wády el Kússis.* The valley of the (Christian) priest.

وادي القطلنا   Jy   *Wâdy el Kŭtŭfa.* The valley of the St. John's wort (plant).

وادي المقاري   My   *Wâdy el Makârch.* · The valley of the hill-tops.

وادي المشخّة   Ly   *Wâdy el Meshukhkhah.* The valley of the staling place. *cf. Shukhkh ed Dŭbâ, p.* 315.

وادي المِلح   Ja   *Wâdy el Milh.* The valley of el Milh. *See Khŭrbet el Milh, p.* 431.

وادي المعلّق   Ky   *Wâdy el Muâllak.* The overhanging valley.

وادي مغائر العبيد   My   *Wâdy Mŭghâir el 'Abîd.* The valley of the caves of the slaves.

وادي المغرقة   Lz   *Wâdy el Mughrakah.* The valley of the immersed rock or flooded place.

وادي المقيبرات   Lz   *Wâdy el Mukeiberât.* The valley of the cemetery.

وادي مطعن منجد   My   *Wâdy Mŭtân Mŭnjid.* The valley where the ally was stabbed.

وادي النار   Ky   *Wâdy en Nâr.* The valley of fire.

وادي رأفات   Ky   *Wâdy Râ-fât.* The valley of Râ-fât. q.v.

وادي الرخيم   {Ly}{Ky}   *Wâdy er Rakhîm.* The soft valley.

وادي رابض   Ny   *Wâdy Rubdhah.* The valley of the lair.

وادي رجم الخليل   My   *Wâdy Rujm el Khŭlîl.* The valley of the cairn of Khŭlîl. p.n. (Abraham).

وادي سلنطيح   Jz   *Wâdy Salantâh.* The valley of Salantâh. p.n.

وادي سعوة   Jz   *Wâdy Sâweh.* The valley of Sâweh. *See Khŭrbet Sâweh,* p. 431.

وادي سيال   Mz   *Wâdy Seiyâl.* The valley of the acacia (*Acacia seiyal*).

وادي السموع   Ky   *Wâdy es Semûâ.* The valley of Semûâ. *See Khŭrbet es Semûâ, p.* 433.

وادي الشرقي   Ny   *Wâdy esh Sherky.* The eastern valley.

وادي الشقفان   Jy   *Wâdy esh Shukfân.* The valley of clefts.

وادي السيميا   Ky   *Wâdy es Sîmia.* The valley of the sign. *See Khŭrbet es Sîmia, p.* 402.

وادي السمسم Jy   *Wâdy es Simsim.* The valley of sesame.

وادي السنطة My   *Wâdy es Sunta.* The valley of the acacia.

وادي السنين My   *Wâdy es Sennein.* The valley of the teeth (serrated ridge).

وادي التبّان My   *Wâdy et Tebbân.* The valley of the straw-cutter.

وادي تراب الخليسي Ly   *Wâdy Trâb el Haleisi.* The valley of the verdant soil.

وادي ام عش My   *Wâdy Umm 'Osh.* The valley with the nests.

وادي ام الحيران Jy   *Wâdy Umm el Hirân.* The valley with the maze.

وادي ام جمعات My   *Wâdy Umm Jem'ât.* The valley of the gatherings.

وادي ام خروبة Kz   *Wâdy Umm Kharrûbeh.* The valley with the locust-tree (*Ceratonia siliqua*).

وادي ام مراضيف Ny   *Wâdy Umm Merâdhif.* The valley of the stones heated for cooking with.

وادي ام سدري Ly   *Wâdy Umm Sidry.* The valley with the lotus-trees.

وادي ام زيتونة Ly   *Wâdy Umm Zeitûneh.* The valley with the olive-trees.

وادي زانوتا Jy   *Wâdy Zânûta.* The valley of Zânûta. *See Khûrbet Zânûta, p.* 431.

SHEET XXVI.

عين عنيبة   Ny   *'Ain 'Oneibeh.* The spring of the grapes.

بركة الخليل   Ny   *Birket el Khŭlil.* The pool of Khŭlil (Abraham, or Hebron).

ظهرة القطاعي   Nz   *Dhahret el Kutââl.* The ridge of the quarryman.

سبة   Nz   *Sebbeh.* Sebbeh. p.n. ; ('reviling').

وادي الخشيبة   Ny   *Wâdy el Khasheibeh.* The valley of mud.

وادي الخبيرة   Ny   *Wâdy el Khubera.* The valley of soft soil.

وادي محرس   Ny   *Wâdy Mahras.* The valley of the guard-house.

وادي سيال   Ny   *Wâdy Seiyâl.* The valley of the acacia.

وادي الصفاصف   Ny   *Wâdy Sufeisif.* Valley of the osier willows.

LONDON :
PRINTED BY HARRISON AND SONS, PRINTERS IN ORDINARY TO HER MAJESTY,
ST. MARTIN'S LANE.